# CELLULAR PEPTIDASES IN IMMUNE FUNCTIONS AND DISEASES 2

# ADVANCES IN EXPERIMENTAL MEDICINE AND BIOLOGY



Recent Volumes in this Series

Volume 468
THE FUNCTIONAL ROLES OF GLIAL CELLS IN HEALTH AND DISEASE: Dialogue between Glia and Neurons
Edited by Rebecca Matsas and Marco Tsacopoulos

Volume 469
EICOSANOIDS AND OTHER BIOACTIVE LIPIDS IN CANCER, INFLAMMATION, AND RADIATION INJURY, 4
Edited by Kenneth V. Honn, Lawrence J. Marnett, and Santosh Nigam

Volume 470
COLON CANCER PREVENTION: Dietary Modulation of Cellular and Molecular Mechanisms
Edited under the auspices of the American Institute for Cancer Research

Volume 471
OXYGEN TRANSPORT TO TISSUE XXI
Edited by Andras Eke and David T. Delpy

Volume 472
ADVANCES IN NUTRITION AND CANCER 2
Edited by Vincenzo Zappia, Fulvio Della Ragione, Alfonso Barbarisi, Gian Luigi Russo, and Rossano Dello Iacovo

Volume 473
MECHANISMS IN THE PATHOGENESIS OF ENTERIC DISEASES 2
Edited by Prem S. Paul and David H. Francis

Volume 474
HYPOXIA: Into the Next Millennium
Edited by Robert C. Roach, Peter D. Wagner, and Peter H. Hackett

Volume 475
OXYGEN SENSING: Molecule to Man
Edited by Sukhamay Lahiri, Naduri R. Prabhakar, and Robert E. Forster, II

Volume 476
ANGIOGENESIS: From the Molecular to Integrative Pharmacology
Edited by Michael E. Maragoudakis

Volume 477
CELLULAR PEPTIDASES IN IMMUNE FUNCTIONS AND DISEASES 2
Edited by Jürgen Langner and Siegfried Ansorge

# CELLULAR PEPTIDASES IN IMMUNE FUNCTIONS AND DISEASES 2

Edited by

Jürgen Langner

*Martin Luther University*
*Halle, Germany*

and

Siegfried Ansorge

*Otto von Guericke University*
*Magdeburg, Germany*

Springer Science+Business Media, LLC

Library of Congress Cataloging-in-Publication Data

Cellular peptidases in immune functions and diseases 2/edited by Jürgen Langner and Siegfried Ansorge.

p. cm.—(Advances in experimental medicine and biology: v.477)

"Proceedings of an International Conference on Cellular Peptidases in Immune Functions and Diseases (II), held September 12–14, 1999, in Magdeburg, Germany"—T.p. verso

Includes bibliographical references and index.

DOI 10.1007/978-0-306-46826-1

1. Peptidase—Immunology—Congresses. 2. Peptidase—Pathophysiology—Congresses. 3. Endopeptidases—Immunology—Congresses. 4. Endopeptidase—Pathophysiology—Congresses. I. Langner, Jürgen. II. Ansorge, Siegfried. III. International Conference On Cellular Peptidases in Immune Functions and Diseases (2nd: 1999: Magdeburg, Germany) IV. Series.

[DNLM: 1. Peptide Hydrolases—Immunology—Congresses. 2. Peptide Hydrolases—Physiology—Congresses. QU 136 C3935 2000]

QP609.P45 C452 2000
616.07'9—dc21 00-022919

Proceedings of an International Conference on Cellular Peptidases in Immune Functions and Diseases (II), held September 12–14, 1999, in Magdeburg, Germany

ISSN: 0065-2598


Originally published by Kluwer Academic / Plenum Publishers, New York in 2000
MyCopy version of the original edition 2000

http://www.wkap.nl/

10 9 8 7 6 5 4 3 2

A C.I.P. record for this book is available from the Library of Congress

# PREFACE

The good response and high scientific output of the first meeting in 1996 encouraged us to continue with a second one, "Cellular peptidases in Immune Functions and Diseases (II)" which was held September 12–14th, 1999 again at Magdeburg-Herrenkrug, Germany. This meeting again was organised by the Sonderforschungsbereich 387 of the Deutsche Forschungsgemeinschaft "Zelluläre Proteasen, Bedeutung für Immunmechanismen und entzündliche Erkrankungen". This field has expanded continuously and has become an established and highly effective area during the last years. Since the first meeting in 1996 new proteases have been detected and new topics have been explored where proteolysis plays an important role. Increasingly it is becoming clear that many—if not all—pathomechanisms of infectious and inflammatory diseases and malignancies are closely associated with dysregulation of proteolytic enzymes. The active participation of colleagues from about twenty laboratories in Germany and from other European countries—in addition to the invited speakers—provides telling evidence of the attraction of this field, also outside the SFB.

The present volume contains most of the presentations of the invited speakers. The three parts of the contents each are introduced by a review of the topic of the respective section. We understand this as a helpful introduction for those readers who are not so close to the special topics of proteolysis in relation to immune functions covered by the meeting. In addition, short communications are included from the members of the SFB showing the variety of projects as well as their development. Included are also some contribution from participants of groups outside the SFB, originally presented as posters.

We are glad to have again the opportunity to publish this volume in the series "Advances in Experimental Medicine and Biology" of Kluwer Academic / Plenum Publishers, and we are especially grateful to Joanna Lawrence, London, for her invaluable help and all cooperation in the production of this book.

Financial support of the meeting was provided by the Deutsche Forschungsgemeinschaft, the Ministerium für Kultur des Landes Sachsen-Anhalt, Otto-von-Guericke-Universität Magdeburg, Martin-Luther-Universität Halle-Wittenberg and the companies: Becton Dickinson GmbH Heidelberg, Biochrom Beteiligungs GmbH & Co, Centeon/Liederbach, Karl Roth GmbH & Co KG, NEN Life Science Products GmbH, R & D Systems, Roche Diagnostics GmbH.

The organization of the meeting was greatly managed by Gudrun Plexnies and Barbara Schotte. The editorial work was excellently performed by Christl Walcker. Invaluable help with all editorial computer problems was provided by Cornelius Hempel. Our deep gratitude goes to all of them.

Jürgen Langner
Siegfried Ansorge

## CONTENTS

**Part I: Membrane Ectopeptidases with Influence on Immune Functions**

1. Review: The Role of Membrane Peptidases in Immune Functions ..... 1
*Uwe Lendeckel, Thilo Kähne, Dagmar Riemann, Klaus Neubert, Marco Arndt, and Dirk Reinhold*

2. Structure and Function of Aminopeptidase N ..... 25
*Hans Sjöström, Ove Norén, and Jørgen Olsen*

3. Modulation of WNT-5a Expression by Actinonin: Linkage of APN to the WNT-Pathway? ..... 35
*Uwe Lendeckel, Marco Arndt, Karin Frank, Antje Spiess, Dirk Reinhold, and Siegfried Ansorge*

4. Enzymatic Activity is not a Precondition for the Intracellular Calcium Increase Mediated by mAbs Specific for Aminopeptidase N/CD13 ..... 43
*Alexander Navarrete Santos, Jürgen Langner, and Dagmar Riemann*

5. Transforming Growth Factor-β Increases the Expression of Aminopeptidase N/CD13 mRNA and Protein in Monocytes and Monocytic Cell Lines ..... 49
*Astrid Kehlen, Jürgen Langner, and Dagmar Riemann*

6. Cell-Cell Contact Between Lymphocytes and Fibroblast-like Synoviocytes Induces Lymphocytic Expression of Aminopeptidase N/CD13 and Results in Lymphocytic Activation ..... 57
*Dagmar Riemann, Jana Röntsch, Bettina Hause, Jürgen Langner, and Astrid Kehlen*

7. Natural Substrates of Dipeptidyl Peptidase IV ..... 67
*Ingrid De Meester, Christine Durinx, Gunther Bal, Paul Proost, Sofie Struyf, Filip Goossens, Koen Augustyns, and Simon Scharpé*

8. Relating Structure to Function in the Beta-Propeller Domain of Dipeptidyl Peptidase IV: Point Mutations that Influence Adenosine Deaminase Binding, Antibody Binding and Enzyme Activity ......... 89
*Mark D. Gorrell, Catherine A. Abbott, Thilo Kähne, Miriam T. Levy, W. Bret Church, and Geoffrey W. McCaughan*

9. Development of a Tertiary-Structure Model of the C-Terminal Domain of DPP IV ............................................ 97
*Wolfgang Brandt*

10. Post Proline Cleaving Peptidases Having DP IV Like Enzyme Activity: Post-Proline Peptidases ........................................ 103
*Catherine A. Abbott, Denise Yu, Geoffrey W. McCaughan, and Mark D. Gorrell*

11. A New Type of Fluorogenic Substrates for Determination of Cellular Dipeptidyl Peptidase IV (DP IV/CD26) Activity .................. 111
*Susan Lorey, Jürgen Faust, Frank Bühling, Siegfried Ansorge, and Klaus Neubert*

12. Potent Inhibitors of Dipeptidyl Peptidase IV and their Mechanisms of Inhibition ............................................ 117
*Angela Stöckel-Maschek, Beate Stiebitz, Ilona Born, Jürgen Faust, Werner Mögelin, and Klaus Neubert*

13. N-terminal HIV-1 TAT Nonapeptides as Inhibitors of Dipeptidyl Peptidase IV. Conformational Characterization ..................... 125
*Carmen Mrestani-Klaus, Annett Fengler, Jürgen Faust, Wolfgang Brandt, Sabine Wrenger, Dirk Reinhold, Siegfried Ansorge, and Klaus Neubert*

14. Signal Transduction Events Induced or Affected by Inhibition of the Catalytic Activity of Dipeptidyl Peptidase IV (DP IV, CD26) ...... 131
*Thilo Kähne, Dirk Reinhold, Klaus Neubert, Ilona Born, Jürgen Faust, and Siegfried Ansorge*

15. Specific Inhibitors of Dipeptidyl Peptidase IV Suppress mRNA Expression of DP IV/CD26 and Cytokines ........................ 139
*Marco Arndt, Dirk Reinhold, Uwe Lendeckel, Antje Spiess, Jürgen Faust, Klaus Neubert, and Siegfried Ansorge*

16. Dipeptidyl Peptidase IV in Inflammatory CNS Disease .............. 145
*Andreas Steinbrecher, Dirk Reinhold, Laura Quigley, Ameer Gado, Nancy Tresser, Leonid Izikson, Ilona Born, Jürgen Faust, Klaus Neubert, Roland Martin, Siegfried Ansorge, and Stefan Brocke*

17. Dipeptidyl Peptidase IV (CD26): Role in T Cell Activation and Autoimmune Disease ..... 155
*Dirk Reinhold, Bernhard Hemmer, Bruno Gran, Andreas Steinbrecher, Stefan Brocke, Thilo Kähne, Sabine Wrenger, Ilona Born, Jürgen Faust, Klaus Neubert, Roland Martin, and Siegfried Ansorge*

18. Effects of Nonapeptides Derived from the N-Terminal Structure of Human Immunodeficiency Virus-1 (HIV-1) TAT on Suppression of CD26-Dependent T Cell Growth ..... 161
*Sabine Wrenger, Dirk Reinhold, Jürgen Faust, Carmen Mrestani-Klaus, Wolfgang Brandt, Annett Fengler, Klaus Neubert, and Siegfried Ansorge*

19. DNA Synthesis in Cultured Human Keratinocytes and HACAT Keratinocytes is Reduced by Specific Inhibition of Dipeptidyl Peptidase IV (CD26) Enzymatic Activity ..... 167
*Robert Vetter, Dirk Reinhold, Frank Bühling, Uwe Lendeckel, Ilona Born, Jürgen Faust, Klaus Neubert, Siegfried Ansorge, and Harald Gollnick*

20. Attractin: A CUB-Family Protease Involved in T Cell-Monocyte/Macrophage Interactions ..... 173
*Jonathan S. Duke-Cohan, Wen Tang, and Stuart F. Schlossman*

21. Analogs of Glucose-Dependent Insulinotropic Polypeptide with Increased Dipeptidyl Peptidase IV Resistance ..... 187
*Kerstin Kühn-Wache, Susanne Manhart, Torsten Hoffmann, Simon A. Hinke, R. Gelling, Raymond A. Pederson, Christopher H.S. McIntosh, and Hans-Ullrich Demuth*

22. Dipeptidyl Peptidase IV (DPP IV, CD26) in Patients with Mental Eating Disorders ..... 197
*Martin Hildebrandt, Matthias Rose, Christine Mayr, Petra Arck, Cora Schüler, Werner Reutter, Abdulgabar Salama, and Burghard F. Klapp*

23. The Membrane-Bound Ectopeptidase CPM as a Marker of Macrophage Maturation in Vitro and in Vivo ..... 205
*Michael Rehli, Stefan W. Krause, and Reinhard Andreesen*

24. Matrix Metalloproteinases (MMP-8, -13, and -14) Interact with the Clotting System and Degrade Fibrinogen and Factor XII (Hagemann Factor) ..... 217
*Harald Tschesche, Andrea Lichte, Oliver Hiller, André Oberpichler, Frank H. Büttner, and Eckart Bartnik*

25. The Neprilysin Family in Health and Disease ..................... 229
*Anthony J. Turner, Carolyn D. Brown, Julie A. Carson, and Kay Barnes*

**Part II: Cellular Endopeptidases: New Cathepsins; Results from Knock-out-mice; Regulatory Aspects**

26. Review: Novel Cysteine Proteases of the Papain Family ............. 241
*Frank Bühling, Annett Fengler, Wolfgang Brandt, Tobias Welte, Siegfried Ansorge, and Dorit K. Nägler*

27. Development and Validation of Homology Models of Human Cathepsins K, S, H, and F ..................................... 255
*Annett Fengler and Wolfgang Brandt*

28. The Function of Propeptide Domains of Cysteine Proteinases ......... 261
*Bernd Wiederanders*

29. Human Cathepsins W and F Form a New Subgroup of Cathepsins that is Evolutionarily Separated from the Cathepsin B- and L-like Cysteine Proteases ............................................ 271
*Thomas Wex, Brynn Levy, Heike Wex, and Dieter Brömme*

30. Cathepsin K Expression in Human Lung .......................... 281
*Frank Bühling, Nadine Waldburg, Annegret Gerber, Carsten Häckel, Sabine Krüger, Dirk Reinhold, Dieter Brömme, Ekkehard Weber, Siegfried Ansorge, and Tobias Welte*

31. Expression of Cathepsins B and L in Human Lung Epithelial Cells is Regulated by Cytokines ....................................... 287
*Annegret Gerber, Tobias Welte, Siegfried Ansorge, and Frank Bühling*

32. Functions of Cathepsin K in Bone Resorption: Lessons from Cathepsin K Deficient Mice ................................................ 293
*Paul Saftig, Ernst Hunziker, Vincent Everts, Sheila Jones, Alan Boyde, Olaf Wehmeyer, Anke Suter, and Kurt von Figura*

33. Ceramide as an Activator Lipid of Cathepsin D .................... 305
*Michael Heinrich, Marc Wickel, Supandi Winoto-Morbach, Wulf Schneider-Brachert, Thomas Weber, Josef Brunner, Paul Saftig, Christoph Peters, Martin Krönke, and Stefan Schütze*

34. Human Cathepsin X: A Novel Cysteine Protease with Unique Specificity ................................................... 317
*Robert Ménard, Dorit K. Nägler, Rulin Zhang, Wendy Tam, Traian Sulea, and Enrico O. Purisima*

35. A Novel Proteolytic Mechanism for Termination of the $Ca^{2+}$ Signalling Evoked by Proteinase-Activated Receptor-1 (PAR-1) in Rat Astrocytes 323
*Joachim J. Ubl and Georg Reiser*

36. Natural and Synthetic Inhibitors of the Tumor-Associated Serine Protease Urokinase-Type Plasminogen Activator .................. 331
*Viktor Magdolen, Nuria Arroyo de Prada, Stefan Sperl, Bernd Muehlenweg, Thomas Luther, Olaf G. Wilhelm, Ulla Magdolen, Henner Graeff, Ute Reuning, and Manfred Schmitt*

37. Processing of Interleukin-18 by Human Vascular Smooth Muscle Cells 343
*Elena Westphal, Ivar Friedrich, Rolf-Edgar Silber, Karl Werdan, and Harald Loppnow*

**Part III: Peptidases and Peptidase Inhibitors in Pathogenesis of Diseases**

38. Review: Peptidases and Peptidase Inhibitors in the Pathogenesis of Diseases: Disturbances in the Ubiquitin-Mediated Proteolytic System. Protease-Antiprotease Imbalance in Inflammatory Reactions. Role of Cathepsins in Tumour Progression .............................. 349
*Ute Bank, Sabine Krüger, Jürgen Langner, and Albert Roessner*

39. The Role of Proteolysis in Alzheimer's Disease ................... 379
*Nigel M. Hooper, Alison J. Trew, Edward T. Parkin, and Anthony J. Turner*

40. Observing Proteases in Living Cells ............................ 391
*Kamiar Moin, Lisa Demchik, Jianxin Mai, Jan Duessing, Christoph Peters, and Bonnie F. Sloane*

41. The Role of Cysteine Proteases in Intracellular Pancreatic Serine Protease Activation ............................................ 403
*Markus M. Lerch, Walter Halangk, and Burkhard Krüger*

42. Peptidases in the Asthmatic Airways ............................ 413
*Vincent H.J. van der Velden, Brigitta A.E. Naber, Peter Th. W. van Hal, Shelley E. Overbeek, Henk C. Hoogsteden, and Marjan A. Versnel*

43. Inactivation of Interleukin-6 by Neutrophil Proteases at Sites of Inflammation: Protective Effects of Soluble IL-6 Receptor Chains ... 431
*Ute Bank, Bernadette Küpper, and Siegfried Ansorge*

44. Antisense Inhibition of Cathepsin B in a Human Osteosarcoma Cell Line ........................................................ 439
*Sabine Krueger, Carsten Haeckel, and Frank Buehling*

45. Protease-Protease Inhibitor Balance in the Gastric Mucosa: Influence of *Helicobacter pylori* Infection ..... 445
*Manfred Nilius, Thomas Vahldieck, Ina Repper, Armin Sokolowski, Paul Janowitz, and Peter Malfertheiner*

46. The Role of Bacterial and Host Proteinases in Periodontal Disease ..... 455
*James Travis, Agnieszka Banbula, and Jan Potempa*

47. Multifunctional Role of Proteases in Rheumatic Diseases ..... 467
*Jörn Kekow, Thomas Pap, and Susanne Zielinski*

48. Evidence of Proteolytic Activation of Transforming Growth Factor β in Synovial Fluid ..... 477
*Susanne Zielinski, Kathleen Bartels, Kristin Cebulski, Cornelia Kühne, and Jörn Kekow*

49. Matrix Metalloproteinases and Tace Play a Role in the Pathogenesis of Endometriosis ..... 483
*Constanze Gottschalk, Kurt Malberg, Marco Arndt, Johannes Schmitt, Albert Roessner, Dagmar Schultze, Jürgen Kleinstein, and Siegfried Ansorge*

50. Influence of Proliferation, Differentiation and Dedifferentiation Factors on the Expression of the Lysosomal Cysteine Proteinase Cathepsin L (CL) in Thyroid Cancer Cell Lines ..... 487
*Alexander Plehn, Dagmar Günther, Hendryk Aurich, Chuong Hoang-Vu, and Ekkehard Weber*

Contributors ..... 497

Index ..... 515

# CELLULAR PEPTIDASES IN IMMUNE FUNCTIONS AND DISEASES 2

# REVIEW: THE ROLE OF MEMBRANE PEPTIDASES IN IMMUNE FUNCTIONS

Uwe Lendeckel, Thilo Kähne, Dagmar Riemann[1], Klaus Neubert[2], Marco Arndt and Dirk Reinhold

*Institute of Experimental Internal Medicine, Center of Internal Medicine, Otto-von-Guericke University Magdeburg, Leipziger Strasse 44, D-39120 Magdeburg, Germany*

*[1]Institute of Medical Immunology,Martin-Luther-University Halle-Wittenberg, Strasse der Opfer des Faschismus 6, D-06097 Halle, Germany*

*[2]Institute of Biochemistry, Dept. Biochemistry & Biotechnology, Martin-Luther-University Halle-Wittenberg, Kurt-Mothes-Strasse 3, D-06120 Halle, Germany*

## 1. INTRODUCTION

Cell-surface proteolytic enzymes (ectopeptidases) comprise a growing family which over the last years attracted increasing numbers of both scientists in several disciplines and physicians. These enzymes differ in structure, catalytic mechanisms and substrate specificity, and, as far as the immune system is concerned, they are generally expressed in a cell-specific and highly regulated manner. For example, carboxypeptidase M (CPM) is expressed on macrophages only (Rehli 1995, Krause 1998) and dipeptidyl peptidase IV (DP IV, CD26) surface expression is increased significantly upon activation of NK and T cells (Kähne 1999). The nomenclature and terminology of ectopeptidases still lack complete coherence, but valuable attempts have been made to impose some order (Barrett 1998, Kenny 1997, Hooper 1994, Rawlings 1993). Roughly, peptidases requiring a free N- or C-terminus in a substrate are exopeptidases, those that do not are endopeptidases. Exopeptidases are classified according to their capability to release single amino acids, dipeptides or tripeptides. Endopeptidases are commonly subdivided into metallo-, serine, cysteine, and aspartic endopeptidases, dependent of the catalytic mechanisms they use. Principally, exopeptidases use the same catalytic mechanisms, allowing further subdividing of e. g. carboxypeptidases.

A growing number of cell-surface peptidases have been identified as leukocyte differentiation antigens (CD antigens): CD10 (CALLA, the common acute lymphoblastic leukaemia antigen) is identical to endopeptidase 24.11 (Letarte 1988), CD13 to aminopeptidase N (Look 1989), CD26 to dipeptidyl peptidase IV (Ulmer 1990) CD143 to angiotensin-converting enzyme (ACE), CD156 to A Disintegrin And Metalloproteinase 8 (ADAM 8), and the murine BP-1/6C3 antigen is aminopeptidase A (Wu 1990). Furthermore, there is evidence for cell-surface expression of CPM on monocytes (Rehli 1995), aminopeptidase B on T cells (Belhacene 1994), aminopeptidase A on early B cells (Welch 1995, Wang 1998), and aminopeptidase P on endothelial cells and activated T cells (Lasch 1998, Hendriks 1991), although no CDs have been assigned to these enzymes as yet.

Cell-surface peptidases of leukocytes and endothelial cells play essential roles in the regulation of immune function and they control physiological processes such as growth, differentiation, activation, cell-cell-interactions and transformation. A growing body of evidence suggests that even slight disturbances of the normal proteolytic balance promote the development of inflammatory and autoimmune diseases, metastasis, and e.g. Alzheimer's disease (cf. Hooper *et al* this book).

This subtle balancing of the proteolytic activity is defined firstly by the repertoire of ectopeptidases expressed on a single cell and, secondly, by the actual composition of different cell types at a local site. Thirdly, endopeptidases released from cells upon appropriate "triggering" and, of course, the local inhibitor concentration determine the net proteolytic activity. Strict regulatory mechanism act at all these levels, but may be disturbed under certain pathological conditions, such as infection and inflammation. Different proteases compete for or complement each other in the processing of a substrate. At the cellular level, ectopeptidase expression is regulated both differentiation- and activation-dependent with special combinations of cell-surface enzymes forming the "functional repertoire" of a cell. Notably, this may require contrary expression mechanisms as observed e.g. with the PMA-induced differentiation of K562 cells which results in abolishment of Kell blood group mRNA and protein, but induces mRNA, protein and enzymatic activity of the CD10 antigen (NEP) (Belhacene 1998). Similarly, an up-regulation of APN and DP IV, but not of aminopeptidase A, in response to IL-4, or a down-regulation of DP IV and aminopeptidase A, but not of APN, by TGF-β1 have been observed (Kehlen 1998).

Cell-surface peptidases are believed to proteolytically activate or inactivate susceptible hormones, growth factors, and transmitters. Loss of biological activity is observed e.g. after cleavage of a number of biologically active peptides, e.g. of the chemokine SDF-1 by DP IV (CD26) (Proost 1998b), substance P by APN and DPIV (Ahmad 1992, Kato 1978), and enkephalins by neutral endopeptidase (Malfroy 1982, Fulcher 1982). In contrast, the strong hypertonic angiotensin II is generated from the inactive angiotensin I by ACE. Proteolytic processing of a substrate may also switch its receptor-specifity: bradykinin binds with high affinity to the BK2 receptor, whereas des-arg-bradykinin resulting from cleavage by CPM has higher affinity to the BK1 receptor.

Our current knowledge of ectopeptidase substrates (Tab 1) is mainly derived from *in vitro* studies. Thus, showing the *in vivo* cleavage of potential substrates remains a challenging task. Insight into mechanisms of enzyme-substrate interaction at the molecular level are hampered by the limited availability of three-dimensional structures of ectopeptidases. The complex structure of these enzymes resulting from their large size, intense glycosylation and homo- or hetero-oligomerization probably accounts for this fact. Alternatively, computer-based structure modelling may help to circumvent crystallisation and, in the case of DP IV (CD26) already reached a promising stage (Brandt this book, ch. 9, Gorell this book, ch. 8). DNA-sequencing, biochemical studies and site-directed mutagenesis were helpful in disclosing primary structures and the extent of glycosylation and oligomerisation, and also identified active site residues. The assumed structure of representative cell-surface peptidases is schematically shown in Fig 1.

As outlined in a number of excellent recent reviews, cell-surface peptidases fulfil various functions dependent both on their substrate specificity and the site of expression (Kenny 1997). In this chapter the focus is on ectopeptidases of immune cells and several members of this group are discussed on this background below. In particular, the roles of dipeptidyl peptidase IV (DP IV, CD26) and alanine aminopeptidase (aminopeptidase N, APN, CD13) for normal and pathogenic immune responses will be emphasised.

## 2. AMINOPEPTIDASE N (CD13)

Aminopeptidase N (APN; CD13, EC 3.4.11.2) is a zinc-dependent metallopeptidase of the superfamily of gluzincins (Hooper 1994). Sequence comparisons of the cloned cDNA showed that APN is identical with the CD13 molecule (Look 1989). APN is a homodimer with a single helical transmembrane region and only a short N-terminal cytoplasmic tail, AKGFYISK (Olsen 1988). APN structure is described in more by H. Sjöström *et al* (this book, ch. 2). The enzyme hydrolyses a broad spectrum of

*Table 1.* Properties and functions of ectopeptidases in immune cells[1]

| enzyme names and synonyms | protease family substrate specificity[2] | mammalian expression | endogenous substrates | established inhibitors |
|---|---|---|---|---|
| **NEP (CD10)**<br>EC. 3.4.24.11<br>Neprilysin<br>CALLA<br>Neutral endopeptidase<br>Enkephalinase | family M13<br><br>polypeptides between hydrophobic residues, particularly with Phe or Tyr at P1' | kidney, leukemic lymphoblasts, CNS, lung, intestine, male genital tract, lymph nodes, polymorphonuclear neutrophils, placenta, fibroblasts, epithelial cells, glandular tissue | enkephalins, bradykinin, substance P, neurotensin, chemotactic peptide, gastrin, atrial natriuretic factor, luliberin | phosphoramidon, acetorphan, thiorphan, retrothiorphan, bestatin, carbaphethiol, kelathorphan, 1,10-phenanthroline, EDTA, DTT, phenylglyoxal, butanedione, puromycine, glutathione, N-(1(R,S)-carboxy-2-phenylethyl)-Phe-p-aminobenzoate |
| **APN (CD13)**<br>EC 3.4.11.2<br>Alanine aminopeptidase<br>Microsomal aminopeptidase<br>Aminopeptidase M<br>Aminopeptidase N<br>Amino-oligopeptidase | family M1<br>↓<br>H - Xaa - Xaa - ...<br>**Ala**<br>Phe<br>Tyr<br>~~Pro~~ | ubiquitous<br>high in: epithelial cells of the proximal tubuli of kidneys, small intestine, placenta, liver, fibroblasts, endothelial cells, monocytes, dendritic cells, granulocytes, plasma | MCP-1, Met-enkephalin, somatostatin, MSH, neurokinin A, tuftsin, kallidin, angiotensins, antigenic peptides, extracellular matrix | actinonin, bestatin, amastatin, chelating agents, p-hydroxymercuribenzoate, $Co^{2+}$, $Zn^{2+}$, $Mn^{2+}$, $Ca^{2+}$, $Ni^{2+}$, leucine, proline, L-alanine, L-arginine, L-glutamine, L-methionine, puromycin, $Hg^{2+}$, N-ethylmaleimide, 8-hydroxyquinoline, 2,2'-bipyridine, $NaN_3$, KCN, ammonium oxalate |
| **DP IV (CD26)**<br>EC 3.4.14.5<br>T-cell activation antigen<br>Tp103<br>Adenosine deaminase complexing protein-2<br>ADAbp | family S9B<br>↓<br>H - Xaa – Xaa - Xaa - ...<br>Gly **Pro** ~~Pro~~<br>Ala **Ala** ~~Hyp~~<br>Val ...<br>Leu Hyp<br>... Ser | ubiquitous<br>high in: kidney, liver, small intestine, placenta, T-lymphocytes, Morris hepatoma cells, epidermis (keratinized cells), fibroblasts, serum, hepatocytes, meconium, pancreas, submaxillary gland | substance P, ß-casomorphin, neuropeptide Y, collagen, peptide YY, fibrin (α-chain, monomeric), chemotactic peptides (RANTES, SDF-1, LIF....) | N-Ala-Pro-O-(4-nitrobenzoyl)-hydroxylamine, Xaa-Pro $^{(p)}$[O-Aryl]$_2$, Xaa-thiazolides, Xaa-pyrrolidides, Xaa-piperidides, Xaa-boroPro, DFP, PMSF, p-chloromercuribenzoate, N-ethylmaleimide, HIV-Tat, diprotin A, diprotin B, Xaa-Pro-dipeptides, proline in position 3 of peptide sequence |
| **ACE (CD143)**<br>EC 3.4.15.1<br>Dipeptidyl carboxypeptidaseI<br>Peptidyl dipeptidase I<br>Angiotensin I-converting enzyme<br>Kininase II<br>Peptidase P | family M2<br>↓<br>... – Xaa - Xaa - Xaa - OH<br>~~Pro~~ ~~Asp~~<br>~~Pro~~ | most or all organs<br>lung, serum, liver, blood vessels, kidney, intestine (brush border ), macrophages, brain, blood plasma, heart, skeletal muscle, adrenal, pancreas, spleen, placenta, seminal plasma, lymph nodes, testis | angiotensin I<br>bradykinin<br>enkephalins | bradykinin-potentiating peptides, angiotensin II, angiotensin III, fibrinopeptide A, tricyclic peptides, N-[1(S)-carboxy-5-aminopeptidyl]-glycylglycine, gastrin I, secretin, captopril, human albumin, acetyltryptophan, enalapril maleate, N-[(S)-1-carboxy-3-phenylpropyl]-L-Ala-L-Pro, 3,6-dihydroxy-1-phenazine-carboxylic acid (phenacein), p-chloromercuribenzoate, chelating agents, lisinopril |

| **ADAM 8 (CD156)**<br>EC 3.4.24.-<br>a disintegrin and<br>metalloprotease 8 | family M12B | neutrophils, monocytes | collagen, extracellular matrix | (chelating agents ?) |
|---|---|---|---|---|
| **MT-MMP1**<br>EC 3.4.24.-<br>MMP-14<br>Membrane-Type Matrix<br>Metalloproteinase 1 | family M10A | stromal cells of colon, breast, head and neck, monocytes | pro-gelatinase A (activation) fibronectin, vitronectin, laminin ß-chain, dermatan sulfate proteoglycan, gelatin, casein, elastin | TIMP-2, TIMP-3, (chelating agents ?) |
| **CPM**<br>EC 3.4.17.12<br>Carboxypeptidase M | family M14<br>↓<br>... - Xaa - Xaa - Xaa - OH<br>**Arg**<br>**Lys** | placenta (microvilli), kidney, macrophages, lung, | bradykinin, anaphylatoxins (C3a, C4a, C5a), fibrino-peptides, EGF, EGF-like proteins (1-7), enkephalins, HGF, MSP, erythropoietin, hemoglobin α-chain | guanidinoethyl mercaptosuccinic acid, 1,10-phenanthroline, 2-mercaptomethyl-3-guanidinoethyl thiopropanoic acid, $Cd(CH3COO)_2$, bathophenanthroline disulfonic acid |
| **ECE**<br>EC 3.4.24.71<br>Endothelin-converting<br>enzyme 1 | family M13 | vascular endothelial cells, lung, adrenal cortex, pancreas, placenta, peripheral blood leukocytes, prostate, testis | big endothelin 1, opioids, tachykinins, natriuretic peptides | phosphoramidon, EDTA, 1,10-phenanthroline, FR9015133 |
| **APA**<br>EC 3.4.11.7<br>Aminopeptidase A<br>Aspartate aminopeptidase<br>Angiotensinase A<br>Glutamyl aminopeptidase<br>Membrane aminopeptidase II | family M1<br>↓<br>H - Xaa - Xaa - ...<br>**Glu**<br>**Asp** | serum, kidney, intestine, intestinal mucosa, adrenal cortex, duodenum, B-cells, bone marrow cells, thymic stromal cells | angiotensin, angiotensin II | $Mn^{2+}$, $Ni^{2+}$, $Cd^{2+}$, $Cu^{2+}$, $Hg^{2+}$, p-hydroxy-mercuribenzoate, dithiothreitol, EDTA, 1,10-phenanthroline, sodium azide, L-aspartic acid, L-glutamic acid, puromycin, amastatin, EGTA, mercaptoethanol, L-α-glutaryl-L-phenylalanine |
| **APP**<br>EC 3.4.11.9<br>Aminopeptidase P<br>X-Pro aminopeptidase.<br>Proline aminopeptidase | family M24B<br>↓<br>H - Xaa - Xaa - ...<br>Arg **Pro**<br>Lys<br>~~Pro~~ | lung, kidney, heart, placenta, liver, small intestine, colon, leukocytes, platelets | bradykinin, collagen neuropeptide Y | apstatin, 2-hydroxy-3-aminoacyl-Pro, 2-hydroxy-3-aminoacid-pyrrolidide, 2-hydroxy-3-aminoacid-thiazolidide, EDTA, p-chloro-mercuribenzoate, 1,10-phenanthroline, $Co^{2+}$, heavy metal ions |

[1]In the table appear only those proteinases which – regardless of their presence in other cells and tissues – have been detected also in immune cells

[2]Arrows indicate bonds cleaved, ~~Pro~~ etc. denotes amino acid residues at the respective position that inhibit attack, cf. text

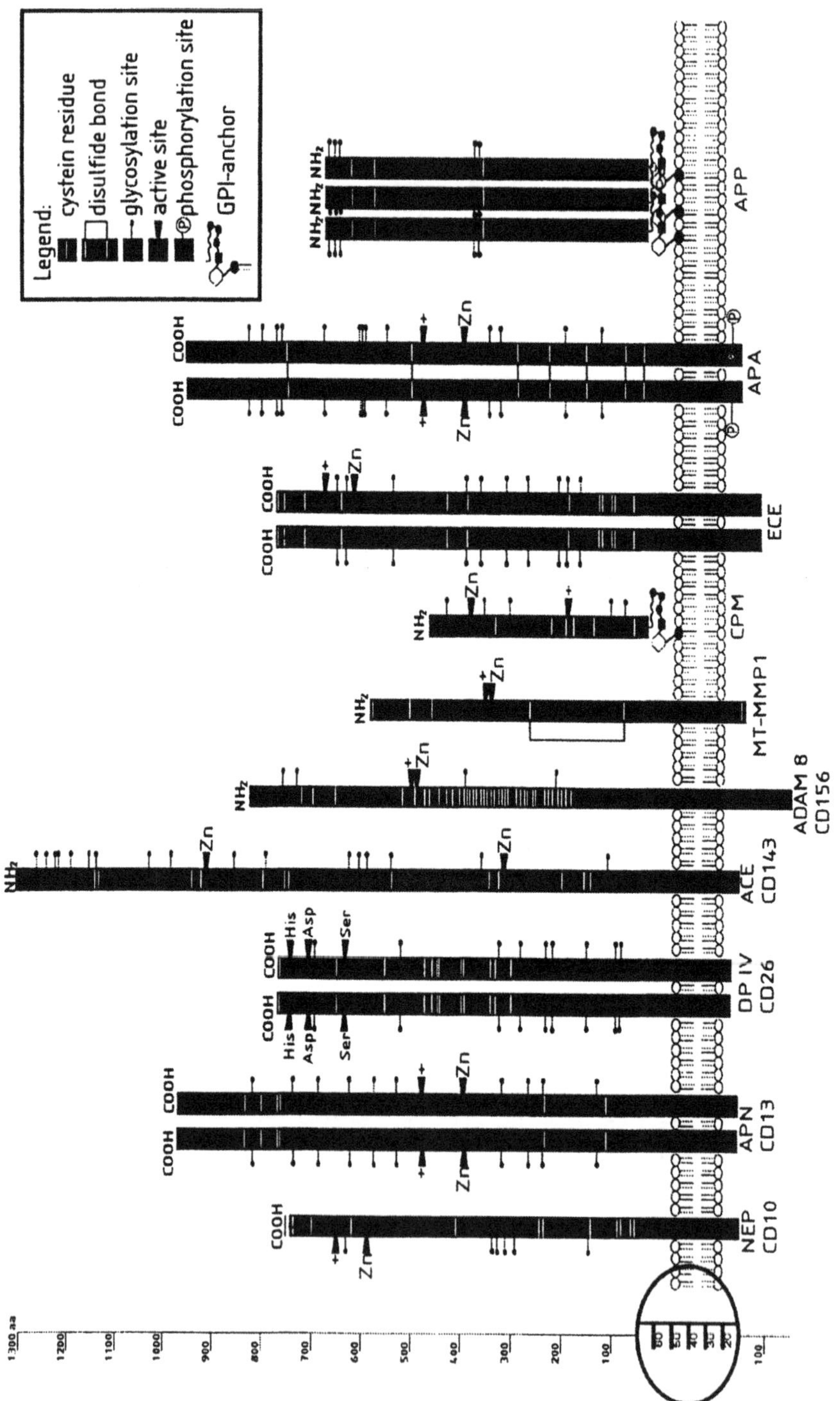


Fig.1 Schematic presentation of the protein structure of the membrane ectoenzymes covered in this review (extended and modified acc. to Shipp and Look, 1993)

oligopeptides. It cleaves preferentially neutral amino acids from the unsubstituted N-terminus of oligopeptides. APN has a widespread distribution, occurring on fibroblasts, epithelial cells, and endothelial cells, with main sources being brush border membranes of kidney proximal tubule cells (George 1973) and of enterocytes (Louvard 1973). With respect to hemopoietic cells, APN expression has been considered to be specific for the myeloid lineage, since monocytes/macrophages and granulocytes but not peripheral blood lymphocytes express this enzyme. However, CD13 surface expression has been described in some of the CD34+ B cell precursor cells (Syrjälä 1994) as well as early stages of T cell development (Spits 1995). APN specific mRNA has been detected in leukaemic T cell lines as well as in activated peripheral T cells (Lendeckel 1994, 1996). Expression of the myeloid marker molecule CD13 on otherwise typical lymphoblasts is often associated with specific genetic abnormalities, such as the t (12;21) translocation which fuses the TEL and AML1 genes (Pui 1998). Furthermore, CD13+ T cells can be found in the synovial fluid of patients with various forms of arthritis (Riemann 1993), on tumor-infiltrating lymphocytes of lung cancer (Riemann 1997a) and renal cancer (Riemann 1994a) or on pericardial fluid T cells, especially of patients undergoing thoracic surgery for heart valve replacement (Riemann 1994b).

The function of APN/CD13 varies depending on its location. In gut and kidneys, APN has been discussed to be involved in the terminal peptide degradation and amino acid scavenging (Kenny 1987). Otherwise, APN inactivates biologically active peptide substrates, or has been implicated to be involved in antigen presentation, trimming peptides protruding out of the binding groove of MHC class II molecules (Larsen 1996). Furthermore, similar as dipeptidyl peptidase IV/CD26, APN has been considered as an auxiliary adhesion molecule localised at sites of cell-cell contact in melanoma cells (Menrad 1993) and at places tightly associated with extracellular matrix components.

Since APN possesses no zymogen pro-form, its enzymatic activity is modulated either by regulation of synthesis, cellular transport and compartmentalisation, degradation, and by endogeneous inhibitors. Bradykinin and substance P are natural peptides inhibiting the enzyme in micromolar concentrations (Xu 1995). Otherwise, T-cell-derived cytokines, such as interleukin-4 and interferon-γ can up-regulate APN protein and mRNA expression in a variety of cells, among them monocytes and endothelial cells (Van Hal 1994). Similar as with other proteases, cell-cell contact seems to be involved in the regulation of APN/CD13 (see for review: Riemann 1999). Thus, cell-cell contact with bone marrow stromal cells causes a down-regulation of the enzyme on maturing B cells *in vitro* (Saito 1995). Otherwise, co-culture with various cells expressing APN (fibroblast-like synoviocytes, endothelial cells, monocytes/macrophages and others) induces mRNA synthesis as well as surface expression of catalytically active APN on T- and B lymphocytes (Riemann 1997b). It has been shown that direct cell-cell contact is required to induce lymphocytic APN expression, a finding further substantiated by the data of Riemann *et al* obtained from co-culture experiments with lymphocytes and synoviocytes (Riemann 1999).

Cell-surface peptidases might play a key role in the control of growth and differentiation of various cellular systems by modulating the activity of peptide factors and by regulating their access to adjacent cells. As an example, differentiation, e.g. of synovial macrophages, goes along with an augmentation of APN expression (Koch 1991, Emmrich 1990). Additionally, mitogenic activation of cells can be associated with an up-regulation of APN expression (Kohno 1985, Kunz 1993, Lendeckel 1996, 1994). It has been suggested that another ectopeptidase, namely neutral endopeptidase 24.11/CD10 can modulate proliferation of bronchial epithelial cells by cleavage of mitogenic bombesins (Ganju 1994). Expression of neutral endopeptidase 24.11 is inversely correlated with proliferation in bronchial epithelial cells and lung cancer cells (Shipp 1991, Cohen 1994). Otherwise, inhibiting APN enzyme activity by inhibitors, mAbs (Löhn 1997) or antisense strategies (Wex 1997) can inhibit cell growth of APN-expressing leukocytes (Review: Lendeckel *et al* 1999). This points to a role of this enzyme also in processes essential for proliferation. A possible explanation for the growth-diminishing effects of aminopeptidase inhibitors give Lendeckel and co-workers (this book, ch. 3) describing the actinonin-mediated prevention of the early activation-dependent increase of Wnt-5a in T cells. Furthermore, the selective inhibitors of APN, actinonin and probestin, have been shown to induce an increased expression and an activation of MAP kinases p42/p44 (Erk1/2) (Lendeckel 1998).

Biologically active peptide substrates cleaved by APN can be neuropeptides such as enkephalins and endorphins, vasoactive peptides such as kallidin and angiotensin III, or chemotactic peptides such as MCP-1 (Tab. 1). Interestingly, most of these substrates signal via G-protein-coupled hepta-helical receptors. Signal transduction initiated at these receptors involves kinase cascades commonly used by growth factors or during adhesion. One can consider that APN – by activating or inactivating biologically active peptides – could indirectly influence these signalling pathways. However, first observations point to a more complex picture with APN possibly directly involved in signal transduction: CD13-specific mAbs cannot only inhibit cell proliferation but can trigger an increase in the concentration of free cytoplasmic $Ca^{2+}$ in monocytic cell lines (McIntyre *et al* 1987). Furthermore, CD13 specific mAbs provoke phosphorylation of the MAP kinases ERK1/2, JNK and p38 in U937 cells (Navarrete Santos, submitted), results comparable to those obtained after cross-linking of CD26 (Hegen 1997b). Although the *in vivo* ligand as well as possible co-operating membrane molecules remain to be identified these results suggest that APN is a novel signal transduction molecule in human monocytes and, probably, in T cells.

## 3. AMINOPEPTIDASE P

Aminopeptidase P (APP; X-Pro-aminopeptidase; EC 3.4.11.9) is a widely distributed zinc metallopeptidase that is specific for N-terminal X-Pro peptide bonds in both short and longer peptides. The enzyme lacks the typical HEXXH motif common to many zinc peptidases. APP belongs to the methionine aminopeptidase family (M24) together with methionine aminopeptidase, creatinase and prolidase (Rawlings and Barrett 1995). These peptidases share a "pita bread" fold (Bazan *et al* 1994). Recent investigations showed that APP from E.coli contains two divalent metal ions within the active site and the biological function, amino acid sequence and metal-ion dependence of APP are very similar to those of human prolidase (Wilce *et al* 1998, Zhang *et al* 1998). Histidine residues crucial for APP enzymatic activity have been identified recently (Lim 1996). A soluble and a membrane form of APP have been isolated from mammalian tissues. Complementary DNA clones encoding the human membrane APP (mAPP, XPNPEP2) were obtained first from kidney and lung (Venema 1997). The mAPP gene has been mapped to chromosome Xq25 (Sprinkle 1998). Vanhoof *et al* reported on the sequence of a putative soluble APP (sAPP, XPNPEPL) cDNA, obtained from PHA-stimulated T cells (Vanhoof 1997). The high degree of homology and identity at both the nucleotide and amino acid level with the subsequently cloned and functionally expressed form of rat cytoplasmic form of APP (Czirjak 1999) confirmed the existence of two genetically distinct APP forms.

In humans, soluble APP is expressed in platelets (Vanhoof 1992), leukocytes (Rusu 1992), and testis (Czirjak 1999), whereas the expression of the membrane form (mAPP) is highest in the kidney, followed by other tissues in the order liver/small intestine > heart > lung > colon/placenta (Venema 1997). mAPP-mRNA could not be detected in human brain, skeletal muscle, leukocytes, pancreas, spleen, thymus, prostate, ovary or testis (Venema 1997). However, both aminopeptidase P enzymatic activity and immunoreactivity have been located at the cell-surface of human lymphocytes and endothelial cells (Lasch 1998, Ryan 1996).

The membrane APP is one of the few GPI-anchored cell-surface peptidases (Hooper 1988). A soluble form found in serum might be released from the membrane form by the action of endogenous phospholipases (Ryan 1992). The mAPP purified from pig kidney has a molecular weight of 91.000 with about 25 % by weight being due to N-linked glycosylation (Hooper 1990). Studies from Lloyd and Turner (1995) showed that the effect of activation of APP with divalent metal ions depends on the substrate. Activation is essential for the hydrolysis of Gly-Pro-Hyp, the hydrolysis of Arg-Pro-Pro or bradykinine, however, is inhibited by divalent cations.

APP is relatively resistant to inhibition by aminopeptidase inhibitors, such as actinonin, bestatin, puromycin and amastatin (Hooper 1990, 1992). However, some inhibitors of the angiotensin-converting enzyme (ACE) (L155.212, enalaprilat, cilazaprilat) significantly inhibit APP. The most efficient inhibitors currently available are apstatin (Orawski 1995) and analogous compounds containing a 2-hydroxy-3-amino acid in P1-position and a proline residue or proline analogue in P1'-position (Stöckel 1997).

A large number of biologically active peptides share a N-terminal X-Pro-sequence and, thus, are potential substrates of both APP and, as discussed below, DP IV. Cleaving the N-terminal Arg-Pro bond of the potent vasodilator bradykinin by APP is a key regulatory mechanism in the lung. Other APP substrates identified so far are substance P, neuropeptide Y and peptide YY (Orawski 1987, Yoshimoto 1994, Medeiros 1994).

All blood cells contain high activity of APP. A comparison of serum and thrombocytes showed, that APP activity in thrombocytes is more than 100 times higher than in serum (Scharpe 1990). Hendriks *et al* (1991) detected in mitogen activated human lymphocytes besides the increase of the specific activity of DP IV also an increase of APP activity.

## 4. DIPEPTIDYL PEPTIDASE IV (CD26)

The dipeptidyl peptidase IV (DP IV, EC 3.4.14.5) is an exopeptidase catalysing the release of N-terminal dipeptides from oligo- and polypeptides preferentially with proline, hydroxyproline and, with less efficiency, alanine in the penultimate position (Vanhoof 1995, Yaron 1993). This enzyme has a type II membrane topology with an extracellularly oriented catalytic domain (ectoenzyme). The cDNA encoding the human DP IV predicts a protein of 766 amino acids with 9 potential glycosylation sites (Fig 1). In the plasma membrane, DP IV occurs as a homodimer with a total molecular mass of 220 - 240 kD (Schön 1984). Abbott *et al* (1994) reported the detailed genomic organisation of the human DP IV gene, which is localised on chromosome 2q24.3 and is composed of 26 small exons. DP IV was firstly described by Hopsu-Havu and 33 years ago. Subsequent investigations outlined the ubiquitous distribution of this enzyme with exceptional high expression in intestine, kidney and liver (Fleischer 1994).

In 1977 Lojda *et al* (Lojda 1977) firstly demonstrated DP IV in human peripheral blood lymphocytes. 11 years later was discovered that DP IV is identical with the leukocyte surface antigen CD26 and at the 4th Workshop on

Leukocytes Differentiation Antigens a number of monoclonal antibodies recognizing DP IV were subsumed under the term CD26 (Ulmer 1990, Hegen 1990). In the haematopoietic system, DP IV/CD26 is expressed on the surface of resting and activated T cells, activated B, and activated NK cells (Schön 1984, Bühling 1995, Bühling 1994). The expression of CD26 is upregulated following mitogenic, antigenic, anti-CD3 or IL-2 stimulation of T cells, St. aureus protein stimulation of B cells and IL-2 stimulation of NK cells (Ansorge 1991, Schön 1984, Bühling 1994, Bühling 1995). Antigen-specific, CD4+ T cell clones (TCC) also express high levels of DP IV/CD26 (Reinhold 1998 and this book, ch. 17).

Within the gastrointestinal system the main function of DP IV is probably the delivering of proline containing dipeptides for reutilization by final digestion of nutrients (Tiruppathi 1993). Recently, Pauly *et al* (1996) expanded this function by reporting a DP IV-mediated processing of gastrointestinal hormones such as glucagon-like peptide (GLP) and glucose-dependent insulinotropic polypeptide (GIP) (Kühn-Wache *et al* this book, ch. 21). The function of DP IV/CD26 within the haematopoetic system seems to be much more complex.

Firstly, DP IV exopeptidase activity is supposed to play a role in the activation or inactivation of biologically active peptides (Vanhoof 1995, Yaron 1993, De Meester 1999). A number of peptides originating from the neuroendocrine system (e.g. substance P, ß-casomorphin, neuropeptide Y, peptide YY, growth hormone releasing factor) were described to be substrates of DP IV (Yaron 1993). Various cytokines and growth factors, such as IL-3, IL-5, IL-10, IL-11, IL-13, GM-CSF, G-CSF, RANTES, LIF and thrombopoietin, are characterized by the particular N-terminal structure with proline in the second position which is probably susceptible for DP IV (Kähne 1999, De Meester 1999 and this book, ch. 7).

Hoffmann *et al* (1993) found that DP IV alone, or in combination with the aminopeptidase N, is capable of hydrolyzing oligopeptides analogous to the N-terminal structure of different cytokines (IL-1ß, IL-2, murine IL-6, TNF-ß), rather than the corresponding intact full length cytokines. The rate of DP IV-catalyzed hydrolysis was negatively correlated with their chain length (Hoffmann 1993). Nevertheless, in recent publications it was shown, that some chemokines like RANTES and SDF-1 represent good substrates of DP IV in vitro (Proost 1998a, 1998b, Schols 1998, Oravecz 1997, De Meester 1999 and this book, ch. 7). Whether these chemokines are also substrates of DP IV in vivo remains to be established. Furthermore, it cannot be excluded that the binding of other cytokines and growth factors to their receptors may alter the DP IV susceptibility for these peptides due to conformational changes and unmasking of their N-terminal part.

Secondly, DP IV/CD26 could be a receptor or ligand for different proteins. The enzyme is reportedly associated with CD45 (Torimoto 1991) and has been described as a collagen receptor (Hanski 1985, Löster 1995) and an adenosine deaminase (ADA) binding protein (Kameoka 1993, Morrison 1993). Recently, Dong *et al* could assign the detailed ADA binding domain to the CD26 amino acids $L^{340}$, $V^{341}$, $A^{342}$ and $R^{343}$ (Dong 1997). ADA binding seems not to influence the enzymatic activity of DP IV (Kähne, 1999). Gutheil *et al* (1994) demonstrated that HIV-1 Tat binds to DP IV/CD26 and inhibits its dipeptidyl peptidase activity. Detailed investigations of Tat-mediated inhibition of DP IV pointed out the requirement of the N-terminal nonapeptide Tat(1-9) for this process (Wrenger 1997 and this book, ch.18).

Thirdly, different groups have shown a key role of DP IV in the regulation of differentiation and growth of T lymphocytes (Ansorge 1991, Dang 1990a, Dang 1990b, Dang 1991, Schön 1991, Schön 1984, Schön 1989, Schön 1987, Torimoto 1991, Torimoto 1992). Because triggering of the T cell receptor (TcR)/CD3 complex alone cannot induce T cell proliferation, accessory molecules are required for generation of costimulatory signals. Various studies provided evidence that besides the known accessory molecules, e. g. CD2 or CD28, the CD26 antigen also can provide a costimulatory signal in human T cells (Dang 1990a, Mittrücker 1995, Fleischer 1994, Morimoto 1998, Ansorge 1991). Certain anti-CD26 monoclonal antibodies are capable of enhancing the submitogenic activity of low doses of phorbol myristate acetate (PMA), cytokines, or anti-CD2/CD3 monoclonal antibodies which trigger T lymphocyte proliferation (Morimoto 1989, Dang 1991, Ansorge 1997, Morimoto 1998, Plana 1991, Torimoto 1991, Reinhold 1997c).

Using a CD26-transfected Jurkat T cell line which lacked DP IV enzymatic activity but still reacted with CD26 antibodies, Tanaka *et al* (1993) showed that DP IV enzymatic activity enhances IL-2 production induced via CD26-related and -unrelated pathways. They found that transfectants expressing DP IV activity on their surface produced substantially more IL-2 than did transfectants with proteolytically inactive DP IV after triggering by a combination of anti-CD26 and anti-CD3 or anti-CD3 antibodies plus PMA (Tanaka 1993). Hower, using a mouse $TCR^+$ T cell hybridoma, other authors have shown that the enzymatic activity of CD26 is not required for its stimulatory and costimulatory activity (Steeg 1995, Von Bonin 1998).

One of the most common approaches to examine the cellular function of an enzyme is to use specific inhibitors. In 1985 Schön *et al* (1985) showed for the first time that N-Ala-Pro-O-(nitrobenzoyl-)hydroxylamine, which irreversibly inhibits DP IV, is capable of suppressing the proliferation of human PBMC stimulated with mitogens. This was the first evidence that DP IV plays a critical role in the regulation of DNA synthesis and that the enzymatic activity of DP IV is involved in this process. In the mean time a multitude of different synthetic inhibitors of DP IV have been studied under different stimulation conditions in different human and mouse T cell models.

Flentke *et al* (Flentke 1991, Subramanyam 1993) demonstrated that Ala-boro-Pro and Pro-boro-Pro are potent and specific inhibitors of DP IV which suppress the antigen-induced proliferation and IL-2 production in murine T cell lines, and the antigen- and CD3-induced DNA synthesis of human PBMC.

The reversible DP IV inhibitors Lys[Z($NO_2$)]-thiazolidide, -piperidide, and -pyrrolidide inhibit DNA synthesis and production of IL-2, IL-10, IL-12, and IFN-$\gamma$ of pokeweed mitogen (PWM)-stimulated PBMC and purified T cells (Reinhold 1997a). Moreover, these inhibitors suppress the DNA synthesis as well as production of IL-2, IL-6 and IL-10 of PHA-stimulated mouse splenocytes and Con A-stimulated mouse thymocytes (Reinhold 1997b). As shown by competitive RT-PCR, in the presence of the DP IV inhibitor Lys[Z($NO_2$)]-thiazolidide, levels of IL-2 mRNA were found to be significantly decreased in PHA-stimulated T cells (Kähne 1999, Arndt *et al* this book, ch. 15).

Interestingly, after DP IV inhibitor treatment, an enhanced production and secretion of latent Transforming Growth Factor-ß1(TGF-ß1) was found on pokeweed mitogen (PWM)-stimulated PBMC and purified T cells (Reinhold, 1997a) as well as on PHA-stimulated mouse splenocytes and Con A-stimulated mouse thymocytes (Reinhold 1997b). Moreover, levels of TGF-ß1 mRNA were shown to be 2- to 3-fold increased in PHA- and PWM-stimulated PBMC and T cells in presence of Lys[Z($NO_2$)]-thiazolidide (Kähne 1998). TGF-ß1 exhibits the same inhibitory effects as DP IV inhibitors on DNA synthesis and cytokine production (Reinhold 1994, Reinhold 1995). A neutralising chicken anti-TGF-ß1 antibody was capable of abolishing the DP IV inhibitor-induced suppression of DNA synthesis of PWM-stimulated PBMC and T cells, suggesting that TGF-ß1 might have key functions in the molecular action of DP IV/CD26 concerning the regulation of DNA synthesis and cytokine production.

Data of different groups provided evidence that anti-CD26 binding or inhibitor treatment can provoke the transduction of cellular signals. Whether these signal transduction events can be generated by DP IV alone, by recruiting of accessory molecules or by the disturbance of a delicately balanced steady state of bioactive peptides unknown so far, but trimmed by DP IV, remains to be clarified. Crosslinking of CD26 by monoclonal antibodies causes tyrosine phosphorylation of several intracellular proteins involved in TCR/CD3-mediated signal transduction, including the $\zeta$ chain, $p56^{lck}$, $p59^{fyn}$, ZAP-70 and mitogen-activated kinase (MAPK) (Hegen 1997b). Recent findings indicated that besides anti-CD26 antibodies also DP IV inhibitors directly induce a panel of newly tyrosine phosphorylated proteins (Kähne 1998). A temporary activation of the MAP kinase p38 HOG was found after DP IV inhibition, whereas other MAP kinase pathways including Erk1/2 and SAP/Jun kinase remained unaffected (Kähne 1999). These findings support the hypothesis of a p38 MAPK-mediated pathway of DP IV inhibitor-induced TGF-ß1 production and secretion (Kähne 1999).

Moreover, DP IV inhibitors are found to suppress early activation signals of T cells induced by PMA or anti-CD3, like the PMA-induced hyper-phosphorylation of $p56^{lck}$ (Kähne 1995) and the anti-CD3-induced $Ca^{2+}$ flux in these cells (Kähne 1998). Recently, it was demonstrated that the temporarily anti-CD3-mediated activation of two other signal cascades, namely the phospholipid kinase pathway initiated by PI3-kinase PKB (akt) and the ras-erk1/erk2 pathway can be also suppressed by DP IV inhibitors (Kähne 1999 and this book, ch. 14).

Taken together, the data suggest that immune intervention via DP IV/CD26 could have potential practical consequences in autoimmune diseases, allograft rejection and malignancies. Korom *et al* (1997) reported about the usage of the irreversible DP IV inhibitor prodipine in a rat transplantation model. This treatement prevented the increase in serum DP IV normally seen during the first days after cardiac allotransplantation. Both allograft rejection and the concomitant increase of DP IV activity in transplantant tissue were delayed. In two other studies, arthritis symptoms in rats could be suppressed by several DP IV inhibitors (Tanaka 1997, Tanaka 1998). Studies are also in progress to investigate the effect of DP IV inhibitors in experimental autoimmune encephalomyelitis (EAE) and Multiple Sclerosis (Steinbrecher *et al,* this book, ch. 16).

## 5. CARBOXYPEPTIDASES

Carboxypeptidases (CPs) catalyse the hydrolysis of the C-terminal peptide bond, releasing a single amino acid. For many bioactive peptides this results in profound changes of biological activity or receptor specificity. Basically, CPs fall into the two groups of metallo-CPs or serine CPs. On the basis of substrate specificity metallo-CPs can be subdivided into enzymes of CPA- or CPB-type. The A-type enzymes preferentially cleave off hydrophobic residues from the C-terminus of a substrate, whereas the action of B-type enzymes is directed towards Lys or Arg residues. Among the CPs only CPM and CPD have been clearly shown to be expressed on the plasma membrane (Skidgel 1996, Kuroki 1994).

CPM, a member of the group of basic carboxypeptidases (B-type), was initially isolated from human placenta (Skidgel 1989), where it is expressed at high level. Its cDNA was cloned in 1989 (Tan 1989). The 62 kDa monomeric, GPI-anchored enzyme is Zn-dependent and has a neutral pH-optimum. It is involved in the processing of various substrates including bradykinin, enkephalin hexapeptides and anaphylatoxins (for review see Krause 1998).

Two groups proved the expression of CPM on blood cells: CPM is not detectable on monocytes, but highly expressed upon their differentiation into macrophages (Rehli 1995 and this book, ch. 23). Furthermore, pre-B cells and activated germinal center cells seem to express CPM, whereas circulating B cells do not (de Saint-Vis 1995). Stimulation of resting B cells by CD40 ligation or CD19 ligation specifically enhances CPM expression on these cells (Galibert 1996). Expression of the other membrane-bound CP, CPD, was reported for lymphoid cells (Xin 1997) and the mouse monocyte/macrophage-like cell line J774.A1 (McGwire 1997). The function of leukocyte CPD remains to be established.

It is intriguing, however, that the two cell-surface CPs, CPM and CPD, are expressed on immune cells, where they are regulated in a differentiation-dependent manner. Based on the substrates identified so far, CPM expressed on macrophages presumably plays an anti-inflammatory role. Inactivation of bradykinin, anaphylatoxins C5a, C4a and C3a all may contribute to this effect. As discussed extensively by others (Krause 1998, Rehli this book), effector-mechanisms employing activated macrophages as in allograft rejection, tumour cytotoxicity and sarcoidosis are associated with enhanced CPM expression.

## 6. ADAMs

ADAM proteins, highly conserved among different species perform several functions in a variety of cells. These proteins form a subfamily of the metzincin superfamily of Zn-dependent metalloproteinases, containing a metalloproteinase attached to a disintegrin domain and were therefore named ADAMs (A Disintegrin And Metalloproteinase) or referred to as MDCs (Metalloproteinase Disintegrin Cysteine rich proteins) (Wolfsberg and White 1995, 1996, 1998, Black 1998). They share this structure and the catalytic site sequence, HEXGHXXGXXHD, with the other members of the subfamily, the PIII class of snake venom metalloproteinases (SVMPs). The ADAM family members contain an additional cytoplasmatic domain, a transmembrane region and an EGF repeat (Black and White 1998). At present, 29 different ADAMs are known, which are expressed mainly in reproductive organs and all are cell surface adhesion proteins (Yamamoto 1999).

So far, our knowledge of the function of ADAMs is very limited. The majority of ADAMs (e.g. ADAMs 1-7) play a role in spermatogenesis and/or sperm-egg cell fusion, but are also expressed in some other cells (e.g. ADAM 1 or 5) (Yamamoto 1999). The meltrins ADAM 9 (γ), 12 (α) and 19 (β) may be involved in myogenesis and/or osteogenesis and are expressed mainly in lung, breast, bone and fetal muscle (Yagami-Hirosama 1995, Inoue 1998,

Gilpin 1998, Weskamp 1996). The cytoplasmatic tail of ADAM 9 (MDC9) contains two proline-rich Src homology 3 (SH3)-binding sequences that are capable of binding biotinylated Src protein and might function in signal transduction (Weskamp 1996). SH3-binding sequences are also found in ADAM 10, 12, 15-17 and 19 (Wolfsberg and White 1996, Yamamoto 1999). ADAM 11 could be found in several organs and MDC-524 appears to be a truncated form of it (Katagiri 1995, Yamamoto 1999). Metargidin (MDC15; ADAM 15) contributes in regulating blood vessel function (Herren 1997) and ADAM 13 is involved in neuronal crest migration and/or somatogenesis (Alfandari 1997).

The human homologue of mouse MS2 (ADAM 8) is identical with the myeloid antigen CD156 which was mapped at the sixth International Congress on Human Differentiation Antigens (Higuchi 1998). Similar to ADAM 9, CD156 contains a SH3- and an additional Abl SH3-binding sequence suggesting a role in signal transduction. CD156 is involved in immune function, where it reportedly facilitates tissue infiltration by neutrophils (reviewed in Yamamoto 1999). Recently, a novel member of ADAMs, ADAM 28 was identified, existing in two forms, a transmembrane (MDC-Lm) and a secreted one (MDC-Ls). The mRNA of MDC-Ls is mainly expressed in lymphoid tissue (e.g. spleen and lymph node). MDC-Lm protein is expressed in human peripheral blood lymphocytes, lymphoid tissue and transformed B- and T-lymphocyte cell lines. Therefore, a possible role of ADAM 28 in the interacting of lymphocytes with their constantly changing environment (Roberts 1999) is discussed.

Two ADAMs are well characterised: ADAM 10 (MADM; kuzbanian; SUP-17) and ADAM 17 (TNF-α converting enzyme, TACE) (for reviews see Blobel 1997, Black and White 1998, Yamamoto 1999, and references therein). ADAM 10, a 62 kDa protein was first identified as kuzbanian required for neuronal development in Drosophila in the process of lateral inhibition. Kuzbanian is directly or indirectly involved in the Notch-signalling pathway. The Notch-signalling pathway, highly conserved from invertebrates to humans, is employed in lineage decisions in both vertebrates and invertebrates, e. g. in intrathymic T-cell differentiation into single-positive $CD4^+$ or $CD8^+$ cells (reviewed in Robey 1997, Hayday 1999). The identity of the Notch activating protease, however, is still a matter of debate. In addition, ADAM 10 is a candidate for TNF-α processing.

The well known protease that specifically cleaves the cell surface TNF-α precursor to release the mature form is the 85 kDa protein ADAM 17 (TACE). TNF-α is a key player in inflammatory processes, but it causes damage when expressed in large excess. Among other cell types monocytes, T-cells, neutrophils and myocytes express TACE mRNA. Patel and co-workers (1998) show that TACE is upregulated in arthritis-affected cartilage and this may influence TNF-α expression or activity. The role of TACE as a α-secretase for amyloid precursor protein is still discussed controversially.

In addition to the functions described above, 17 of the 29 ADAM proteins are putative active proteases, which cleave both extracellular matrix (ECM) molecules as well as cell surface proteins (Hooper 1997). Candidates for such substrates are: pro-cytokines, pro-growth factors, their receptors, ligands for "death receptors", membrane-bound enzymes and adhesion molecules. It should be emphasised that ADAMs 8-10, 15, 17, and 28 are expressed in macrophages and/or T-cells (Yamamoto and Roberts 1999), suggesting a role in T-cell development, activation or immune effector mechanisms.

## REFERENCES


Abbott, C.A., *et al.*, 1994, Genomic organization, exact localization, and tissue expression of the human CD26 (dipeptidyl peptidase IV) gene. *Immunogenetics* **40:** 331-338.

Alfandari, D., *et al.*, 1998, ADAM 13: a novel ADAM expressed in somitic mesoderm and neural crest cells during *Xenopus laevis* development. *Dev. Biol.* **182:** 314-330.

Ansorge, S., *et al.*, 1991, Membrane-bound peptidases of lymphocytes: Functional implications. *Biomed. Biochem.. Acta* **50:** 799-807.

Ansorge, S., *et al.*, 1997, CD26 dipeptidyl peptidase IV in lymphocyte growth regulation. *Adv. Exp. Med. Biol.* **421:** 127-140.

Barrett, A.J., Rawlings, N.D., and Woessner, J.F., 1998, Handbook of proteolytic enzymes. Academic Press,

Bazan, J.F., *et al.*, 1994, Sequence and structure comparison suggest that methionine aminopeptidase, prolidase, aminopeptidase P, and creatinase share a common fold. *Proc. Natl. Acad. Sci. USA* **91:** 2473-2477.

Belhacene, N., *et al.*, 1993, Characterisation and purification of T-lymphocyte aminopeptidase B: a putative marker of T cell activation. *Eur. J. Biochem..* **23:** 1948-1955.

Belhacene, N., *et al.*, 1998, Differential expression of the Kell blood group and CD10 antigens: two related membrane metallopeptidases during differentiaton of K562 cells by phorbol ester and hemin. *FASEB J.* **12:** 531-539.

Black, R.A., 1998, In *Handbook of Proteolytic Enzymes* (A.J. Barrett, N.D. Rawlings and J.F. Woessner, eds), Academic Press, San Diego/London, pp. 1315-1317.

Black, R.A. and White, J.M., 1998, ADAMs: focus on the protease domain. *Curr. Opin. Cell Biol.* **10:** 654-659.

Blobel C.P., 1997, Metalloprotease-Disintegrins: links to cell adhesion and cleavage of TNFα and Notch. *Cell* **90:** 589-592.

Bühling, F., *et al.*, 1994, Expression and functional role of dipeptidyl peptidase IV on human natural killer cells. *Nat. Immun.* **13:** 270-279.

Bühling, F., *et al.*, 1995, Functional role of CD26 on human B lymphocytes. *Immunol. Lett.* **45:** 47-51.

Cohen A.J., *et al.*, 1996, Neutral endopeptidase : variable expression in human lung, inactivation in lung cancer, and modulation of peptide-induced calcium flux. *Cancer Res.* **56:** 831-839.

Cottrell, G.S., *et al.*, 1998, The cloning and functional expression of human pancreatic aminopeptidase P. *Biochem. Soc. Trans.* **26:** S248.

Czirjak, G., *et al.*, 1999, Cloning and functional expression of the cytoplasmic form of rat aminopeptidase P. *Biochim. Biophys. Acta* **1444:** 326-336.

Dang, N.H., *et al.*, 1990a, Comitogenic effect of solid-phase immobilized anti-1F7 on human CD4 T cell activation via CD3 and CD2 pathways. *J. Immunol.* **144:** 4092-4100.

Dang, N.H., *et al.*, 1990b, Cell surface modulation of CD26 by anti-1F7 monoclonal antibody. Analysis of surface expression and human T cell activation. *J. Immunol.* **145:** 3963-3971.

Dang, N.H., *et al.*, 1991, 1F7 (CD26): A marker of thymic maturation involved in the differential regulation of the CD3 and CD2 pathways of human thymocyte activation. *J. Immunol.* **147:** 2825-2832.

De Meester, I., *et al.*, 1999, CD26, let it cut or cut it down. *Immunol. Today* **20:** 367-375

de Saint-Vis, B., *et al.*, 1995, Distribution of carboxypeptidase-M on lymphoid and myeloid cells parallels the other zinc-dependent proteases CD10 and CD13. *Blood* **86:** 1098-1105.

Dong, R.P., *et al.*, 1997, Determination of adenosine deaminase binding domain on CD26 and its immunoregulatory effect on T cell activation. *J. Immunol.* **159:** 6070-6076.

Emmrich, F., *et al.*, 1990, Phenotype analysis of "inflammatory" macrophages in rheumatoid arthritis by two-colour cytofluorometry and immunohistology. Agents Actions **29:** 95-97.

Fleischer, B., 1994, CD26: a surface protease involved in T-cell activation. *Immunol. Today* **15:** 180-184.

Flentke, G.R., *et al.*, 1991, Inhibition of dipeptidyl aminopeptidase IV (DP IV) by Xaa-boroPro dipeptides and use of these inhibitors to examine the role of DP IV in T-cell function. *Proc. Natl. Acad. Sci.USA* **88:** 1556-1559.

Fulcher I.S., *et al.*, 1982, Kidney neutral endopeptidase and the hydrolysis of enkephalin by synaptic membranes show similar sensitivity to inhibitors. *Biochem. J.* **203:** 519-22.

Galibert, L., *et al.*, 1996, CD40 and B cell antigen receptor dual triggering of resting B lymphocytes turns on a partial germinal center phenotype. *J. Exp. Med.* **183:** 77-85.

Ganju, R.K., *et al.*, 1994, CD10/NEP in non-small cell lung carcinomas. Relationship to cellular proliferation. *J. Clin. Invest.* **94:** 1784-1791.

George, S.G. and Kenny, A.J., 1973, Studies on the enzymology of purified preparations of brush border from rabbit kidney. *Biochem. J.* **134:** 43-57.

Gilpin, B.J., *et al.*, 1998, A novel secreted form of human ADAM 12 (meltrin $\alpha$) provokes myogenesis *in vivo*. *J. Biol. Chem.* **273:** 157-166.

Gutheil, W.G., *et al.*, 1994, Human immunodeficiency virus 1 Tat binds to dipeptidyl aminopeptidase IV (CD26): A possible mechanism for Tat s immunosuppressive activity. *Proc. Natl. Acad. Sci. USA* **91:** 6594-6598.

Hanski, C., *et al.*, 1985, Involvement of plasma membrane dipeptidyl peptidase IV in fibronectin-mediated adhesion of cells on collagen. *Biol. Chem. Hoppe-Seyler* **366:** 1169-1176.

Hayday, A.C., *et al.*, 1999, Signals involved in gamma/delta T cell versus alpha/beta T cell lineage commitment. *Semin. Immunol.* **11:** 239-249.

Hegen, M., *et al.*, 1990, The T cell triggering molecule Tp103 is associated with dipeptidyl aminopeptidase IV activity, J. Immunol. **144:** 2908-2914.

Hegen, M., *et al.*, 1997a, CD26 workshops panel report. In *Leucocyte Typing VI*. (T. Kishimoto, M. Kikutani, and A.G.E. Kr. von dem Borne, eds), Garland Publishing, Inc., New York/London, pp. 478-81.

Hegen, M., *et al.*, 1997b, Cross-linking of CD26 by antibody induces tyrosine phosphorylation and activation of mitogen-activated protein kinase. *Immunology* **90:** 257-264.

Hendriks, D., *et al.*, 1991, Aminopeptidase P and dipeptidyl peptidase IV activity in human leukocytes and in stimulated lymphocytes. *Clin. Chimica Acta* **196**: 87-96.

Herren, B., *et al.*, 1997, Expression of a disintegrin-like protein in cultured human vascular cells and in vivo, *FASEB J.* **11**: 173-180.

Higuchi, Y., *et al.*, 1997, CD156 workshop panel report. In *Leucocyte typing IV.* (T. Kishimoto, M. Kikutani, and A.G.E. Kr. von dem Borne, eds), Garland Publishing Press, New York/London, pp.1083-1085.

Hoffmann, T., *et al.*, 1993, Dipeptidyl peptidase IV (CD26) and aminopeptidase N (CD13) catalyzed hydrolysis of cytokines and peptides with N-terminal cytokine sequences. *FEBS Lett.* **336**: 61-64.

Hooper, N. M., *et al.*, 1988, Ectoenzymes of the kidney microvillar membrane. Aminopeptidase P is anchored by a glycosyl-phosphatidylinositol moiety. *FEBS Letters* **229**: 340-344.

Hooper, N. M., and Turner, A.J., 1990, Purification and characterisation of pig kidney aminopeptidase P. A glycosyl-phosphatidylinositol-anchored ectoenzyme. *Biochem. J.* **267**:509-515.

Hooper, N. M., *et al.*, 1992, Inhibition by converting enzyme inhibitors of pig kidney aminopeptidase P. *Hypertension* **19**: 281-285.

Hooper, N. M., 1994, Families of zinc metalloproteinases. *FEBS Letters* **354**: 1-6.

Hooper, N. M., *et al.*, 1997, Membrane protein secretases. *Biochem. J.* **321**: 265-279.

Hopsu-Havu, K.V., and Glenner, G.G., 1966, A new dipeptide naphthylamidase hydrolyzing glycyl-prolyl-beta-naphthylamide. *Histochemie* **7**: 197-201.

Inoue, D., *et al.*, 1998, Cloning and initial characterization of mouse meltrin beta and analysis of the expression of four metalloprotease-disintegrins in bone cells. *J. Biol. Chem.* **273**: 4180-4187.

Kameoka, J., *et al.*, 1993, Direct association of adenosine deaminase with a T cell activation antigen, CD26. *Science* **261**: 466-469.

Katagiri, T., *et al.*, 1995, Human metalloprotease/disintegrin-like (MDC) gene: exon-intron organization and alternative splicing. *Cytogen. Cell. Genet.* **68**: 39-44.

Kato, T., *et al.*, 1978, Successive cleavage of N-terminal Arg1-Pro2 and Lys3-Pro4 from substance P but no release of Arg1-Pro2 from bradykinin, by X-Pro dipeptidyl-aminopeptidase. *Biochim. Biophys. Acta* **525**: 417-422.

Kähne, T., *et al.*, 1995, Enzymatic activity of DP IV/CD26 is involved in PMA-induced hyperphosphorylation of p56lck. *Immunol. Lett.* **46**: 189-193.

Kähne, T., *et al.*, 1998, Early phosphorylation events induced by DPIV/CD26-specific inhibitors. *Cell. Immunol.* **189**: 60-66.

Kähne, T., *et al.*, 1999, Dipeptidyl peptidase IV: A cell surface peptidase involved in regulating T cell growth (Review). *Int. J. Mol. Med.* **4**: 3-15.

Kehlen, A., *et al.*, 1998, Regulation of the expression of aminopeptidase A, aminopeptidase N/CD13 and dipeptidyl peptidase IV/CD26 in renal cell carcinomas cells and renal tubular epithelial cells by cytokines and cAMP-increasing mediators. *Clin. Exp. Immunol.* **111**: 435-441.

Kenny, A.J., *et al.*, 1987, Cell surface peptidases. In *Mammalian Ectoenzymes* (A.J. Kenny and A.J. Turner, eds), Elsevier Science Publishers, Amsterdam, pp. 169-210.

Kenny, A.J., 1997, Ectoenzymes, ectopeptidases and CD antigens. In *Cell-surface peptidases in health and disease* (A.J. Kenny and C.M. Boustead, eds.), BIOS Scientific Publishers Ltd., Oxford, 1-9.

Koch, A.E., *et al.*, 1991, Monoclonal antibodies detect monocyte/macrophage activation and differentiation antigens and identify functionally distinct subpopulations of human rheumatoid synovial tissue macrophages. *Am. J. Pathol.* **138**: 165-173.

Kohno, H., and Kanno, T., 1985, Properties and activities of aminopeptidases in normal and mitogen-stimulated human lymphocytes. *Biochem. J.* **226:** 59-65.

Korom, S., *et al.*, 1997, Inhibition of CD26/dipeptidyl peptidase IV activity in vivo prolongs cardiac allograft survival in rat recipients. *Transplantation* **63:** 1495-1500.

Krause, S. W., Rehli, M., and Andreesen, R., 1998, Carboxypeptidase M as a marker of macrophage maturation. *Immunol. Reviews* **161:** 119-127.

Kunz, D., *et al.*, 1993, Aminopeptidase N (CD13, EC 3.4.11.2) occurs on the surface of resting and concanavalin A-stimulated T-lymphocytes. *Biol. Chem. Hoppe-Seyler* **374:** 291-296.

Kuroki, K., *et al.*, 1994, A cell surface protein that binds avian hepatitis B virus particles. *J. Virol.* **68**: 2091-2096.

Larsen, S.Ll *et al.* (1996, T cell responses affected by aminopeptidase N (CD13)-mediated trimming of major histocompatibility complex class II-bound peptides. *J. Exp. Med.* **184:** 183-189.

Lasch, J., *et al.*, 1998, Aminopeptidase P - a cell-surface antigen of endothelial and lymphoid cells: catalytic and immuno-histotopical evidences. *Biol. Chem.* **379:** 705-709.

Lendeckel, U., *et al.*, 1999, Role of alanyl aminopeptidase in growth and function of human T cells (Review). *Int. J. Mol. Med.* **4:** 17-27.

Lendeckel, U., *et al.*, 1998, Inhibition of alanyl aminopeptidase induces MAP-kinase p42/ERK2 in the human T cell line KARPAS-299. *Biochem. Biophys. Res. Commun.* **252:** 5-9.

Lendeckel, U., *et al.*, 1996, Induction of the membrane alanyl aminopeptidase gene and surface expression in human T-cells by mitogenic activation. *Biochem. J.* **319:** 817-823.

Lendeckel, U., *et al.*, 1994, Expression of the aminopeptidase N (CD13) gene in the human T cell lines HuT78 and H9. *Cell. Immunol.* **153:** 214-226.

Lerche, C., *et al.*, 1996, Human aminopeptidase N is encoded by 20 exons. *Mammalian Genome* **7:** 712-713.

Lim, J., and Turner, A.J., 1996, Chemival modification of porcine kidney aminopeptidase P indicates the involvement of two critical histidine residues. *FEBS Letters* **381:** 188-190.

Look, A.T., *et al.*, 1986, Molecular cloning, expression, chromosomal localization of the gene encoding a human myeloid membrane antigen (gp150). *J. Clin. Invest.* **78:** 914-921.

Look, A.T., *et al.*, 1989, Human myeloid plasma membrane glykoprotein CD13 (gp150) is identical to aminopeptidase N. *J. Clin. Invest.* **83:** 1299-1307.

Loehn, M., *et al.*, 1997, Aminopeptidase N-mediated signal transduction and inhibition of proliferation of human myeloid cells. *Adv. Exp. Med. Biol.* **421:** 85-91.

Lojda, Z., 1977, Studies on glycyl-proline naphthylamidase. I. Lymphocytes. *Histochemistry* **54:** 299-309.

Lloyd, G.S., and Turner, A.J., 1995, Aminopeptidase P: cation activation and inhibitor sensitivity are substrate-dependent. *Biochem. Soc. Trans.* **23:** 60S.

Löster, K., *et al.*, 1995, The cysteine-rich region of dipeptidyl peptidase IV (CD26) is the collagen-binding site. *Biochem. Biophys. Res. Commun.* **217:** 341-348.

Louvard D *et al.*, 1973, On the distribution of enterokinase in porcine intestine and on its subcellular localization. *Biochim Biophys Acta*; **309:** 127-137.

Malfroy, B. and Schwartz, J.C. (1982, Properties of "enkephalinase" from rat kidney: comparison of dipeptidyl-carboxypeptidase and endopeptidase activities. *Biochem. Biophys. Res. Commun.* **106:** 276-285.

McGwire, G. B., *et al.*, 1997, Identification of a membrane-bound carboxypeptidase as the mammalian homolog of duck gp180, a hepatitis B-virus binding protein. *Life Sci.* **60:** 715-724.

McIntyre, E.A., *et al.*, 1987, In *Leucocyte Typing III: White cell differentiation antigens* (A.J. McMichael, P.C.L. Beverly, S. Cobbold, and M.J. Crumpton, eds.), University Press, Oxford, pp. 685-688.

Medeiros, M., Dos, S., and Turner, A.J., 1994, Processing and metabolism of peptide YY: pivotal roles dipeptidyl peptidase IV, aminopeptidase P and endopeptidase-24.11. *Endocrinology* **134:** 2088-2094.

Menrad, A.D., *et al.*, 1993, Biochemical and functional characterization of aminopeptidase N expressed by human melanoma cells. *Cancer Res.* **53:** 1450-1455.

Mittrücker, H., *et al.*, 1995, The cytoplasmic tail of the T cell receptor zeta-chain is requirend for signaling via CD26. *Eur. J. Immunol.* **25:** 295-297.

Morimoto, C., *et al.*, 1989, 1F7, a novel cell surface molecule, involved in helper function of CD4 cells. *J. Immunol.* **143:** 3430-3439.

Morimoto, C., and Schlossman, S.F., 1998, The structure and function of CD26 in the T-cell immune response. *Immunol. Rev.* **161:** 55-70.

Morrison, M.E., *et al.*, 1993, A marker for neoplastic progression of human melanocytes is a cell surface ectopeptidase. *J. Exp. Med.* **177:** 1135-1143.

Olsen, J., *et al.*, 1988, Complete amino acid sequence of human intestinal aminopeptidase N as deduced from cloned cDNA. *FEBS Letters* **238:** 307-314.

Oravecz, T., *et al.*, 1997, Regulation of the receptor specificity and function of the chemokine RANTES (regulated on activation, normal T cell expressed and secreted) by dipeptidyl peptidase IV (CD26)-mediated cleavage. *J. Exp. Med.* **186:** 1865-1872.

Orawski, A.T., Susz, J.P., and Simmons, W.H., 1987, Aminopeptidase P from bovine lung: Solubilization, properties, and potential role in bradykinin degradation. *Mol. Cell. Biochem.* **75:** 123-132.

Orawski, A.T., and Simmons, W.H., 1995, Purification and properties of membrane-bound aminopeptidase P from rat lung. *Biochem.* **34:** 11227-11236.

Pauly, R.P., *et al.*, 1996, Investigation of glucose-dependent insulinotropic polypeptide-(1-42) and glucagon-like peptide-1-(7-36) degradation in vitro by dipeptidyl peptidase IV using matrix-assisted laser desorption/ionization time of flight mass spectrometry - A novel kinetic approach. *J. Biol. Chem.* **271:** 23222-23229.

Patel, I.R., *et al.*, 1998, TNF-alpha convertase enzyme from human arthritis-affected cartilage: isolation of cDNA by differential display, expression of the active enzyme, and regulation of TNF-alpha. *J. Immunol.* **160:** 4570-4579.

Plana, M., *et al.*, 1991, Induction of interleukin 2 and interferon-gamma and enhancement of IL-2 receptor expression by a CD26 monoclonal antibody. *Eur. J. Immunol.* **21:** 1085-1088.

Proost, P., *et al.*, 1998a, Amino-terminal truncation of chemokines by CD26/dipeptidyl-peptidase IV - Conversion of RANTES into a potent inhibitor of monocyte chemotaxis and HIV-1-infection. *J. Biol. Chem.* **273:** 7222-7227.

Proost, P., *et al.*, 1998b, Processing by CD26/dipeptidyl-peptidase IV reduces the chemotactic and anti-HIV-1 activity of stromal-cell-derived factor-1a. *FEBS Letters* **432:** 73-76.

Pui, C.H., *et al.*, 1998, Reappraisal of the clinical and biological significance of myeloid-associated antigen expression in childhood acute lymphoblastic leukemia. *J. Clin. Oncol.* **16:** 3768-3773.

Rawlings, N.D., and Barrett, A.J., 1993, Evolutionary families of peptidases. *Biochem. J.* **290:** 205-218.

Rawlings, N.D., and Barrett, A.J., 1995, Evolutionary families of metallopeptidases. *Methods Enzymol.* **248:** 183-228.

Rehli, M., *et al.*, 1995, Carboxypeptidase M is identical to the MAX.1 antigen and its expression is associated with monocyte to macrophage differentiation. *J. Biol. Chem.* **270:** 15644-15649.

Reinhold, D., *et al.*, 1994, Transforming growth factor-b1 (TGF-b1) inhibits DNA synthesis of PWM stimulated PBMC via suppression of IL-2 and IL-6 production. *Cytokines* **6:** 382-388.

Reinhold, D., *et al.*, 1995, Transforming Growth Factor beta1 inhibits Interleukin-10 mRNA expression and production in pokeweed mitogen-stimulated peripheral blood mononuclear cells and T cells. *J. Interferon Cytokine Res.* **15:** 685-690.

Reinhold, D., *et al.*, 1997a, Inhibitors of dipeptidyl peptidase IV induce secretion of transforming growth factor-b1 in PWM-stimulated PBMC and T cells. *Immunology* **91:** 354-360.

Reinhold, D., *et al.*, 1997b, Inhibitors of dipeptidyl peptidase IV (DP IV, CD26) induces secretion of transforming growth factor-b1 (TGF-b1) in stimulated mouse splenocytes and thymocytes. *Immunol. Lett.* **58:** 29-35.

Reinhold, D., *et al.*, 1997c, The effect of anti-CD26 antibodies on DNA synthesis and cytokine production (IL-2, IL-10 and IFN-gamma) depends on enzymatic activity of DP IV CD26. *Adv. Exp. Med. Biol.* **421:** 149-155.

Reinhold, D., *et al.*, 1998, Inhibitors of dipeptidyl peptidase IV/CD26 suppress activation of human MBP-specific CD4+T cell clones. *J. Neuroimmunol.* **87:** 203-209.

Riemann, D. *et al.*, 1993, Demonstration of CD13/aminopeptidase N on synovial fluid T cells from patients with different forms of joint effusions. *Immunobiol.* **187:** 24-35.

Riemann, D., *et al.*, 1997b, Induction of aminopeptidase N/CD13 on human lymphocytes after adhesion to fibroblast-like synoviocytes, endothelial cells, epithelial cells, and monocytes/macrophages. *J. Immunol.* **158:** 3425-3432.

Riemann, D., *et al.*, 1999, CD13 – not just a marker in leukemia typing. *Immunol Today* **20:** 83-88.

Riemann, D., *et al.*, 1994a, Expression of aminopeptidase N/CD13 in tumour-infiltrating lymphocytes from human renal cell carcinoma. *Immunol. Letters* **42:** 19-23.

Riemann, D., *et al.*, 1997a,. Phenotypic analysis of T lymphocytes isolated from non-small-cell lung cancer. *Int. Arch. Allergy Immunol.* **114:** 38-45.

Riemann, D., *et al.*, 1994b, Immunophenotype of lympocytes in pericardial fluid from patients with different forms of heart disease. *Int. Arch. Allergy Immunol.* **104:** 48-56.

Roberts, C.M., *et al.*, 1999, MDC-L, a novel metalloprotease disintegrin cysteine-rich protein family member expressed by human lymphocytes, *J. Biol. Chem.* **274:** 29251-29259.

Robey, E., 1997, Notch in vertebrates. *Curr. Opin. Genet. Dev.* **7:** 551-557.

Rusu, I., and Yaron, A., 1992, Aminopeptidase P from human leukocytes. *Eur. J. Biochem.* **210:** 93-100.

Ryan, J.W., *et al.*, 1992, Purification and characterization of guinea pig serum aminoacylproline hydrolase (aminopeptidase P). *Biochim. Biophys. Acta* **1119:** 140-147.

Ryan, J.W., *et al.*, 1996, Aminopeptidase P is disposed on human endothelial cells. *Immunopharmacol.* **32:** 149-152.

Saito, M., *et al.*, 1995, Stromal cell-mediated transcriptional regulation of the CD13/ aminopeptidase N gene in leukemic cells. *Leukemia* **9:** 1508-1516.

Scharpe, S.L., *et al.*, 1990, Exopeptidases in human platelets: an indication for proteolytic modulation of biologically active peptides. *Clin. Chim. Acta* **195:** 125-132.

Schols, D., *et al.*, 1998, CD26-processed RANTES(3-68), but not intact RANTES, has potent anti-HIV-1 activity. *Antiviral Res.* **39:** 175-187.

Schön, E., *et al.*, 1984, Dipeptidyl peptidase IV of human lymphocytes. Evidence for specific hydrolysis of glycylprolin in T-lymphocytes, *Biochem. J.* **223:** 255-258.

Schön, E., *et al.*, 1985, The dipeptidyl peptidase IV, a membrane enzyme involved in the proliferation of T lymphocytes, *Biomed. Biochem. Acta* **44**:K9-K15.

Schön, E., *et al.*, 1987, The role of dipeptidyl peptidase IV in human T lymphocyte activation. Inhibitors and antibodies against dipeptidyl peptidase IV suppress lymphocyte proliferation and immunoglobulin synthesis in vitro., *Eur. J. Immunol.* **17:** 1821-1826.

Schön, E., *et al.*, 1989, Dipeptidyl peptidase IV in human T lymphocytes. Impaired induction of interleukin 2 and gamma interferon due to specific inhibition of dipeptidyl peptidase IV., *Scand. J. Immunol.* **29:** 127-132.

Schön, E., *et al.*, 1991, Dipeptidyl peptidase IV in the immune system. Effects of specific enzyme inhibitors on activity of dipeptidyl peptidase IV and proliferation of human lymphocytes, *Biol. Chem. Hoppe Seyler* **372:** 305-311.

Shipp, M.A., *et al.*, 1991, CD10/neutral endopeptidase 24.11 hydrolyzes bombesin-like peptides and regulates the growth of small cell carcinomas of the lung. *Proc. Natl. Acad. Sci. USA* **88:** 10662-10666.

Skidgel, R.A., Davis, R.M., and Tan, F., 1989, Human carboxypeptidase M: Purification and characterization of a membrane-bound carboxypeptidase that cleaves peptide hormones. *J. Biol. Chem.* **264:** 2236-2241.

Skidgel, R.A., McGwire, G.B., and Li, X.-Y., 1996, Membrane anchoring and release of carboxypeptidase M: Implications for for extracellular hydrolysis of peptide hormones. *Immunopharmacol.* **3:** 48-52.

Spits, H., *et al.*, 1995, Development of human T and natural killer cells. *Blood* **85:** 2654-2670.

Sprinkle, T. J., *et al.*, 1998, Assignement of the membrane-bound human aminopeptidase P gene (XPNPEP2) to chromosome Xq25. *Genomics* **50:** 114-116.

Steeg, C., *et al.*, 1995, Unchanged signaling capacity of mutant CD26/Dipeptidyl peptidase IV molecules devoid of enzymatc activity., *Cell. Immunol.* **164:** 311-315.

Stöckel, A., Stiebitz, B., and Neubert, K., 1997, Specific inhibitors of aminopeptidase P. Peptides and pseudopeptides of 2-hydroxy-3-amino acids. *Adv. Exp. Med. Biol.* **421:** 31-35.

Subramanyam, M., *et al.*, 1993, Mechanism of HIV-1 Tat induced inhibition of antigen-specific T cell responsiveness, *J. Immunol.* **150:** 2544-2553.

Syrjälä, M., *et al.*, 1994, A flow cytometrix assay of CD34-positive cell populations in the bone marrow. *Br. J. Haematol.* **88:** 679-684.

Tan, F., *et al.*, 1989, Molecular cloning and sequencing of the cDNA for human membrane-bound carboxypeptidase M. Comparison with carboxypeptidases A, B, H, and N. *J. Biol. Chem.* **264:** 13165-13170.

Tanaka, S., *et al.*, 1997, Suppression of arthritis by the inhibitors of dipeptidyl peptidase IV. *Int. J. Immunopharmacol.* **19:** 15-24.

Tanaka, S., *et al.*, 1998, Anti-arthritic effects of the novel dipeptidyl peptidase IV inhibitors TMC-2A and TSL-225. *Immunopharmacol.* **40:** 21-26.

Tanaka, T., *et al.*, 1993, The costimulatory activity of the CD26 antigen requires dipeptidyl peptidase IV enzymatic activity. *Proc. Natl. Acad. Sci. USA* **90:** 4586-4590.

Tiruppathi, C., *et al.*, 1993, Genetic evidence for role of DPP IV in intestinal hydrolysis and assimilation of prolyl peptides. *Am. J. Physiol.* **265:** G81-G89.

Torimoto, Y., *et al.*, 1991, Coassociation of CD26 (dipeptidyl peptidase IV) with CD45 on the surface of human T lymphocytes. *J. Immunol.* **147:** 2514-2517.

Torimoto, Y., *et al.*, 1992, Biochemical characterization of CD26 (dipeptidyl peptidase IV): Functional comparison of distinct epitopes recognized by various anti-CD26 monoclonal antibodies. *Mol. Immunol.* **29:** 183-192.

Ulmer, A.J., *et al.*, 1990, CD26 antigen is a surface Dipeptidyl peptidase IV (DPPIV) as characterised by monoclonal antibodies clone TII-19-4-7 and 4EL1C1. *Scand. J. Immunol.* **31**:429-435.

Vanhoof, G., *et al.*, 1992, Kininase activity in human platelets: cleavage of the $Arg^1$-$Pro^2$ bond of bradykinin by aminopeptidase P. *Biochem. Pharmacol.* **44:** 479-487.

Vanhoof, G., *et al.*, 1995, Proline motifs in peptides and their biological processing. *FASEB J.* **9:** 736-744.

Vanhoof, G., *et al.*, 1997, Isolation and sequence analysis of a human cDNA clone (XPNPEPL) homologous to X-prolyl aminopeptidase (aminopeptidase P). *Cytogenet. Cell Genet.* **78:** 275-280 .

Venema,R.C., *et al.*, 1997, Cloning and tissue distribution of human membrane-bound aminopeptidase P. *Biochim et Biophys. Acta* **1354:** 45-48.

Von Bonin, A., *et al.*, 1998, Dipeptidyl-peptidase IV/CD26 on T cells: analysis of an alternative T-cell activation pathway. *Immunol. Rev.* **161:** 43-53.

Weskamp, G., *et al.*, 1996, MDC9, a widely expressed cellular disintegrin containing cytoplasmic SH3 ligand domains. *J. Cell Biol.* **132:** 717-726.

Wex, T., *et al.*, 1997, Antisense-mediated inhibition of aminopeptidase N (CD13) markedly decreases growth rates of hematopoietic tumour cells. *Adv. Exp. Med. Biol.* **421:** 67-73.

Wilce, M.C.J., *et al.*, 1998, Structure and mechanism of a proline-specific aminopeptidase from Escherichia coli. *Proc. Natl. Acad. Sci. USA* **95:** 3472-3477.

Wolfsberg, T.G., *et al.*, 1995, ADAM, a novel family of membrane proteins containing A Disintegrin And Metalloprotease domain: multipotential functions in cell-cell and cell-matrix interactions. *J. Cell Biol.* **131:** 275-278.

Wolfsberg, T.G. and White, J.M. (1996, ADAMs in fertilization and development. *Dev. Biol.* **180:** 389-401.

Wolfsberg, T.G. and White, J.M., 1998, In *Handbook of Proteolytic Enzymes* (A.J. Barrett, N.D. Rawlings, and, J.F. Woessner, eds.), Academic Press, San Diego/London, pp. 1310-1313.

Wrenger, S., *et al.*, 1997, The N-terminal structure of HIV-1 Tat is required for suppression of CD26-dependent T cell growth. *J. Biol. Chem.* **272:** 30283-30288.

Xin, X., *et al.*, 1997, Cloning and sequence analysis of cDNA encoding rat carboxypeptidase D. *DNA Cell. Biol.* **16:** 897-909.

Xu Y., *et al.*, 1995, Substance P and Bradykinin are natural inhibitors of CD13/ aminopeptidase N. *Biochem. Biophys. Res. Communicat.* **208:** 664-674.

Yagami-Hiromasa, T., *et al.*, 1995, A metalloprotease-disintegrin participating in myoblast fusion. *Nature* **377:** 652-656.

Yamamoto, S., *et al.*, 1999, ADAM family proteins in the immune system. *Immunol. Today* **20:** 278-284.

Yaron, A., and Naider, F., 1993, Proline-dependent structural and biological properties of peptides and proteins. *Crit. Rev. Biochem. Mol. Biol.* **28:** 31-81.

Yoshimoto, T., Orawski, A.T., and Simmons, W.H., 1994, Substrate specifity of aminopeptidase P from E. coli: comparison with membrane-bound forms from rat and bovine lung. *Arch. Biochem. Biophys.* **311:** 28-34.

Zhang, L., *et al.*, 1998, Spectroscopic identification of a dinuclear metal center in manganese (II)-activated aminopeptidase P from E. coli: implications for human prolidase. *J. Biol. Inorg. Chem.* **3:** 470-483.

# STRUCTURE AND FUNCTION OF AMINOPEPTIDASE N

Hans Sjöström, Ove Norén and Jørgen Olsen
*Department of Medical Biochemistry and Genetics, Biochemistry Laboratory C, The Panum Institute, University of Copenhagen, Copenhagen, Denmark*

## 1. INTRODUCTION

Aminopeptidase N (EC 3.4.11.2) is an enzyme preferentially releasing neutral amino acids from the N- terminal end of oligopeptides. It is a type II membrane glycoprotein with a zinc-dependent catalytic activity and is expressed in many tissues, being most abundant in the enterocytes of the small intestine and in the epithelium of kidney proximal tubules. For a recent review see Norén, O. *et al* (1997). The enzyme has over the years been known under several different names e.g. microsomal aminopeptidase, aminopeptidase M, particle-bound aminopeptidase, p 146, p161 and gp 150. The nomenclature committee of IUBMB (http://www.chem.qmw.ac.uk/iubmb/enzyme/) recommends the name membrane alanyl aminopeptidase. Knowledge of the primary structure of aminopeptidase N (Olsen *et al* 1988) led Look *et al* (1989) to identify CD 13 as the same protein. Furthermore aminopeptidase N serves as a receptor for coronaviruses belonging to a certain genetic subset (Delmas *et al* 1992, Yeager *et al* 1992). It is the aim of the present paper to summarise our present knowledge on the structure of aminopeptidase N and to integrate it into functional data giving a view of structure-function relationships of this ubiquitous enzyme.

## 2. CLASSIFICATION

Aminopeptidase N belongs to the M1 family of the MA clan of peptidases (Rawlings and Barrett 1999) also called gluzincins (Hooper 1994). The M1 family has 20 different entries. Four of them (aminopeptidase N, cystinyl aminopeptidase, EC 3.4.11.3, glutamyl aminopeptidase, EC

3.4.11.7 and pyroglutamyl-peptidase II, EC 3.4.19.6) are membrane-bound type II glycoproteins. Aminopeptidase N has been cloned and sequenced from six different mammalian species (table 1) and from chicken egg yolk (called aminopeptidase Ey). Interestingly an insect aminopeptidase N, which has been cloned and characterised is inserted in the membrane via a C-terminal GPI anchor. There is no known three-dimensional structure in the M1 family but it has been described for thermolysin in the M4 family of the same clan.

Table 1. Cloned and sequenced aminopeptidase N from different species and tissues

| Species | Tissue/cell type | Reference |
|---|---|---|
| Human | Caco-2 cells | Olsen *et al*, 1988 |
| | Myeloid cells | Look *et al* ,1989 |
| Rat | Kidney | Malfroy *et al*, 1989 |
| | | Watt and Yip, 1989 |
| Rabbit | Kidney/intestine | Yang *et al*, 1993 |
| Pig | Small intestine | Delmas *et al*, 1994 |
| Cat | Fcwf-cells | Tresnan *et al*, 1996 |
| | CRFK-cells | Kolb *et al*, 1997 |
| Mouse | CFTL-cells | Chen *et al*, 1996 |
| Chicken | Liver | Midorikawa *et al*, 1998 |

## 3. STRUCTURAL INFORMATIONS

The following structural discussion is based on electron microscopic studies on the purified enzyme incorporated into liposomes, knowledge of the exon-intron organisation of the gene, computer-aided predictions of the secondary structure and comparison of the primary structure with the same enzyme of different species or similar enzymes from the same or different species (see Fig 1).

### 3.1 Electron microscopy

The overall structure of the purified pig intestinal aminopeptidase N was studied by Hussain *et al* (1981) after incorporation into lipid membranes. Electron microscopic observations demonstrated a dimeric symmetrical structure of aminopeptidase N with the dimensions 13,5 x 5,5 nm separated by a 5 nm gap from the membrane. Biochemical studies showed that the catalytically active part was located outside the liposomes and the anchoring peptide associated with the membrane.

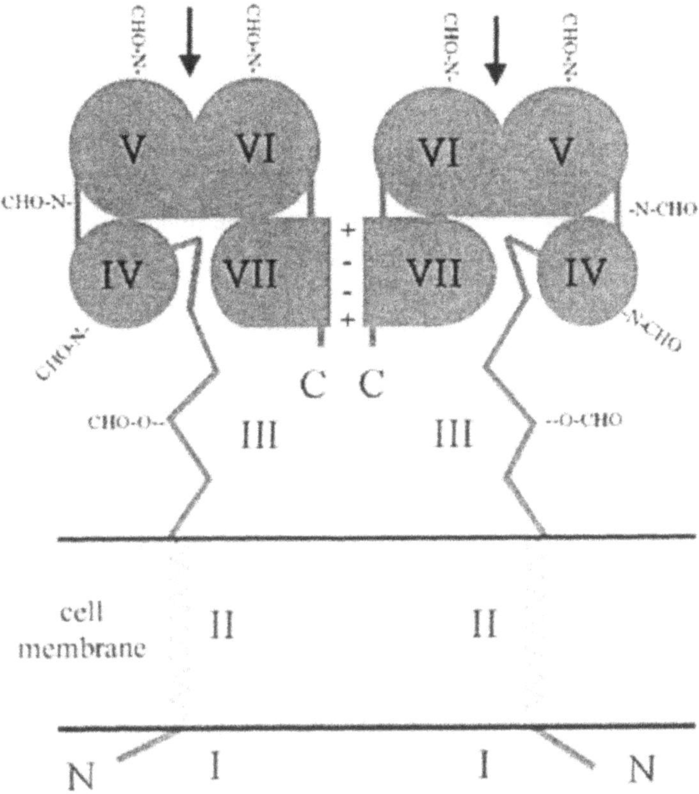


*Figure 1.* Hypothetical model of aminopeptidase N with different domains (I – VII), N- and C-terminus (N, C), active site (arrow), O-glycosylation (CHO-O--), N-glycosylation (CHO-N-) and dimerisation area (+--+) indicated.

## 3.2 Exon-intron structure

The exon-intron structure of the aminopeptidase N gene was mapped by Lerche *et al* (1996). It was originally suggested that exons define different structural domains (Gilbert 1987). This suggestion may not be true (Stoltzfus *et al* 1994), as introns instead may have spread in the genomes rather late in the evolution. However, there is evidence that the occurrence of introns has allowed exons to be shuffled between genes later in the evolution and thereby contribute to the generation of new genes. The aminopeptidase N gene consists of 20 exons encoding segments differing in size from 18 to 205 amino acids.

## 3.3 Secondary structure predictions

For secondary structure predictions the PredictProtein service (Rost *et al* Internet server was used. This program uses a neural network protocol with a stated accuracy of more than 70 %. The algorithm uses comparison of the target sequence with evolutionary related sequences. Proteins from the SwissProtein database were used and some proteins not occurring here were added manually. In all structure comparisons the numbering of human aminopeptidase N (Olsen *et al* 1988) is used.

# 4. DOMAIN STRUCTURE

The considerations mentioned above and further detailed in the following sections led us to suggest a domain structure of seven different domains. Part of this domain model fits with the domain structure of porcine aminopeptidase A, including the generation of a 45 kDa C-terminal fragment, suggested by Hesp and Hooper (1997). Bauer (1994) also demonstrated that trypsin released a fragment from pyroglutamyl-peptidase II, indicating that there is a similar model for all type II membrane bound aminopeptidases.

## 4.1 Domain I

Domain I is the cytosolic part of aminopeptidase N. The size of this domain varies a lot between the four mammalian type II aminopeptidases. Aminopeptidase N contains seven residues, glutamyl aminopeptidase has 17 residues, the rat pyroglutamyl-peptidase II 38 residues, whereas this domain of cystinyl aminopeptidase comprises 137 residues. Domain I of cystinyl aminopeptidase has been suggested to contain a signal for correct intracellular transport in the form of two di-leucine motives (Keller *et al* 1995). It has not been possible to ascribe any function to this domain of aminopeptidase N, although the enzyme is known to cluster in coated pits in connection with TGEV infection. As this clustering also occurs using an aminopeptidase N devoid of domain I (Hansen *et al* 1998), it is tempting to suggest that aminopeptidase N signals over the membrane via an auxiliary membrane protein.

## 4.2 Domain II

Domain II is the membrane-spanning domain probably existing as an $\alpha$-helix. The domain is well conserved within the aminopeptidase Ns, but differs a lot from the other three type II aminopeptidases. It may be that the amino acid sequence of this domain determines the Triton X-100 insolubility of aminopeptidase N at 4 °C (Danielsen 1995). This property has been suggested to form the basis for selective apical transport of newly synthesised aminopeptidase N from the trans-Golgi network.

## 4.3 Domain III

Comparisons between type II membrane aminopeptidases on one side and cytosolic and GPI anchored aminopeptidases of the M1 type on the other side allow a suggestion of a stalk region between residue 40 and residue 70. After this residue there in aminopeptidase N is a sequence WNXXRLP, with homologies also to non-membrane bound aminopeptidases. Aminopeptidase N from human seminal plasma, and normal and pregnancy serum forms have residues 48 and 59 and 69 as N-terminals (Huang *et al* 1997, Watanabe *et al* 1995, 1998). As exon 1 encodes the first 205 amino acids it can be excluded that these forms are generated by an alternative splicing mechanism. Studies on the release of other membrane proteins have suggested that there are several membrane-bound enzymes, which release ectoenzymes. Cleavage between position 68 and 69 demonstrates that this residue is localised within an exposed part of the stalk. The divergence of amino acid composition even within aminopeptidase Ns from different species is striking. A common feature, however, is the frequent occurrence of prolines. It seems reasonable to suggest that a stalk has an ordered structure, even if the secondary structure predictions suggest an unordered structure. The presence of prolines excludes an α-helix but is suggestive of a polyproline II-helix structure, having 3 amino acids per turn.

## 4.4 Domain IV

Domain IV comprises amino acid 70 to 252. The domain corresponds to most of exon 1 and all of exon 2. The end is defined by the start of a domain corresponding to thermolysin. The secondary structure prediction suggests an unordered structure with minor stretches of β-strands. Residues 216-227 represent a conserved region with similarities also to soluble mammalian and bacterial aminopeptidases. This could indicate a function of this region connected to the aminopeptidase enzymatic activity. However mutagenesis experiments with a conserved aspartic acid (residue 225) has shown that this residue is not critically involved in the hydrolytic mechanism (Luciani *et al* 1998). One of the two conserved N-glycosylation sites occurs in this domain (see later).

## 4.5 Domain V and VI

Domain V and VI are distinguished by a thermolysin comparison using the three zinc ligands as fixpoints and correspond to the two spherical domains of thermolysin connected by a loop between helix 2 and 3 of thermolysin. It includes amino acids 253 to 580 corresponding to exons 3 to 10.

Residue 576 is an arginine followed by proline in all examined aminopeptidase N structures except the pig where reside 577 is serine. This allows the bond to be cleaved by trypsin, a proteolysis that has been shown to occur *in vivo* in the pig, providing further evidence that this region is a linkage between two domains (VI and VII) (Sjöström *et al* 1978).

It might be reasonable to substantiate the possible structure similarity between the structure of thermolysin and the secondary structure prediction of APN in domain V and VI. First a comparison between the known three-dimensional structure of thermolysin, having seven α-helices, with the predicted one showed a rather high concordance with some evident mistakes. The first two α-helices are not predicted and three short predicted α-structures (129-132, 155-159 and 230-233) are not found in thermolysin. The four C-terminal α-helices are predicted with high reliability.

In aminopeptidase N the region around the HEXXH motif is not predicted as an α-helix, but this is also the case for thermolysin. It seems therefore reasonable to suggest that the region from positions 382 to 400 is an α-helix, which corresponds to helix 2 in thermolysin. The third zinc ligand (residue 411) is contained in a region, which for both enzymes is suggested to be α-helical. On the N-terminal side residues 315 to 332 in aminopeptidase N are located in a position similar to helix 1 in thermolysin. Two predicted β-strands in aminopeptidase N have counterparts in thermolysin, whereas the predicted α-helix in position 273-290 cannot be found in thermolysin. C-terminally to the active site there are four α-helices in thermolysin and four predicted α-helices in aminopeptidase N (481-490, 493-507, 515-524 and 536-546). Whereas the position of the first α-helix correlates well with that of thermolysin, the other three helices are positioned 12 residues closer to the N-terminus than they are in thermolysin. It is suggested that 12 amino acids are replaced by a short loop between helix 4 and helix 5 as the end of helix 4 and the start of helix 5 are located very close to each other in thermolysin.

Mutation of glutamic acid 355 in an aminopeptidase conserved region (the GAMEN motif) led to an almost completely inactive enzyme (Luciani *et al* 1998) Further characterisation indicated that this glutamic acid belongs to the anionic binding site in aminopeptidase N thereby specifying its aminopeptidase activity. Interestingly, using the above structural comparison, this glutamic acid, (not occurring in thermolysin, which has no aminopeptidase activity) is located in the opening of the active site cleft.

## 4.6 Domain VII

Domain VII constitutes the remaining C-terminal part of the enzyme and includes amino acids 581 to 967 (exons 11-20). This domain has a very high content of predicted α-helices. In the pig enzyme the domain can be dissociated from the rest of the enzyme under non-reducing conditions (Sjöström and Norén 1982). Thus there are no disulfide bridges between domain VII and the rest of the enzyme, in spite of the fact that this domain harbours four of the seven cysteine residues. Of these four residues only residue 761 is conserved in all four type II aminopeptidases. It is therefore difficult to predict which (if any) of the four cysteins that are connected via disulfide linkages.

The final part of aminopeptidase N (residue 936 to 964) is predicted to constitute an extraordinarily long α-helix. It may be hard to accommodate an α-helix of this size within the domain VII structure and it is therefore suggested to be located outside the main part. This segment is furthermore characterised by a high amount of charged amino acids in a regular pattern with most of them on the same side of the putative helix. These properties, together with the high conservation within aminopeptidase N from different species suggest a distinct biological function.

# 5. POST-TRANSLATIONAL MODIFICATIONS

## 5.1 N-glycosylation

Pig aminopeptidase N is heavily N-glycosylated corresponding to a molecular weight of about 25 kDa (Danielsen *et al* 1983). As judged from comparisons of apparent molecular weights in SDS gel electrophoresis with number of amino acids, most of the carbohydrates are bound to residues between number 1 and 580. Only two of the eleven potential N-glycosylation sites (128 and 625) of human aminopeptidase N are conserved within the seven aminopeptidase N molecules compared. Interestingly one of them is located within domain III. Provided this domain is facing the membrane, it may be of significance for the interaction with another membrane protein of importance for signalling/membrane localisation of aminopeptidase N. Alternatively it can be suggested that the conserved sites mark regions which are particularly well suited to monitor the progress of folding during the biosynthesis (Tatu and Helenius 1997).

## 5.2 O-glycosylation

The occurrence and importance of O-glycosylation were studied in a cell-line by transfecting different forms of human and pig aminopeptidase N into a mutant CHO cell line ldl(D), which has a UDP-Gal/UDP-GalNAc 4-epimerase deficiency making it unable to convert UTP-glucose and UTP-N-acetylglucoseamine to the corresponding galactose derivatives. This made it possible to selectively block the O-glycosylation (Norén, K. *et al* 1997). Using $^{3}$H-labelled GalNAc and either wild type aminopeptidase N or aminopeptidase N devoid of the anchor and the stalk it could be demonstrated that the enzyme is O-glycosylated and that this glycosylation mainly or exclusively is confined to the stalk. These results corroborated the outcome of a computer analysis of human aminopeptidase N for the presence of O-glycosylation. This analysis indicated a high probability for O-glycosylation in the stalk region and in the link region between domain VI and domain VII. The experiments with the particular cell line furthermore showed that the absence of O-glycosylation does not affect the intracellular transport to cell surface nor the enzymatic activity.

## 5.3 Dimerisation

The dimerisation of the pig enzyme is due to non-covalent bonds as monomers can be generated by treatment with SDS only (Sjöström and Norén 1982). This is in contrast to glutamyl aminopeptidase which is suggested to be dimerised by an S-S-linkage (Hesp and Hooper 1997). The interaction is in the pig enzyme independent of the cytosolic and the membrane spanning domains, as proteolytically solubilised forms behave as dimers in gel filtration experiments (Sjöström *et al* 1983). An interacting structure - besides being specific for aminopeptidase N - is therefore expected to be of non-covalent character and localised within domains V to VII. The earlier described charged C-terminal helix is a candidate.

## ACKNOWLEDGMENTS

Work performed in the authors' laboratory has been supported by the Danish Research Council of Health Sciences, the Danish Cancer Society, the Lundbeck foundation, the Novo-Nordic Foundation and the Benzon

Foundation. The authors are members of the Biomembrane Research Center, Aarhus University, Denmark.

# REFERENCES

Bauer, K., 1994, Purification and characterisation of the thyrotropin-releasing-hormone degrading ectoenzyme. *Eur. J. Biochem.* **224**: 387-396.

Chen, H., Kinzer, C.A., and Paul, W.E., 1996, p161, a murine membrane protein expressed on mast cells and some macrophages, is mouse CD13/aminopeptidase N. *J. Immunol.* **157**: 2593-2600.

Danielsen, E.M., Cowell, G.M., Norén, O., Sjöström, H., and Dorling, P.R., 1983, Biosynthesis of intestinal microvillar proteins. The effect of swainsonine on post-translational processing of aminopeptidase N. *Biochem. J.* **216**: 325-331.

Danielsen, E.M., 1995, Involvement of detergent-insoluble complexes in the intracellular transport of intestinal brush border enzymes. *Biochem.* **34**: 1596-1605.

Delmas, B., Gelfi, J., L'Haridon, R., Vogel, L.K., Sjöström, H., Norén, O., and Laude, H., 1992, Aminopeptidase N is a major receptor for the enteropathogenic coronavirus TGEV. *Nature* **357**: 417-419.

Delmas, B., Gelfi, J., Kut, E., Sjöström, H., Norén, O., and Laude, H., 1994, Determinants essential for the transmissible gastroenteritis virus-receptor interaction reside within a domain of aminopeptidase-N that is distinct from the enzymic site. *J. Virol.* **68**: 5216-5224.

Gilbert, W., 1987, The exon theory of genes. *Cold Spring Harbor Symposia Quant. Biol.* **52**: 901-905.

Hansen, G.H., Delmas, B., Besnardeau, L., Vogel, L.K., Laude, H., Sjöström, H., and Norén, O., 1998, The coronavirus transmissible gastroenteritis virus causes infection after receptor-mediated endocytosis and acid-dependent fusion with an intracellular compartment. *J. Virol.* **72**: 527-534.

Hesp, J.R., and Hooper, N.M., 1997, Proteolytic fragmentation reveals the oligomeric and domain structure of porcine aminopeptidase A. *Biochem.* **36**: 3000-3007.

Hooper, N.M., 1994, Families of metalloproteases. *FEBS Lett.* **354**: 1-6.

Huang, K., Takahara, S., Kinouchi, T., Takeyama, M., Ishida, T., Ueyama, H., Nishi, K., and Ohkubo, I., 1997, Alanyl aminopeptidase from human seminal plasma: purification, characterization, and immunohistochemical localization in the male genital tract. *J. Biochem.* **122**: 779-787.

Hussain, M.M., Tranum-Jensen, J., Norén, O., Sjöström, H., and Christiansen, K., 1981, Reconstitution of purified amphiphilic pig intestinal aminopeptidase. Mode of membrane insertion and morphology. *Biochem. J.* **199**: 179-186.

Keller, S.R., Scott, H.M., Mastick, C.C., Aebersold, R., and Lienhard, G.E., 1995, Cloning and characterization of a novel insulin-regulated membrane aminopeptidase from Glut4 vesicles. *J. Biol. Chem.* **270**: 23612-23618.

Kolb, A.F., Hegyi, A., and Siddell, S.G., 1997, Identification of residues critical for the human coronavirus 229E receptor function of human aminopeptidase N. *J. Gen. Virol.* **78**: 2795-2802.

Lerche, C., Vogel, L.K., Shapiro, L.H., Norén, O., and Sjöström, H., 1996, Human aminopeptidase N is encoded by 20 exons. *Mammalian Genome* **7**: 712-713.

Look, A.T., Ashmun, R.A., Shapiro, L.H., and Peiper, S.C., 1989, Human myeloid plasma membrane glycoprotein CD13 (gp150) is identical to aminopeptidase N. *J. Clin. Invest.* **83**: 1299-1307.

Luciani, N., Marie-Claire, C., Ruffet, E., Beaumont, A., Roques, B.P., and Fournié-Zaluski, M.-C., 1998, Characterisation of $Glu^{350}$ as a critical residue involved in the N-terminal amine binding site of aminopeptidase N (EC 3.4.11.2): Insights into its mechanism of action. *Biochem.* **37**: 686-692.

Malfroy, B., Kado-Fong, H., Gros, C., Giros, B., Schwartz J.-C., and Helmiss, R., 1989, Molecular cloning and amino acid sequence of rat kidney aminopeptidase M: a member of a super family of zinc-metallohydrolases. *Biochem. Biophys. Res. Comm.* **161**: 236-241.

Midorikawa, T., Abe, R., Yamagata, Y., Nakajima, E., and Ichishima, E., 1998, Isolation and characterization of cDNA encoding chicken egg yolk aminopeptidase Ey. *Comp. Biochem. Phys. Part B* **119**: 513-520.

Norén, K., Hansen, G.H., Clausen, H., Norén, O., Sjöström, H., and Vogel, L.K., 1997, Defectively N-glycosylated and non-O-glycosylated aminopeptidase N (CD13) is normally expressed at the cell surface and has full enzymatic activity. *Exp. Cell Res.* **231**: 112-118.

Norén, O., Sjöström, H., and Olsen, J., 1997, Aminopeptidase N. *In Cell-surface peptidases in health and disease* (A.J. Kenny, and C.M. Boustead, eds.), BIOS Scientific Publishers, Oxford, pp. 175-191.

Olsen, J., Cowell, G.M., Kønigshøfer, E., Danielsen, E.M., Møller, J., Laustsen, L., Hansen, O.C., Welinder, K.G., Engberg, J., Hunziker, W., Spiess, M., Sjöström, H., and Norén, O., 1988, Complete amino acid sequence of human intestinal aminpeptidase N as deduced from cloned cDNA. *FEBS Lett.* **238**: 307-314.

Rawlings, N.D., and Barrett, A.J., 1999, MEROPS: the peptidase database. *Nucl. Acids Res.* **27**: 325-331.

Rost, N., Sander, C., and Schneider, R., 1994,. PHD – an automatic mail server for protein secondary prediction. *Comput. Appl. Biosci.* **10**: 53-60.

Sjöström, H., Norén, O., Jeppesen, L., Staun, M., Svensson, B., and Christiansen, L., 1978, Purification of different amphiphilic forms of a microvillus aminopeptidase from pig small intestine using immunoadsorbent chromatography. *Eur. J. Biochem.* **88**: 503-511.

Sjöström, H., and Norén, O., 1982, Changes of the quaternary structure of microvillus aminopeptidase in the membrane. *Eur. J. Biochem.* **122**: 245-250.

Sjöström, H., Norén, O., Danielsen, E.M., and Skovbjerg, H., 1983, Structure of microvillar enzymes in different phases of their life cycles. In *Brush border membranes,* Ciba Foundation Symposium 95, Pitman, London, pp 50-69.

Stoltzfus, A., Spencer, D.F., Zuker, M., Logsdon, J.M., and Doolittle, W.F., 1994, Testing the exon theory of genes: the evidence from protein structure. *Science* **265**: 202-207.

Tatu, U., and Helenius, A., 1997, Interactions between newly synthesized glycoproteins, calnexin and a network of resident chaperons in the endoplasmic reticulum. *J. Cell Biol.* **136**: 555-565.

Tresnan, D.B., Levis, R., and Holmes, K.V., 1996, Feline aminopeptidase N serves as a receptor for feline, canine, porcine and human coronaviruses in serogroup I. *J. Virol.* **70**: 8669-8674.

Watanabe, Y., Iwaki-Egawa, S., Mizukoshi, H., and Fujimoto, Y., 1995, Identification of an alanine aminopeptidase in human maternal serum as a membrane-bound aminopeptidase N. *Biol. Chem. Hoppe Seyler* **376**: 397-400.

Watanabe, Y., Ito, K., Iwaki-Egawa, S., Yamaguchi, R., and Fujimoto, Y., 1998, Aminopeptidase N in sera of healthy subjects is a different N-terminal processed derivative from the one obtained from maternal serum. *Mol. Gen. Metab.* **63**: 289-294.

Watt, V.M., and Yip, C.C., 1989, Amino acid sequence deduced from a rat kidney cDNA suggests it encodes Zn-peptidase aminopeptidase N. *J. Biol. Chem.* **264**: 5480-5487.

Yang, X.F., Milhiet, P.E., Gaudoux, F., Crine, P., and Boileau, G., 1993, Complete sequence of rabbit kidney aminopeptidase N and mRNA localization in rabbit kidney by *in situ* hybridization. Biochem. *Cell. Biol.* **71**: 278-287.

Yeager, C.L., Ashmun, R.A., Williams, R.K., Cardellichio, C.B., Shapiro, L.H., Look, A.T., and Holmes, K.V., 1992, Human aminopeptidase N is a receptor for human coronavirus 229E. *Nature* **357**: 420-422.

# MODULATION OF WNT-5A EXPRESSION BY ACTINONIN: LINKAGE OF APN TO THE WNT-PATHWAY?

Uwe Lendeckel, Marco Arndt[1], Karin Frank, Antje Spiess[1], Dirk Reinhold, and Siegfried Ansorge

*Institute of Experimental Internal Medicine, [1]Institute of Immunology, Center of Internal Medicine, Otto-von-Guericke University Magdeburg*
*Leipziger Strasse 44, D-39120 Magdeburg, Germany*

Key words alanyl aminopeptidase (APN, CD13, EC 3.4.11.2), T cell, Wnt

Abstract Inhibition of alanyl-aminopeptidase gene expression or enzymatic activity compromises T cell proliferation and function. Molecular mechanisms mediating these effects are not known as yet. Applying the cDNA array technique we identified the proto-oncogen Wnt-5a strongly affected by APN-inhibition. Wnt-5a and other members of the Wnt family of secreted factors are implicated in cell growth and differentiation. Wnt-5a was moderately expressed in resting T cells, but strongly down-regulated in response to activation by OKT3/IL-4/IL-9. Actinonin increased Wnt-5a-mRNA contents as confirmed by RT-PCR. In addition, expression of GSK-3ß, an inherent component of the Wnt-pathway, was found to be increased in response to activation, but suppressed by actinonin at both the mRNA and protein level. These findings may provide a rationale for the strong growth inhibitory effects resulting from an inhibition of alanyl aminopeptidase expression or activity.

## 1. INTRODUCTION

Alanyl aminopeptidase (EC 3.4.11.2, APN, CD13) is an 150 kDa metalloprotease of the M1 family of peptidases (clan MA, gluzincins; Hooper 1994) which is expressed as a homodimer on the surface of various cells. The human APN gene was cloned and mapped to chromosome 15 (q25-q26)

(Look 1986, Watt 1990). Its coding sequence is spread over 20 exons (Lerche 1996). Among leukocytes APN is predominantly expressed on myelo-monocytic lineage cells. Malignant transformation, inflammation or T cell activation may induce APN gene and surface expression on B and T cells (Lendeckel 1996, 1999, and Riemann 1993, 1994, 1995).

Inhibition of either APN gene expression or APN enzymatic activity decrease the proliferation of mononuclear blood cells, T cells, and related cell lines. In the T cell line Karpas-299 inhibition of APN enzymatic activity induces MAP kinase p42/p44 (Erk2/Erk1) expression and activity (Lendeckel 1998).

Here, by applying the cDNA array technique we identified the Wnt-5a proto-oncogen as a second potential downstream target of APN inhibition. Members of the highly conserved Wnt family of glycosylated and secreted factors regulate cell-cell interactions during developmental decisions through activation of receptor-mediated signalling pathways (Moon 1997). Wnt proteins upon binding to the seven-pass transmembrane receptor frizzled (Fz) transduce their signals through dishevelled (dsh) proteins to inhibit glycogen synthase kinase 3β (GSK-3β). Consequently, β-catenin accumulates in the cytosol leading to the activation of TCF/LEF-1 transcription factors (for review see: Nusse, 1999). Furthermore, Wnt-5a is a hematopoietic growth factor (Van Den Berg 1998) and may function as a tumor suppressor (Olson 1997, 1998). The aim of this study was to determine the influence of actinonin on expression levels of Wnt-5a and GSK-3β in activated T cells by quantitative RT-PCR or immunoblot techniques.

# 2. MATERIALS AND METHODS

## 2.1 Reagents

The ATLAS Human Cell Cycle Array was from Clontech. IL-4 and IL-9 were purchased from PBH. Anti-CD3 mab, clone OKT3, was purified form hybridoma supernatant. Actinonin was from Sigma.

## 2.2 Cell culture

Mononuclear cells (MNC) were prepared from peripheral blood of healthy donors by Ficoll-Paque gradient centrifugation (Boyum 1968). T cells were enriched from the MNC fraction by the nylon wool adherence technique (Julius 1973). T cells were kept overnight in IMDM-medium and then cultured at a density of 1 x $10^5$ cells/ml with the additions and for periods of times indicated in the figures.

## 2.3 RNA-isolation and RT-PCR

Total RNA was prepared by means of the RNeasy kit (Qiagen). 2 μg RNA were reverse-transcribed using AMV-RT (Stratagene) and 1/10th of the cDNA was used for RT-PCR.

Quantitative determination of GSK-3β mRNA contents was performed using the Lightcycler LC24 (Idaho Technology) and the primers GSK-3β-US (5′-GAAGAGAGTGATCATGTCAG) and GSK-3β-DS (5′-CTTTCTT CTCACCACTGGAG). 18S mRNA amounts were determined using the RT primer pair commercially available from Ambion and used to normalise sample cDNA content. Wnt-5a mRNA amounts were measured by semi-quantitative RT-PCR using an Autogene II (CLF Laboratory), ThermoScript™ RT-PCR System and *Taq*PCRx DNA polymerase (Gibco BRL), and the primers Wnt-5a-US (5′-ACTGTGCCACTTGTATCAGG) and Wnt-5a-DS (5′-CCTCGTTGTTGTGCAGGTTC). Samples were separated in an 1.4 % agarose gel and images of the ethidium bromide stained gels were densitometrically evaluated using the RFLP-Scan software (Scanalytics).

## 2.4 Immunoblotting

Cells were resuspended in 20 mM HEPES buffer plus 1 volume 4xRotiLoad (Roth). Aliquots of equal protein content were separated on 4 - 20 % gradient NuPage gels (Novex) using MES-SDS running buffer. After blotting on BAS-85 membrane (Schleicher&Schüll) GSK-3 protein was detected by means of anti-GSK-3 mab (Biomol) and goat-anti-mouse-POD/Super Signal Extended Duration chemoluminescence substrate (Pierce). GSK-3β was quantified by densitometric analysis using the RFLP-Scan software (Scanalytics).

# 3. RESULTS

### *Actinonin induces Wnt-5a mRNA expression*

By using the ATLAS Cell Cycle Array (Clontech) to compare the mRNA expression of T cells activated by OKT3/IL-4/IL-9 in the presence or absence of 10 μM actinonin, we found Wnt-5a mRNA levels strongly elevated in cells exposed to the aminopeptidase inhibitor (not shown). This finding was subsequently confirmed by semi-quantitative RT-PCR. Wnt-5a mRNA could be detected in resting T cells. Activation by OKT3/IL-4/IL-9

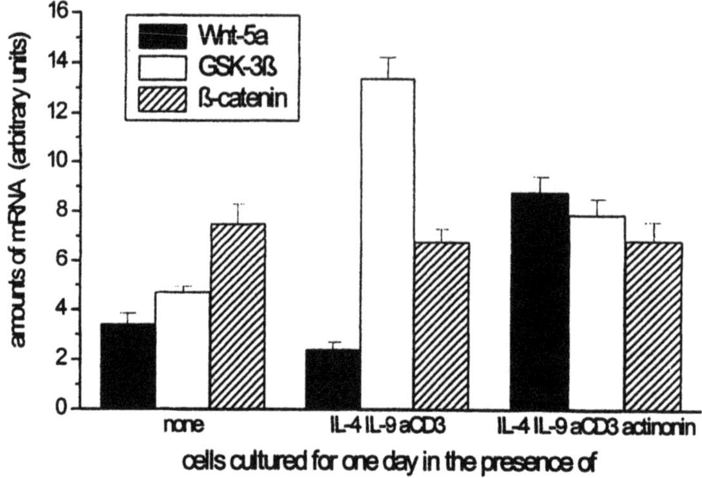


*Figure 1.* Effects of 10 µM actinonin on Wnt-5a, GSK-3β, and β-catenin mRNA amounts of human peripheral T cells. T cells were cultured for 24 hours with the additions of 0.2 ng/ml IL-4, 1 ng/ml IL-9 and 1 µg/ml anti-CD3 mab, clone OKT3 as indicated. β-catenin mRNA amounts did change neither in response to T cell activation nor due to the administration of actinonin (mean of three experiments ± sd).

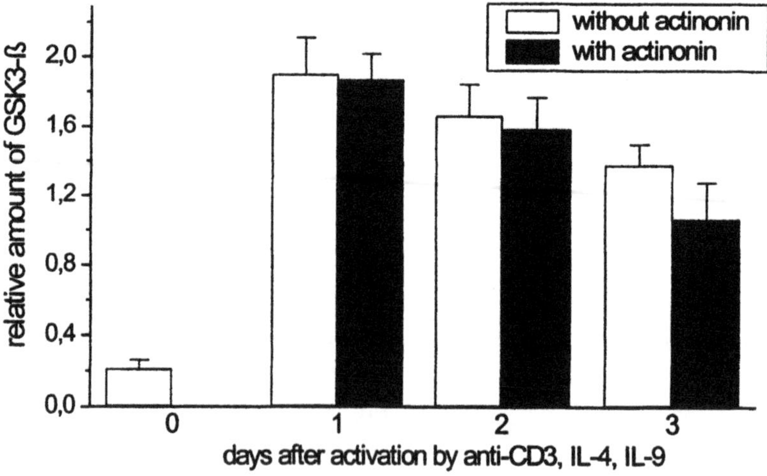


*Figure 2.* Effect of 10 µM actinonin on the amount of GSK-3β protein. T cells were cultured for different periods of time with the additions indicated in the figure. Amounts of GSK-3β were then determined by immunoblot technique using an anti-GSK-3 mab (Upstate Biotechnology) (mean of three experiments ± sd).

leads to a 2-fold decrease in Wnt-5a mRNA amounts after 24 hours (Fig1). On the contrary, activation in the presence of actinonin induced Wnt-5a mRNA amounts about 3-fold compared to resting T cells.

### *Decrease of GSK-3β mRNA expression by actinonin*

Quantitative RT-PCR revealed that 24 hours after T cell activation by OKT3/IL-4/IL-9 there was a 3-fold increase in GSK-3β mRNA amounts, in comparison to resting T cells. In the presence of actinonin, this induction of GSK-3β mRNA was completely abolished (Fig 1).

### *Inhibition of GSK-3β protein amounts by actinonin*

As observed with mRNA levels, there was an activation-dependent increase in GSK-3β protein amounts (Fig 2). Maximum GSK-3β expression was observed one day after activation, thereafter GSK-3β levels were steadily declining. Actinonin provoked a marked reduction of the activation-dependent increase in GSK-3β. Over the period of time this suppressive effect of actinonin became more pronounced, resulting in a 30 % reduction at day 3.

## 4. CONCLUSIONS

Wnt-signalling typically leads to an inhibition of GSK-3β activity. As an increase of Wnt-5a expression was observed 24 hours after administration of actinonin, one could assume that also changes in GSK-3β activity or expression may occur. In this study GSK-3β activity has not been analysed. However, actinonin provoked a suppression of both GSK-3β mRNA and protein. It is tempting to speculate that these changes are due to an increase in Wnt-5a levels and go along with decreased GSK-3β activity. Clearly, this has to be addressed in further studies. Notably, the effect of actinonin on GSK-3β protein amounts was delayed, compared to the rapid repression of corresponding mRNA amounts. This finding might reflect the slow fade-out of pre-existing GSK-3β. Our data show a considerable expression of Wnt-5a in resting T cells. The observed down-regulation of Wnt-5a expression upon T cell activation might be essential for entry and progression through the cell-cycle, thereby enabling clonal expansion. In support of this view, Cyclin D1 proteolysis is regulated through GSK-3β (Diehl 1998). In conclusion, we found an increase in Wnt-5a mRNA expression and an inhibition of GSK-3β protein and mRNA in human T cells after administration of 10 μM actinonin. Beside the previously reported induction of MAP kinase p42/Erk2 mRNA and activity, this is the second T cell signalling pathway found to be affected by inhibition of alanyl-aminopeptidase enzymatic activity. The mechanisms by which actinonin enhances Wnt-5a expression remain to be established. Our data suggest that actinonin excerts its known anti-proliferative activity by preventing the early activation-dependent increase of Wnt-5a expression in T cells.

## ACKNOWLEDGMENTS

This work was supported by the Deutsche Forschungsgemeinschaft, SFB387. We thank Christine Wolf, Ruth Hilde Hädicke, and Bianca Schultze for excellent technical assistance.

## REFERENCES

Boyum, A., 1968, Isolation of mononuclear cells and granulocytes from human blood. *Scand. J. Clin. Lab. Invest.* suppl. **97:** 77-89.

Diehl, J.A., *et al.*, 1998, Glycogen synthase kinase-3beta regulates cyclin D1 proteolysis and subcellular localization. Genes. Dev. **15:** 3499-3551.

Hooper, N.M., 1994, Families of zinc metalloproteases. *FEBS Lett.* **354:** 1-6.

Julius, M.H., *et al.*, 1973, A rapid method for the isolation of functional thymus-derived murine lymphocytes. *Eur. J. Immunol.* **3:** 645.

Lendeckel, U., *et al.*, 1999, Role of alanyl aminopeptidase in growth and function of human T cells (review). *Int. J. Mol. Med.* **4:** 17-27.

Lendeckel, U., *et al.*, 1998, Inhibition of alanyl aminopeptidase induces MAP-kinase p42/ERK2 in the human T cell line KARPAS-299. *BBRC* **252:** 5-9.

Lendeckel, U., *et al.*, 1996, Induction of the membrane alanyl aminopeptidase gene and surface expression in human T-cells by mitogenic activation. *Biochem. J.* **319:** 817-823.

Lerche, C., *et al.*, 1996, Human aminopeptidase N is encoded by 20 exons. *Mammalian Genome* **7:** 712-713.

Look, A.T., *et al.*, 1986, Molecular cloning, expression, chromosomal localization of the gene encoding a human myeloid membrane antigen (gp150). *J. Clin. Invest.* **78:** 914-921.

Moon, R.T., *et al.*, 1997, WNTs modulate cell fate and behavior during vertebrate development. *TIG* **13:** 157-162.

Nusse, R., 1999, WNT targets - repression and activation. *Trends Genet.* **15:** 1-3.

Olson, D.J., *et al.*, 1997, Reversion of uroepithelial cell tumorigenesis by the ectopic expression of human wnt-5a. *Cell Growth Diff.* **8:** 417-423.

Olson, D.J., *et al.*, 1998, Ectopic expression of wnt-5a in human renal cell carcinoma cells suppresses in vitro growth and telomerase activity. *Tumour Biology* **19:** 244-252.

Riemann, D., *et al.*, 1995, Stimulation of the expression and the enzyme activity of aminopeptidase N/CD13 and dipeptidylpeptidase IV/CD26 on human renal cell carcinoma cells and renal tubular epithelial cells by T cell-derived cytokines, such as IL-4 and IL-13. *Clin. Exp. Immunol.* **100:** 277-283.

Riemann, D., *et al.*, 1993, Demonstration of CD13/aminopeptidase N on synovial fluid T cells from patients with different forms of joint effusions. *Immunobiol.* **187:** 24-35.

Riemann, D., *et al.*, 1994, Immunophenotype of lymphocytes in pericardial fluid frompatients with different forms of heart disease. *Int. Arch. Allergy Immunol.* **104:** 48-56.

Van Den Berg, D.J., *et al.*, 1998, Role of members of the Wnt gene family in human hematopoiesis. *Blood* **92:** 3189-3202.

Watt, V.M., and Willard, H.F., 1990, The human aminopeptidase N gene: isolation, chromosome localization, and DNA polymorphism analysis. *Hum.Genet.* **85:** 651-654.

# ENZYMATIC ACTIVITY IS NOT A PRECONDITION FOR THE INTRACELLULAR CALCIUM INCREASE MEDIATED BY mAbs SPECIFIC FOR AMINOPEPTIDASE N/CD13

Alexander Navarrete Santos, Jürgen Langner, and Dagmar Riemann
*Institute of Medical Immunology*
*Martin Luther University Halle*
*Straße der OdF 6, D-06097 Halle, Germany*
*e-mail: alexander.navarrete@medizin.uni-halle.de*

Key words: aminopeptidase N, CD13, intracellular calcium, inhibitors

## 1. INTRODUCTION

Aminopeptidase N (APN)/CD13 similar to dipeptidyl peptidase (DP) IV/CD26 and neutral endopeptidase 24.11/CD10 is a membrane-bound ectoenzyme. Substrates of these enzymes, such as neuropeptides, kinins and chemotactic peptides, signal via G-protein linked heptahelical receptors. Besides their capability to transduce signals via activating or inactivating peptide mediators, for DPIV/CD26 a costimulatory activity has been described which is mediated directly via the enzyme molecule. Similarly, cross-linking with mAbs to APN/CD13 can cause a rise in intracellular free calcium ions ($[Ca^{2+}]_i$) in the cell line U937 (McIntyre *et al.* 1987) as well as phosphorylation of the mitogen activated protein kinase family (Navarrete Santos *et al.* submitted). In this study we show that inhibitors of aminopeptidase N/CD13 are not able to inhibit the increase in $[Ca^{2+}]_i$ evoked by CD13 mAbs.

# 2. MATERIALS AND METHODS

## 2.1 Cells

The human monocytoid leukemia cell line U937 was obtained from the German Collection of Microorganisms and Cell Cultures (Braunschweig, Germany). Cells were maintained as suspension cultures in RPMI 1640 medium supplemented with 10 % heat inactivated FCS, 2 mM L-glutamine, and antibiotics.

## 2.2 $[CA^{2+}]_I$ Measurement

Before loading of the cells with fluorescence dyes, they were incubated with the following APN/CD13 inhibitors: actinonin (Sigma, 100 μM) and bestatin (Sigma, 50 μM). Probestin was kindly provided by Prof. Aoyagi from the Institute of Microbial Chemistry (Tokio) and used at concentration of 50 μM. The incubation in presence of inhibitors was for one hour at 37 °C. U937 cells were then loaded with the $Ca^{2+}$-sensitive dye fluo-3/AM (2 μM) and with the non-$Ca^{2+}$-sensitive dye SNARF-1/AM (0.04 μM) for 30 min at 37 °C. This allows the resolution of the ratio fluo-3/SNARF-1 as a more sensitive parameter for measuring changes in $[Ca^{2+}]_i$. The measurements of $[Ca^{2+}]_i$ were performed by flow cytometry on a FACS Vantage (Becton Dickinson, USA). The absolute values of $[Ca^{2+}]_i$ were determined according to the method described by Grynkiewicz *et al.* (1985).

# 3. RESULTS

In control U937 cells (incubation without APN/CD13 inhibitors) the addition of the CD13 mAb clone MY7 evoked an increase in $[Ca^{2+}]_i$ from 110 ± 20 nM to 370 ± 60 nM. Incubation of U937 cells with actinonin, probestin (Fig 1A, B) and bestatin (data not shown) was not able to reduce the $[Ca^{2+}]_i$ increase evoked by MY7. In the same series of experiments we tested the effect of the CD13 mAb clone WM15 which can inhibit APN/CD13 activity (Razak 1993). Application of WM15 mAb to control cells induced even a more sustained elevation of $[Ca^{2+}]_i$ in comparison with the clone MY7 (Fig 2).

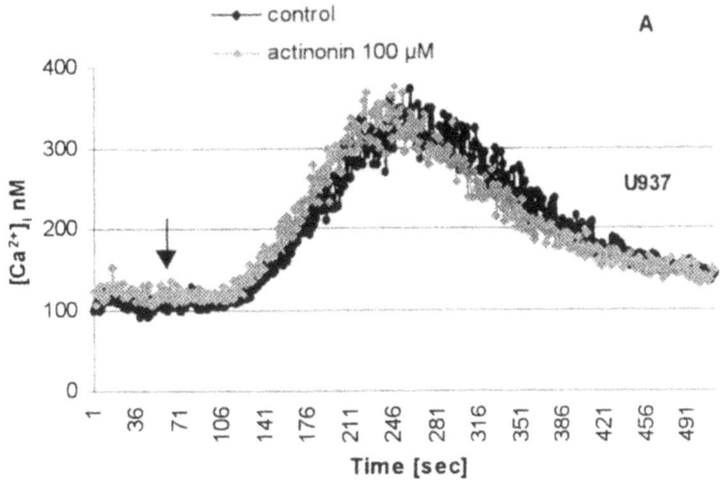


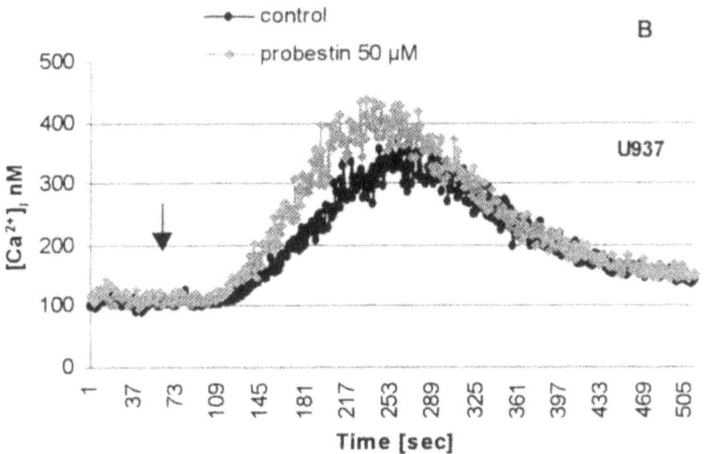


*Figure 1.* Neither actinonin (A) nor probestin (B) are able to reduce $[Ca^{2+}]_i$ increase evoked by MY7. The moment of adding the mAb (final concentration 2 μg/ml) is shown by the arrow.

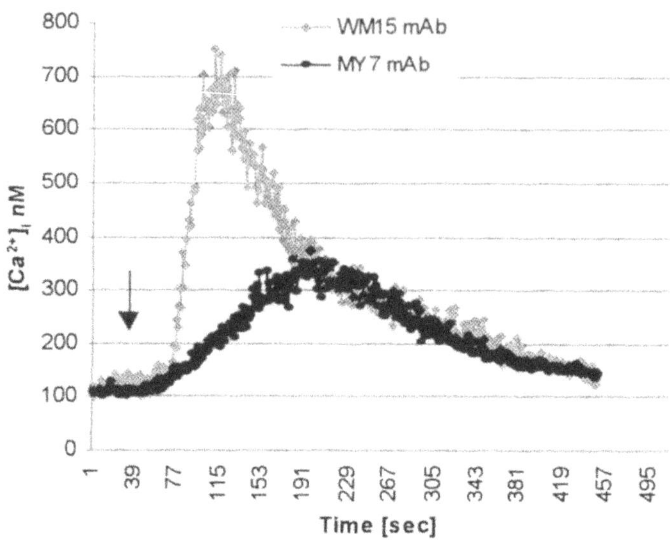


*Figure 2.* The enzyme-activity inhibiting clone WM15 induces a more sustained increase in $[Ca^{2+}]_{i;}$ in comparison with MY7 (final concentration of both mAbs 2 µg/ml)

# 4. DISCUSSION

Inhibition of APN/CD13 activity was shown to comprise cell proliferation in a number of cells (for review see Lendeckel 1999). This points to a role of APN/CD13 substrates in cell proliferation. The aim of the present study was to analyse the role of APN/CD13 enzyme activity in the $[Ca^{2+}]_i$ increase evoked by CD13 ligation. The application of three potent enzyme activity inhibitors (actinonin and probestin inhibit the enzyme activity by more than 91 %, bestatin by 52 %) (Tieku 1992) was not able to reduce the increase in $[Ca^{2+}]_i$ evoked by CD13 mAbs in U937 cells. Furthermore, we show that ligation with the APN/CD13-inhibiting clone WM15 stimulates the cells to increase $[Ca^{2+}]_i$. We conclude that the enzymatic activity is not necessary for the increase in $[Ca^{2+}]_i$ mediated by CD13 mAbs. Further investigations are needed to characterize in vivo ligands capable to induce direct signal transduction via APN/CD13

# ACKNOWLEDGMENTS

We thank Grit Helbing for their excellent technical assistance. This work was funded by the Deutsche Forschungsgemeinschaft (SFB 378).

## REFERENCES

Grynkiewicz G., Poeni M., and Tsien R.Y., 1985, A new generation of $Ca^{2+}$ indicators with greatly improbed fluorescence properties, *J. Biol. Chem.* **260:** 3440-3450.

Lendeckel U., Arndt, M., Frank, K., Wex, T., and Ansorge S.,1999, Role of alanyl aminopeptidase in growth and function of human T cells, *Int. J. Mol. Med.* **4:** 17-27.

McIntyre, E. A., Jones, H.M., Roberts, P.J., Tidman, N., and Linch, D.C., 1987, *Leucocyte Typing III: White cell differentiation antigens* (McMichael A.J., Beverley, P.C.L., Cobbold, S. and Crumpton, M. J. eds. Oxford, University Press. Oxford) pp 685-688.

Razak, K., Collins, P.W., Kelsey, S.M., and Newland, A.C., 1993, Binding of antibodies to different CD13 epitopes on human myeloid leukaemic cell, *Int. J. Oncology* **3:** 749-754.

Tieku, S., and Hooper, N.M., 1992, Inhibition of Aminopeptidase N, A and W. A re-evaluation of the action of bestatin and inhibitors of angiotensin converting enzyme, *Biochem. Pharmacol.* **44:** 1725-1730.

# TRANSFORMING GROWTH FACTOR-β INCREASES THE EXPRESSION OF AMINOPEPTIDASE N/CD13 mRNA AND PROTEIN IN MONOCYTES AND MONOCYTIC CELL LINES

Astrid Kehlen, Jürgen Langner and Dagmar Riemann
*Institute of Medical Immunology, Martin Luther University Halle, Strasse der OdF 6, D-06097 Halle, Germany*
*astrid.kehlen@medizin.uni-halle.de*

Key words Aminopeptidase N (APN)/CD13, TGF-β, monocytic cells

Abstract Aminopeptidase N (APN)/CD13 is a membrane-bound surface ectopeptidase with a ubiquitous distribution. In hematopoiesis, APN/CD13 is expressed on stem cells and during most developmental stages of myeloid cells. Because APN/CD13 has been implicated in the trimming on the cell surface of peptides that protrude out of MHC class II molecules, we wanted to study the regulation of this membrane peptidase in antigen presenting cells by TGF-β. TGF-β is a potent inducer of the maturation of monocyte precursors towards a macrophage phenotype. Using competitive RT-PCR and cytofluorimetric analyses, we quantified the modulation of the APN/CD13 mRNA as well as protein expression by TGF-β1 and -2 and found a stimulation of the APN/CD13 expression in a time- and dose-dependent manner in monocytic cells. In U937 cells, the time course showed a maximum for APN/CD13 mRNA at 24 hours incubation with TGF-β. In experiments with actinomycin D- treated cells was found a stabilization of APN/CD13 mRNA by TGF-β1. Contrary to the IL-4-induced expression of APN/CD13 as well as of MHC class II in monocytic cells, we could show that TGF-β is able to augment the APN/CD13 expression but decreases the MHC class II expression.

# 1. INTRODUCTION

Transforming growth factor-β (TGF-β) is a multifunctional cytokine which in mammals exists in 3 isoforms (TGF-β1, 2 and 3). The biological activities of TGF-β are not species-specific. TGF-β is the most potent known growth inhibitor for normal and transformed epithelial cells, lymphoid cells and other hematopoietic cell types (Wahl 1992). TGF-β has anti-inflammatory actions, such as the down-regulation of MHC class II molecules in different cells.

Members of the TGF-β family signal through serin/threonin kinase transmembrane receptors. On activation by TGF-β or other ligands of this family, signaling from the receptor to the nucleus is mediated by phosphorylation of cytoplasmatic mediators called Smads (for review see Massague 1998, Zhang and Derynck 1999). The receptor-associated Smads, such as Smad 2 and 3 interact directly with and are phosphorylated by activated TGF-β receptor complex. They are ligand specific and form after activation heteromeric complexes with Smad4. Smad 4 is the common mediator of all Smad pathways. These complexes are translocated in the nucleus, where they act as transcription factors, possibly in association with other proteins. One of these genes is Smad 7, an inhibitory Smad, which is involved in a negative feed-back loop.

Aminopeptidase N (APN)/CD13 is a membrane-bound surface ectopeptidase with a ubiquitous distribution. The enzyme acts with a broad substrate specificity on peptides with N-terminal neutral amino acids. In hematopoiesis, APN/CD13 is expressed on stem cells and during most developmental stages of myeloid cells (for review see Sanderink *et al* 1988; Shipp and Look, 1993). Lymphocytes of peripheral blood, spleen or tonsils are APN/CD13 negative but can become positive after cell-cell contact with monocytes and other cells (Riemann *et al* 1997). APN/CD13 has been implicated in the trimming on the cell surface of peptides that protrude out of MHC class II molecules (Larsen *et al* 1996). Therefore, we wanted to study the regulation of this membrane peptidase in antigen presenting cells. Previously, IL-4 could be shown to upregulate MHC class II as well as APN/CD13 expression in monocytes (van Hal *et al* 1994).

# 2. MATERIALS AND METHODS

## 2.1 Preparation and culture of cells

Pericardial fluid (about 4 - 5 ml) was collected from patients undergoing open cardiac operations for different reasons. Pericardial fluid was collected via heparinized syringes and diluted with RPMI 1640 medium. After washing cells with PBS, macrophages were harvested by incubation in culture dishes in RPMI 1640 with 10 % fetal calf serum (FCS) at 37 °C for 16 hours to allow attachment of macrophages. Lymphocytes were removed in two washing steps. The total cell count of pericardial fluid ranged from 500 to 3000 cells/mm³, with, on average, one third lymphocytes and two thirds of $CD14^+$ cells (Riemann *et al.* 1994).

Macrophages were treated for varying times (1 hour to 4 days) with the following recombinant cytokines: TGF-β1 (1 - 2 x $10^7$ U/mg; Strathmann AG Hamburg, Germany) and TGF-β2 (R & D Systems Wiesbaden, Germany). The use of pericardial fluid as a source of monocytes/ macrophages was performed with the permission of the local ethics committee. The human myeloid cell lines U937 (American Type Culture Collection, ATCC) was cultured in RPMI 1640 with 10 % FCS. Mono-Mac 6 cells were purchased from ATCC and grown in RPMI 1640 with 10 % FCS supplemented with 1 mM Na-pyruvic acid and 9 µg/ml bovine insulin.

## 2.2 RNA isolation, standard construction and competitive RT-PCR

Total cellular RNA was isolated according to Chomczynski and Sacchi (1987). Competitive RT-PCR for quantification of APN/CD13 mRNA expression was performed as previously described (Kehlen *et al* 1998). Briefly, the first strand DNA was synthesized using 0.5 µg total RNA in presence of dilutions of internal competitive standard RNA. Ten-percent portions of the cDNA were amplified with the following APN/CD13 specific primers: sense 5' GTC TAC TGC AAC GCT ATC GC 3'; antisense 5'GAT GGA CAC ATG TGG GCA CCT TG 3'. The relative amounts of target and standard products were calculated after densitometric analysis using the Scan Pack software 2.0. For the construction of the internal competitive Smad 7 standard RNA, composite primers were synthesized. Primer 1 was composed of two specific primers complementary to the coding strand of Smad 7 with the sequence: 5' <u>CTT AGC CGA CTC TGC GAA CT</u>C AAC TTC TTC TGG AGC CT 3'. Primer 2 contained a sequence for the SP6 RNA polymerase and one of the specific sequences of primer 1 (underlined): 5' GAT TTA GGT GAC ACT ATA GAA TAC <u>CTT AGC CGA CTC TGC GAA CT</u> 3'. The purified product of the first PCR amplification with primers 1 and 3 (5' ACG CCT TCT CGT AGT CGA AA 3') was template for the second amplification using primers 2 and 3. The amplified DNA was gel purified (QIA quick gel extraction kit, Qiagen, Hilden, Germany) following an in vitro transcription by the SP6 promoter using the transcription system of Boehringer (Roche, Mannheim, Germany). The recombinant RNA was quantified by absorbance at 260 nm and was used as an internal standard in the cDNA synthesis and competitve PCR. Competitive RT-PCR for quantification of Smad 7 mRNA expression was performed with the primers 3 and 4 (5' CTT AGC CGA CTC TGC GAA CT 3') as described for APN/CD13 quantification

## 2.3 Cytofluorimetric analysis

Cells were washed twice with cold PBS, then standard surface membrane immunofluorescence techniques were used. Cells were stained with PE-labelled anti-CD13 (IgG1, Leu-M7) and PE-labelled IgG1 isotype control from Becton Dickinson (Heidelberg, Germany). Fluorescence was analysed in a Becton Dickinson FACScan. Five thousend cells per sample were counted. Differences of mean fluorescence intensity (MFI) were calculated as sample MFI-control antibody MFI.

# 3. STATISTICAL ANALYSIS

Data are expressed as range or as mean ± SD. The Wilcoxon rank sum test was used to determine whether two experimental values were significantly different, * $p < 0.05$.

# 4. RESULTS

Using competitive RT-PCR and cytofluorimetric analyses, we quantified the modulation of the APN/CD13 mRNA as well as protein expression by TGF-β. TGF-β1 and -2 were found to stimulate APN/CD13 expression in a time-and concentration dependent manner. After a 1-day TGF-β1 incubation, we measured a 1.6 fold higher APN/CD13 mRNA expression in monocytes obtained from pericardial fluid and in the monocytic cell line Mono Mac 6. In the cell line U937, even a 3.9 fold increased expression under similar conditions could be found (Fig 1). Mono Mac6 is a monocytic cell line and has the phenotypic and functional characteristics of mature peripheral blood monocytes. U937 cells represent early stages of monocytic activation. The time course of the APN/CD13 mRNA expression increased by TGF-β1 shows a maximum at 24 hours in U937 cells. In these cells, we detected a concentration-dependent stimulation of the APN/CD13 mRNA by TGF-β1 with a maximum effect at 20 ng/ml TGF-β1.

TGF-β treatment augmented also APN/CD13 protein expression (135 ± 14 % of the control after 2 – 3 days) in monocytes obtained from pericardial fluid. In U937 cells, we detected an 2.6 ± 0.5 –fold higher APN/CD13 Pro-

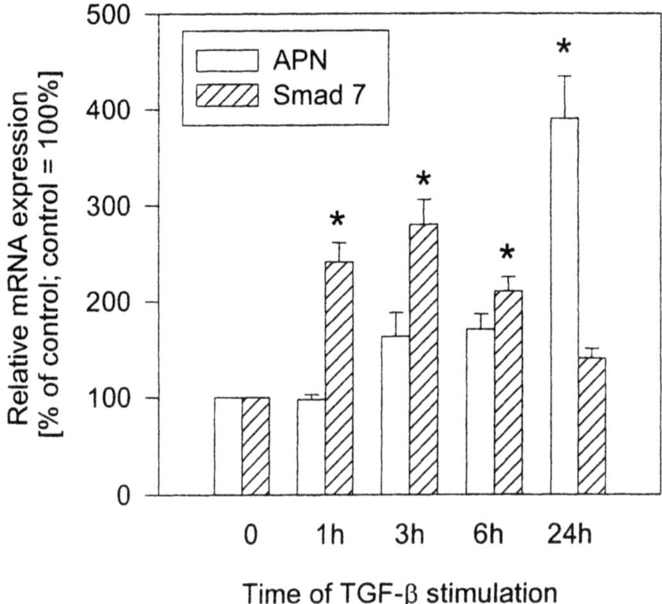


*Figure 1.* Different kinetics of APN/CD13 mRNA and Smad 7 mRNA expression in U937 cells after addition of 20 ng/ml TGF-β1. Data represent mean ± SD of five separate experiments.

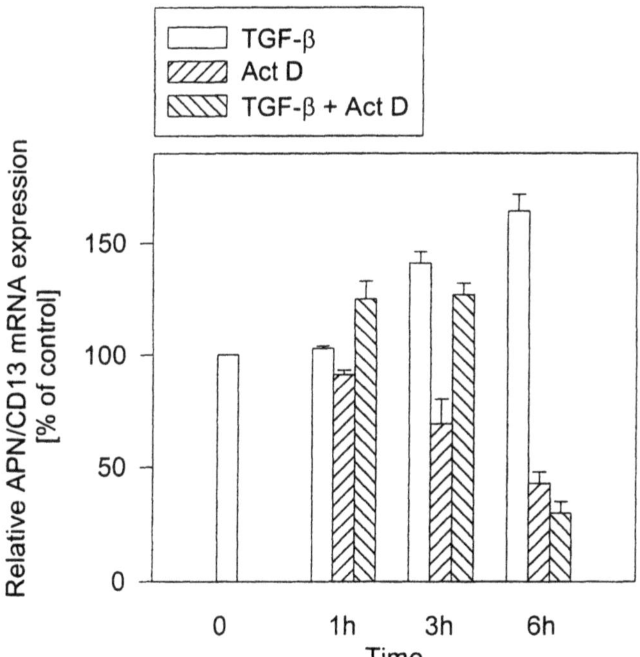


*Figure 2.* Effect of Actinomycin D (10 ng/ml) on the TGF-β1 (20 ng/ml) induced APN/CD13 mRNA expression in U937 cells. Data represent mean ± SD of three separate experiments.

tein expression compared to the untreated control after 2 days. Furthermore, we measured the Ala-pNA cleaving activity after TGF-β1 treatment and found 190 ± 11 % activity compared to the control after 2 days in U937 cells.

Also in U937 cells, we followed the Smad7 mRNA in parallel to TGF-β1 induced APN/CD13 mRNA expression. Using competitive RT-PCR we detected an increase in Smad 7 mRNA expression with a maximum time at 3 hours. For APN/CD13 mRNA expression, the time course showed a maximum at 24 hours (Fig 1).

It is known that TGF-β1 can regulate gene expression by a posttranscriptional mechanism (McGowan *et al* 1997). We investigated the TGF-β1 induced APN/CD13 mRNA as well as Smad7 mRNA expression in Actinomycin D-treated U937 cells. We found a stabilization of APN/CD13 mRNA due to TGF-β1 during the first 3 hours of the treatment (Fig 2).

## 5. CONCLUSION

We describe for the first time that TGF-β1 as a potent inducer of the maturation of monocyte precursors towards a macrophage phenotype augments APN/CD13 expression in different monocytic cells. Since APN/CD13 might be involved in antigen processing, we wanted to study the regulation of this membrane peptidase. The first insights into APN/CD13 expression suggest that it is regulated by a variety of different mechanisms. In addition to developmentally regulated expression, preliminary studies suggest a varying expression of APN/CD13 during cell-growth and differentiation. Thus, maturation and differentiation of monocytes is accompanied by increasing APN/CD13 expression (Laouar *et al* 1993). Cell surface APN/CD13 activity of several cell types increases in the presence of phorbol esters, as shown among others for human myeloid cells (Gregg *et al* 1984). Also the antiinflammatory cytokine IL-4 increases APN/CD13 activity of monocytes/macrophages (van Hal *et al* 1994). IL-4 is considered to induce monocytic maturation, to inhibit production of proinflammatory cytokines by macrophages and to up-regulate MHC class II molecules (van Hal *et al* 1992). Contrary to the IL-4 effect in monocytic cells, we could show that TGF-β is able to augment the APN/CD13 expression but decreases the MHC class II expression.

The time course showed a maximum time for APN/CD13 mRNA at 24 hours incubation with the mediator. Much faster is the increase in Smad 7 mRNA expression after TGF-β treatment with a maximum time at 3 hours in U937 cells (Fig 1), and we observed that TGF-β1 can stabilize the APN/CD13 mRNA (Fig 2).

TGF-β treatment augmented APN/CD13 protein expression and Ala-pNA cleaving activity in monocytes and monocytic cell lines. With peripheral blood monocytes we observed no significant effect of TGF-β on APN/CD13 (data not shown). This could be the result of the high endogeneous TGF-β production by peripheral blood monocytes or the presence of platelets as TGF-β source in the culture well.

The expression of other peptidases has been found inhibited by TGF-β, the TGF-β effects on peptidase expression, however, are cell type dependent. TGF-β down-regulates dipeptidylpeptidase IV/CD26 as well as aminopeptidase A expression in renal tubular epithelial cells and renal cell carcinoma cells in culture, but not APN/CD13 expression. In these cells, we detected no significant effects of TGF-β on APN/CD13 expression (Kehlen *et al* 1998).

The induced APN/CD13 protein synthesis represents a potentially increased cellular ability to inactivate inflammatory mediators as shown for enkephalins or the chemokine MCP-1 (Amoscato *et al* 1993, Weber *et al* 1996). So we postulate that the up-regulation of APN/CD13 in monocytes and monocytic cell lines may be an indirect mechanism of TGF-β for modulating the action of bioactive peptides. Further studies have to deal with the question of signal transduction pathways involved in the TGF-β-induced APN/CD13 expression.

## ACKNOWLEDGMENTS

This work has been supported by Deutsche Forschungsgemeinschaft SFB 387, Teilprojekt A9.

## REFERENCES

Amoscato, A. A., Spiess, R.R., Sansoni, S.B., *et al.*, 1993, Degradation of enkephalins by rat lymphocyte and purified rat natural killer cell surface aminopeptidases. *Brain. Behav. Immun.* **7:** 176-187.

Chomczynski, P., and Sacchi, N., 1987, Single-step method of RNA isolation by acid guanidinium thiocyanate-phenol-chloroform extraction. *Anal. Biochem.* **162:** 156-159.

Gregg, S.L., Gajl-Peczalska, K.J., LeBien, T.W., *et al.*, 1984, Monoclonal antibody MCS-2 as the marker of phorbol diester-induced myeloid differentiation in acute undifferentiated leukemia. *Cancer. Res.* **44:** 2724-2730.

Kehlen, A., Gohring, B., Langner, J., *et al.*, 1998, Regulation of the expression of aminopeptidase A, aminopeptidase N/CD13 and dipeptidylpeptidase IV/CD26 in renal carcinoma cells and renal tubular epithelial cells by cytokines and cAMP-increasing mediators. *Clin. Exp. Immunol.* **111:** 435-441.

Laouar, A. Wietzerbin, J., and Bauvois, B., 1993, Divergent regulation of cell surface protease expression in HL-60 cells differentiated into macrophages with granulocyte macrophage colony stimulating factor or neutrophils with retinoic acid. *Int. Immunol.* **5:** 965-973.

Larsen, S.L., Pedersen, L.O., Buus, S. *et al.*, 1996, T cell responses affected by aminopeptidase N (CD13)-mediated trimming of major histocompatibility complex class II-bound peptides. *J. Exp. Med.* **184:** 183-189.

Massague, J., 1998, TGF-β signal transduction. *Ann. Rev. Biochem.* **67:** 753-791.

McGowan, S.E., Jackson, S.K., Olson, P.J., *et al.*, 1997, Exogenous and endogenous transforming growth factors-beta influence elastin gene expression in cultured lung fibroblasts. *Am. J. Respir. Cell Mol. Biol.* **17:** 25-35.

Riemann, D., Kehlen, A., Thiele, K., *et al.,* 1997, Induction of aminopeptidase N/CD13 on human lymphocytes after adhesion to fibroblast-like synoviocytes, endothelial cells, epithelial cells, and monocytes/macrophages. *J. Immunol.* **158:** 3425-3432.

Sanderink, G.J., Artur, Y., and Siest, G., 1988, Human aminopeptidases: a review of the literature. *J. Clin. Chem. Clin. Biochem.* **26:** 795-807.

Shipp, A., and Look, A.T., 1993, Hematopoietic differentiation antigens that are membrane-associated enzymes: cutting is the key! *Blood* **82:** 1052-1070.

van Hal, P.T., Hopstaken-Broos, J.P., Prins, A., *et al.*, 1994, Potential indirect anti-inflammatory effects of IL-4. Stimulation of human monocytes, macrophages, and endothelial cells by IL-4 increases aminopeptidase N activity (CD13; EC 3.4.11.2). *J. Immunol.* **153:** 2718-2728.

Wahl, S.M., 1992, Transforming growth factor beta (TGF-beta) in inflammation: a cause and a cure. *J. Clin. Immunol.***12:** 61-74.

Weber, M., Uguccioni, M., Baggiolini, M., *et al.*, 1996, Deletion of the NH2-terminal residue converts monocyte chemotactic protein 1 from an activator of basophil mediator release to an eosinophil chemoattractant. *J. Exp. Med.* **183:** 681-685.

Zhang, Y., and Derynck, Y., 1999, Regulation of Smad signalling by protein association and signalling crosstalk. *Trends Cell. Biol.* **9:** 274-279.

# CELL-CELL CONTACT BETWEEN LYMPHOCYTES AND FIBROBLAST-LIKE SYNOVIOCYTES INDUCES LYMPHOCYTIC EXPRESSION OF AMINOPEPTIDASE N/CD13 AND RESULTS IN LYMPHOCYTIC ACTIVATION

Dagmar Riemann, Jana Röntsch, *Bettina Hause, Jürgen Langner, Astrid Kehlen

*Institute of Medical Immunology, Martin Luther University Halle-Wittenberg, Strasse der OdF 6, 06097 Halle; *Institute of Plant Biochemistry Halle, Germany*

Key words cell-cell contact, aminopeptidase, CD13, synoviocytes

## 1. INTRODUCTION

Direct cell-cell contact seems to regulate the expression not only of adhesion molecules but also of different kinds of proteases which can be up- and down-regulated (review in Riemann 1999). Our group described for the first time the induction of the transmembrane ectoenzyme aminopeptidase N (APN)/CD13 on lymphocytes after direct contact with various cells expressing APN, e.g. monocytes, endothelial cells, renal tubular epithelial cells, and fibroblast-like synoviocytes (SFC) (Riemann 1997). The time course of APN induction on lymphocytes is rapid and differs from the well-described temporal characteristics of an increase in APN expression on monocytes elicited by interleukin-4 (Van Hal 1994). Exposure of tonsillar T and B cells to synoviocytes for as little as 1 hour sufficed to induce mRNA and protein expression of the enzymatically active peptidase on lymphocytes. Mitogens could be shown to enhance lymphocytic APN expression in short-time cell-cell contact assays, whereas soluble APN prepared from human kidneys was not able to induce lymphocytic APN expression (Riemann 1997).

To identify genes differentially expressed in T cells with and without direct contact with SFC, differential display analysis was performed (Thiele 2000): In Jurkat T cells, mRNA coding for M-type pyruvat kinase, for the cytoskeletal protein tropomyosin TM30 and for interleukin-17 receptor was found up-regulated after 24 hours coculture with fibroblasts. Otherwise, mRNA for the p54nrb gene product was down-regulated.

Now, we studied lymphocytic APN induction after cell contact with SFC in more detail, using confocal laser scanning microscopy, quantitative RT-PCR and Western blotting. Lymphocytes migrating into a three-dimensional collagen lattice with SFC were taken as a model of lymphocytic APN induction after cell-cell contact with stromal cells.

# 2. MATERIAL AND METHODS

## 2.1 Cell isolation and Coculture Experiments

Tonsillar and peripheral blood lymphocytes were prepared as described by Riemann *et al* (1997). The human lymphoblastic cell line Jurkat was kindly provided by O. Werdelin (Panum Institute Kopenhagen) and cultured in RPMI 1640 medium with 10 % FCS and antibiotics. SFC were isolated from synovectomy tissues of patients suffering from rheumatoid arthritis and passaged as reported (Schwachula 1994). For microscopy, SFC were seeded into chamber slides (Nunc). A 2-hour coculture between lymphocytes and SFC was performed as described (Riemann 1997). Collagen gels were prepared by combining 0.5 ml collagen solution (type I from rat tail, 3.56 mg/ml), 8.9 µl 1 N NaOH and 0.491 ml ITS supplement (Becton Dickinson) diluted 1 : 100 with RPM1640. 2 x $10^5$ SFC/ml were mixed with the gel, when indicated, and cultured for 24 hours. Lymphocytes (2 x $10^6$ cells/ml RPMI medium containing 10 % FCS and 100 U/ml IL-2) were allowed to migrate into an empty collagen matrix overnight, and released from the gel with 0.05 % collagenase 1a (Sigma). These lymphocytes were divided and used for coculture on SFC-monolayers and for migration into a SFC-containing collagen matrix. For fixation experiments, we used 0.05 % glutaraldehyde or 2 - 4 % paraformaldehyde in PBS, pH 7.2.

## 2.2 Functional assay for aminopeptidase activity

Enzyme activity was assayed using 1.5 mM p-nitroanilide (Sigma) as described previously (Riemann 1997).

## 2.3 Immunofluorescence staining

Double-staining of T lymphocytes was done according to standard procedures using mAbs specific for CD3-FITC and CD13-phycoerythrin (Leu-M7, Becton Dickinson). Flow cytometry was performed on a Becton Dickinson FACScan. Indirect immunofluorescence staining of SFC/T cell-cocultures was done using clone Leu-M7 and a Cy-3-labeled secondary Ab. Microscopical examinations were performed on an inverted confocal scanning microscope (Carl Zeiss, Jena).

## 2.4 Standard construction and competitive RT-PCR

Competitive RT-PCR for quantification of mRNA expression was performed as previously described (Kehlen 1998). Briefly, the first strand DNA was synthesized using 0.5 µg total RNA in the presence of dilutions of internal competitive standard RNA. Ten-percent portions of the cDNA were amplified with the specific primer 3 and 4 (Tab 1). The relative amounts of target and standard products were calculated after densitometric analysis (Scan Pack software 2.0). For the construction of the internal competitve standard RNA, composite primers were synthesized. Primer 1 was composed of two specific primers complementary to the coding strand of the appropriate transcription factor. Primer 2 contained a sequence for the SP6 RNA polymerase and one of the specific sequences of primer 1 (underlined). The purified product of the first PCR amplification with primers 1 and 3 was template for the second amplification using primers 2 and 3. The amplified DNA was gel purified (QIA quick gel extraction kit, Qiagen, Hilden) followed by in vitro transcription by the SP6 promoter using the transcription system of Boehringer (Roche, Mannheim). The recombinant RNA was quantified by absorbance at 260 nm and was used as an internal standard in the cDNA synthesis and competitve PCR.

## 2.5 Western Blotting

Cells were washed twice and harvested into ice-cold RIPA buffer (PBS, 1 % Nonidet P-40, 0.5 % sodium deoxycholate) with an inhibitor cocktail (Sigma). The cell lysate was transferred to microcentrifuge tubes, incubated on ice for 60 min and centrifuged at 14,000 rpm at 4 °C for 20 min. Protein was quantified using the BCA Protein Assay Reagent Kit (KMF, Leipzig). Western blot analysis was carried out using 40 µg cell lysate protein.

*Table 1.* Sequences of primers used for standard construction and PCR amplification

| Name | Number | Sequence 5' - 3' |
|---|---|---|
| c-myb | 1 | TACGGTCCGAAACGTTGGTCCAAAGCTACTGCCTGGACGA |
| | 2 | GATTTAGGTGACACTATAGAATACTACGGTCCGAAACGTTGGTC |
| | 3 | TAGCTCATTCTCGGTTGAC |
| | 4 | TACGGTCCGAAACGTTGGTC |
| c-ets1 | 1 | ACTCCTGGCACCATGAAGGAACAGCAACGACTGGGGATC |
| | 2 | GATTTAGGTGACACTATAGAATAC ACTCCTGGCACCATGAAG |
| | 3 | GAACACACTGGGCATGCTCA |
| | 4 | ACTCCTGGCACCATGAAG |
| c-ets2 | 1 | TGCAGCAAGGCTGTGATGAGCTGGTGAACGTGAATCTGCAG |
| | 2 | GATTTAGGTGACACTATAGAATACTGCAGCAAGGCTGTGATGAG |
| | 3 | CTCAGAGCTGAGTACGCTGG |
| | 4 | TGCAGCAAGGCTGTGATGAG |

Proteins were electroblotted from NuPAGE gels (NOVEX, Frankfurt-Hoechst) onto Hybond ECL membrane (Amersham). The membrane was blocked with 5 % dry milk in TBST. Blots were incubated with the primary antibody (pERK1/2 - Santa Cruz; 1 : 1,000 dilution) in TBST with 1 % BSA at 20 °C for 1.5 hours. After 3 washing steps, blots were incubated with the secondary Ab coupled with horseradish peroxidase (1 : 1,000 dilution, Dianova) for 1 hour. Immunodetection was accomplished using the ECL Western blotting detection reagents for chemiluminescent detection (Amersham). Immunoreactivity was quantified by scanning densitometry using the Scan Pack 2.0 software (Biometra).

## 3. RESULTS

1. After coculture of lymphocytes with SFC, crawling movements of lymphocytes on and beneath SFC can be observed (Fig 1). Immunofluorescence staining of cocultured synoviocytes and T cells reveals a homogeneous APN expression of SFC with vesicular structures, as well as a patchy APN staining of lymphocytes (Fig 2).

2. Fixation of SFC with paraformaldehyde (PFA, 2-4 %) cannot prevent APN induction after cell-cell contact. This fixation does, furthermore, not change Ala-pNA cleaving activity of SFC (95 % of native SFC). However, mild glutaraldehyde fixation of SFC partially destroys Ala-pNA cleaving activity (72 %) and prevents lymphocytic induction of APN after coculture.

3. SFC detached from culture flasks by trypsin are able to induce a lymphocytic APN expression after coculture of both kinds of cells (data not shown). A 60-min pretreatment of SFC with cytochalasin B (disrupting cytoskeleton) attenuates CD13 expression of PBL and tonsillar lymphocytes after contact with SFC (n = 6-8) ,

| | |
|---|---|
| with 0 μM | 100 % |
| with 1 μM | 77 ± 11 % |
| with 10 μM | 80 ± 6 % |
| with 100 μM | 48 ± 12 % |

relative lymphocytic CD13 expression. A cytochalasin concentration of 10 μM (and higher) results in rounding and detachment of SFC beginning already after 30 min. With leukemic cells, coculture in the presence of small amounts of cytochalasin B often resulted in an inverse effect, thus we observed an up-regulation of APN expression in Jurkat cells after a 1-hour coculture with SFC (data not shown).

4. After contact of lymphocytes with SFC, we could not detect changes in the lymphocytic mRNA expression of the transcription factors Ets-1 and -2 (Fig 3) and Myb (data not shown). However, c-ets-1 and -2 could be induced

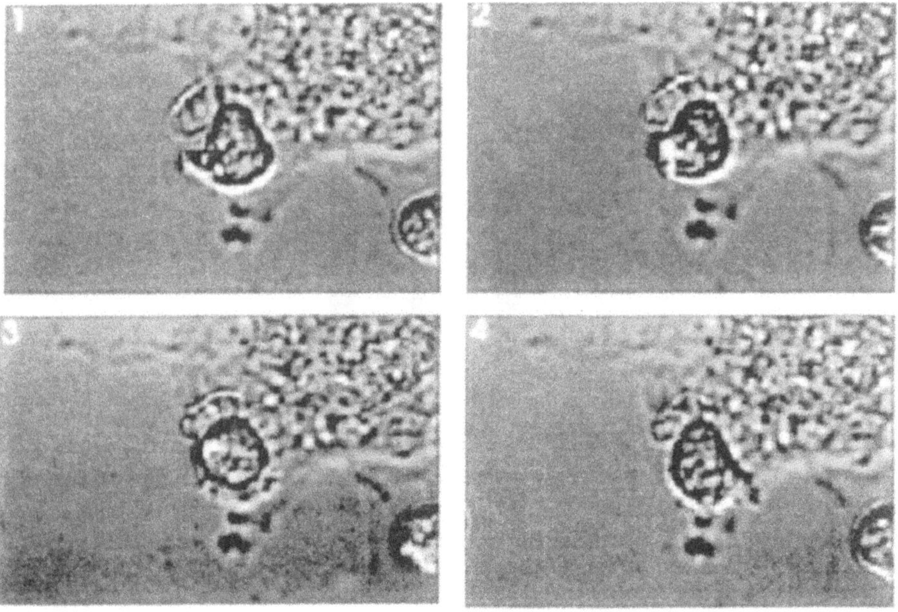

*Figure 1.* Series of phase contrast micrographs (time-lags of 15 s) showing crawling movements of a T lymphocyte on adherent SFC.

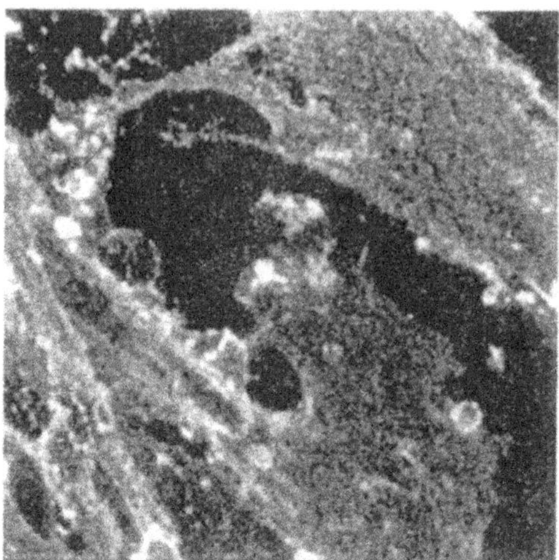

*Figure 2.* Confocal laser scanning micrograph of T cells adhering to SFC. CD13 staining(Cy3, red) is shown in grey color. One lymphocyte with a patchy CD13 staining is visible.

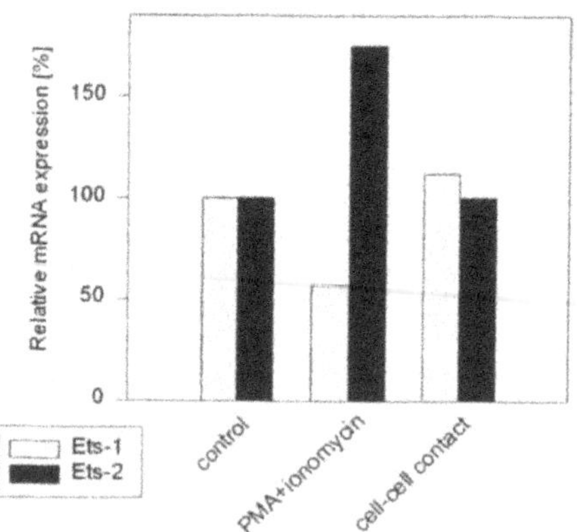


*Figure 3.* Results of quantitative RT-PCR illustrating that coculture with SFC for 4 hours does not increase lymphocytic mRNA expression of Ets 1 and –2. As a positive control, lymphocytes were cultured with PMA (10 ng/ml, Sigma) and ionomycin (1 μM).

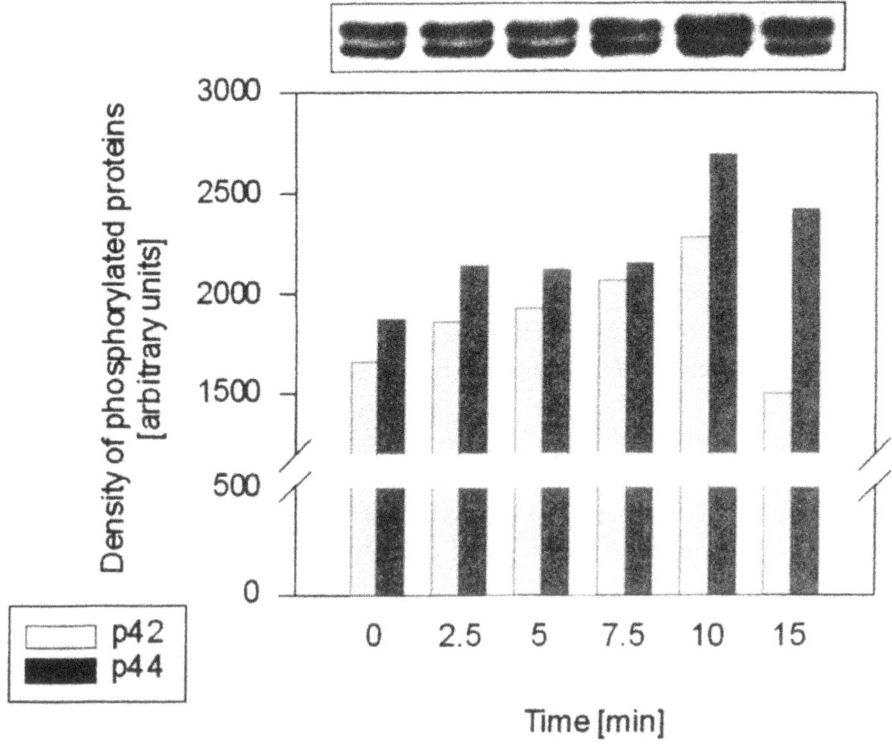


*Figure 4. Cell-cell contact of lymphocytes with SFC increases the phosphorylation of the MAP kinases ERK 1/2 in a time-dependent manner*

by PMA/ionomycin treatment of lymphocytes, as already described (Bhat 1989). Western blot analysis illustrated that activation of the mitogen-activated protein (MAP) kinases ERK1/2 in lymphocytes occurs already 10 min after cell-cell contact with SFC (Fig 4).

5. T lymphocytes (tonsillar, peripheral blood) migrate into a 3-dimensional collagen matrix with SFC and can be found APN positive thereafter. Contact of T cells with collagen type I augments the coculture-induced APN expression whereas collagen I and collagenase alone does not induce APN expression on lymphocytes. Lymphocytes were allowed to migrate into an empty collagen gel overnight, and were then cocultured with confluent SFC-monolayers or migrated into a collagen matrix with SFC for 2 hours. In the latter case, we observed a lymphocytic APN expression of 196 ± 46 % of the APN expression after coculture on adherent SFC. Furthermore, in most cases we observed that migration into an empty collagen gel augments further the lymphocytic APN expression induced with monolayers of SFC.

## 4. DISCUSSION

Our data give first hints to the complexity of signal transduction pathways involved in cell-cell contact between lymphocytes and SFC:

1. Within the rheumatoid pannus, lymphocytes are found in close contact with HLA class II-expressing SFC (Kobayashi 1973). In vitro studies of T cell-SFC interactions have shown that lymphocytes with their pseudopods form close contact areas with fibrillar and microtubular structures oriented perpendicular to the plane of the interface of both cell types, similar to desmosomal structures (Holoshitz 1991).

2. Glutaraldehyde fixation is known to eliminate enzymatic activity of cells with only a few exceptions, such as alkaline phosphatase (Maranto 1988). Further investigations will have to show why glutaraldehyde disturbs lymphocytic APN expression after cell-cell contact with SFC.

3. We show that the loss of cytoskeletal structures after treatment of SFC with cytochalasin B is associated with an attenuated capability to induce APN on lymphocytes (e.g., of peripheral blood) after cell-cell contact. Disruption of the actin cytoskeleton disturbs signal transduction pathways since phosphorylation of both MAP kinases and focal adhesion kinase (e.g. upon integrin engagement) depends on cytoskeletal formation.

4. The myeloid promoter of the APN gene is GC rich, lacks a TATA box and contains binding sites for members of the transcription factor families Myb and Ets (Shapiro 1991; Shapiro 1995). For c-Myb and c-Ets1 a synergistic effect on the APN promoter was observed (Shapiro 1995). Members of at least six subfamilies of Ets proteins are nuclear targets of the Ras-Raf-MAP kinase cascade (Wasylyk 1998). However, though the coculture of T cells with SFC results in Erk1/2 activation, the expression of the transcription factors Ets 1 and –2 is not up-regulated by cell-cell contact in our experiments.

5. We describe that migration into a collagen gel augments the cell-cell contact-induced APN expression in lymphocytes. This result could be due to lymphocytic activation caused by contact with extracellular matrix, or due to a change of synoviocytic properties after seeding into the 3-dimensional collagen matrix. About 25 % of PBL migrating in a collagen gel develop a spontaneous locomotory activity accompanied by enhanced tyrosine phosphorylation of focal adhesion kinase a process which has been claimed to be independent of PKC activity (Entschladen 1997). Also fibroblasts

cultured within 3-dimensional collagen matrix show a specific reprogramming of biosynthetic activity, resulting in a strong induction of proteolytic activity (Mauch 1989). Whereas transient adhesion to fibronectin has been described to induce the production of membrane-type matrix metalloproteinase-1 (MMP-14) in T cells (Esparza 1999), we did not find an induction of APN after adhesion of lymphocytes to fibronectin and collagen I.

## ACKNOWLEDGMENTS

This work was supported by the Deutsche Forschungsgemeinschaft (SFB 387) and the Thyssen Stiftung (Az. 926 98 003).

## REFERENCES

Bhat, N.K. et al., 1989, Expression of ets genes in mouse thymocyte subsets and T cells. *J. Immunol* **142**: 672-678.

Entschladen, F. et al., 1997, Differential requirement of protein tyrosine kinases and protein kinase C in the regulation of T cell locomotion in three-dimensional collagen matrices. *J. Immunol* **159**, 3203-3210.

Esparza J, et al., 1999, Fibronectin upregulates gelatinase B (MMP-9) and induces coordinated expression of gelatinase A (MMP-2) and its activator MT1-MMP (MMP-14) by human T lymphocyte cell lines. A process repressed through RAS/MAP kinase signaling pathways. *Blood* **94**:2754-66.

Holoshitz, J. et al., 1991, T lymphocyte-synovial fibroblast interactions induced by mycobacterial proteins in rheumatoid arthritis. *Arthritis Rheum.* **34:** 679-686.

Kehlen, A. et al., 1998, Regulation of the expression of aminopeptidase A, aminopeptidase N/CD13 and dipeptidylpeptidase IV/CD26 in renal carcinoma cells and renal tubular epithelial cells by cytokines and cAMP-increasing mediators. *Clin. Exp. Immunol.* **111:** 435-441.

Kobayashi, I., and Ziff, M., 1973, Electron microscopic, studies of lymphoid cells in the rheumatoid synovial membrane. *Arthritis Rheum.* **16**: 471-486.

Maranto, A.R. and Schoen, F.J., 1988, Phosphatase enzyme activity is retained in glutaraldehyde treated bioprosthetic heart valves. *ASAIO Trans* **34**: 827-830.

Mauch, C. et al., 1989, Collagenase gene expression in fibroblasts is regulated by a three-dimensional contact with collagen. *FEBS Lett.* **250**: 301-305.

Riemann, D. et al., 1997, Induction of aminopeptidase N/CD13 on human lymphocytes after adhesion to fibroblast-like synoviocytes, endothelial cells, epithelial cells, and monocytes/macrophages. *J. Immunol.* **158:** 3425-3432.

Riemann, D., et al., 1999; CD13 – not just a marker in leukemia typing. *Immunol. Today* **20:** 83-88.

Schwachula, A. et al., 1994, Characterization of the phenotype and functional properties of fibroblast-like synoviocytes in comparison to skin fibroblasts and umbilical vein endothelial cells. *Immunobiology* **190**: 67-92.

Shapiro, L. H. et al., 1991, Separate promoters control transcription of the human aminopeptidase N gene in myeloid and intestinal epithelial cells. *J. Biol. Chem.* **266:** 11999-2007.

Shapiro, L. H .et al., 1995, Myb and Ets proteins cooperate to transactivate an early myeloid gene. *J. Biol. Chem.* **270:** 8763-8771.

Thiele, K. et al., 2000, Cell-cell contact of human T cells with fibroblasts changes lymphocytic mRNA expression: Increased mRNA expression of interleukin-17 and interleukin-17 receptor.,*Eur. Cytokine Network* **11:** in press.

Van Hal, T.T. et al., 1994, Potential indirect anti-inflammatory effects of IL-4. Stimulation of human monocytes, macrophages, and endothelial cells by IL-4 increases aminopeptidase N activity (CD13; EC 3.4.11.2). *J. Immunol.* **153**: 2718-2728.

Wasylyk, B. et al., 1998, Ets transcription factors: nuclear effectors of the Ras-MAP-kinase signaling pathway. *TIBS* **23**: 213-216.

# NATURAL SUBSTRATES OF DIPEPTIDYL PEPTIDASE IV

Ingrid De Meester[1], Christine Durinx[1], Gunther Bal[2], Paul Proost[3], Sofie Struyf[3], Filip Goossens[1], Koen Augustyns[2] and Simon Scharpé[1]
[1]*Laboratory of Clinical Biochemistry*
*University of Antwerp, Wilrijk, Belgium*
[2]*Laboratory of Pharmaceutical Chemistry*
*University of Antwerp, Wilrijk, Belgium*
[3]*Laboratory of Molecular Immunology*
*Rega Institute for Medical Research*
*University of Leuven, Leuven, Belgium*

## 1. INTRODUCTION

Dipeptidyl peptidase IV (DPP IV, EC 3.4.14.5) belongs to a group of cell-membrane-associated peptidases. Scientists in several disciplines have independently investigated various aspects of this molecule for over three decades now. The present knowledge in the DPP IV/CD26 area is summarized in a number of recent reviews (De Meester *et al* 1999, Augustyns *et al* 1999, Morimoto *et al* 1998, von Bonin *et al* 1998). As is the case for the majority of cell-surface peptidases, DPP IV is a type II integral membrane protein, being anchored to the plasma membrane by its signal sequence. This glycoprotein is expressed on a variety of differentiated epithelia, endothelia and hemapoetic cells, mainly T cells. A soluble form that lacks the transmembrane and intracellullar segment is found in plasma and seminal fluid. DPP IV has been identified as the leukocyte differentiation marker CD26. Other cell-surface peptidases that are well-known to immunologists and hematologists are endopeptidase -24.11(EC 3.4.24.11) or CD10 and aminopeptidase N (EC 3.4.11.2) or CD13 (Riemann *et al* 1999). The role of DPP IV/CD26 in the activation of immune cells and the molecular mechanisms underlying its involvement have been the focus of intense research (De Meester *et al* 1999, Augustyns *et al* 1999, Morimoto *et al* 1998, Kähne *et al* 1999).

In a continuing attempt to unravel the in vivo function of this abundant but highly regulated enzyme, the hydrolysis of natural substrates by DPP IV has been studied by several research groups. At this moment, a number of naturally occuring peptides are proven substrates for DPP IV in vitro (De Meester *et al* 1999). In vivo studies using potent and selective inhibitors are needed to elicit which intact peptide levels increase upon eliminating DPP IV enzymatic activity and to investigate the feasibility of DPP IV inhibition in clinical settings. Here we want to focus on a number of natural peptides that may serve as substrates for DPP IV.

## 2. SUBSTRATE SPECIFICITY OF DPP IV

DPP IV/CD26 has a preference for peptides bearing a proline at the second position and it releases dipeptides from their $NH_2$-terminus (Yaron *et al* 1993). The P1 Pro residue can be replaced by Ala and to a lesser extend by Hyp, Ser, Gly, Val and Leu. The available data support a strong selectivity towards Pro or Ala, indicating that physiologically relevant substrates most probably possess one of these amino acids at the second position. The specificity towards various chromogenic substrates has been reviewed before (Yaron *et al* 1993).

DPP IV is exceptional in combining a serine-type protease mechanism with an exopeptidase activity whereas most exopeptidases are metalloproteases. It belongs to the prolyl oligopeptidase family (Barrett *et al* 1998). Another member of this distinct family of serine-type peptidases is prolyl oligopeptidase (Goossens *et al* 1995) (PO, EC 3.4.21.26) of which the 3D-structure was resolved (Fulöp *et al* 1998). Similarly to DPP IV, PO hydrolyses peptide bonds at the carboxyl side of internal proline residues and acts as an oligopeptidase.

## 3. OLD AND NEW SUBSTRATES OF VARYING SIZE: A FEW EXAMPLES

In the late seventies, Heymann and Mentlein reported on the degradation of the undecapeptide Substance P (Arg-Pro-Lys-Pro-Gln-Glu-Phe-Phe-Gly-Leu-Met-$NH_2$) by liver DPP IV (Heymann *et al* 1978). By a stepwise release of Arg-Pro and Lys-Pro, substance P(1-11) in plasma is converted to the C-terminal heptapeptide which exerts increased activity (Conlon *et al* 1983). Substance P signals by activating G-protein-coupled heptahelical receptors (Jarpe *et al* 1998). Remarkably, although the function and structure of the natural peptides that are truncated by DPP IV is very diverse (see below), most of them have in common the fact that they bind to and signal through prototypic G protein-coupled heptahelical receptors.

In contrast to the above described activation, DPP IV can also cause inactivation of peptide substrates. ßcasomorphin-5 (Tyr-Pro-Phe-Pro-Gly), a pentapeptide with potent opioid activity has been isolated from bovine milk. It can be quickly hydrolyzed by DPP IV to inactive fragments (Hartrodt 1982). Only very recently, it is shown that endomorphin-2, an endogenous tetrapeptide with high affinity for the $\mu$ opioid receptor can be inactivated by DPP IV in vivo (Shane *et al* 1999).

The first report on the DPP IV mediated cleavage of dipeptides from larger peptides describes the release of the $NH_2$-terminal glycylproline from the $\alpha$ chain of monomeric fibrin (Mentlein *et al* 1982). The truncation reduces the clotting capacity of fibrin. The presence of DPP IV at the surface of endothelial cells lining the blood vessels suggest a physiological involvement of DPP IV in this process. If there is an active participation of this enzyme, DPP IV deficiency would favor clot formation. To date no further info on the in vivo relevance of the above described finding is reported in the literature.

## 4. PROCOLIPASE AND ENTEROSTATIN: A ROLE FOR DPP IV IN THEIR METABOLISM?

Procolipase, a protein of about 100 amino acids occuring in the pancreatic juice, is activated to colipase by the loss of the 5 $NH_2$-terminal amino-acids. Active colipase is a cofactor of triacylglycerol lipase (EC 3.1.1.3) and forms a 1:1 stoichiometric complex with it, enabling lipase to hydrolyze its substrate at the lipid/water interface (Thomson *et al* 1993, van Tilbeurgh *et al* 1992). Without colipase, the enzyme is washed off by bile salts.

Heymann and coworkers reported an efficient release of the $NH_2$-terminal dipeptide (Phe-Pro) from porcine procolipase (Heymann *et al* 1986). Under certain conditions a second dipeptide Asp-Pro was generated more slowly, resulting in partial activation. Fig 1 depicts how a combined action of DPP IV and Aminopeptidase B, which also occurs in an active form in the pancreas and which prefers Arg as P1 residue, could fully activate procolipase (Heymann *et al* 1986). This stepwise processing mimics the trypsin-mediated activation of procolipase, a process that generates a pentapeptide called 'enterostatin' from the $NH_2$-terminus of procolipase (Okada *et al* 1991) (Fig 1). Enterostatin is produced in the mucosal epithelia of the small intestine and in the gastric mucosa. It appears in the lymph and in the circulation after feeding. In the intestine, immunoreactive enterostatin is present in a 1 : 1 molar ratio to pancreatic colipase after a test meal, suggesting a complete cleavage of procolipase to colipase and enterostatin in that organ (van Tilbeurgh *et al* 1992).

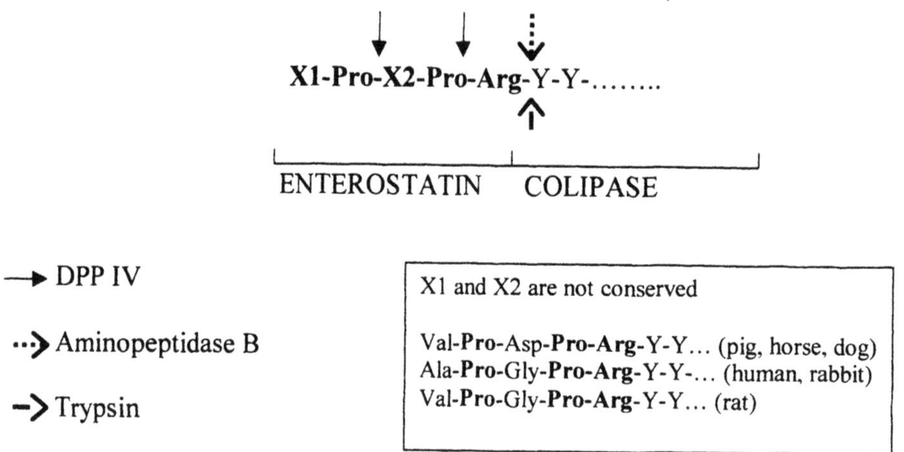


*Figure 1.* The $NH_2$-terminal sequence of mammalian procolipase is shown as well as its processing by trypsin cleavage, resulting in the formation of the active lipase cofactor COLIPASE and a pentapeptide ENTEROSTATIN. The Pro residues at position 2 and 4 as well as the Arg at position 5 are conserved among mammalian species while at positions 1 and 3 (X1 and X2) residues are different. A number of known $NH_2$-terminal sequence of mammalian procolipases are given in the insert. An alternative processing pathway for procolipase involving dipeptidyl peptidase IV and aminopeptidase B is indicated with arrows above the sequence.

The interpretation of data concerning the structure activity relationship of enterostatins and the importance of different peptidases in their metabolisation (if not excreted as such) are hampered (i) by the occurrence of sequence variation at the $NH_2$-terminus of procolipase in different species. The major forms of $NH_2$-termini of procolipase (and hence the derived enterostatin sequences) are shown in Fig 1. The variation implies that different peptidases can play a role in the procolipase activation and/or enterostatin degradation depending on the species studied (Heymann *et al* 1986, Bouras *et al* 1995) (ii) by the lack of sensitive and specific assays to determine intact enterostatin levels in biological systems (Bowyer *et al* 1991, Dezan *et al* 1994) and (iii) finally by the lack of defined receptors, mediating the observed effects of enterostatins. The role of DPP IV in the metabolization of human enterostatin is rather limited (Huneau *et al* 1994), although another report demonstrates that DPP IV inhibition stabilizes enterostatin (VPDPR) and hence improves its intestinal absorption (Bouras *et al* 1996). When injected intraperitoneally into rats, enterostatin produces a dose-dependent reduction in fat intake (Erlanson-Albertsson *et al* 1997). Chronically it also reduces body weight and body fat. Both peripheral and central sites of action have been proposed. The multiple metabolic effects include a reduction of insulin secretion (Ookuma *et al* 1998). When administered centrally, Val-Pro-Asp-Pro-Arg and Ala-Pro-Gly-Pro-Arg were equally potent, but after peripheral injection only Val-Pro-Asp-Pro-Arg affected fat intake. Intraduodenal administration of Ala-Pro-Gly-Pro-Arg also reduced fat intake. Prasad and coworkers evaluated levels of 2 different enterostatins in obese and lean individuals. The mean fasting level of Val-Pro-Asp-Pro-Arg was increased in obese vs lean subjects, whereas the rise in Ala-Pro-Gly-Pro-Arg after a meal was significantly higher in lean individuals than in obese persons (Prasad *et al* 1999).

The analysis of enterostatin from pig showed Asp-Pro as a critical sequence. Indeed, when cyclized into the diketopiperazine peptide cyclo-Asp-Pro it has all the biological activities of enterostatin (Lin *et al* 1994). The human sequence however, lacks this aspartic acid and the minimal structure for the active human form(s) as well as the enzymes involved in the metabolization are not yet defined.

## 5. PANCREATIC POLYPEPTIDE FAMILY AS WELL CHARACTERIZED SUBSTRATES

An important and remarkably conserved group of substrates are members of the pancreatic polypeptide family including pancreatic polypeptide, neuropeptide Y (NPY) and peptide YY (PYY) (Mentlein *et al* 1993a, Medeiros *et al* 1994). NPY and PYY are 36 amino acid regulatory peptides with COOH-terminal amidated ends (Fig 2). The tertiary structure appears to be characteristic for the whole family of peptides and has been termed the 'Pancreatic Polypeptide-fold' or 'PP-fold'. The proline residues at positions 2, 5 and 8 are highly conserved among species as well as among the members of the pancreatic polypeptide family (Fig 2). $NH_2$-terminal deletion experiments on porcine NPY revealed that sequential deletion of Tyr1, Pro2 and Ser3 had no effect on the structural (as determined by circular dichroism) or aggregation properties of NPY. Additional removal of Lys4-Pro5 decreased the helical content and abolished aggregation to a dimeric form. The residues around Pro5 seem important for the formation of the compact, PP-fold structure (Hu *et al* 1994). DPP IV rapidly cleaves the $NH_2$-terminal Tyr-Pro from NPY and PYY, while almost no release of Ala-Pro from pancreatic polypeptide is observed (Mentlein *et al* 1993a). Mentlein and coworkers determined turnover rates for NPY, PYY, β casomorphin and substance P in parallel: The kcat/Km value obtained was about 20 times higher for NPY compared to substance P (Mentlein *et al* 1993a). Studies using specific inhibitors demonstrated the involvement of DPP IV in the metabolization of NPY and PYY in human serum. Also cultivated endothelial cells liberate the $NH_2$-terminal dipeptide from PYY and NPY. The relevance of DPP IV activity in the modulation of NPY and PYY is further supported by the presence of NPY(3-36) and PYY(3-36) in tissue extracts and body fluids, e. g. PYY(3-36) accounts for 30 – 60 % of total PYY immunoreactivity in plasma. Upon incubation of human cerebrospinal fluid with exogenous NPY, the most important NPY-immunoreactive components were NPY(1-36) and NPY(3-36) (Nilsson *et al* 1998).

| **Neuropeptide Y** | |
|---|---|
| human | **YP**SKPDNPGEDAPAEDMARYYSALRHYINLITRQRY |
| rat | **YP**SKPDNPGEDAPAEDMARYYSALRHYINLITRQRY |
| pig | **YP**SKPDNPGEDAPAEDLARYYSALRHYINLITRQRY |
| sheep | **YP**SKPDNPGDDAPAEDLARYYSALRHYINLITRQRY |
| **Peptide YY** | |
| human | **YP**IKPEAPGEDASPEELNRYYASLRHYLNLVTRQRY |
| rat | **YP**AKPEAPGEDASPEELSRYYASLRHYLNLVTRQRY |
| pig | **YP**AKPEAPGEDASPEELSRYYASLRHYLNLVTRQRY |
| rabbit | **YP**SKPEAPGEDASPEELNRYYASLRHYLNLVTRQRY |

*Figure 2.* The amino acid sequences of members of the pancreatic polypeptide family from different species are given. Neuropeptide Y (NPY) and peptide YY (PYY) are efficiently truncated by DPP IV in vitro and in vivo.

NPY is widely distributed in the nervous system and displays a large range of functions. It serves as an important cardiovascular regulator, primarily as a vascular growth factor, which in conditions of stress also becomes a vasoconstrictor (Zukowska-Grojec 1997, Zukowska-Grojec 1998). Aside from these direct cardiovascular activities, NPY exerts potent metabolic and endocrine effects, including stimulation of the appetite (mainly for carbohydrates), lipogenesis and anxiolysis/sedation. The above effects might contribute to its importance in cardiovascular physiology and in the pathogenesis of cardiovascular diseases. Recent data indicate that

NPY can directly affect T cell function by inducing integrin-mediated adhesion to ECM components and by inducing a distinct pattern of cytokine secretion (Levite 1998). PYY is localized in the endocrine cells of the gastrointestinal mucosa and has a variety of actions in the digestive system: it inhibits exocrine pancreatic secretion, has a vasoconstrictory action and inhibits jejunal and colonic mobility (Leiter *et al* 1987).

Several neuropeptide Y/peptide YY receptors (NPY-R) have been cloned and characterized. They all belong to the group of G-protein-coupled seven-transmembrane proteins (Michel *et al* 1998). The typical signaling responses of NPY receptors are similar to those of other G-protein-coupled receptors. Inhibition of adenylyl cyclase is found in almost every tissue and cell type investigated. Additional signaling responses such as mobilization of Ca++ from intracellular stores are seen in certain cell or tissue types. The affinity of a receptor for various intact or truncated pancreatic peptides varies from one receptor type to the other. Removal of $NH_2$-terminal dipeptides from PYY or NPY by DPP IV transforms the unselective agonists into the more selective short peptides PYY(3-36) and NPY(3-36), Zukowska-Grojec 1998). In contrast to their mature forms, which activate vasoconstrictive Y1 receptors, $NH_2$-truncated NPY/PYY(2-36) and NPY/PYY(3-36) are inactive at Y1 receptors but bind to other NPY receptors such as Y2 or Y5 receptors that are involved in the non-vasoconstrictive cardiovascular actions of NPY. The NPY effects on the regulation of feeding behavior are mediated by the Y5 in combination with Y1 receptor subtype (Michel *et al* 1998). These conclusions are drawn from feeding experiments in NPY1- or NPY5-R knockout animals in combination with specific receptor antagonists (Innis 1999). In addition, NPY inhibits insulin secretion from the islets of Langerhans. Competition-binding studies using NPY fragments and use of specific receptor antagonists point to a Y1 receptor mediated mechanism. NPY inhibits the stimulation of cAMP by glucagon-like peptide-1(7-36)amide (tGLP-1) and it also blocks insulin secretion stimulated by tGLP-1 (Morgasn *et al* 1998).

Despite the great number of studies on NPY, a lot of questions in the field of human NPY activity profile and metabolization remain to be answered partially because of the species differences that are observed in NPY receptor distribution (Dumont *et al* 1998).

# 6. THE GLUCAGON/VASOINTESTINAL PEPTIDE FAMILY OPENS THERAPEUTIC POSSIBILITIES

The glucagon superfamily includes the polypeptides glucagon, secretin, vasoactive intestinal peptide (VIP), gastric inhibitory peptide (GIP, also referred to as glucose-dependent insulinotropic peptide) and growth hormone releasing factor (GRF). Although the sizes and sequences of their precursor peptides are distinct, the characterization of the cDNA clones reveals a similar structural organisation. Each has a signal peptide, an $NH_2$-terminal peptide and one, two or three peptides whose sequences are related to glucagon (Bell 1986).

The peptides of this family share a considerable sequence similarity at their $NH_2$-terminus. They either start with Tyr-Ala, His-Ala or His-Ser. The peptides with a penultimate Ala are substrates for DPP IV (Mentlein *et al* 1993b). As an intact $NH_2$-terminus is necessary for their biological activity, metabolization by DPP IV inactivates them.

One of these peptides, proglucagon (Fig 3) is cleaved by tissue-specific prohormone convertases to proglucagon-derived peptides. The major pancreatic product of proglucagon-processing is glucagon, whereas the intestinal products include glicentin, oxyntomodulin, glucagon-like peptide-1(GLP-1) and glucagon-like peptide -2. The active form of GLP-1, truncated glucagon-like peptide-1 (GLP-1(7-36) $NH_2$, tGLP-1) is a 30 residue, COOH-terminally amidated hormone. The release of tGLP-1 and the other proglucagon-derived peptides from the ileal L-cells is under complex neural and hormonal regulation (Rocca *et al* 1999). The main biological effect of tGLP-1 is its action as an incretin, stimulating the secretion of insulin in glucose-dependent manner. Other metabolic effects include the suppression of glucagon secretion and enhancement of glucose disposal. This spectrum of activities offers great potential for use of tGLP-1 in type 2 diabetes (Deacon *et al* 1999). The major problem in realizing this goal is its

**Glucagon precursor (1-180)**

1 *MKSIYFVAGL FVMLVQGSWQ* RSLQDTEEKS RSFSASQADP LSDPDQMNED KR**HSQGTFTS**
*signal sequence* (1-20) **proglucagon** (21-180)

61 **DYSKYLDSRR AQDFVQWLMN T** KRNRNNIAK R*HDEFER* **HA E GTFTSDVSSY LEGQAAKEFI**
**glucagon** (53-81) **tGLP-1** (98-127)

121 **AWLVKGR**GRR DFPEEVAIVE ELGRR**HA DGS FSDEMNTILD NLAARDFINW LIQTKITDRK**
**GLP-2** (146-178)

*Figure 3.* The preproglucagon sequence is depicted and some of the peptides encoded within this sequence are indicated below the codes. The shaded sequences are peptide substrates for DPP IV. Cleavage sites are indicated with an arrow. tGLP1, truncated glucagon-like peptide 1(7-36) $NH_2$, GLP-2, glucagon like peptide 2.

metabolic instability, the most important degradation product found in circulation is GLP-1(9-36) $NH_2$ DPP IV turned out to be the responsible enzyme for the rapid inactivation (half-life of tGLP-1 in vivo reported to be less than 5 min). As the $NH_2$-terminal histidine plays a crucial role in receptor activation, most analogues have modified penultimate residues. The introduction of D-amino acids in the second position largely stabilizes tGLP-1 against degradation by DPP IV. In vivo this resulted in an increased potency compared to native tGLP-1, but surprisingly not a significantly extended biological activity (Ritzel *et al* 1998). Des-amino-tGLP-1 showed both an increased potency and a longer biological action. Other modifications at position 2 include the introduction of threonine, glycine, serine or $\alpha$-aminobutyric acid (Siegel *et al* 1999).

Characterization of the cDNAs encoding proglucagon identified a second glucagon-like peptide, GLP-2, situated COOH-terminal to the GLP-1 sequence in mammalian proglucagon (see Fig 3). It concerns a 33-amino acid peptide with a highly conserved sequence among different mammalian species (Drucker 1998). GLP-2 is the proglucagon-derived peptide with significant intestinal growth factor activity. GLP-2 is trophic for the small as well as the large bowel epithelium as is suggested by its distribution in enteroendocrine cells of both the small and the large intestine. The predominant tissue and circulating GLP-2 appears to be the GLP-2(1-33), but degradation by DPP IV is also observed yielding GLP-2(3-33). Consistent with the observed inactivation by DPP IV, GLP-2 analogues that were made DPP IV resistant were more intestinotrophic (Drucker 1998).

Another member of the glucagon superfamily of peptides, the glucose-dependent insulinotropic peptide or gastric inhibitory peptide (GIP), is an important insulin-releasing hormone secreted from endocrine cells in the intestinal tract in response to feeding. Like glucagon-like peptide-1, GIP supports the pancreatic islets in the control of blood glucose homeostasis and nutrient metabolism (O'Harte *et al* 1999). In plasma, it is inactivated mainly by DPP IV which releases the $NH_2$-terminal Tyr-Ala. Both peptides not only lose their incretin activity, they even serve as antagonists at their respective receptors upon truncation by DPP IV. Although GLP-1 is considered as a more potent insulin secretagogue compared to GIP, recent studies with $NH_2$-terminally modified GIP seem promising. O'Harte and colleagues showed that glycation at the aminoterminal Tyr of GIP markedly limits its degradation and prolongs its half-life in vivo (O'Harte *et al* 1999). This results in an enhanced antihyperglycemic activity and increased insulin concentrations in vivo, making a further evaluation of modified GIP worthwile. It is interesting to note that a similar glycation of the tGLP-1 $NH_2$-terminus impaired its insulinotropic potency. The proposed use of DPP IV inhibitors in type II diabetes mainly aims to optimize the effect of tGLP-1 but the involvement of other mechanisms in the glucose lowering properties of these products cannot be ruled out at this moment. Several other peptides that contain a X-Ala or X-Pro $NH_2$-terminus are possibly involved in the glucose homeostasis. Gastrin-releasing peptide is such a peptide that not only influences meal-stimulated exocrine pancreas secretion, but also glucose-induced insulin-secretion. The importance of the $NH_2$-terminus is not thoroughly studied, but the conservation of the penultimate Pro during the evolution is striking (Table 1).

*Table 1.* $NH_2$-terminal amino-acid sequences of gastrin releasing peptide from different species

| | |
|---|---|
| Human | **vplp**agg gtvltkmypr gnhwavghlm |
| Sheep | **ap**vtagr agalakmytr gnhwavghlm |
| Rat | **ap**vstga gggtvlakmy prgshwavgh lm |
| Dog | **ap**vpggqgtv ldkmyprgnh wavghlm |
| Cavia | **ap**vsvgggtv lakmyprgnh wavghlm |
| Chick | **ap**lqpggspa ltkiyprgsh wavghlm |
| Pig | **ap**vsvgggtv lakmyprgnh wavghlm |

Growth hormone releasing hormone (GRF) is yet another peptide of the glucagon family. This 44-amino acid long amidated peptide is released by the hypothalamus and acts on the adenohypophyse to stimulate the secretion of growth hormone. As early as in 1986, Frohman reported on the rapid enzymatic degradation of growth hormone-releasing hormone by plasma in vitro as well as in vivo (Frohman *et al* 1986). The major metabolite appeared to be the inactive GRF(3-44) $NH_2$, a product of DPP IV mediated inactivation of GRF. Analogues of GRF(1-44) $NH_2$ that are truncated at their COOH-terminus exert excellent biological activity when the chains are 29 to 44 amino acids long and when having an intact $NH_2$-terminus. A further shortening causes a sudden loss in biological activity. The effect of $NH_2$-terminal substitutions introduced to stabilize the molecule (Sato *et al* 1987, Martin *et al* 1993, Coy *et al* 1985, Bongers *et al* 1992), also depends on the chain length, suggesting reciprocal conformational effects of COOH and $NH_2$-terminal residues in GRF (Coy *et al* 1987).

## 7. CHROMOGRANIN A-VASOSTATINS

In the early nineties the idea that DPP IV could only truncate relatively short peptides was reinforced by the failure to process the intact interleukins 2 and 6 at their $NH_2$-terminus (Hoffmann *et al* 1993). Apart from some early reports mentioned above, the largest well-documented substrate was indeed growth-hormone releasing hormone, GRF(1-44) $NH_2$. At that moment a study was started to examine the influence of protein chain length on DPP IV activity. We evaluated whether the enzyme cleaves a Leu-Pro dipeptide from the structurally related proteins chromogranin A, vasostatin I and vasostatin II, that share the same NH2-terminal part but differ in the number of amino acid residues (431, 76 and 113 residues respectively). Chromogranin A belongs to a family of acidic proteins, widely distributed in the secretory granules of endocrine and neuroendocrine tissues, which secrete peptide hormones and neurotransmitters. It is the precursor of several biological active peptides including vasostatin I and vasostatin II. Vasostatin I has suppressive effects on the endothelin-induced contraction in human blood vessels, while vasostatin II inhibits parathyroid hormone secretion stimulated by a low calcium concentration (Winkler *et al* 1992).

Using electrospray mass spectrometry, it was possible to demonstrate that vasostatin I is a substrate for DPP IV, while intact vasostatin II is not. It was only truncated at its $NH_2$-terminus after degradation to a peptide of 78 amino acids. Also for chromogranin A, no $NH_2$-terminal truncation was observed (Zhang *et al* 1999). Studying the 3D structure of chromogranin A will reveal whether steric hindrance is the cause of inability to release the dipeptide or whether it is a matter of peptide mass.

## 8. CALCITONIN FAMILY: POSSIBLE SUBSTRATES?

At present 4 genes having nucleotide sequence homologies are grouped in the 'calcitonin gene family'. The precursor protein preprocalcitonin (Fig 4) is encoded by the CALC-I gene. The same gene may be responsible for the generation of the procalcitonin (PCT) that is strongly increased during bacterial infections (Karzai *et al* 1997). The CALC-I gene undergoes alternative splicing ; the primary RNA-transcript is processed into different mRNA by inclusion or exclusion of different exons belonging to the primary transcript. Calcitonin-encoding mRNA is the main product of CALC-I transcription in the C-cells of the thyroid, whereas calcitonin-gene related peptide-mRNA is produced in nervous tissue of the central and peripheral nerve system. In contrast to calcitonin, calcitonin gene-related peptide has no influence on calcium and phosphate, but it acts as a vasoactive peptide with vasodilatative properties (Jonas *et al* 1985).

Calcitonin-mRNA encodes a preprocalcitonin of 141 amino acids (Fig 4). It comprises (i) a signal peptide, (ii) the $NH_2$-terminal region of procalcitonin (N-PCT), (iii) calcitonin and (iv) the COOH-terminal region of procalcitonin, called katacalcin. The signal sequence is degraded in the endoplasmatic reticulum and the remaining protein is procalcitonin (PCT).

The mid-portion comprises the sequence for calcitonin, flanked by basic amino acids that serve as clieving site for prohormone convertase. This processing results in the main cleavage products of PCT : N-PCT, calcitonin and katacalcin. At normal physiologic conditions, the processing occurs intracellularly and mature calcitonin is secreted into the circulation, regulated by Ca-dependent stimuli. Its half-life in circulation is very short (Jonas *et al* 1985). With systemic bacterial infections, in sepsis and multi-organ dysfunction, very high concentrations of calcitonin-precursor peptides are found in the circulation without increase of calcitonin levels. Among these peptides, procalcitonin is the main peptide with a plasma half life of 25 to 30 h in vivo (as measured by immuo-assay) (Karzai *et al* 1997).

The aminoterminus of PCT is well conserved between different species, especially at its penultimate position (Pro). Thus PCT and its $NH_2$-terminal region N-PCT are hypothetical substrates for DPP IV. Also the pro-hormone of calcitonin-gene related peptide contains a similar aminoterminus. To date the physiological role of the propeptides or their remaining $NH_2$-terminal region after liberation of the calcitonin or related peptide, remains obscure. As will become clear from the following example, the presence of a penultimate Pro residue in the propeptide seems characteristic for this family of hormones.

Islet amyloid has been recognized as a pathological entity in type 2 diabetes since many years (Kah *et al* 1999). Amyloid consists of small proteins that ultrastructurally form fibrils with a β-pleated sheet structure. The classification of these deposits is based essentially on the nature of the precursor protein that forms the backbone of the fibril deposits. The major and unique component of islet amyloid is islet amyloid polypeptide (IAPP) or amylin, a 37 amino-acid peptide. Encoded by the CALC-IV gene (Fig 4A), it belongs to the calcitonin family and has some homology with calcitonin-gene-related peptide, but it differs significantly in its mid-portion (Kah *et al* 1999, Sanket *et al* 1988). IAPP is expressed by islet B-cells and co-released with insulin. IAPP is synthesized as an 89 amino acid long prepropeptide having a typical signal peptide and IAPP flanked by 2 short peptides. They are ultimately cleaved at basic residues. The $NH_2$-terminal flanking region consists of a peptide which contains a possible cleavage site for DPP IV. The sequence is (Tyr-Pro-Ile-Glu-Ser-His-Gln-Val-Glu-Lys-Arg) (van Hilst *et al* 1999). Using specific antisera, it was shown that not only the mature IAPP but also the $NH_2$-terminal flanking peptide is present in islet amyloid deposits. It remains to be shown (1) if the propeptide has any physiological function or (2) if it plays a role in the pathogenesis of amyloid deposits or (3) whether it is only a side product of IAPP formation without any biological significance (Westermark *et al.* 1989).

**A Islet Amyloid Polypeptide precursor (1-89)**

1 *MGILKLQVFL IVLSVALNHL KA* **TP IESHQV EKR** KCNTATC ATQRLANFLV
*signal sequence* (1-22) **propeptide(23-33)** **islet amyloid-**

51 HSSNNFGAIL SSTNVGSNTY GKRNAVEVLK REPLNYLPL
**polypeptide** (34-70)

**B Calcitonin precursor (1-141)**

1 *MGFQKFSPFL ALSILVLLQA GSLHA* **AP FRS ALESSPADPA TLSEDEARLL**
*signal sequence* (1-25) **procalcitonin** (26-141) N-PCT(26-82)

51 **LAALVQDYVQ MKASELEQEQ EREGSSLDSP RS** KR CGNLST CMLGTYTQDF
**calcitonin** (85-116)

101 NKFHTFPQTA IGVGAP GKKR DMSSDLERDH RPHVSMPQNA N
**katacalcin** (121-141)

**C α-type Calcitonin gene related peptide I precursor (1-128)**

1 *MGFQKFSPFL ALSILVLLQA GSLHA* **AP FRS ALESSPADPA TLSEDEARLL**
*signal sequence* (1-25) **propeptide**

51 **LAALVQDYVQ MKASELEQEQ EREGSRIIAQ** KR ACDTATCV THRLAGLLSR
**calcitonin gene related**

101 SGGVVKNNFV PTNVGSKAF G RRRRDLQA
**peptide I** (83-119)

*Figure 4.* The amino acid sequences of 4 members of the calcitonin-gene related family of proteins is given:
A) Islet amyloid polypeptide precursor, the propeptide of which is a putative DPP IV substrate
B) Calcitonin precursor containing the 116 amino acid long procalcitonin (PCT) with $NH_2$-terminal Ala-Pro as putative DPP IV recognition site. In case of calcitonin formation the $NH_2$-terminal region of PCT (N-PCT) is formed, which might also serve as a DPP IV substrate.
C) Calcitonin gene related peptide precursor. The propeptide of calcitonin gene related peptide also contains an $NH_2$-terminal X-Pro motif at its terminus. The physiological function of the propeptides or the remaining parts after liberation of the mature peptide as well as their further truncation by DPP IV remains to be investigated.

# 9. CHEMOKINE PROCESSING BY DPP IV: AN ARRAY OF CONSEQUENCES

Chemokines were first described as inducible mediators of leukocyte attraction in inflammation, but later several constitutively produced chemokines were identified that fulfill homeostatic functions. Chemokines form a large family of small proteins with four cysteines linked by disulfide bridges. The main subfamilies are distinguished depending on the sequence around the first cysteine residue, resulting in chemokines of the C, CC, CXC or CX3C family. They act via heptahelical G-protein-coupled receptors (designated CXR, CCR, CXCR and CX3CR and CCR followed by a number) and are unusually large ligands for this type of receptors (Luster 1998, Baggiolini *et al* 1997). Changes in intracellular $Ca^{++}$ are often used as a measure of signaling through their receptor(s) and chemotaxis assays as an in vitro estimate of their biological function.

*Table 2*. $NH_2$-terminal sequence of chemokines with X-Pro or X-Ala motifs

| | Chemokine | $NH_2$-terminal sequence |
|---|---|---|
| CXC | IL-8 (monocyte derived form) | **SAK**ELRCQC |
| | GCP-2 | **GP***VSAVLTELRCTC |
| | PF4 | **EA**EEDGDLQCLC |
| | IP-10 | **VPL**SRTVRCTC |
| | MIG | **TP**VVRKGRCSC |
| | SDF-1 | **KP**VSLSYRCPC |
| | GRO-□ | **AP**LATELRCQC |
| | I-TAC | **FP**MFKRGRCLC |
| CC | RANTES | **SP**YSSDTTPC C |
| | LD78□ | **AP**LAADTPTAC C |
| | MIP-1□ | **AP**MGSDPPTAC C |
| | MCP-1 | **QP**DAINAPVTC C |
| | MCP-2 | **QP**DSVSIPITC C |
| | MCP-3 | **QP**VGINTSTTC C |
| | MCP-4 | **QP**DALNVPSTC C |
| | Eotaxin | **GP**ASVPTTC C |
| | MDC | **GPYG**ANMEDSVC C |

* underlined dipeptides are released by CD26/DPP IV

Analysis of the primary structure of the more than 40 chemokines identified so far, reveals a low overall sequence homology. However, comparison of the known three-dimensional structures indicates a remarkable similarity in the backbone fold. The motif formed by the two cysteines is followed by an extended solvent exposed loop region. The further structure consists of 3 $\beta$ strands and a COOH-terminal $\alpha$ helix.

Although physical methods can describe the structure into great detail, most of these structural aspects give only little insight into their mode of action. Indeed, the $NH_2$-terminal region before the first cysteine (usually less than 11 amino acids) has almost maximal flexibility and at the same time the $NH_2$-terminus contains critical residues for receptor triggering (Baggiolini *et al* 1997, Crump *et al* 1998). Removal or modification of recognition motifs in the short aminoterminal stretch yields molecules that still recognize the receptor but do not induce functional responses and act as antagonists.

The above described importance of the $NH_2$-region in chemokine signaling, the occurrence of an $NH_2$-terminal X-Pro or X-Ala motif in a great number of chemokines (listed in table 2) and the observation that naturally isolated peptides are often truncated (Wnyts *et al* 1999), prompted a study to the role of DPP IV in these processes. The consequences of chemokine processing by DPP IV are reviewed recently (van Damme *et al* 1999) and are summarized in table 3. Examples of significant biological effects of dipeptide removal by DPP IV include RANTES (Oravecz *et al* 1997, Proost *et al* 1998a, Schols *et al* 1998), SDF-1 (Ohtsuki *et al* 1998, Proost *et al* 1998b), and eotaxin (Struyf *et al* 1999). As a consequence of a decreased signaling through CCR1 and CCR3, RANTES(3-68) is a poor chemoattractant for monocytes and eosinophils, while an increased binding to CCR5 clearly enhances its anti-HIV properties73. In contrast, the removal of Lys-Pro from the CXC chemokine SDF-1 reduces both its chemotactic and anti-viral activity. DPP IV-mediated processing of eotaxin, another chemokine that like SDF-1 activates a single receptor, also results in a loss of chemotactic activity while binding to its receptor CCR3 remains intact (Struyf *et al* 1999). That enables the truncated eotaxin(3-74) to inhibit chemotactic responses towards intact eotaxin. In order to reveal the biological significance of chemokine processing by soluble or cell-surface DPP IV/CD26, more detailed kinetic and comparative studies are required to identify the preferred chemokine substrates. Even with that knowledge, it will be hard to predict the in vivo scenario, as the regulation of chemokine secretion and receptor expression is complex with redundancy at some but not all levels.

The rather unexpected removal of an additional Tyr-Gly from MDC under circumstances where other chemokines only lost their aminoterminal X-Pro upon incubation with purified DPP IV/CD26 suggests that the substrate selectivity may not be as restricted as generally accepted (Proost *et al* 1999). This finding confirms early reports on the cleavage of analogues of GRF (Martin *et al* 1993, Bongers *et al* 1992) and it serves as a trigger to elaborate further on the extended substrate recognition of DPP IV (preferred P2, P1', P2' etc). A number of candidate-substrates indeed still await further examination. The new data will at least increase our understanding of the extended substrate specificity of DPP IV.

*Table 3.* Biological effects of chemokine processing by DPP IV

| | Chemotaxis | Receptor activation | HIV-inhibition |
|---|---|---|---|
| SDF-1α (3-68) | lymphocytes↓↓ | CXCR4 ↓↓ | ↓↓ |
| RANTES (3-68) | monocytes<br>lymphocytes<br>macrophages = | CCR1 ↓↓<br>CCR3 ↓↓<br>CCR5 ↑ | ↑↑ |
| MDC (5-69) | MDDC ↓<br>monocytes = | CCR4 ↓↓ | (=) |
| Eotaxin (3-74) | eosinophils ↓ | CCR3 ↓ | (=) |
| GCP-2 (3-77) | neutrophils = | CXCR1 =<br>CXCR2 = | ↑ |
| MCP-2 | no cleavage | | |

## 10. CONCLUSIONS

During the last decade it has become clear that DPP IV may have various substrates in vivo and that the preferred peptide will depend on the localization and physiological circumstances. It is at present impossible to depict a certain chain length as the maximal acceptable substrate size as it turns out that the immediate surrounding and surface accessibility of the $NH_2$-terminal dipeptide are determining the susceptibility for cleavage of a peptide.

From the above, it is clear that the result of dipeptide removal by DPP IV may vary from no effect over activation or change in receptor selectivity to inactivation of the substrate. Therefore, biological interpretation of assays that do not distinguish intact and modified peptides, should be interpreted with caution. Furthermore, collection and conservation of samples for peptide analysis should occur in the cold and if possible in the presence of appropriate protease inhibitors.

Clinical implications of peptide processing by DPP IV include that the therapeutic potential of peptides that are degraded by DPP IV may largely be enhanced by creating DPP IV-resistant, active analogues, and that the most suitable $NH_2$-terminal modification may vary from one peptide to another. Most exciting are the observations that the in vivo introduction of DPP IV specific inhibitors can enhance the levels of intact endogeneous peptides creating therapeutical perspectives (Holst *et al* 1998). Extensive in vivo experiments to reveal whether DPP IV is a powerful and safe pharmaceutical target, are awaited with interest.

# REFERENCES


Augustyns, K., Bal, G., Thonus, G., Belyaev, A., Zhang, X.M., Bollaert, W., Lambeir, A.M., Durinx ,C., Goossens, F., and Haemers, A., 1999, The unique properties of dipeptidyl-peptidase IV (DPP IV/CD26) and the therapeutic potential of DPP IV inhibitors. *Curr. Med.. Chem.* **6:** 307-317.

Baggiolini, M., Denwald, B., and Moser, B., 1997, Human Chemokines: an update. *Ann. Rev. Immunol.* **15:** 675-705.

Barrett, A., Rawlings, N., and Woesner, J., 1998, Handbook of Proteolytic enzymes

Bell, G.I., 1986, The glucagon superfamily: precursor structure and gene organization. *Peptides* **7:** 27-36.

Bongers, J., Lambros, T., Ahmad, M., and Heimer, E.P., 1992, Kinetics of dipeptidyl peptidase IV proteolysis of growth hormone-releasing factor and analogs. *Biochim. Biophys. Acta* **1122:** 147-153.

Bouras, M., Huneau, J.F., and Tome, D., 1996, The inhibition of intestinal dipeptidylaminopeptidase-IV promotes the absorption of enterostatin and des-arginine-enterostatin across rat jejunum in vitro. *Life Sci.* **59:** 2147-2155.

Bouras, M., Huneau, J.F., Luengo, C., Erlanson-Albertsson, C., and Tomé, D., 1995, Metabolism of enterostatin in rat intestine, brain membranes, and serum : Differential involvement of proline-specific peptidases. *Peptides* **16:** 399-405.

Bowyer, R.C., Jehanli, A.M.T., Patel, G., Hermon-Taylor, J., 1991, Development of enzyme-linked immunosorbent assay for free human pro-colipase activation peptide (APGPR). *Clin. Chim. Acta* **200:** 137-152.

Conlon, J.M., and Sheehan, L., 1983, Conversion of substance P to C-terminal fragments in human plasma. *Regulatory peptides* **7:** 335-345.

Coy, D.H., Murphy, W.A., Lance, V.A., and Heiman, M.L., 1987, Differential effects of N-terminal modifications on the biological potencies of growth hormone releasing factor analogues with varying chain lengths. *J. Med. Chem.* **30:** 219-222.

Coy, D.H., Murphy, W.A., Sueiras-Diaz, J., Coy, E.J., and Lance, V.A., 1985, Structure-activity on the N-terminal region of growth hormone releasing factor. *J. Med. Chem.* **28:** 181-185.

Crump, M., Rajarathnam, K., Kim, K-S, Clark-Lewis, I., and Sykes, B., 1998, Solution structure of eotaxin, a chemokine that selectively recruits eosinophils in allergic inflammation. *J. Biol. Chem.* **273:** 22471-22479.

De Meester, I., Korom, S., Van Damme, J., and Scharpé, S., 1999, CD26, let it cut or cut it down. *Immunol. Today* **20:** 367-375.

Deacon, C.F., Holst, J.J., and Carr, R.D., 1999, Glucagon-like peptide-1: A basis for new approaches to the management of diabetes. *Drugs Today* **35:** 159-170.

Dezan, C., Daniel, C., Hirn, J., Sarda, L., and Bellon, B., 1994, Monoclonal antibodies to human pancreatic procolipase : Production and characterization by competitive binding studies. *Hybridoma* **13:** 509-517.

Drucker, D.J., 1998, Glucagon-like peptides. *Diabetes* **47:** 159-169.

Drucker, D.J., Shi, Q., Crivici,A., Sumner-Smith, M., Tavares, W., Hill, M., DeForest, L., Cooper; S., and Brubaker, P.L., 1997, Regulation of the biological activity of glucagon-like peptide 2 *in vivo* by dipeptidyl peptidase IV. *Nat. Biotechnol.* **15:** 673-677.

Dumont, Y., Jacques, D., Bouchard, P., and Quirion, R., 1998, Species differences in the expression and distribution of the neuropeptide Y Y1, Y2, Y4 and Y5 receptors in rodents, guinea pig and primate brains. *J. Comp. Neurol.* **402:** 372-384

Erlanson-Albertsson, C., and York, D., 1997, Enterostatin – A peptide regulating fat intake. *Obes. Res.* **5:** 360-372.

Frohman, L.A., Downs, T.R., Williams, T.C., Heimer, E.P., Pan Y.C., and Felix, A.M., 1986, Rapid enzymatic degradation of growth hormone-releasing hormone by plasma *in vitro* and *in vivo* to a biologically inactive product cleaved at the $NH_2$-terminus. *J. Clin. Invest.* **78:** 906-913.

Fulöp, V., Böcskei, Z., and Polgár, 1998, Prolyl oligopeptidase : An unusual β-propeller domain regulates proteolysis. *Cell* **94:** 161-170.

Goossens, F., De Meester, I., Vanhoof, G., Hendriks, D., Vriend, G., and Scharpé, S., 1995, The purification, characterization and analysis of primary and secondary-structure of prolyl oligopeptidase from human lymphocytes. Evidence that the enzyme belongs to the α/β hydrolase fold family. *Eur. J. Biochem.* **233:** 432-441.

Hartrodt, B., Neubert, K., Fischer, G., Demuth, U., Yoshimoto, T., and Barth, A., 1982, Degradation of β casomorphin-5 by PSE and PPCE. Comparative studies of the β casomorphin-5 cleavage by DPPIV. *Pharmazie* **37:** 72-73.

Heymann, E., and Mentlein, R., 1978, Liver dipeptidyl aminopeptidase IV hydrolyzes substance P. *FEBS Lett.* **91:** 360-364.

Heymann, E., Mentlein, R., Nausch, I., Erlanson-Albertsson, C., Yoshimoto, T. and Feller, A.C., 1986, Processing of pro-colipase and trypsinogen by pancreatic dipeptidyl peptidase IV. *Biomed. Biochim. Acta* **45:** 575-584.

Hoffmann, T., Faust, J., Neubert, K., and Ansorge, S., 1993, Dipeptidyl peptidase IV (CD 26) and aminopeptidase N (CD 13) catalyzed hydrolysis of cytokines and peptides with N-terminal cytokine sequences. *FEBS Lett.* **336:** 61-64.

Holst, J.J., and Deacon, C.F., 1998, Inhibition of the activity of dipeptidyl-peptidase IV as a treatment for type 2 diabetes. *Diabetes* **47:** 1663-1670.

Hu, L., Balse, P., and Doughty, M.B., 1994, Neuropeptide Y N-terminal deletion fragments : Correlation between solution structure and receptor binding activity at Y1 receptors in rat brain cortex. *J. Med. Chem.* **14:** 3622-2629.

Huneau, J.F., Erlanson-Albertsson, C., Beauvallet, C., and Tomé, D., 1994, The *in vitro* intestinal absorption of enterostatin is limited by brush-border membrane peptidases. *Regul. Pept.* **54:** 495-503.

Inuis, A., 1999, Neuropeptide Y feeding receptors: Are multiple subtypes involved? *TIPS* **20:** 43-46.

Jarpe, M.B., Knall, C., Mitchell, F.M., Buhl, A.M., Duzic, E., and Johnson, G.L., 1998, [D-Arg1,D-$Phe^5$,$Trp^{7,9}$-$Leu^{11}$]substance P acts as a biased agonist toward neuropeptide and chemokine receptors. *J. Biol. Chem.* **273:** 3097-3104.

Jonas, V, Lin, C.R., Kawashima E., Semon, D., Swanson, L., Mermod, J., Evans, R., and Rosenfeld M., 1985, Alternative RNA processing events in human calcitonin/calcitonin gene-related peptide gene expression. *Proc. Natl. Acad. Sci USA* **82:** 1994-1998.

Kah, S.E., Andrikopoulos, S., and Verchere, C.B., 1999, Islet amyloid : A long-recognized but underappreciated pathological feature of type 2 diabetes. *Diabetes* **48:** 241-253.

Kähne, T., Lendeckel, U., Wrenger, S., Neubert, K., Ansorge, S., Reinhold, D., 1999, Dipeptidyl peptidase IV: A cell surface peptidase involved in regulating T cell growth. *Int. J. Mol. Med.* **4:** 3-15.

Karzai, W., Oberhoffer, A., Meier-Hellmann, A., Reinhart, K., 1997, Procalcitonin – A new indicator of the systemic response to severe infections. *Infect.* **25:** 329-334.

Leiter, A.B., Toder, A., Wolfe, H.J., Taylor, I.L., Cooperman, S., Mandel, G., and Goodman, R.H., 1987, Peptide YY : Structure of the precursor and expression in exocrine pancreas. *J. Biol. Chem.* **262:** 12894-12988.

Levite, M., 1998, Neuropeptides, by direct interaction with T cells, induce cytokine secretion and break the commitment to a distinct T helper phenotype. *Proc. Natl. Acad. Sci.* **95:** 12544-12549.

Lin, L., Okada, S., York, D.A., and Bray, G.A., 1994, Structural requirements for the biological activity of enterostatin. *Peptides* **15:** 849-854.

Luster, A., 1998, Chemokines-chemotactic cytokines that mediate inflammation. *N. Engl. J. Med.* **338:** 436-445.

Martin, R.A., Cleary, D.L., Guido, D.M., Zurcher-Neely, H.A., and Kubiak, T.M., 1993, Dipeptidyl peptidase IV (DPP-IV) from pig kidney cleaves analogs of bovine growth hormone-releasing factor (bGRF) modified at position 2 with Ser, Thr or Val. Extended DPP-IV substrate specificity? *Biochim. Biophys. Acta* **1164:** 252-260.

Medeiros, M.S., and Turner, A.J., 1994, Post-secretory processing of regulatory peptides : The pancreatic polypeptide family as a model example. *Biochim.* **76:** 283-287.

Mentlein, R., and Heymann, E., 1982, Dipeptidyl peptidase IV inhibits the polymerization of fibrin monomers. *Arch. Biochem. Biophys.* **217:** 748-750.

Mentlein, R., Dahms, P., Grandt, D., and Krüger, R., 1993, Proteolytic processing of neuropeptide Y and peptide YY by dipeptidyl peptidase IV. *Regul. Pept.* **49:** 133-144.

Mentlein, R., Gallwitz, B., and Schmidt, E., 1993, Dipeptidyl-peptidase IV hydrolyses gastric inhibitory polypeptide, glucagon-like peptide-1(7-36)amide, peptide histidine methionine and is reponsible for their degradation in human serum. *Eur. J. Biochem.* **214:** 829-855.

Michel, M.C., Beck-Sickinger, A., Cox, H., Doods, H.N., Herzog, H., Larhammar, D., Quirion, R., Schwartz, T., Westfall, T., 1998, XVI. International union of pharmacology recommendations for the nomenclature of neuropeptide Y, peptide YY, and pancreatic polypeptide receptors. *Pharmacol. Rev.* **50:** 143-150.

Morgan, D.G., Kulkarni, R.N., Hurley, J.D., Wang, Z.L., Wang, R.M., Ghatei, M.A., Karlsen, A.E., Bloom, S.R., and Smith, D.M., 1998, Inhibition of glucose stimulated insulin secretion by neuropeptide Y is mediated via the Y1 receptor and inhibition of adenylyl cyclase in RIN 5AH rat insulinoma cells. *Diabetologia* **41:** 1482-1491.

Morimoto, C., and Schlossman, S.F., 1998, The structure and function of CD26 in the T-cell immune response. *Immunological Rev.* **161:** 55-70.

Nilsson, C., Westman, A., Blennow, K, and Ekman, R., 1998, Processing of neuropeptide Y and somatostatin in human cerebrospinal fluid as monitored by radioimmunoassay and mass spectrometry. *Peptides* **19:** 1137-1146.

O'Harte, F.P., Mooney, M.H., Flatt, P.R., 1999, $NH_2$-Terminally modified gastric inhibitory polypeptide exhibits amino-peptidase resistance and enhanced antihyperglycemic activity. *Diabetes* **48:** 758-765.

Ohtsuki, T., Hosono, O., Kobayashi, H., Munakata, Y., Souta, A., Shioda, T., and Morimoto, C., 1998, stromal cell-derived factor 1α by CD26/dipeptidyl peptidase IV. *FEBS Lett.* **431:** 236-240.

Okada, S., York, D.A., Bray, G.A., and Erlanson-Albertsson, C., 1991, Enterostatin (Val-Pro-Asp-Pro-Arg), the activation peptide of procolipase, selectivity reduces fat intake. *Physiol. Behav.* **49:** 1185-1189.

Ookuma, M. and York, D.A., 1998, Inhibition of insulin release by enterostatin. *Int. J. Obes. Relat.. Metab. Disord.* **22:** 800-805.

Oravecz, T., Pall, M., Roderiquez, G., Gorrell, M.D., Ditto, M., Nguyen, N., Boykins, R., Unsworth, E., and Norcross, M.A., 1997, Regulation of the receptor specificity and function of the chemokine RANTES (Regulated on activatin, normal T cell expressed and secreted) by dipeptidyl peptidase IV (CD26)-mediated cleavage. *J. Exp. Med.* **186:** 1865-1872.

Prasad, C., Immamura, M., Debata, C., Svec, F., Sumar, N., Hermon-Taylor, J., 1999, Hyperenterostatinemia in premenopausal obese women. *J. Clin. Endocrinol. Metabol.* **84:** 937-941.

Proost, P., De Meester, I., Schols, D., Struyf, S., Lambeir, A.M., Wuyts, A., Opdenakker, G., De Clercq, E., Scharpé, S., and Van Damme, J., 1998, Amino-terminal truncation of chemokines by CD26/dipeptidyl-peptidase IV. *J. Biol. Chem.* **273:** 7222-7227.

Proost, P., Struyf, S., Schols, D., Durinx, C., Wuyts, A., Lenaerts, J.P., De Clercq, E., De Meester, I., and Van Damme, J., 1998, Processing by CD26/dipeptidyl-peptidase IV reduces the chemotactic and anti-HIV-1 activity of stromal-cell-derived factor-1α. *FEBS Lett.* **432:** 73-76.

Proost, P., Struyf, S., Schols, D., Opdenakker, G., Sozzani, S., Allavena, P., Mantovani, A., Augustyns, K., Bal, G., Haemers, A., Lambeir, A.M., Scharpé, S, Van Damme, J., and De Meester, I., 1999, Truncation of macrophage-derived chemokine by CD26/dipeptidyl-peptidase IV beyond its predicted cleavage site affects chemotactic activity and CC chemokine receptor 4 interaction. *J. Biol. Chem.* **274:** 3988-3993.

Riemann, D., Kehlen, A. and Langner, J. (1999) CD13 – not just a marker in leukemia typing. Immunol. Today 20, 83-88.

Ritzel, U., Leonhardt, U., Ottleben, M., Ruhmann, A., Eckart, K., Spiess, J., and Ramadori, G., 1998, A synthetic glucagon-like peptide-1 analog with improved plasma stability. *J. Endocrinol.* **159:** 93-102.

Rocca, A.S., and Brubaker, P.L., 1999, Role of the vagus nerve in mediating proximal nutrient-induced glucagon-like peptide-1 secretion. *Endocrinology* **140:** 1687-1694.

Sanket, T., Bell, G.I., Sample, C., Rubenstein, A.H., and Steiner, D.F., 1988, An islet amyloid peptide is derived from an 89-amino acid precursor by proteolyltic processing. *J. Biol. Chem.* **263:** 17243-17246.

Sato, K., Hotta, M., Kageyama, J., Chiang, T.C., Hu, H.Y., Dong, M.H., and Long, N., 1987, Synthesis and *in vitro* bioactivity of human growth hormone-releasing factor analogs substituted with a single D-amino acid. *Biochem. Biophys. Res. Commun.* **16:** 531-517.

Schols, D., Proost, P., Struyf, S., Wuyts, A., De Meester, I., Scharpé, S., Van Damme, J., and De Clercq, E., 1998, CD26-processed RANTES(3-68), but not intact RANTES, has potent anti-HIV-1 activity. *Antivir. Res.* **39:** 175-187.

Shane, R., Wilk, S., Bodnar, R.J., 1999, Modulation of endomorphin-2-induced analgesia by dipeptidyl peptidase IV. *Brain Res.* **815:** 278-286.

Siegel, E.G., Gallwitz, B., Scharf, G., Mentlein, R., Morys-Wortmann, C., Fölsch, U.R., Schrezenmeir, J., Drescher, K., and Schmidt, W.E., 1999, Biological activity of GLP-1 analogues with N-terminal modifications. *Regul. Pept.* **79:** 93-102.

Struyf, S., Proost, P., Schols, D., De Clercq, E., Opdenakker, G., Lenaerts, J.P., Detheux, M., Parmentier, M., De Meester, I., Scharpé, S., and Van Damme, J., 1999, CD26/Dipeptidyl-peptidase IV down-regulates the eosinophil chemotactic potency, but not the anti-HIV activity of human eotaxin by affecting its interaction with CC chemokine receptor 3. *J. Immunol.* **162:** 4903-4909.

Thomson, A., Schoeller, C., Keelan, M., et al., 1993, Lipid absorption: Passing through the unstirred layers, brush border membrane, and beyond. *Can. J. Physiol. Pharmacol.* **71:** 531-555.

Van Damme, J., Struyf, S., Wuyts, A., Van Coillie, E., Menten, P., Schols, D., Sozzani, S., De Meester, I., and Proost, P., 1999, The role of CD26/DPP IV in chemokine processing. *Chem. Immunol.* **72:** 42-56.

van Hulst, K.L., Oosterwijk, C., Born, W., Vroom, Th.M., Nieuwenhuis, M.G., Blankenstein., M.A., Lips, C.J.M., Fischer, J.A, and Höppener, J.W.M., 1999, Islet amyloid polypeptide/amylin messenger RNA and protein expression in human insulinomas in relation to amyloid formation. *Eur. J. Endocrinol.* **140:** 69-78.

van Tilbeurgh, H., Sarda, L., Verger, R., and Cambillau, C., 1992, Structure of the pancreatic lipase-procolipase complex. *Nature* **359:** 159-162.

von Bonin, A., Hühn J. and Fleisher, B. (1998) Dipeptidyl peptidase IV/CD26 alternative T-cells: analysis of an alternative activation pathway. *Immunological Rev.* **161:** 43-53.

Westermark, P., Engstrom, U., Westermark, G.T., Kohnson, K.H., Permerth, J., Betsholtz, C. (1989) Islet amyloid polypeptide (IAPP) and pro-IAPP immunoreactivity in human islets of Langerhans. *Diab. Res. Clin. Pract.* **18:** 219-226.

Winkler, H., and Fischer-Colbrie, 1992, The chromogranins A and B: the first 25 years and future perspectives. *Neuroscience* **49:** 497-528.

Wuyts, A., Govaerts, C., Struyf, S., Lenaerts, J.P., Put. W., Conings, R., Proost, P., and Van Damme, J., 1999 Isolation of the CXC chemokines ENA-78, GRO alpha and GRO gamma from tumor cells and leukocytes reveals NH2-terminal heterogeneity. Functional comparison of different natural isoforms. *Eur. J. Biochem.* **260:** 421-429.

Yaron, A., and Naider, F., 1993, Proline-dependent structural and biological properties of peptides and proteins. *Crit. Rev. Biochem. Mol. Biol.* **28:** 31-81.

Zhang, X.Y., De Meester, I., Lambeir, A.M., Dillen, L., Van Dongen, W., Esmans, E.L., Haemers, A., Scharpé, S., Claeys, M., 1999, Study of the enzymatic degradation of vasostatin I and II and their precursor chromogranin A by dipeptidyl peptidase IV using high-performance liquid chromatography/electrospray mass spectrometry. *J. Mass Spectrom.* **34:** 255-263.

Zukowska-Grojec, Z., 1997, Neuropeptide Y: Implications in vascular remodeling and novel therapeutics. *DN & P* **10:** 587-595.

Zukowska-Grojec, Z., Karwatowska-Prokopczuk, E., Rose, W., Rone, J., Movafagh, S., Ji, H., Yeh, Y., Chen, W.T., Kleinman, H.K., Grouzmann, E., and Grant, D.S., 1998, Neuropeptide Y: A novel angiogenic factor from the sympathetic nerves and endothelium. *Circ. Res.* **83:** 187-195.

# RELATING STRUCTURE TO FUNCTION IN THE BETA-PROPELLER DOMAIN OF DIPEPTIDYL PEPTIDASE IV

*Point mutations that influence adenosine deaminase binding, antibody binding and enzyme activity*

Mark D. Gorrell, Catherine A. Abbott, Thilo Kähne[1], Miriam T. Levy, W. Bret Church[2] and Geoffrey W. McCaughan

*A.W. Morrow Gastroenterology and Liver Centre, Centenary Institute of Cell Biology and Cancer Medicine, Royal Prince Alfred Hospital and The University of Sydney, NSW, Australia. [1]Otto-von-Guericke University, Magdeburg. [2]Garvan Institute of Medical Research, Sydney.*

Key words Dipeptidyl peptidase IV, beta propeller, adenosine deaminase, epitopes

Abstract Point mutations in human CD26/DP IV were analysed for adenosine deaminase (ADA) binding, monoclonal antibody (mAb) binding and DP IV enzyme activity. Point mutations at either Leu294 or Val341 ablated ADA binding. Binding by mAbs that inhibit ADA binding was found to involve both Leu340 to Arg343 and Thr440/Lys441. Glu205 and Glu206 were found to be essential for enzyme activity. All residues of interest were mapped onto a model of the β-propeller domain of DP IV. These data led us to suggest that in DP IV and related peptidases ligand and antibody binding sites are non-linear and that enzyme activity depends on charged sidechains that surround the entrance to the central tunnel of the β-propeller.

## 1. INTRODUCTION

Dipeptidyl peptidase IV (DP IV; EC 3.4.14.5) is a post-proline dipeptidyl aminopeptidase also called adenosine deaminase binding protein (ADAbp) or CD26. DP IV/CD26 binding to the soluble extracellular enzyme adenosine deaminase (ADA; EC 3.5.4.4) is independent of the catalytic activities of either enzyme (De Meester *et al* 1994). CD26/DP IV is a type II glycoprotein of 766 amino acids in which 738 amino acids are extracellular. The C-terminal, hydrolase domain of DP IV forms an α/β hydrolase fold similar to that of prolyl endopeptidase (PEP; EC 3.4.21.26) (Fülop *et al* 1998). The structure of the glycosylation-rich and cysteine-rich regions of DP IV, upstream from the α/β hydrolase domain, could include a β-propeller fold, which is a torus formed by repeated 4-stranded anti-parallel β sheets of about 50 amino acids, as occurs in PEP (Fülop *et al* 1998). The hydrolase and propeller domains of PEP both contribute to the surface of the inner cavity in which catalysis occurs.

MAbs to DP IV can be grouped into five epitopes (Gorrell *et al* 1997; Dong *et al* 1998). ADA binding and the epitope of the group of mAbs that inhibit ADA binding are known to involve the region Leu340 to Arg343 (Dong *et al* 1997, Dong *et al* 1998).

## 2. MATERIALS AND METHODS

### *Peptide library*

A set of 92 15mer peptides with a 7-residue overlap and representing the 738 extracellular amino acids of DP IV was synthesised with a N-terminal 4 amino acid spacer and biotin on each peptide (Chiron Mimotopes, Melbourne, Australia). Peptides were bound to streptavidin-coated EIA plates then reacted with mAbs, ADA or BSA in NaCl/Pi at pH 7 or 8.2. Binding was assessed using the immunoperoxidase method, anti-mouseIg-horseradish peroxidase (HRPO; P0260, Dako, Santa Barbara), a rabbit anti-ADA antibody (Abbott *et al* 1999a) and protein G-HRPO (Pierce, Rockford). The mAbs tested were PEG2/C3, EF5/A3, EF6/B10, F11, A10 and G11 (Kähne *et al* 1996), OX61 and 236.3.

### *Analyses of transfected cells*

Specific binding of mAbs (listed elsewhere; Abbott *et al* 1999b) and ADA to transfected COS-7 cells was detected by immunocytochemistry, Western blot and flow cytometry. Adherent monolayers were fixed in cold ethanol prior to staining. Suspensions of trypsin/EDTA-harvested cells were washed in Hank's buffer prior to enzyme assay or flow cytometry. These methods are described elsewhere (Gorrell *et al* 1991, Abbott *et al* 1999b, Levy *et al* 1999).

# 3. RESULTS

## Peptide library

No specific binding to any peptide fragment of DP IV by ADA or any anti-DP IV mAb was detected. However, 5 peptides produced non-specific reactions with rabbit anti-mouseIg-HRPO and proteinG-HRPO in the presence or absence of mouse Ig. The most intense reaction was with peptide 621 to 635, followed in intensity by peptides 453 to 467, 310 to 323, 69 to 83 and 165 to 180. Interestingly, peptide 621 to 635 includes the catalytic serine and peptide 310 to 323 includes the sequence 313LQWLRRI that inhibits the binding of fibronectin to DP IV (Gonzalez-Gronow *et al* 1998). The failure to identify a linear epitope suggested that epitopes of DP IV might be discontinuous and depend on tertiary structure.

## Mutagenesis strategy

ADA-DP IV binding occurs in humans but not rats. ADA and DP IV dissociate in very low ionic strength buffer, suggesting that hydrophobic residues are crucial (De Meester *et al* 1996). Therefore, we selected for study 8 hydrophobic residues non-conserved between rat and human DP IV. Antibody binding often requires charged amino acids, so non-conserved, charged residues were targeted.

## Analysis of point mutations and single amino acid point mutations

All the point mutations clearly retained structural integrity, indicated by cell surface expression, dimerisation, catalytic activity and antibody and ADA binding. The point mutations Leu294Arg and Val341Lys displayed no detectable ADA binding. Binding by the 6 mAbs that inhibit ADA binding was either significantly diminished or ablated in the mutations Arg343Ala, Arg343Gln, 340LeuValAla→ProLysThr and 440ThrLys→ProProSer. The mutations Glu205Lys and Glu206Leu showed no enzyme activity by cytochemistry and little activity in whole-cell enzyme assays of cell surface expressed DP IV.

## A structure prediction of the ADA binding domain of DP IV.

The 354-residue β-propeller domain of PEP (PDB code 1qfm) was used as a template for the model of the 369 amino acids Asp133 to Asp501 of human DP IV (Figure 1). The overall model closely followed a PEP - DP IV sequence alignment output from Threader {v.2.1 (Jones *et al* 1992)} apart from an N terminal extension of the propeller to incorporate the observation that Asp133 to Lys190 (blade 1) has 33 % identity and 43 % similarity with Asp192 to Arg253 (blade 2), concordant with the repeating nature of propeller blades. The model of DP IV placed the mutations influential in ligand binding (Leu294 and Val341) and antibody binding (Leu340/Val341/Arg343 and Thr440/Lys441) on loops between the third and fourth β strands of β sheets 3, 4 and 6 and placed these loops on the outside of the propeller lowerface, distant from both the catalytic site and the propeller central pore. An interesting feature of PEP retained in the DP IV model was that charged residues surround the central pore of the propeller lowerface. The glutamates implicated in enzymatic function, Glu205 and Glu206, are two such residues.

## 4. CONCLUSION

We have shown that residues Leu294 and Val341 of DP IV are essential for ADA binding, that the epitope of 6 mAbs that inhibit ADA binding includes the regions of Leu340 to Arg343 and Thr440/Lys441 and that Glu205 and Glu206 are essential for enzyme activity. The locations of these amino acids concords with our β-propeller fold model of the ADA binding domain of DP IV. The 7-blade β-propeller fold forms a torus of 7 β sheets of about 50 amino acids each. This tertiary structure allows amino acids 47 and 97 positions distant in primary structure to contribute to binding sites (Oxvig and Springer 1998). Therefore, we predict that the epitopes of all antibodies that map to residues 133 to 501 of DP IV, which includes most mAbs (Dong *et al* 1998, Hühn *et al* 1999), are discontinuous in the primary structure. The propensity of antibody binding to depend on charged residues suggests that the basic residues Arg343 and Lys441 may be directly involved in antibody binding.

In PEP, 5 pairs of charged amino acids, located at the ends of the first β strands of blades 1, 2, 3, 4 and 7, surround the central pore of the propeller lower face. The model placed pairs of charged amino acids in similar positions in DP IV. The pairs are Glu145-Glu146, Glu205-Glu206, Lys258-Ala259, Arg317-Arg318 and Lys463-Glu464 in the DP IV model. It is likely that some of these residues surrounding the propeller central pore, about 20 Å from the catalytic site, are involved in substrate entry into this hollow enzyme (Abbott *et al* 1999a).

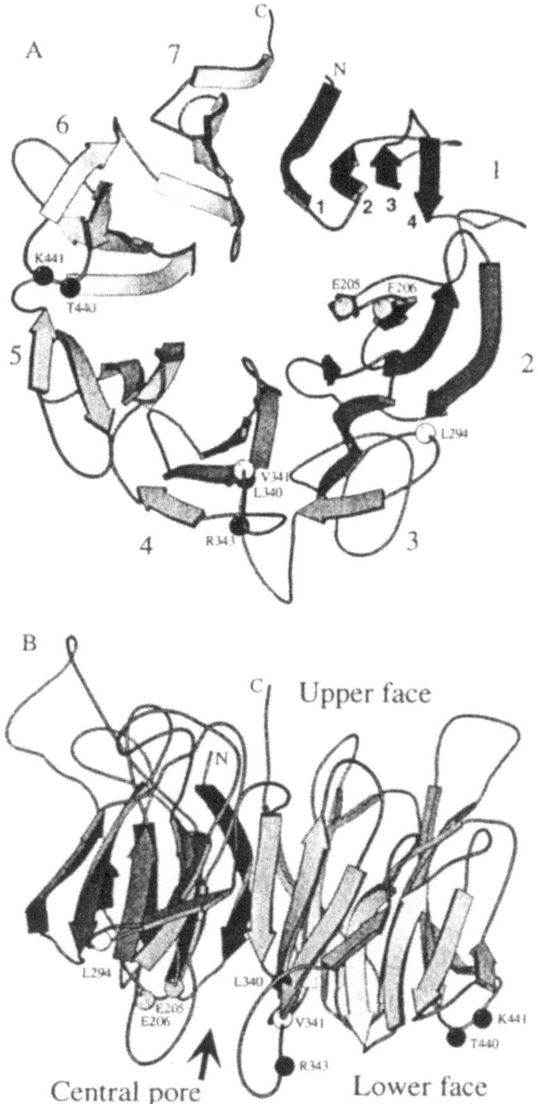


*Figure 1.* A model of DP IV residues Asp133 to Asp501 is depicted in ribbon representation coloured from dark grey to pale grey as the polypeptide chain is traversed from the N to the C terminus of this domain. Locations of the mutated amino acids of interest are displayed as spheres: Sites of point mutations that diminished ADA binding (white), mAb binding (black), and enzyme activity (grey) are shown. A. View down the pseudo-symmetry axis from below the propeller lower face. B. View perpendicular to (A) in which the catalytic domain is anticipated to be above the propeller. Figure generated using MOLSCRIPT (Kraulis 1991).

These findings provide significant insight into the ADA binding domain of DP IV and a foundation for future detailed structure-function studies of the relative roles of ADA binding and DP IV-mediated catalysis in lymphocyte biology.

# ACKNOWLEDGMENTS

The National Health and Medical Research Council of Australia and the Deutsche Forschungsgemeinschaft supported this work. The authors are grateful to C. Morimoto and D. Doyle for gifts of DP IV cDNA, the many laboratories that provided antibodies, D. Yu, B. Creighton and J. Ho for technical assistance and C. Collyer, N. Barclay and K. Stanley for helpful advice and discussions.

# REFERENCES

Abbott, C.A., McCaughan, G.W., and Gorrell, M.D., (1999a) Two highly conserved glutamic acid residues in the predicted beta propeller domain of dipeptidyl peptidase IV are required for its enzyme activity. *FEBS Lett.* **458:** 278-284.

Abbott, C.A., *et al.*, 1999b, Binding to human dipeptidyl peptidase IV by adenosine deaminase and antibodies that inhibit ligand binding involves overlapping, discontinuous sites on a predicted b propeller domain. *Eur. J. Biochem.* in Press.

De Meester, I., *et al.*, 1994, Binding of adenosine deaminase to the lymphocyte surface via CD26. *Eur. J. Immunol.* **24:** 566-570.

De Meester, I., *et al.*, 1996, Use of immobilized adenosine deaminase (EC 3.5.4.4) for the rapid purification of native human CD26 dipeptidyl peptidase IV (EC 3.4.14.5). *J. Immunol. Methods* **189:** 99-105.

Dong, R.P., *et al.*, 1997, Determination of adenosine deaminase binding domain on CD26 and its immunoregulatory effect on T cell activation. *J. Immunol.* **159:** 6070-6076.

Dong, R.P., *et al.*, 1998, Correlation of the epitopes defined by anti-CD26 mAbs and CD26 function. *Mol. Immunol.* **35:** 13-21.

Fülop, V., Bocskei, Z., and Polgar, L., 1998, Prolyl oligopeptidase - an unusual beta-propeller domain regulates proteolysis. *Cell* **94:** 161-170.

Gonzalez-Gronow, M., *et al.*, 1998, Plasmin(ogen) carbohydrate chains mediate binding to dipeptidyl peptidase IV (CD 26) in rheumatoid arthritis human synovial fibroblasts. *Fibrinol. Proteol.* **12:** 366-374.

Gorrell, M.D., *et al.*, (1997) CD26: Adenosine deaminase binding, immunoblotting and species cross reactivity examined using recombinant proteins. In *Leucocyte Typing VI* (T. Kishimoto, ed), New York, Garland Publishing Inc., pp. 484-485.

Gorrell, M.D., Wickson, J., and McCaughan, G.W., 1991, Expression of the rat CD26 antigen (dipeptidyl peptidase IV) on subpopulations of rat lymphocytes. *Cell. Immunol.* **134:** 205-215.

Hühn, J., *et al.*, 1999, The adenosine deaminase-binding region is distinct from major anti-CD26 mAb epitopes on the human dipeptidyl peptidase IV(CD26) molecule. *Cell. Immunol.* **192:** 33-40.

Jones, D.T., Taylor, W.R., and Thornton, J.M., 1992, A new approach to protein fold recognition. *Nature* **358:** 86-89.

Kähne, T., *et al.*, 1996, Alterations in structure and cellular localization of molecular forms of DPPIV/CD26 during T cell activation. *Cell. Immunol.* **170:** 63-70.

Kraulis, P., 1991, MOLSCRIPT: a program to produce both detailed and schematic plots of protein structures. *J. Appl. Crystallogr.* **24:** 946-950.

Levy, M.T., *et al.*, 1999, Fibroblast activation protein: A cell surface dipeptidyl peptidase and gelatinase expressed by stellate cells at the tissue remodelling interface in human cirrhosis. *Hepatology* **29:** 1768-1778.

Oxvig, C., and Springer, T.A., 1998, Experimental support for a beta-propeller domain in integrin alpha-subunits and a calcium binding site on its lower surface. *Proc. Natl. Acad. Sci. USA* ***95:*** 4870-4875.

# DEVELOPMENT OF A TERTIARY-STRUCTURE MODEL OF THE C-TERMINAL DOMAIN OF DPP IV

Wolfgang Brandt
*Department of Biochemistry and Biotechnology, Institute of Biochemistry, Martin-Luther-University Halle-Wittenberg, Kurt-Mothes-Str. 3, 06120 Halle (Saale), Germany*

Key words dipeptidyl peptidase IV, molecular modelling, active site, tertiary structure

Abstract Based on the recently published structure of prolyl oligopeptidase (POP) a model of the C-terminal part of dipeptidyl peptidase IV (DPP IV) which contains the active site has been developed. The structure of the model of DPP IV shows considerable similarity to the structure of POP particularly in the active site. A hydrophobic pocket (Tyr666, Tyr670, Tyr 631, Val556) forms the S1-binding site for recognition of proline. Tyr547 may stabilise the oxyanion formed in the tetrahedral intermediates by a strong hydrogen bond. The positively charged N-terminus of ligands of DPP IV is recognised by forming a salt bridge with the acidic side chain Glu668. A second hydrophobic pocket (S2′ to S5′) may represent an important binding site for HIV-1 Tat-protein derivatives, chemokines and others.

## 1. INTRODUCTION

Dipeptidyl peptidase IV (DPP IV) is a proline-specific serine peptidase of broad medical and biochemical significance. It has been described that DPP IV plays a role in regulation mechanisms of the immunological system (Von Bonin *et al* 1998). Among others, it could be shown that the HIV-1 Tat protein and the related N-terminal nonapeptide (MDPVDPNIE) are DPP IV inhibitors (Wrenger *et al* 1997).

This enzyme sequentially removes dipeptides off the N-terminus of a peptide chain whenever a proline or alanine residue is present in the penultimate position. An X-ray structure of DPP IV does not exist up to now. The recently published structure of the related enzyme prolyl oligopeptidase (POP) is now public available (Fülöp *et al* 1998). The detailed knowledge of the binding sites of ligands to DPP IV may not only serve to understand the substrate specificity but also as a basis to design new highly specific inhibitors. By means of homology modelling approaches a model of the C-terminal part of DPP IV which contains the active site has been developed.

## 2. METHODS

The recently published structure of prolyl oligopeptidase (Brookhaven Protein Data Bank entry 1qfm) was used as a target to model the C-terminal catalytically active region of DPP IV. COMPOSER (Blundell *et al* 1988) a program for homology modelling which is included in the molecular graphics program package SYBYL (TRIPOS Associates Inc.) was used to prepare the model of DPP IV. The modelling procedure consisted of the following steps: the amino acid sequences of POP and DPP IV were aligned (Fig 1), structurally conserved regions (SCRs) were identified and a framework of conserved regions was defined as mean positions of structurally equivalent Cα atoms. Structurally varable regions SVRs (loops) were selected from a database of peptide fragments extracted from a protein database in order to satisfy end-to-end distances of the SCRs already positioned in the framework.

The obtained tertiary structure was minimised with the TRIPOS force field (Clark *et al* 1989) including electrostatic interactions based on Gasteiger partial charge distributions (Gasteiger and Marsili 1980) by using a dielectric constant with a distance dependent function $\varepsilon = 4r$ and a nonbonded cut-off of 8 Å. Low temperature (100 K) molecular dynamics simulations were performed for refinement of the structure particularly of the loop regions. HIV-1 Tat nonapeptides were docked to the active site of DPP IV using FLEXIDOCK (Judson 1997), a program of SYBYL that allows the simulation of induced fit mechanisms. The quality of the model was checked and proven with PROCHECK (Laskowski *et al* 1993) and PROSA (Suppl. 1993).

## 3. RESULTS AND DISCUSSION

The structure of POP is characterised by two major domains, a so called propeller domain and the C-terminal region which contains the catalytically active site. In this paper we focus on the modelling of the C-terminal domain of DPP IV only. For this purpose the co-ordinates of the amino acid residues of the propeller domain together with the N-terminal chain of POP were removed. The remaining sequence was aligned with the likely corresponding sequence of human DPP IV starting with residue 502. The resulting optimal alignment is presented in Fig 1.

```
    502      10         20         30         40        50
    KMLQNVQMPSKKLDFIILNETKFWYQMILPPHFDKSKKYPLLLDVYAGP
    SDYQTVQIF-----YPSKDGTKIPMFIVHKKGIKLDGSHPAFLYGYGG-

CSQKADTVFRLNWATYLASTENIIVASFDGRGSGYQGDKIMHAI-NRRLGT
FNISITPNYSVSRLIFVRHMGGVLAVANIRGGGEYGETWHKGGILANKQNC
                                    630
FEVEDQIEAARQFSKMGFVDNKRIAIWGWSYGGYVTSMVLGSGSGVFKCGI
---FDDFQCAAEYIKEGYTSPKRLTINGGSNGGLLVATCANQRPDLFGCVI

AVAPV-SRWEY--Y--DSVYTERYMGLPTPE--DNLDHYR--NSTVMSRAE
AQVGVMDMLKFHKYTIGHAWTTDYGCSDSKQHFEWLIKYSPLHNVKLPEAD
                 708
NFKQVEYLLIHGTADDNVHFQQSAQISKALVDVGV----DFQAMWY---TD
DIQYPSMLLLTADHDDRVVPLHSLKFIATLQYIVGRSRKQNNPLLIHVDTK
  740
EDHGIASSTAH-QHIYTHMSHFIKQCFSFP
AGHGAGKPTAKVIEEVSDMFAFIARCLNIDWIP
```

*Figure 1.* Alignment of the amino acid sequences of DPP IV (upper row) and POP (lower row). Residues forming the catalytic triad are labelled.

Using several procedures as described in the Methods Section a model of the C-terminal region of DPP IV could be obtained that passes all necessary criteria for a native fold indicated by PROCHECK (i.e. more than 99 % of all backbone dihedral angles in most and additional favoured regions) and PROSA (the whole structure has negative relative energies).

The structure of the model of DPP IV (Fig 1) shows considerable similarity to the structure of POP particularly in the active site. The S1-binding site of DPP IV responsible for the recognition of proline is mainly formed by side chains of tyrosine residues (Tyr666, Tyr670, Tyr 631) and by Val556 representing the bottom of the pocket (Fig 3). In the case of POP phenylalanine instead of tyrosine residues in DPP IV occupy nearly the same spatial arrangement. In high similarity to POP the phenolic hydroxyl group of an additional tyrosine residue points to the S1-pocked. This Tyr547 may stabilise the oxyanion formed in the tetrahedral intermediates by a strong hydrogen bond (Fig 4). This may explain for instance the unusual substrate hydrolysis of N-peptidyl-O-hydroxylamines (Demuth *et al* 1989).

The protonated and positively charged N-terminus of substrates of DPP IV is recognised by the acidic side chain Glu668 which is a striking difference to POP. All these residues forming the binding site for ligands are conserved in DPP IV sequences or related enzymes.

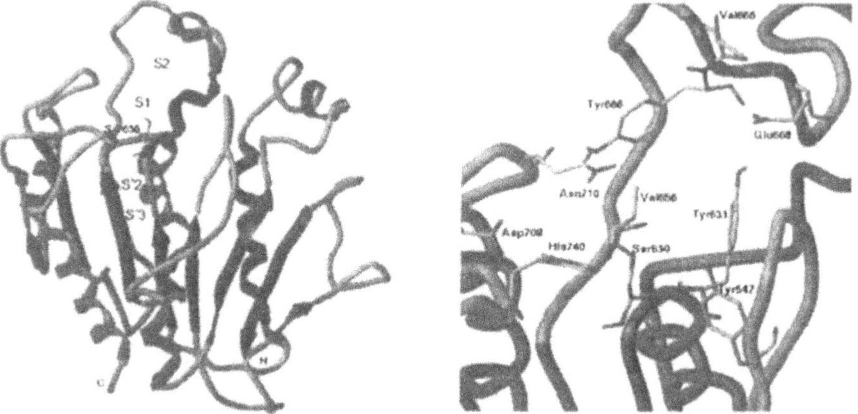


*Figure 2.* Model of the tertiary structure of the C-terminal domain of DPP IV.

*Figure 3.* The catalytically active site of DPP IV.

*Figure 4.* Stabilisation of the first oxyanion by hydrogen bond to the phenolic hydroxyl group of Tyr547.

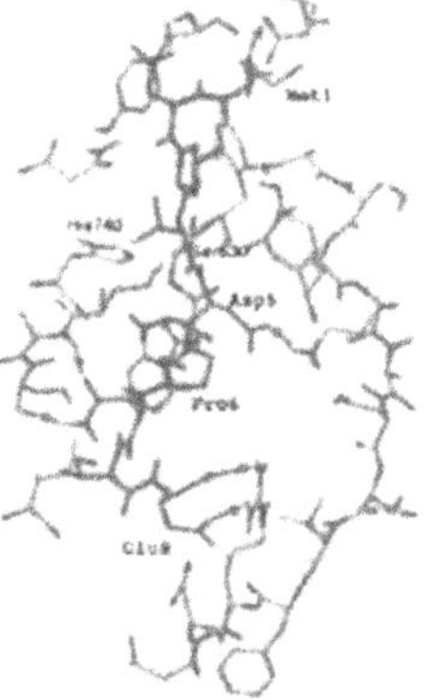

*Figure 5.* Interaction of HIV-Tat(1-9) with both binding pockets.

Calculating the molecular electrostatic as well as the lipophilic surface potentials of DPP IV, a second deep mostly hydrophobic pocked became visible (see: http//www.biochemtech.uni-halle.de/~mdqba/dpivFigurehtm). This pocket can be described as S2′ to S5′ binding sites for ligands. Tat(1-9) may interact with DPP IV with either the S1-S2 or S2′-S5′ binding pocket or with both together. The HIV1 Tat(1-9) (MDPVDPNIE) was docked in the active site of DPP IV (Fig 5). A multitude of attractive hydrophibic interactions and several hydrogen bonds are formed with the enzyme.

## 4. CONCLUSIONS

The structure of the C-terminal domain of DPP IV in principal seems to be correct since important criteria for a native fold were passed. By means of this model not only the substrate specificity of DPP IV can be understood but also the unusual interaction behaviour of of N-peptidyl-O-hydroxylamines. Furthermore, a second pocked close to the active site seems to be of high importance for the interaction of several ligands of DPP IV such as N-terminal HIV1-Tat peptides, chemokines and others. These results offer a new insight in structure and function relationships of DPP IV.

## ACKNOWLEDGMENT

We are grateful for research support by Deutsche Forschungsgemeinschaft (SFB 387/TP A8).

## REFERENCES

Blundell, T., et al., 1988, Knowledge-based protein modelling and design. *Eur. J. Biochem.* **172:** 513-520.

Clark, M., et al., 1989, Validation of the General Purpose Tripos 5.2 Force Field. *Comput. Chem.* **10:** 982-991.

Demuth, H.-U., et al., 1989, Reactions between dipeptidyl peptidase IV and diacyl hydroxylamines: mechanistic investigations. *J.Enyme Inhib.* **2:** 239-248.

Fülöp, V., et al., 1998, Prolyl Oligopeptidase: An unusual ß-propeller domain regulates proteolysis. *Cell* **94:** 161-170.

Gasteiger, J., and Marsili, M., 1980, Iterative partial equalization of orbital electronegativity - a rapid access to atomic charges. *Tetrahedron* **36:** 3219-3328.

Judson, R., 1997, In *Reviews in Computational Chemistry 10* (K.B. Lipkowitz, and D.B. Boyd, eds.), VCH Publishers, Inc. pp. 1-73,

Laskowski, R.A., et al., 1993, PROCHECK: a program to check the stereochemical quality of protein structures. *J. Appl. Cryst.* **26:** 283-289.

Sippl, M.J., 1993, Recognition of errors in three-dimensional structures of proteins. *Proteins* **17:** 355-362.

von Bonin, A., et al., 1998, Dipeptidyl-peptidase IV/CD26 on T cells: analysis of an alternative T-cell activation pathway. *Immunol. Rev.* 161, 43-53.

Wrenger, S., et al., 1997, The N-terminal structure of HIV-1 Tat is required for suppression of CD26-dependent T cell growth. *J. Biol. Chem.* **272:** 30283-30288.

# POST PROLINE CLEAVING PEPTIDASES HAVING DP IV LIKE ENZYME ACTIVITY

*Post-proline peptidases*

Catherine A. Abbott, Denise Yu, Geoffrey W. McCaughan and Mark D. Gorrell.

*A.W. Morrow Gastroenterology and Liver Centre, Centenary Institute of Cell Biology and Cancer Medicine, Royal Prince Alfred Hospital and The University of Sydney, NSW, Australia.*

Key words Dipeptidyl peptidase IV, prolyl oligopeptidase, serine protease

Abstract DP IV has been studied extensively in disease and in the immune system by the use of enzyme assays which detect hydrolysis of Gly-Pro or Ala-Pro substrate. In addition many studies have used inhibitors of DP IV enzyme activity. The characterisation of a novel DP IV like protein, DPP4R, and of other proteases which have a substrate specificity similar to DP IV or that bind DP IV inhibitors suggests that these studies require further evaluation.

## 1. INTRODUCTION

Very few enzymes are able to cleave the prolyl bond. The most widely studied of these enzymes is the type II integral membrane protein dipeptidyl peptidase IV (DP IV; EC 3.4.14.5). Recently, novel enzymes have been reported whose substrate specificities are similar to DP IV. We discuss here the relationship between these proteases and DP IV. The C-terminus of DP IV has high homology with several other non-classical serine proteases including the prolyl endopeptidases (PEP; EC 3.4.21.26) and the acylaminoacyl peptidases (ACPH; 3.4.19.1). These three enzymes belong to the enzyme clan SC as they contain a serine nucleophile and the catalytic residues in the order Ser-Asp-His and have been classified in the prolyl oligopeptidase S9 family (Rawlings and Barrett 1999). The three dimensional structure of pig PEP has been solved and it has two domains, one with a α/β hydrolase fold and one with a β propeller fold (Fülop *et al* 1998). We have predicted that DP IV, FAP and DPPX have similar tertiary structures including an atypical β propeller domain of 4-stranded β sheets as occurs in PEP (Abbott et al. 1999b; Gorrell et al. 2000).

We recently suggested that the dipeptidyl aminopeptidase short and long forms (DPPX-s and DPPX-l) (Wada et al. 1993), and fibroblast activation protein (FAP) (Scanlan et al. 1994) form a distinct sub-class of the POP family, a DP IV-related gene family (Abbott et al. 1999a, Table 1). DPPX has no peptidase activity due to a mutation in the serine recognition site. FAP has both a dipeptidyl aminopeptidase activity and a gelatinase/collagenase activity (Levy et al. 1999; Park et al. 1999). The distinction between DP IV, PEP and APCH also has been recognised by the Enzyme Nomenclature Committee in allocating members of the DP IV-related gene family to subfamily B of the S9 family (http://www.bi.bbsrc.ac.uk) (Rawlings and Barrett 1999).

*Table 1.* Comparison of structural and functional characteristics of members of the DP IV-related family. √ = yes, X= no

| | DP IV | FAP | DPP6 | DPP4R |
|---|---|---|---|---|
| AA size | 766 | 760 | 803 & 865 | 882 |
| AA identity with DP IV | | 52 % | 33 % | 27 % |
| AA similarity with DP IV | | 71 % | 55 % | 51 % |
| Gene location | 2q24.3 | 2q23 | 7 | 15q22 |
| N-Glycosylation sites | 9 | 6 | 6 | None |
| Monomer | 110 kDa | 95 kDa | 97 kDa | 101 kDa |
| Dimer | 150 kDa | 180 kDa | Unknown | None |
| Transmembrane domain | √ | √ | √ | X |
| Cleaves H-Gly-Pro | √ | X | X | X |
| Cleaves H-Ala-Pro | √ | √ | X | √ |

The role of DP IV enzyme activity in the immune system has primarily been studied using synthetic inhibitors of DP IV both *in vitro* and *in vivo*. DP IV enzyme activity has been suggested to be essential for the role of DP IV in modulating T cell proliferation (Schön *et al* 1987, Flentke *et al* 1991; Reinhold *et al* 1993) but some data is contradictory. For example, inhibitors of DP IV were shown to suppress arthritis in an animal model, but this suppression was also achieved in DP IV deficient rats (Tanaka *et al* 1997), suggesting that this DP IV inhibitor may bind to a DP IV-related protein. Therefore, the aim of recent work by us and others has been to clone novel proteins related to DP IV that could have a DP IV-related enzyme activity.

DP IV-β is a cell surface glycoprotein of 82 kDa that cleaves Gly-Pro which was purified from a DP IV-negative cell line (C8166) (Jacotot *et al* 1996). DP IV-β binds DP IV inhibitors but with less affinity than DP IV

(Blanco *et al* 1998). Until a cDNA for DP IV-β is cloned it will be unclear to which gene or enzyme family this protein belongs. DPPT-L is a 175 kDa protein purified from human serum reported to hydrolyse Gly-Pro and also to bind DP IV monoclonal antibodies (Duke-Cohan *et al* 1995). DPPT-L has been cloned recently and renamed attractin (Duke-Cohan *et al* 1998). Attractin shares no significant sequence homology with any member of Clan SC, or any other known peptidase. Human attractin contains a serine within the consensus sequence required for serine proteases (Table 2) however this is not conserved in mouse attractin (GRKGR) (Gunn *et al* 1999). In addition, the 200 aa immediately following this motif contain a CUB domain and two EGF domains, so it is improbable that this serine residue is within an α/β hydrolase fold. The Gly-Pro hydrolytic activity reported for attractin may therefore come from a novel mechanism of cleaving prolyl bonds or from another protein that binds or is activated by attractin.

## 2. A NOVEL DP IV RELATED PROTEIN

We have recently cloned a novel human serine protease using an expressed sequence tag homologous to DP IV and FAP as a probe (Abbott *et al* in preparation). Analysis of the cDNA sequence predicted a 101 kDa protein of 882 amino acids having 27 % identity and 51 % similarity with DP IV, therefore we named this DP IV-related gene DPP4-related protein (DPP4R, Table 1). The catalytic residues in DPP4R that potentially form the charge-relay system are $Ser^{739}$, $Asp^{817}$ and $His^{849}$. The serine residue is found in the same serine recognition sequence as DP IV (Table 2). Therefore DPP4R is a serine protease, a member of Clan SC, and a member of the S9 enzyme family, subfamily B. Northern blot hybridisation showed that DPP4R is highly expressed in testis and placenta but has a ubiquitous tissue expression similar to DP IV. Another major difference between DPP4R and DP IV is that DPP4R contains no N-linked glycosylation and no transmembrane domain and was expressed in the cytoplasm of transfected COS-7 cells. Transfected COS-7 cells hydrolysed the DP IV substrate Ala-Pro, but not Gly-Pro. Therefore the range of potential natural substrates of DPP4R is different to that of DP IV.

## 3. POST-PROLINE PEPTIDASES UNRELATED TO DP IV

The inhibitor L-valinyl-L-boroproline (VbP) has been used extensively to inhibit DP IV expressed by T lymphocytes (Flentke *et al* 1991, Snow *et al*

1994, Coutts *et al* 1996). Recently a novel apoptotic pathway in quiescent lymphocytes has been identified through use of this and similar inhibitors. A target of this inhibitor is the recently cloned protease quiescent peptidyl peptidase (QPP) which is found in lysosomes but is also secreted in an active form (Chiravuri *et al* 1999, Underwood *et al* 1999). The catalytic residues in QPP are ordered Ser-Asp-His as in DP IV. However QPP has 42 % amino acid identity with prolyl carboxypeptidase (PCP, Angiotensinase C, EC 3.4.16.2) and little with DP IV. The serine residue in both QPP and PCP is found in a serine recognition site, similar to that of DP IV (Table 2). PCP is thought to be the evolutionary link between the serine carboxypeptidases (S10 family) and the S9 family and has been placed in its own family S28. Another lysosomal exopeptidase that cleaves the prolyl bond is dipeptidyl peptidase II (DPPII; EC 3.4.14.4). DPP II has a similar substrate specificity to DP IV but is active only at acidic pH (Mentlein and Struckhoff 1989). DPP II has been assigned to the S28 family based upon amino terminal sequence similarity to PCP but is an aminopeptidase related QPP (Rawlings and Barrett 1996). A novel MHC-encoded serine peptidase highly expressed by cortical epithelial cells of the thymus (Genbank accession AAC33563) probably belongs to this family as it has 48 % aa similarity with PCP and 53 % similarity with QPP. The enzyme activity of this novel protein is unknown.

*Table 2.* Some post-proline cleaving enzymes and their attributes. Superscript numbers next to each protein are related to the arrangement around the serine.

| Classification | Clan SC FamilyS9 SubfamilyA | Clan SC FamilyS9 Subfamily B | Clan SC Family S28 | Clan MH Family M28 Subfamily B | |
|---|---|---|---|---|---|
| Enzyme | PEP | DP IV, FAP, DPP4R | [1]QPP , [1]PCP, DPPII, and [2]AAC33563 | [3]NAALDaseI, [4]NALDaseII [5]NAALDaseL | Attractin |
| Order of Catalytic residues | Ser, Asp, His | All have Ser, Asp, His | All have Ser, Asp, His | All have Ser, Asp and His | Ser$^{26}$ Asp ? His ? |
| Arrangement around serine GxSxG | GGSNG | GWSYG | [1]GGSYG [2]GGSYA | [3]SVSFD [4]GVSFD [5]SISLG | GRSGG |

Three novel type II integral membrane proteins human glutamate carboxypeptidases, *N-a*cetlyated *a*lpha-*l*inked *a*cidic *d*eptid*ase*s (NAALADase I, NAALADase II and NAALADase L) have been cloned and are reported to hydrolyse Gly-Pro as well N-acetyl-L-aspartyl-L-(3,4)-glutamate (Pangalos *et al* 1999). NAALADases are in family M28 (Table 2) and are homologous to the transferrin receptor and the bacterial co-catalytic zinc metallopeptidases. Like the S28 family, the M28 family contains both aminopeptidases and carboxypeptidases. It is hard to reconcile the DP IV activity of these proteases as sequence analyses did not find a serine in the consensus sequence required for serine protease activity (Table 2). It is possible that the hydrolysis of Gly-Pro by transfected cells is due to upregulation of a prolyl aminopeptidase activity. It has been observed that recombinant expression of DP IV in melanoma cells upregulates endogenous FAP expression (Wesley *et al* 1999).

## 4. CONCLUSION

Many diverse biological roles have been suggested for the post-proline dipeptidyl aminopeptidases DP IV, particularly in the immune system, and FAP in tumour growth and tissue remodelling. It will be interesting to investigate physiological substrates and roles of the novel peptidases, DPP4R and QPP, in disease and immune functions. QPP has been shown to mediate a novel apoptotic pathway. The discoveries of proteins other than DP IV that cleave DP IV substrates indicate that DP IV inhibitors require re-evaluation. Indeed, new DP IV specific inhibitors are being developed for use in type II diabetes (Hughes *et al* 1999). The identification of a novel member of the DP IV-related gene family, DPP4R, will assist understanding the perhaps separate roles of individual members of this gene family in various biological processes.

## REFERENCES


Abbott, C.A., McCaughan, G.W., and Gorrell, M.D., 1999a, Two highly conserved glutamic acid residues in the predicted beta propeller domain of dipeptidyl peptidase IV are required for its enzyme activity. *FEBS Lett.* **458:** 278-284.

Abbott, C.A., et al., 1999b, Binding to human dipeptidyl peptidase IV by adenosine deaminase and antibodies that inhibit ligand binding involves overlapping, discontinuous sites on a predicted β propeller domain. *Eur. J. Biochem.* in Press.

Blanco, J., et al., 1998, Dipeptidyl-peptidase IV-beta. Further characterization and comparison to dipeptidyl-peptidase IV activity of CD26. *Eur. J. Biochem.* **256:** 369-378.

Chiravuri, M. et al., 1999, A novel apoptotic pathway in quiescent lymphocytes identified by inhibition of a post-proline cleaving aminodipeptidase: A candidate target protease, quiescent cell proline dipeptidase. *J. Immunol.* **163:** 3092-3099.

Coutts, S.J. et al., 1996, Structure-activity relationships of boronic acid inhibitors of Dipeptidyl Peptidase IV. 1. variation of the P-2 position of X(Aa)-Boropro dipeptides. *J. Medicinal Chem.* **39:** 2087-2094.

Duke-Cohan, J.S., et al., 1998, Attractin (DPPT-L), a member of the CUB family of cell adhesion and guidance proteins, is secreted by activated human T lymphocytes and modulates immune cell interactions. *Proc. Natl. Acad. Sci. USA* **95:** 11336-11341.

Duke-Cohan, J.S., et al., 1995, A novel form of Dipeptidylpeptidase IV found in human serum - isolation, characterization, and comparison with T lymphocyte membrane Dipeptidylpeptidase IV (CD26). *J. Biol. Chem.* **270:** 14107-14114.

Flentke, G.R., et al., 1991, Inhibition of dipeptidyl aminopeptidase IV (DP-IV) by Xaa-boroPro dipeptides and use of these inhibitors to examine the role of DP-IV in T-cell function. *Proc. Natl. Acad. Sci. USA* **88:** 1556-1559.

Fülop, V., Bocskei, Z., and Polgar, L., 1998, Prolyl oligopeptidase - an unusual beta-propeller domain regulates proteolysis. *Cell* **94:** 161-170.

Gorrell, M.D., et al., 2000, Relating structure to function in the beta-propeller domain of dipeptidyl peptidase IV. *Adv. Exp. Med. Biol.* (This volume)

Gunn, T.M., et al., 1999, The mouse mahogany locus encodes a transmembrane form of human attractin. *Nature* **398:** 152-156.

Hughes, T.E., et al., 1999, NVP-DPP728 - (1-[[[2-[(5-cyanopyridin-2-yl) amino]ethyl] amino] acetyl]-2-cyano-(S)-pyrrolidine), a slow-binding inhibitor of dipeptidyl peptidase IV. *Biochemistry* **38:** 11597-11603.

Jacotot, E., et al., 1996, Dipeptidyl-peptidase IV-beta, a novel form of cell-surface-expressed protein with dipeptidyl-peptidase IV activity. *Eur. J. Biochem.* **239:** 248-58.

Levy, M.T., et al., 1999, Fibroblast activation protein: A cell surface dipeptidyl peptidase and gelatinase expressed by stellate cells at the tissue remodelling interface in human cirrhosis. *Hepatology* **29:** 1768-1778.

Mentlein, R., and Struckhoff, G., 1989 Purification of two dipeptidyl aminopeptidases II from rat brain and their action on proline-containing neuropeptides. *J. Neurochem.* **52:** 1284-1293.

Pangalos, M.N., et al., 1999, Isolation and expression of novel human glutamate carboxypeptidases with N-acetylated alpha-linked acidic dipeptidase and dipeptidyl peptidase IV activity. *J. Biol. Chem.* **274:** 8470-83.

Park, J.E., et al., 1999, Fibroblast activation protein: A dual-specificity serine protease expressed in reactive human tumor stromal fibroblasts. *J. Biol. Chem.* in press.

Rawlings, N.D., and Barrett, A.J., 1996, Dipeptidyl-peptidase II is related to lysosomal Pro-X carboxypeptidase. *Biochim. Biophys. Acta* **1298:** 1-3.

Rawlings, N.D., and Barrett, A.J., 1999, MEROPS: the peptidase database. *Nuc. Acids Res.* **27:** 325-331.

Reinhold, D., et al., 1993, Dipeptidyl peptidase IV (CD26) on human lymphocytes. Synthetic inhibitors of and antibodies against dipeptidyl peptidase IV suppress the proliferation of pokeweed mitogen-stimulated peripheral blood mononuclear cells, and IL-2 and IL-6 production. *Immunobiology* **188:** 403-414.

Scanlan, M.J., et al., 1994, Molecular cloning of fibroblast activation protein alpha, a member of the serine protease family selectively expressed in stromal fibroblasts of epithelial cancers. *Proc. Natl. Acad. Sci. USA* **91:** 5657-5661.

Schön, E., et al., 1987, The role of dipeptidyl peptidase IV in human T lymphocyte activation. Inhibitors and antibodies against dipeptidyl peptidase IV suppress lymphocyte proliferation and immunoglobulin synthesis in vitro. *Eur. J. Immunol.* **17:** 1821-1826.

Snow, R.J., et al., 1994, Studies on proline boronic acid dipeptide inhibitors of dipeptidyl peptidase IV: identification of a cyclic species containing a B-N bond. *J. Am. Chem. Soc.* **116:** 10860-10869.

Tanaka, S., et al., 1997, Suppression of arthritis by the inhibitors of dipeptidyl peptidase IV. *Int. J. Immunopharmacol.* **19:** 15-24.

Underwood, R., et al., 1999, Sequence, purification and cloning of an intracellular serine protease, quiescent cell proline dipeptidase (QPP). *J. Biol. Chem.* in Press.

Wada, K., et al., 1993, Genetic mapping of the mouse gene encoding dipeptidyl aminopeptidase-like proteins. *Mamm. Genome* **4:** 234-237.

Wesley, U.V., et al., 1999, A role for dipeptidyl peptidase IV in suppressing the malignant phenotype of melanocytic cells. *J. Exp. Med.* **190:** 311-322.

# A NEW TYPE OF FLUOROGENIC SUBSTRATES FOR DETERMINATION OF CELLULAR DIPEPTIDYL PEPTIDASE IV (DP IV/CD26) ACTIVITY

Susan Lorey[1], Jürgen Faust[1], Frank Bühling[2], Siegfried Ansorge[2], and Klaus Neubert[1]

*[1] Martin-Luther-University Halle-Wittenberg, Institute of Biochemistry and Biotechnology, Kurt-Mothes Str. 3, D-06120, Halle, Germany; [2] Otto-von-Guericke-University Magdeburg, Institute of Experimental Medicine, Leipziger Str. 44, D-39120, Magdeburg, Germany*

Key words dipeptidyl peptidase IV, rhodamine 110, fluorogenic substrates

Abstract The stability of cell associated fluorescence is an essentiell requirement for measurements of cellular enzymatic activity via enzyme catalyzed liberation of fluorophores. Rhodamine 110 (R110), a highly fluorescent xanthene dye, was used to synthesize nonfluorescent dipeptidyl peptidase IV (DP IV) substrates Xaa-Pro-R110-Y allowing the stable covalent binding of the enzymatically released fluorescent R110-Y on cells. All compounds have been characterized as substrates of isolated DP IV with $k_{cat}/K_m$ values of about $10^6$ $M^{-1}\cdot s^{-1}$. The hydrophobicity of the residue Y affects the affinity of the substrate to the catalytic site of DP IV. The compounds are characterized as sensitive substrates of cell surface associated DP IV of DP IV rich U-937 cells. The binding of the enzymatically released R110-Y on cells results in a stable cellular fluorescence. This way, the quantitative determination of cell surface associated DP IV activity is possible.

## 1. INTRODUCTION

Dipeptidyl peptidase IV (DP IV, CD26) is a membrane bound type II glycoprotein present on most mammalian epithelial cells. DP IV catalyzes the cleavage of dipeptides from the N-terminus of oligo- and polypeptides provided that the penultimate residue is proline (Heins *et al* 1988). In the immune system DP IV is described as activation marker on the surface of immune cells (Ansorge *et al* 1995, Bühling *et al* 1995). Furthermore, DP IV seems to be involved in signal transduction processes (Kähne *et al* 1998). Rhodamine 110 (R110, Fig 1) is a highly fluorescent xanthene dye. It possesses two amino groups suitable for coupling of the protease substrate structure (Leytus *et al* 1983).

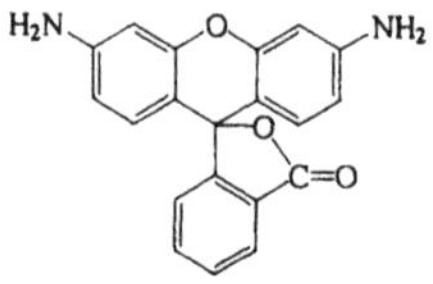

*Figure 1.* Structure of Rhodamine 110 (R110)

Recently we synthesized R110 compounds of the structure $(Xaa\text{-}Pro)_2$-R110 (Xaa as proteinogenic amino acid) as substrates of isolated DP IV displaying $k_{cat}/K_m$ values between $1.2 \cdot 10^6$ and $4.3 \cdot 10^6\ M^{-1} \cdot s^{-1}$ (Lorey *et al* 1997). The compounds were sensitive substrates of cell surface associated DP IV. However, a quantification of cellular enzymatic activity is difficult because the R110 diffuses into the buffer after enzymatic hydrolysis of the substrate. Hence we synthesized a new type of fluorescent substrates Xaa-Pro-R110-Y (Xaa = Gly, Ala and Y = functional group of different length and reactivity) anchoring the fluorescent moiety at the cell surface and in this way allowing the determination of cellular DP IV activity.

## 2. MATERIALS AND METHODS

The substrates Xaa-Pro-R110-Y (Fig 2) were synthesized by coupling Xaa-Pro-R110 with the corresponding carboxylic acid (carbodiimid method) or using the acid chloride of Y. Dipeptidyl peptidase IV was isolated from pig kidney. All substrates were tested for cleavage by DP IV at 30 °C at an enzyme concentration of $6.8 \cdot 10^{-10}$ or $1.4 \cdot 10^{-9}$ M monitoring the enzymatically released R110-Y at 494 nm. Measurements of the cleavage of the substrates by cellular DP IV were done with U-937 cells and CD26 transfected as well as wildtype CHO cells after an incubation time of 20 to 30 min at 37 °C.

# 3. RESULTS

## 3.1 Isolated DP IV

The synthesized DP IV substrates Xaa-Pro-R110-Y (Xaa = Gly or Ala) contained a residue Y of different length, hydrophobicity and reactivity (Fig 2). Xaa-Pro determined the substrate specificity of DP IV and Y acted as reactive anchor group fascilitating the stable fixation of the fluorophor on the cell surface mainly by reaction with cell surface localized thiol or amino groups. The compounds were colorless and nonfluorescent. The DP IV catalyzed substrate hydrolysis resulted in the release of R110-Y with an absorption maximum at 494 nm and an emission maximum at 522 nm.

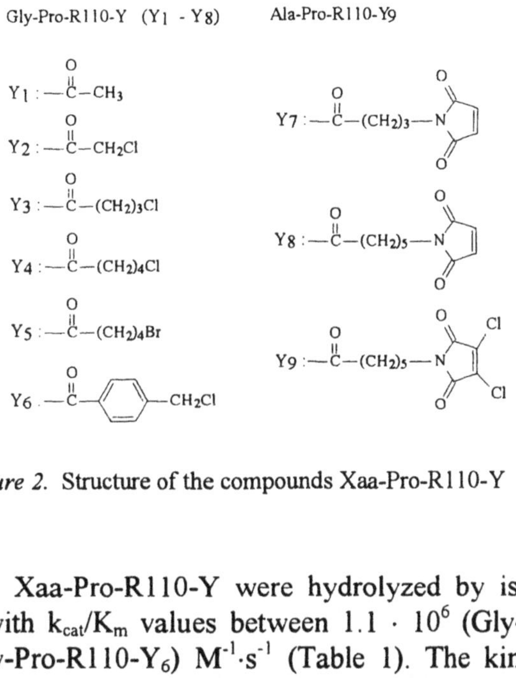

*Figure 2.* Structure of the compounds Xaa-Pro-R110-Y

All compounds Xaa-Pro-R110-Y were hydrolyzed by isolated DP IV from pig kidney with $k_{cat}/K_m$ values between $1.1 \cdot 10^6$ (Gly-Pro-R110-$Y_2$) and $3.3 \cdot 10^6$ (Gly-Pro-R110-$Y_6$) $M^{-1} \cdot s^{-1}$ (Table 1). The kinetic constants were in the same range as the $k_{cat}/K_m$ values of the hydrolysis of the known dipeptide-*p*-nitroanilides ($10^6$ $M^{-1} \cdot s^{-1}$, Heins *et al* 1988).

The compounds were hydrolyzed according to the model of substrate inhibition in a non hyperbolic dependency of initial velocity on substrate concentration. A longer and more hydrophobic residue Y resulted in a decrease of the $K_m$ value reflecting a higher affinity of the substrate to the active site of DP IV, which can be explained by hydrophobic interactions.

*Table 1.* Kinetic constants of the hydrolysis of Xaa-Pro-R110-Y by isolated DP IV

| compound | $10^5$ $K_m$ (M) | $k_{cat}$ ($s^{-1}$) | $10^{-6}$ $k_{cat}/K_m$ ($M^{-1} \cdot s^{-1}$) | $10^4$ $K_i$ (M) |
|---|---|---|---|---|
| Gly-Pro-R110-Y | | | | |
| Y = $Y_1$ | 17.40 ± 6.24 | 235.60 ± 43.25 | 1.35 ± 0.56 | 2.95 ± 1.04 |
| Y = $Y_2$ | 6.23 ± 0.61 | 70.84 ± 4.21 | 1.14 ± 0.13 | 2.30 ± 0.25 |
| Y = $Y_3$ | 3.77 ± 0.40 | 67.49 ± 3.46 | 1.79 ± 0.21 | 6.06 ± 0.90 |
| Y = $Y_4$ | 3.83 ± 0.31 | 59.65 ± 2.55 | 1.56 ± 0.14 | 2.24 ± 0.19 |
| Y = $Y_5$ | 1.54 ± 0.17 | 42.39 ± 1.78 | 2.75 ± 0.33 | 7.87 ± 1.58 |
| Y = $Y_6$ | 1.10 ± 0.20 | 36.64 ± 3.74 | 3.33 ± 0.69 | 1.02 ± 0.27 |
| Y = $Y_7$ | 2.67 ± 0.21 | 60.12 ± 2.08 | 2.25 ± 0.19 | 6.32 ± 0.77 |
| Y = $Y_8$ | 1.25 ± 0.10 | 40.41 ± 1.23 | 3.23 ± 0.28 | 3.69 ± 0.39 |
| Ala-Pro-R110-Y | | | | |
| Y = $Y_9$ | 0.58 ± 0.07 | 14.67 ± 0.92 | 2.55 ± 0.44 | 0.80 ± 0.13 |

## 3.2 Cell surface associated DP IV

All compounds Xaa-Pro-R110-Y were characterized as substrates of DP IV of DP IV rich U-937 cells. Gly-Pro-R110-$Y_1$ with $Y_1$ as acetyl group is not able to anchor the fluorophore on the cell. After one wash step more than 95 % of the fluorescent moiety was removed from the cells. The substrates with $Y_2$ - $Y_9$ as anchor group were able to link the fluorescent moiety R110-Y with the cell surface indicating a stable cellular fluorescence between 30 % and 93 % after 4 wash steps. The cell associated fluorescence varied depending on the reactivity of the anchor Y. The compounds Gly-Pro-R110-Y with Y = $Y_2$ (93 %) and $Y_6$ (71 %) and Ala-Pro-R110-$Y_9$ (77 %) were characterized as the substrates with the most reactive anchor group.

Furthermore, the DP IV catalyzed hydrolysis of the substrates was influenced by steric effects of the anchor group. It was shown that a large ($Y_7$, $Y_8$) as well as short residue Y ($Y_2$) resulted in a low fluorescence release after cleavage of the substrates by cell associated DP IV. Compounds with an alkyl halogenide anchor ($Y_5$, $Y_3$, $Y_4$) were characterized as the best substrates of cell associated DP IV. This way, the cellular fluorescence was determined by the anchor characteristics and the substrate properties of the compounds. The stability of cellular fluorescence was determined by the reactivity of the anchor Y as decisive factor for an improved detection of cellular DP IV activity. This way, according to the differentiation of the CD26 expression by antibody staining, a differentiation between CD26 high expressing cells and CD26 low expressing cells by DP IV activity was possible.

# 4. CONCLUSIONS

Protease substrates with fluorophoric hydrolyzable groups are described as tools for the characterization of isolated and cell associated enzymatic activity. The main problem using these substrates exists in the high background fluorescence resulting from the diffusion of the enzymatically released fluorophores into the medium, thereby complicating their quantification.

The new type of bifunctional DP IV substrates Xaa-Pro-R110-Y described in the present study improves the quantitative determination and differentiation of cell surface associated DP IV activity by fluorescence microscopy and flow cytometry preventing a high background fluorescence via the stable fixation of the fluorophore on the cell surface.

# ACKNOWLEDGMENTS

This work was supported by the Deutsche Forschungsgemeinschaft, SFB 387. We thank Dr. J. Rahfeld for the gift of dipeptidyl peptidase IV and Dr. M. Gorrel and Dr. C. Abbott for the gift of transfected CHO cells.

# REFERENCES

Ansorge, S., Bühling, F., Hoffmann, T., Kähne, T., Neubert, K., and Reinhold, D., 1995, In *Dipeptidyl peptidase VI (CD26) in Metabolism and the Immune Response* (B. Fleischer, ed), Springer-Verlag, Heidelberg, pp. 163-184.

Bühling, F., Junker, U., Neubert, K., Jäger, L., and Ansorge, S., 1995, Functional role of CD26 on human B lymphocytes. *Immunol. Lett.* **45:** 47-51.

Heins, J., Welker, P., Schönlein, C., Born, I., Hardrodt, B., Neubert, K., Tsuru, D. and Barth, A., 1988, Mechanism of proline-specific proteinases: (I) Substrate specificity of dipeptidyl peptidase IV from pig kidney and proline specific endopeptidase from *Flavobacterium meningosepticum*. *Biochim. Biophys. Acta* **954:** 161-169.

Kähne, T., Neubert, K., Faust, J., and Ansorge, S., 1998, Early phosphorylation events induced by DP IV/CD26-specific inhibitors. *Cell. Immunol.* **189:** 60-66.

Leytus, S.P., Melhado, L.L., and Mangel, W.F., 1983, Rhodamine-based compounds as fluorogenic substrates for serine proteinases. *Biochem. J.* **209:** 299-307.

Lorey, S., Faust, J., Bühling, F., Ansorge, S., and Neubert, K., 1997, New fluorogenic Dipeptidyl peptidase IV/CD26 substrates and inhibitors. *Adv. Exp. Med. Biol.* **241:** 157-160.

# POTENT INHIBITORS OF DIPEPTIDYL PEPTIDASE IV AND THEIR MECHANISMS OF INHIBITION

Angela Stöckel-Maschek, Beate Stiebitz, Ilona Born, Jürgen Faust, Werner Mögelin and Klaus Neubert
*Martin-Luther-University Halle-Wittenberg, Department of Biochemistry and Biotechnology, Institute of Biochemistry, Kurt-Mothes-Str.3, 06120 Halle, Germany.*

Key words dipeptidyl peptidase IV, reversible inhibitor, irreversible inhibitor

Abstract Dipeptidyl peptidase IV (DP IV) is a proline specific serine protease which cleaves Xaa-Pro-dipeptides from the N-terminus of longer peptides. A series of product analogous amino acid amides containing different structure modifications like substitution of a ring atom, variation of the ring size and/or the introduction of a thioxo amide bond, phosphono amide bond or reduced amide bond were done to characterize these compounds as inhibitors of DP IV. These compounds are mostly classical reversible inhibitors of DP IV. In contrast amino acyl-2-cyanopyrrolidides inhibit DP IV according to a slow-binding mechanism with inhibition constants in the nanomolare range. On the other hand, diaryl dipeptide phosphonates inhibit irreversibly. In conclusion, this work shows, that the mechanism of inhibition of DP IV depends on the structure of the investigated compounds.

## 1. INTRODUCTION

Dipeptidyl peptidase IV (DP IV, CD26) is a serine protease. It cleaves dipeptides from the amino terminus of peptides behind a penultimate proline or alanine residue. The enzyme exists as dimer and is anchored in the membrane. DP IV hydrolyzes a variety of biologically active peptides. Many efforts have been done to develop inhibitors for DP IV. Previously, it was found that DP IV is inhibited by amino acyl proline dipeptides, which are the products of substrate hydrolysis (Yaron and Naider 1993). Subsequently, we have found that amino acid pyrrolidides represent more potent inhibitors with inhibition constants in the higher nanomolar up to the micromolar range depending on the amino acid (Born *et al* 1994). We started to modify systematically these structures and investigated e.g. different amino acid amides, thioxo amino acid amides, phosphonamides and dipeptide phosphonates.

## 2. MATERIALS AND METHODS

(S)-(+)-1-(2-pyrrolidinylmethyl)-pyrrolidine was purchased from Aldrich. The other investigated compounds were synthesized using different methods of organic chemistry and peptide chemistry. The synthesis will be published elsewhere. DP IV from pig kidney was isolated according to Wolf *et al* (1978). The mechanism of inhibition and the $K_i$ values were determined from the enzyme-catalyzed hydrolysis of Gly-Pro-4NA (4NA: 4-nitroanilide) in the absence and the presence of inhibitor at 390 nm and at 30 °C on a Beckman DU-650 UV/VIS spectrophotometer. The incubation mixture (1 ml) contained 40 mM Tris buffer pH 7.6, 0.125 mM NaCl, various concentrations of Gly-Pro-4NA and various concentrations of inhibitor around the expected inhibition constant. In cases of the slow-binding and the irreversible inhibitors the concentration of substrate was 0.2 mM. The final concentration of DP IV was 1.4 nM for competitive and mixed-type inhibition, 0.7 nM for slow-binding and 0.14 nM for irreversible inhibition. The reaction was initiated by adding the enzyme. Generally, the reaction was followed over a time interval in which less than 10 % cleavage of substrate occurred. The steady state kinetics were analyzed using the following equation

$$\frac{1}{v}=\frac{(1+\frac{\alpha K_m}{[S]})}{\alpha K_m V_{max}}[I]+\frac{1}{V_{max}}(1+\frac{K_m}{[S]}) \qquad (1)$$

$K_i$ is the competitive inhibition constant and factor $\alpha$ multiplied with $K_i$ represents the uncompetitive inhibition constant. For the distinction between competitive and linear mixed type inhibition slopes and intercepts were replotted versus 1/[S] to complete the inhibition mechanism and the constants $K_i$ as well as $\alpha$ (equations 2 and 3) (Segel 1993).

$$slope=\frac{K_m}{K_i V_{max}}\frac{1}{[S]}+\frac{1}{\alpha K_i V_{max}} \qquad (2)$$

$$int\,ercept = \frac{K_m}{V_{max}} \frac{1}{[S]} + \frac{1}{V_{max}} \tag{3}$$

Slow-binding inhibition was monitored over 5 min and fitted to the equation (4) to obtain $v_i$ (initial velocity), $v_s$ (steady-state velocity) and $k_{obs}$ (first order rate constant for the approach of the steady state). These values were obtained for each progress curve and subsequently used to obtain the initial inhibitory constant $K_i$, the overall inhibition constant $K_i$', the association constant $k_{on}$ and the dissociation constant $k_{off}$ (Morrison and Walsh 1988).

$$P = v_s \cdot t + (v_i - v_s) \cdot (1 - e^{-k_{obs}t}) / k_{obs} + d \tag{4}$$

It is possible to describe inactivation of proteases by acylating reagents using models of slow-binding inhibition (Neumann *et al* 1991). Progress-curves were fitted to equation (4) and apparent pseudo-first-order inactivation rate constants $k_{obs}$ were used to determine $k_{inact}$ and $K_i$.

## 3. RESULTS

Firstly, starting from the amino acid pyrrolidides the ring size of the amide residue was changed (Fig 1). The data in table 1 show that the extension as well as the reduction of the ring size result in higher affinity constants compared to the amino acid pyrrolidides. In both cases the inhibition constants are about 10 times higher than for the amino acid pyrrolidides. Surprisingly, Lys[Z($NO_2$)]-piperidide is a slow-binding inhibitor. The overall-inhibition constant amounts to 1.6 μM. The substitution of a $CH_2$-group within the pyrrolidine ring for sulfur results in thiazolidine. Generally, amino acid thiazolidides are more potent inhibitors than amino acid pyrrolidides (Demuth 1990). The data show that the inhibition constants are in the higher nanomolare range (Table 1). In contrast a substitution of a $CH_2$-group for oxygen within the six-membered piperidine ring does not have an effect. The inhibition of DP IV by Ile-morpholide is with an inhibition constant about 4 μM in the same range as the inhibition constant of Ile-piperidide.

Another possibility of structure modifications is the exchange of the amide bond. Firstly, we introduced a thioxo amide bond and synthesized thioxo amino acid pyrrolidides (Xaa-$\Psi$[CS-N]-Pyrr) and thiazolidides (Xaa-$\Psi$[CS-N]-Thia) of the amino acids alanine, valine and isoleucine. The comparison of the $K_i$ values showed that these compounds are less potent inhibitors of DP IV than the corresponding amino acid pyrrolidides and thiazolidides (Table 1). Just as it is known from amino acid amides, the thioxo amino acid thiazolidides are more potent inhibitors than the thioxo amino acid pyrrolidides. Nevertheless, the best inhibitors Val-Ψ[CS-N]-Thia and Ile-Ψ[CS-N]-Thia have inhibition constants of about 0.2 μM like Val-Thia and Ile-Thia.

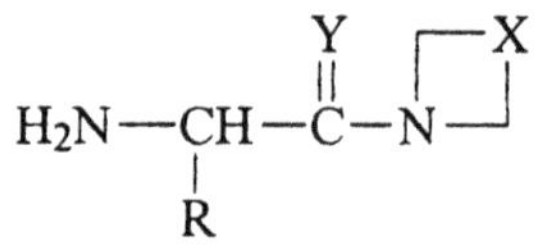

*Figure 1.* General structure of amino acid amides and thioxo amino acid amides (R: $CH_3$, $CH(CH_3)_2$, $CH(CH_3)C_2H_5$, $(CH_2)_4NH[Z(NO_2)]$; X: $(CH_2)_n$ (n=1-3), $S\text{-}CH_2$, $CH_2\text{-}O\text{-}CH_2$; Y: O, S)

*Table 1.* Inhibition of DP IV by amino acid amides and thioxo amino acid amides

| Compound | R | X | Y | $K_i$ [μM] |
|---|---|---|---|---|
| Ile-azetidide | $CH(CH_3)C_2H_5$ | $CH_2$ | O | 2.38±0.14 |
| Lys[Z(NO$_2$)]-azetidide | $(CH_2)_4NH[Z(NO_2)]$ | $CH_2$ | O | 2.28±0.10 |
| Ala-pyrrolidide | $CH_3$ | $CH_2\text{-}CH_2$ | O | 9.06±0.40 |
| Val-pyrrolidide | $CH(CH_3)_2$ | $CH_2\text{-}CH_2$ | O | 0.255±0.017 |
| Ile-pyrrolidide | $CH(CH_3)C_2H_5$ | $CH_2\text{-}CH_2$ | O | 0.218±0.008 |
| Lys[Z(NO$_2$)]-pyrrolidide | $(CH_2)_4NH[Z(NO_2)]$ | $CH_2\text{-}CH_2$ | O | 0.271±0.012 |
| Ile-piperidide | $CH(CH_3)C_2H_5$ | $CH_2\text{-}CH_2\text{-}CH_2$ | O | 3.73±0.19 |
| Lys[Z(NO$_2$)]-piperidide | $(CH_2)_4NH[Z(NO_2)]$ | $CH_2\text{-}CH_2\text{-}CH_2$ | O | 4.01[1)] |
| Ala-thiazolidide | $CH_3$ | $S\text{-}CH_2$ | O | 0.339[2)] |
| Val-thiazolidide | $CH(CH_3)_2$ | $S\text{-}CH_2$ | O | 0.267[2)] |
| Ile-thiazolidide | $CH(CH_3)C_2H_5$ | $S\text{-}CH_2$ | O | 0.126[2)] |
| Lys[Z(NO$_2$)]-thiazolidide | $(CH_2)_4NH[Z(NO_2)]$ | $S\text{-}CH_2$ | O | 0.0923±0.0036 |
| Ile-morpholide | $CH(CH_3)C_2H_5$ | $CH_2\text{-}O\text{-}CH_2$ | O | 4.31±0.20 |
| Ala-Ψ[CS-N]-pyrrolidide | $CH_3$ | $CH_2\text{-}CH_2$ | S | 47.6±4.1 |
| Val-Ψ[CS-N]-pyrrolidide | $CH(CH_3)_2$ | $CH_2\text{-}CH_2$ | S | 1.07±0.15 |
| Ile-Ψ[CS-N]-pyrrolidide | $CH(CH_3)C_2H_5$ | $CH_2\text{-}CH_2$ | S | 1.02±0.04 |
| Ala-Ψ[CS-N]-thiazolidide | $CH_3$ | $S\text{-}CH_2$ | S | 7.88±0.74 |
| Val-Ψ[CS-N]-thiazolidide | $CH(CH_3)_2$ | $S\text{-}CH_2$ | S | 0.208±0.012 |
| Ile-Ψ[CS-N]-thiazolidide | $CH(CH_3)C_2H_5$ | $S\text{-}CH_2$ | S | 0.203±0.010 |

[1)] slow-binding inhibition: $K_i' = 1.59$ μM, $k_{on} = 0.017\ s^{-1}$. [2)] Ref. Demuth 1990

Another replacement of the amide bond is a reduced amide bond. (S)-(+)-1-(2-pyrrolidinylmethyl)-pyrrolidine (Pro-Ψ[$CH_2$-N]-Pyrr) is the proline pyrrolidide analogue with a reduced amide bond (Fig 2). Surprisingly, this compound was able to inhibit DP IV, but with very low affinity (Table 2).

Finally, the amide bond was exchanged by the phosphonamide bond. The synthesis of α-aminoisobutanephosphonic acid methylester pyrrolidide (Val-Ψ[PO($OCH_3$)-N]-Pyrr) yielded two diastereomers which were separated by HPLC (Fig 2). Both diastereomers are able to inhibit DP IV but not very efficiently (Table 2). Interestingly enough, both diastereomers show different inhibition mechanisms. Whereas diastereomer 2 inhibits in accordance with a classical competitive mechanism, diastereomer 1 is a mixed-type inhibitor of DP IV with a $K_i$ value of 180 μM and an α value of 10. That means, that the inhibitor is able to bind to the free enzyme as well as to the enzyme-inhibitor complex and that the substrate is able to form a complex with the free enzyme or the enzyme-inhibitor complex.

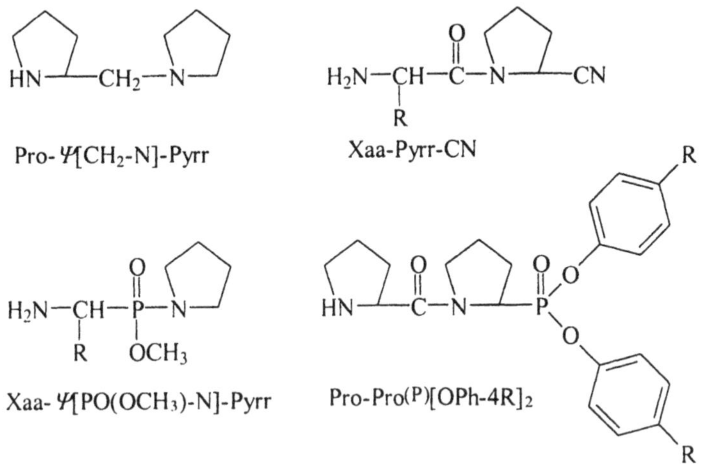

*Figure 2.* Structure of reduced proline-pyrrolidide, aminoacyl-2-cyanopyrrolidide, phosphonamide and dipeptide phosphonates

The investigation of amino acid-2-cyanopyrrolidides (Xaa-Pyrr-CN) showed that DP IV is inhibited according to a slow-binding mechanism. The overall-inhibition constants are in the nanomolare range. Data from the literature are in the same range, but up to recently such compounds are described as competitive inhibitors of DP IV (Li *et al* 1995, Ashworth *et al* 1996, Hughes *et al* 1999).

In contrast to all these compounds, diaryl aminoacyl pyrrolidine-2-phosphonates (Xaa-Pro$^{(P)}$[OPh-4R]$_2$) inhibit DP IV irreversibly (Boduszek *et al* 1994, Lambeir *et al* 1996, Stiebitz *et al* 1999). We were able to separate the diastereomers of bis(4-acetylphenyl)prolyl pyrrolidine-2-phosphonate (Pro-Pro$^{(P)}$[OPh-4Ac]$_2$) and bis(4-chlorophenyl)prolyl pyrrolidine-2-phosphonate (Pro-Pro$^{(P)}$[OPh-4Cl]$_2$) by HPLC (Fig 2). According to their appearance the diastereomers were called diastereomer 1 and diastereomer 2.

In both cases diastereomer 1 is the better inactivator compared to diastereomer 2 (Table 2). Both diastereomers of Pro-Pro$^{(P)}$[OPh-4Ac]$_2$ are more potent inactivators of DP IV than Pro-Pro$^{(P)}$[OPh-4Cl]$_2$. The reason is that the 4-acetylphenoxy residue is a better leaving group than the 4-chlorophenoxy residue. This results in a very efficient inactivation of DP IV by diastereomer 1 of Pro-Pro$^{(P)}$[OPh-4Ac]$_2$ with a second order rate constant of 864 000 $s^{-1}M^{-1}$.

*Table 2.* Potency and inhibition mechanism of different DP IV inhibitors

| compound | mechanism | Inhibition constants |
|---|---|---|
| Pro- $\Psi$[$CH_2$-N]-Pyrr | competitive | $K_i$ = 889 μM |
| Val- $\Psi$[PO($OCH_3$)-N]-Pyrr | | |
| diastereomer 1 | mixed-type | $K_i$=180 μM; α = 10 |
| diastereomer 2 | competitive | $K_i$=215 μM |
| Ile-Pyrr-CN | slow-binding | $K_i$=99.0 nM; $K_i$' = 4.35 nM; $k_{on}$ = 0.35 $s^{-1}$ |
| Lys[Z($NO_2$)]-Pyrr-CN | slow-binding | $K_i$=23.7 nM; $K_i$' = 2.96 nM; $k_{on}$ = 0.23 $s^{-1}$ |
| Pro-Pro$^{(P)}$[OPh-4Cl]$_2$ | | |
| diastereomer 1 | irreversible | $K_i$=10.4 μM; $k_{inact}$=0.17 $s^{-1}$; $k_{inact}/K_i$=16100 $s^{-1}M^{-1}$ |
| diastereomer 2 | irreversible | $K_i$=66.9 μM; $k_{inact}$=0.068 $s^{-1}$; $k_{inact}/K_i$=1010 $s^{-1}M^{-1}$ |
| Pro-Pro$^{(P)}$[OPh-4Ac]$_2$ | | |
| diastereomer 1 | irreversible | $K_i$=0.0538 μM; $k_{inact}$=0.047 $s^{-1}$; $k_{inact}/K_i$=864000 $s^{-1}M^{-1}$ |
| diastereomer 2 | irreversible | $K_i$=5.29 μM; $k_{inact}$=0.15 $s^{-1}$; $k_{inact}/K_i$=28800 $s^{-1}M^{-1}$ |

## 4. CONCLUSION

These investigations show that starting from the dipeptides there exist different possibilities of structural modifications to provide inhibitors of DP IV. These inhibitors are characterized by displaying very different inhibition mechanisms. Besides the classical competitive inhibitors amino acid pyrrolidides and thiazolidides, amino acid-2-cyanopyrrolidides as efficient slow-binding inhibitors and diaryl dipeptide phosphonates as irreversible inhibitors are the most useful compounds in this series.

## ACKNOWLEDGMENTS

We thank Dr. C. Mrestani-Klaus and Dr. A. Schierhorn for NMR spectroscopic and mass spectrometric investigations, respectively. We thank S. Kaufmann for excellent technical assistance. Financial support was provided by the Deutsche Forschungsgemeinschaft, SFB 387, the Fond der Chemischen Industrie and the DECHEMA e.V..

## REFERENCES

Ashworth, D.M., Atrash, B., Baker, G.R., Baxter, A.J., Jenkins, P.D., Jones, D.M. and Szelke, M., 1996, 2-Cyanopyrrolidides as potent, stable inhibitors of dipeptidyl peptidase IV. *Bioorg. Med. Chem. Lett.* **6:** 1163-1166.

Born, I., Faust, J., Heins, J., Barth, A. and Neubert, K., 1994, Potent inhibitors of dipeptidyl peptidase IV. *Eur. J. Cell Biol. Suppl.* **40:** 23.

Boduszek, B., Oleksyszyn, J., Kam, C.-M., Selzler, J., Smith, R.E. and Powers, J.C., 1994, Dipeptide phosphonates as inhibitors of dipeptidyl peptidase IV. *J. Med. Chem.* **37:** 3969-3976.

Demuth, H.-U., 1990, Recent development in inhibiting cysteine and serine proteases, J. Enzyme Inhibition **3:** 249-278.

Hughes, T.E., Mone, M.D., Russell, M.E., Weldon, S.C. and Willhauer, E.B., 1999, NVP-DPP728 (1-[[[2-[(5-Cyanopyridin-2-yl)amino]ethyl]amino]acetyl]-2-cyano-(S)-pyrrolidine), a slow-binding inhibitor of dipeptidyl peptidase IV. *Biochemistry* **38:** 11597-11603.

Lambeir, A.-M., Borloo, M., De Meester, I., Belyaev, A., Augustyns, K., Hendriks, D., Scharpe, S. and Haemers, A., 1996, Dipeptide-derived diphenyl phosphonate esters: Mechanism-based inhibitors of dipeptidyl peptidase IV. *Biochim. Biophys. Acta* **1290:** 76-82.

Li, J., Wilk, E. and Wilk, S., 1995, Aminoacylpyrrolidine-2-nitriles: Potent and stable inhibitors of dipeptidyl peptidase IV. *Arch. Biochem. Biophys.* **323:** 148-154.

Morrison, J.F. and Walsh, C.T., 1988, The behavior and significance of slow-binding enzyme inhibitors. *Adv. Enzymol.* **61:** 201-301.

Neumann, U., Stürzebecher, J., Leistner, S. and Vieweg, H., 1991, Inhibition of Chymotrpsin and pancreatic elastase by 4H-3, 1-benzoxazin-4-ones. *J. Enzyme Inhibition* **4:** 227-232.

Segel, I.H., 1993, Enzyme kinetics. Behavior and analysis of rapid equilibrium and steady-state enzyme systems. John Wiley & Sons, London.

Stiebitz, B., Stöckel-Maschek, A. and Neubert, K., 1999, Dipeptide phosphonates as inhibitors of prolyl oligopeptidase and dipeptidyl peptidase IV. Symposium Peptido- and Proteinomimetics, Houffalize, Belgium.

Wolf, B., Fischer, G. and Barth, A., 1978, Kinetische Untersuchungen an der Dipeptidyl-peptidase IV. *Acta Biol. Med. Germ.* **37:** 409-420.

Yaron, A. and Naider, F., 1993, Proline-dependent structural and biological properties of peptides and proteins. *Crit. Rev. Biochem. Mol. Biol.* **28:** 31-81.

# N-TERMINAL HIV-1 TAT NONAPEPTIDES AS INHIBITORS OF DIPEPTIDYL PEPTIDASE IV. CONFORMATIONAL CHARACTERIZATION

Carmen Mrestani-Klaus[1], Annett Fengler[1], Jürgen Faust[1], Wolfgang Brandt[1], Sabine Wrenger[2], Dirk Reinhold[2], Siegfried Ansorge[2] and Klaus Neubert[1]

*[1]Department of Biochemistry and Biotechnology, Institute of Biochemistry, Martin-Luther-University Halle-Wittenberg, Kurt-Mothes-Str. 3, 06120 Halle; [2]Department of Internal Medicine, Institute of Experimental Internal Medicine, Otto-von-Guericke-University Magdeburg, Leipziger Str. 44, 39120 Magdeburg, Germany*

Key words HIV-1 Tat, dipeptidyl peptidase IV, $^1$H NMR spectroscopy, MD calculations, DP IV inhibitors

Abstract Compared to the N-terminal nonapeptide of the HIV-1 Tat protein as inhibitor of activity of DP IV which is supposed to mediate the immunosuppressive effects of HIV-1 Tat, the Ile$^5$ and Leu$^6$ analogues showed strongly reduced inhibitory activity. Interestingly, replacement of Asp$^2$ with Gly or Lys led to compounds with considerably enhanced inhibition. Therefore, we have applied $^1$H NMR spectroscopy and restrained molecular dynamics calculations to elucidate the molecular conformation of a series of Tat nonapeptides. Conformational backbone differences of these peptides as well as the nature and the arrangement of the side chains *per se* at significant positions preventing effective binding to DP IV might explain their different inhibitory activity on DP IV.

## 1. INTRODUCTION

The membrane-bound serine exopeptidase dipeptidyl peptidase IV (DP IV, CD26) cleaves dipeptides from the N-terminus of oligo- and polypeptides provided that the penultimate residue is proline or alanine.

In the immune system DP IV was found to be expressed on the surface of T and B lymphocytes and of NK cells. Due to the strong expression enhancement of CD26 on peripheral blood T lymphocytes upon T cell activation the enzyme is characterized to be a specific in vivo marker for those immune cells (Ansorge 1995). DP IV seems to play a central role in signal transduction processes via interaction with other membrane expressed antigens and, therefore, in activation processes of immune cells (Torimoto 1991).

DP IV has been reported to mediate the immunosuppressive effect of HIV-1 Tat, the transactivator of the HIV-1 virus (Gutheil 1994). It was shown that the N-terminal Xaa-Xaa-Pro sequence of this protein is important for DP IV inhibition and for suppression of CD26-dependent T cell growth (Wrenger 1996).

As part of our ongoing research program on the investigation of the role of DP IV in the immune system, we characterized the inhibition behaviour of a series of N-terminal HIV-1 Tat nonapeptides (Wrenger 1997; Mrestani-Klaus 1998). Of the compounds studied, $G^2$-Tat(1-9) and $K^2$-Tat(1-9) turned out to be the best inhibitors compared to the parent peptide Tat(1-9), whereas both $Ile^5$ and $Leu^6$ analogues showed strongly reduced inhibitory activity. Knowledge of the solution conformations of these peptides can yield an explanation of the origin of their different DP IV inhibition. We therefore present a conformational study, as obtained from combined use of NMR data and restrained MD calculations.

# 2. MATERIALS AND METHODS

## 2.1 NMR Experiments

NMR spectra were acquired on Bruker ARX 500 and Varian UNITY 500 spectrometers. Samples of 5 mg of the appropriate peptide were dissolved in 0.8 ml $H_2O$ containing 10 % $D_2O$. $d_4$-TSPA was used as internal standard. Water signal suppression was achieved using presaturation during relaxation delay. Spectra were acquired at 303 K. Peaks were completely assigned by using 2D COSY, TOCSY, NOESY and ROESY spectra. Homonuclear coupling constants were extracted from 1D proton spectra. TOCSY, NOESY and ROESY spectra were recorded in the phase-sensitive mode. The ROE cross-peak intensities were measured using the TRIAD software package and were classified strong, medium and weak, corresponding to interproton distances of 1.8-2.5, 1.8-3.3 and 1.8-4.0 Å, respectively.

## 2.2 Structure calculations

The starting structures were put into a box of water molecules (TIP3P model) and calculated with the AMBER 4.1 force field. From ROESY spectra a set of interresidue interproton distances was obtained and employed as distance restraints with a force constant of 32 kcal/Å. After the restrained molecular dynamics (MD) run at 300 K a free MD simulation for 800 ps was performed to check the quality of the structures.

# 3. RESULTS

In water we observed one predominant conformation for each of the five peptides characterized by *all-trans* peptide bonds. The observation of strong $d_{\alpha N}(i,i+1)$ ROEs in the absence of $d_{NN}(i,i+1)$ ROEs indicates that the conformations of each peptide exist mainly in "random coil" forms. In addition, the $d_{\alpha N}(i,i)$ connectivity is weaker than the sequential $d_{\alpha N}(i,i+1)$ one as is characteristic for unfolded peptides. However, several amino acids such as $Val^4$ and $Ile^8$ in both Tat(1-9) and $L^6$-Tat(1-9), $Val^4$, $Ile^5$ and $Ile^8$ in $I^5$-Tat(1-9), $Gly^2$, $Asp^5$ and $Ile^8$ in $G^2$-Tat(1-9) and $Ile^8$ in $K^2$-Tat(1-9) had substantial deviations from the tabulated random-coil values for the $C^{\alpha}H$ chemical shifts as highly sensitive structural parameter, for the $^3J_{\alpha H\text{-}NH}$ coupling constants as well as for the temperature coefficients of the amide proton chemical shifts. These results indicate that the backbone conformations of the peptides studied were not fully extended or random possessing some propensity to form turn-like structuring at the centre amino acids.

Besides the major conformation sets of low intensity signals (less than 5 %) appeared arising from *cis-trans* isomerization of one or more Xaa-Pro peptide bonds (Figure 1). The very low concentration of these less populated structures did not allow a detailed study. The major conformations in water were studied with 15 ROEs for Tat(1-9), 17 ROEs for $G^2$-Tat(1-9), 18 ROEs for both $K^2$-Tat(1-9) and $I^5$-Tat(1-9) and 20 ROEs for $L^6$-Tat(1-9) as restraints in a box of 1400 water molecules. The energetically optimized solution conformations of Tat(1-9) are presented in Figure 2 (a). The main differences of the conformations of $I^5$-Tat(1-9) and $L^6$-Tat(1-9) (Mrestani-Klaus 1998) are changes of the backbone torsion angles $\phi$ and $\psi$. In contrast, the replacement of Asp in position 2 led to less pronounced backbone conformational differences compared to Tat(1-9) (Figure 2 (b)).

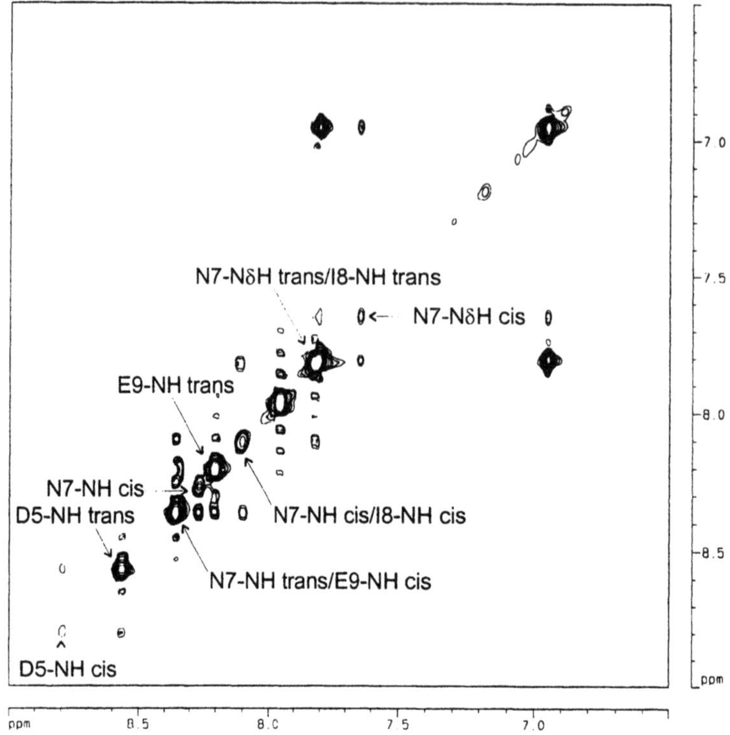


*Figure 1.* The NH region in a NOESY spectrum of Tat(1-9) (400 ms mixing time) after addition of a catalytic amount of cyclophilin 18 shows exchange cross peaks between NHs of the major (*trans*) and minor (*cis*) conformations.

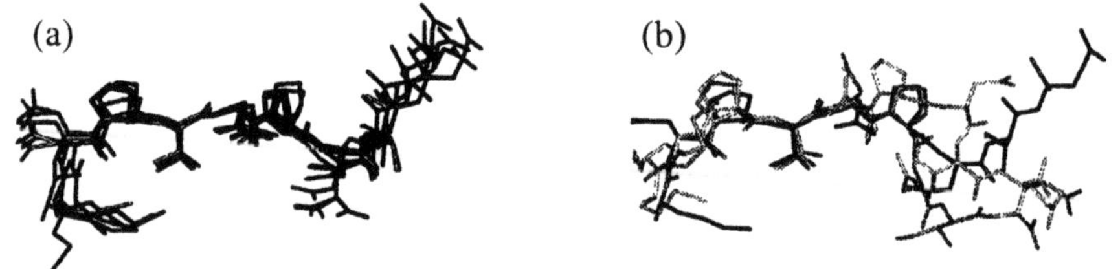


*Figure 2.* (a) Bundle of 8 best TAT(1-9) structures of lowest energy obtained by MD calculations. (b) Comparison of the structures of $G^2$-Tat(1-9) (light) and $K^2$-Tat(1-9) (light) to the parent Tat(1-9) (dark) presented as superposition. The structures of all peptides fit best the NMR data such as ROEs, torsion angles $\phi$ and temperature dependence studies.

## 4. CONCLUSION

By performing a detailed conformational analysis we were able to define the solution conformations of five N-terminal nonapeptides of the HIV-1 Tat protein. Our results indicate structural restraints at certain positions in all analogues forming turn-like structures around the centre amino acids. The conformational differences of Tat(1-9) sequences caused by amino acid substitutions in positions 5 or 6 possibly prevent effective binding to DP IV and might explain the lack of their inhibitory activity on DP IV. Since no major conformational differences were found between those peptides substituted in the second position and the parent peptide Tat(1-9), the considerable enhancement of inhibition can only be due to the replacement of Asp in the second position. This indicates that the negatively charged $Asp^2$ does not represent the most favoured amino acid residue for optimal binding to DP IV. These results will aid us in disclosing the mode of action of the Tat(1-9) peptide analogues and the reason for their different inhibitory activity and will help to gain insight in the ligand-DP IV interaction.

## ACKNOWLEDGMENTS

The authors wish to thank the Deutsche Forschungsgemeinschaft for financial support through SFB 387. Funding from the Fonds der Chemischen Industrie is also acknowledged.

## REFERENCES

Ansorge, S., Bühling, F., Hoffmann, T., Kähne, T., Neubert, K., and Reinhold, D., 1995, In *Dipeptidyl Peptidase IV (CD26) in Metabolism and the Immune Response* (B. Fleischer, ed), Springer-Verlag, Heidelberg, pp. 163-184.

Gutheil, W.G., et al., 1994, Human immunodeficiency virus 1 Tat binds to dipeptidyl aminopeptidase IV (CD26): A possible mechanism for Tat's immunosuppressive activity. *Proc. Natl. Acad. Sci. USA* **91:** 6594-6598.

Mrestani-Klaus, C., et al., 1998, $^1$H NMR conformational study on N-terminal nonapeptide sequences of HIV-1 Tat protein: A contribution to structure-activity relationships. *J. Peptide Sci.* **4:** 400-410.

Torimoto, Y., et al., 1991, Coassociation of CD26 (dipeptidyl peptidase IV) with CD45 on the surface of human T lymphocytes. *J. Immunol.* **147:** 2514-2517.

Wrenger, S., et al., 1996, The N-terminal X-X-Pro sequence of the HIV-1 Tat protein is important for the inhibition of dipeptidyl peptidase IV (DP IV/CD26) and the suppression of mitogen-induced proliferation of human T cells. *FEBS Lett.* **383:** 145-149.

Wrenger, S., et al., 1997, The N-terminal structure of HIV-1 Tat is required for suppression of CD26-dependent T cell growth. *J. Biol. Chem.* **272:** 30283-30288.

# SIGNAL TRANSDUCTION EVENTS INDUCED OR AFFECTED BY INHIBITION OF THE CATALYTIC ACTIVITY OF DIPEPTIDYL PEPTIDASE IV (DP IV, CD26)

Thilo Kähne[1]*, Dirk Reinhold[1], Klaus Neubert[2], Ilona Born[2], Jürgen Faust[2], and Siegfried Ansorge[1]

*[1] Institute of Experimental Internal Medicine, Department of Internal Medicine, Otto-von-Guericke University Magdeburg, D-39120 Magdeburg, Germany, [2] Institute of Biochemistry, Department of Biochemistry and Biotechnology, Martin-Luther-University Halle-Wittenberg, D-06120 Halle(Saale), Germany*

Key words dipeptidyl peptidase IV, CD26, inhibitors, signal transduction

Abstract DP IV (CD26) represents an accessory surface molecule playing an important role in the process of activation and proliferation of human lymphocytes. The molecular events mediated by this ectoenzyme are only partly established and the necessity of DP IV enzymatic activity for its signalling capacity has been discussed controversial. Focusing on the putative role of the catalytic domain of this peptidase, it could be shown that inhibition of the catalytic activity can provoke many cellular effects, including induction of tyrosine phosphorylations and p38 MAP kinase activation as well as suppression of DNA synthesis and reduced production of various cytokines. TGF-ß1, the production and secretion of which is increased after DP IV inhibition, supposedly mediates the observed suppressive effects by maintaining $p27^{kip}$ expression levels which leads to a cell cycle arrest in $G_1$. Moreover, anti-CD3-induced signalling pathways, including $Ca^{2+}$ mobilisation, MEK1-, Erk1/2- and PKB-activation, can be strongly affected by DP IV inhibition. Thus, the enzymatic activity or at least the interaction of effectors with the catalytic domain of CD26 seems to be important for crucial functions of this cell surface antigen.

# 1. INTRODUCTION

Although DP IV was firstly described in 1966 and subsequently a lot of research activities have been concentrated on this enzyme, now, 33 years later the functional role of this widely distributed peptidase is still poorly understood (Hopsu-Havu 1966, De Meester 1999).

Developing and utilising specific inhibitors of DP IV helped to investigate especially the role of the enzymatic activity within the manifold putative functions of DPIV/CD26 in the immune system (Kähne 1999). Whereas various studies have provided evidence that binding of certain CD26 monoclonal antibodies to cell surface DPIV can generate a costimulatory signal in human T cells, the role of the catalytic domain of DPIV regarding the process of DPIV mediated regulation of the immune response has been controversial (Dang 1990, Hegen 1993, Reinhold 1994, Fleischer 1994, Kähne 1997).

The signalling mechanisms underlying the costimulatory effects of CD26 are poorly understood and the early cellular events after DPIV inhibition are completely unknown (Hegen 1997, von Bonin 1998). Here we want to focus on the early cellular events after inhibition of DPIV enzyme activity.

# 2. MATERIALS AND METHODS

## 2.1 Detection of activated kinases (MEK 1/2, ERK 1/2, p38 HOG, PKB)

Sample preparation was performed as previously described (Kähne 1999). 10 µl aliquots of each sample were separated in a 10 % SDS-polyacrylamide gel. After transferring the proteins onto nitrocellulose activated kinases were detected by means of the following primary antibodies (all were from New England BioLabs, USA): MEK 1/2: rabbit-anti-P-MEK1/2 (specificity: Phospho-Ser217/221), ERK1/2 (p42/44): rabbit-anti-P-p42/44 (specificity: Phospho-Thr202/Tyr204), p38 HOG: rabbit-anti-P-p38 (specificity: Phospho-Thr180 and Phospho-Tyr182), PKB (akt): rabbit-anti-P-akt (specificity: Phospho-Ser473). Blots were subsequently incubated with a POD conjugated goat-anti-rabbit-antibody and proteins were visualised using a chemiluminescent substrate (UltraSignal, Pierce, The Netherlands).

## 2.2 Detection of $p27^{kip}$

Resting human T cells were stimulated with PHA (1μg/ml) for 48 h in presence and absence of simultaneously given DP IV inhibitors or TGF-ß1.The level of cdk inhibitor $p27^{kip}$ expression was analysed by Western blot (rabbit anti $p27^{kip}$, Calbiochem) at times indicated in the figure. Blots were subsequently incubated with a POD conjugated goat-anti-rabbit-antibody and proteins were visualised using a chemiluminescent substrate (UltraSignal, Pierce, The Netherlands).

# 3. RESULTS

## 3.1 DP IV signal transduction

One strategy to clarify the functional role of CD26 is to discover the molecular mechanisms underlying the well characterised DP IV-mediated phenomena such as suppression of proliferation and cytokine production as well as enhanced TGF-ß1 secretion.

Data of our and other groups have provided evidence that anti-CD26 binding or inhibitor treatment can provoke the transduction of cellular signals. Whether these signal transduction events can be generated by DP IV alone, by recruiting of accessory molecules or by the disturbance of a delicately balanced steady state of bioactive peptides unknown so far, but trimmed by DP IV, remains to be clarified.

### 3.1.1 Putative signalling mechanisms resulting in TGF-ß1 secretion

Reinhold et al. have described the enhanced production and secretion of TGF-ß1 after exposure of human T cells to DPIV inhibitors (Reinhold 1997). The mechanisms underlying this phenomenon remain unclear. We suppose the occurrence of signalling pathways activated after DPIV inhibition and resulting in the induction of TGF-ß1 expression.

Recent findings of our group indicated that besides anti-CD26 antibodies also DP IV inhibitors directly induce a panel of new tyrosine phosphorylated proteins (Kähne 1998). The nature of the induced phosphoproteins remains unclear at present. Surprisingly, however, a temporary activation of the MAP kinase p38 HOG was found after DP IV inhibition, whereas other MAP kinase pathways including Erk1/2 and SAP/Jun kinase remained unaffected (Fig 1). These data were confirmed by selective blocking of intracellular p38 MAPK using the inhibitor SB 202190 which diminishes the suppressive effect of DP IV inhibitors on DNA synthesis (data not shown). These findings let us speculate about a p38 MAPK-mediated pathway of DP IV inhibitor-induced TGF-ß1 production and secretion.

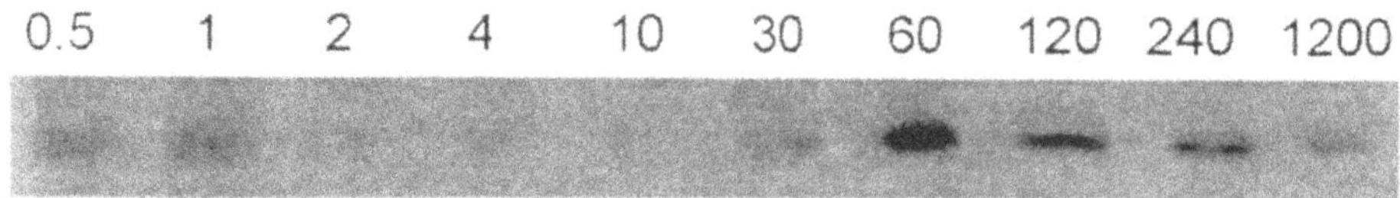


*Figure 1.* Temporary activation of p38 MAP kinase in DP IV inhibitor treated resting human T cells. T cells were incubated with the DP IV inhibitor Lys[Z($NO_2$)]-thiazolidide ($10^{-5}$M) and the time (in minutes)dependent activation of p38 MAP kinase was analysed by Western blot.

Further work was designed to study whether the enhanced secreted TGF-ß1 is responsible for the found suppressive effects of DPIV inhibitors regarding DNA synthesis and cytokine production.

TGF-ß1 is known as an anti-mitogen blocking cell cycle progression by stopping the late G1 transition. This effect is most probably either due to an upregulation or a prevention of degradation of the cdk2 inhibitor $p27^{kip}$ (Polyak 1994). DP IV inhibitor-mediated suppression of cell proliferation is also known to be associated with a cell cycle arrest in late G1 (Schön 1985). These findings let us suggest that this effect might be mediated by TGF-ß1. As shown in Fig 2, both DP IV inhibitors and TGF-ß1 simultaneously given

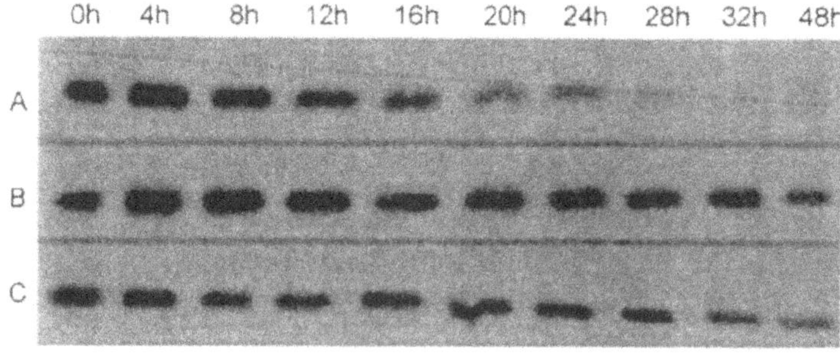


*Figure 2.* The level of cdk inhibitor $p27^{kip}$ expression remains constant in presence of DP IV inhibitor (C) or TGF-ß1 (B). Resting human T cells were stimulated with PHA for 48h in presence (B,C) and absence (A) of simultaneously given DP IV inhibitors or TGF-ß1. $p27^{kip}$ expression was analysed by Western blot at times indicated in the figure.

with a mitogen (PHA) maintain the level of cdk inhibitor $p27^{kip}$ in comparison to PHA-stimulated controls where the amount of inhibitor continuously decreases within the first day after stimulation.

These findings support the hypothesis of a putative mechanism underlying the well studied suppression of T cell proliferation by DP IV inhibition based on the autocrine action of lately (4 to 6 hrs) released immunosuppressive TGF-ß1.

### 3.1.2 DPIV inhibitors affect early T cell receptor generated signal cascades

Besides the TGF-ß1-mediated effects described above very fast processes after DP IV inhibitor treatment were found, particularly with regard to anti-CD3 stimulation of human T cells. Obviously, these effects are not linked to the enhanced TGF-ß1 secretion but do reflect a second signalling mechanism located proximal to the cell membrane and CD3 /TcR complex.

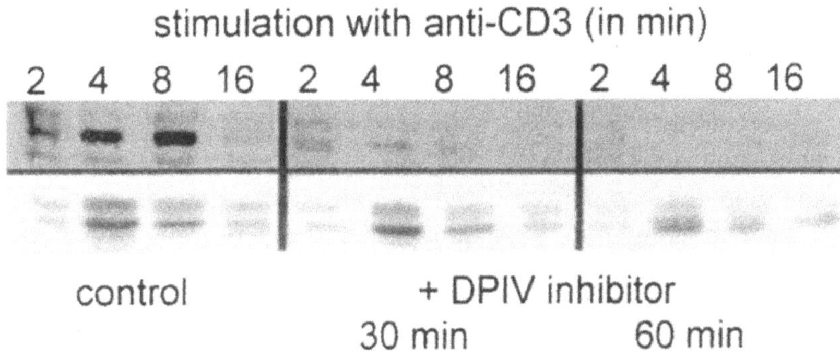


*Figure 3.* DP IV inhibitor preincubation diminish anti-CD3 (OKT3) induced temporary activation of: PKB, akt (top) and ERK1/2 (bottom). T cells were preincubated with DP IV inhibitors as indicated and subsequently stimulated with anti-CD3. The time-dependent activation of PKB and ERK1/2 was analysed by Western blot.

Besides the dose-dependent and reversible inhibition of PMA-induced hyperphosphorylation of $p56^{lck}$ (Kähne 1995) a suppression of anti-CD3-induced $Ca^{2+}$ flux in these cells was found (Kähne 1998). This led us to investigate in more detail whether other anti-CD3-induced activation mechanisms can also be affected by DP IV inhibition. As shown in Fig 3, the temporarily anti-CD3-mediated activation of two other signal cascades can be suppressed by DP IV inhibitors. First, measuring one of the first downstream substrates of the PI3-kinase namely PKB (akt), an inhibition of the phospholipid kinase pathway initiated by PI3-kinase was found. As shown in Fig 3 (top), anti-CD3-induced phosphorylation of akt is strongly suppressed by DP IV inhibitor pre-incubation of the cells. The same holds true for a second well characterised signal cascade involved in CD3/TcR signalling, the ras-erk1/erk2 pathway. The MAP kinases Erk1 and Erk2 (Fig 3, bottom) as well as their upstream regulator MEK1/2 kinase (data not shown) have been analysed representing three members of this important pathway.

## 4. CONCLUSION

The data summarized here clearly point out that the enzymatic activity or at least the catalytic domain of DP IV/CD26 is indeed involved in regulating T cell proliferation. These results led us speculate about supposed cellular mechanisms and consequences of DP IV inhibition:

Inhibition of DP IV enzyme activity results in a subsequent induction of new tyrosine phosphorylated proteins. Moreover, immediately after DP IV inhibition blocking effects with regard to many anti-CD3-induced signal cascades including calcium mobilisation, phospholipid kinase-activation as well as MAPKK- and MAPK-activation were found. These suppressive effects might diminish a CD3/TcR generated activation signal as strong as sufficient to abolish cell cycle progression by preventing $G_0$ exit and in that way leading to the suppressed DNA synthesis and cytokine production. Whether these blocking effects are transmitted by one or more of the primary induced tyrosine phosphorylated proteins rather than by primary interaction of DP IV and other surface molecules forming multimeric cell surface complexes remains unclear at present.

DP IV inhibition is furthermore linked to a temporary activation of MAP kinase p38 HOG. The direct involvement of p38 HOG in DP IV inhibitor signalling is strengthened by the fact that the highly p38 specific inhibitor SB202190 is capable of diminishing the suppressive effects of DP IV inhibitors on DNA synthesis. We suppose that these signalling mechanisms finally lead to the induction of TGF-ß1 mRNA expression and enhanced secretion of TGF-ß1. The released TGF-ß1 might act in an autocrine feedback and induce among other things the expression of the cdk inhibitor $p27^{kip}$ resulting in a cell cycle arrest in late $G_1$. This cell cycle arrest is then, similar to the previously described hypothesis, strongly associated with a suppression of proliferation (DNA synthesis) and perhaps also with an inhibition of mRNA expression and production of different immunostimulatory cytokines like IL-2, IL-10, IL-12 and IFN-$\gamma$.

Within the supposed cellular mechanisms which might be induced by DP IV inhibition, a lot of question marks occur. Further studies will be designed to give deeper insights into the molecular events of CD26 signal transduction and might substantially contribute to the comprehension of the functional role of CD26 in the immune system.

# ACKNOWLEDGMENTS

This work was supported by the Deutsche Forschungsgemeinschaft, SFB 387.

# REFERENCES

Dang, N.H., et al., 1990, Comitogenic effect of solid-phase immobilized anti-1F7 on human CD4 T cell activation via CD3 and CD2 pathways. *J.Immunol.* **144:** 4092-4100.

De Meester, I., et al., 1999, CD26, let it cut or cut it down. *Immunol.Today* **20:** 367-375.

Fleischer, B., 1994, CD26: a surface protease involved in T cell activation. *Immunology Today* **15:** 180-184.

Hegen, M., et al., 1993, Enzymatic activity of CD26 (Dipeptidyl-peptidase IV) is not required for its signalling function in T cells. *Immunobiol.* **189:** 483-493.

Hegen, M., et al., 1997, Cross-linking of CD26 by antibody induces tyrosine phosphorylation and activation of mitogen-activated protein kinase. *Immunology* **90:** 257-264.

Hopsu-Havu, V.K., and Glenner, G.G., 1966, A new peptide naphtylamidase hydrolyzing glycylprolyl-naphtylamide. *Histochemistry* **7:** 197-201.

Kähne, T., et al., 1995, Enzymatic activity of DPIV/CD26 is involved in PMA-induced hyperphosphorylation of p56lck. *Immunol. Lett.* **46:** 189-193.

Kähne, T., et al., 1997, In *Leukocyte Typing VI.* (T. Kishimoto, et. al., eds), Garland Publishing.

Kähne, T., et al., 1998, Early phosphorylation events induced by DPIV/CD26-specific inhibitors. *Cell Immunol.* **189:** 60-66.

Kähne, T., et al., 1999, Dipeptidyl peptidase IV: a cell surface peptidase involved in regulating T cell growth. *Int. J. Mol. Med.* **4:** 3-15.

Polyak, K., et al., ,1994, p27Kip1, a cyclin-Cdk inhibitor, links transforming growth factor-beta and contact inhibition to cell cycle arrest. *Genes Dev.* **8:** 9-22.

Reinhold, D., et al., 1994, Inhibitors of dipeptidyl peptidase IV (DPIV, CD26) specifically suppress proliferation and modulate cytokine production of strongly CD26 expressing U937 cells. *Immunobiology* **192:** 121-136.

Reinhold, D., et al., 1997, Inhibitors of dipeptidyl peptidase IV (DP IV, CD26) induces secretion of transforming growth factor-beta 1 (TGF-beta 1) in stimulated mouse splenocytes and thymocytes. *Immunol. Lett.* **58:** 29-35.

Schön, E., et al., 1985, The dipeptidyl peptidase IV, a membrane enzyme involved in the proliferation of T lymphocytes. *Biomed. Biochim. Acta* **44:** K9-K15.

von Bonin, A., et al., 1998, Dipeptidyl-peptidase IV/CD26 on T cells: analysis of an alternative T- cell activation pathway. *Immunol.Rev.* **161:** 43-53.

# SPECIFIC INHIBITORS OF DIPEPTIDYL PEPTIDASE IV SUPPRESS mRNA EXPRESSION OF DP IV/CD26 AND CYTOKINES

Marco Arndt, Dirk Reinhold,[1] Uwe Lendeckel,[1] Antje Spiess, Jürgen Faust,[2] Klaus Neubert,[2] and Siegfried Ansorge[1]
*Institute of Immunology, [1]Institute of Experimental Internal Medicine, Center of Internal Medicine, Otto von Guericke University Magdeburg*
*Leipziger Strasse 44, D-39120 Magdeburg, Germany*
*[2]Institute of Biochemistry, Department of Biochemistry and Biotechnology, Martin Luther University Halle-Wittenberg*
*Kurt Mothes Str. 3, D-06120 Halle (Saale), Germany*

Key words dipeptidyl peptidase IV (DP IV, CD26), cytokines, PBMC, quantitative RT-PCR

Abstract The dipeptidyl peptidase IV is an activation marker on T, B and NK cells. Specific inhibitors of DP IV suppress DNA synthesis, as well as cytokine protein production. Here, we describe for the first time the quantitative changes of mRNA expression of IFN-γ, IL-2, IL-12 and DP IV after inhibition of DP IV. Due to the stimulation of human peripheral blood mononuclear cells (PBMC) with pokeweed mitogen (PWM) both cytokine as well as DP IV mRNA expression is increased significantly. Treatment with DP IV inhibitor suppresses dose-dependently these changes. Importantly, mRNA expression of DP IV itself was inhibited. The presented data are fully compatible with our hypothesis that inhibition of DP IV leads to cell cycle arrest in late G1 due to enhanced TGF-β1 expression.

## 1. INTRODUCTION

The cell surface ectopeptidase dipeptidyl peptidase IV (DP IV, E.C 3.4.14.5), a type II membrane protein, is identical with the CD26 antigen (Ulmer 1990). The gene encoding for DP IV has been localised to chromosome 2q24.3 (Abbott 1994) and predicts a protein of 766 amino acids

(Hegen 1997). DP IV is an exopeptidase, catalysing the release of N-terminal dipeptides from peptides, preferentially with proline or hydroxyproline in the penultimate position (Rahfeld 1991).

Numerous cytokines and other bioactive peptides are potential substrates of the enzyme, but our knowledge of the *in vivo* substrates is still very limited. It has been shown that CD26 plays a crucial role in B, NK and T cell activation (Ansorge 1991, Bühling 1994, 1995). Inhibition of enzymatic activity of dipeptidyl peptidase IV provokes many cellular effects such as suppression of DNA synthesis and production of T cell activating cytokines (see review Kähne 1999). On the other hand, TGF-β1 production and secretion is increased (Reinhold 1997a, 1997b, Kähne 1999).

Here, we describe for the first time the influence of the DP IV inhibitor Lys[Z($NO_2$)]-thiazolidide on the mRNA expression of different cytokines and dipeptidyl peptidase IV itself.

# 2. MATERIALS AND METHODS

## 2.1 Cell culture and stimulation experiments

Human peripheral blood mononuclear cells (PBMC) were isolated from healthy donors using density gradient centrifugation (Boyum 1968) and T cells were enriched by the nylon wool adherence technique (Julius 1973). Cells were seeded into a 6-well plate to a density of $10^6$ cells/ml, grown in CG-medium (VITROMEX), stimulated by the addition of PWM (Sigma, 2 μg/ml) and incubated with the amount of DP IV inhibitor Lys[Z($NO_2$)]-thiazolidide indicated.

## 2.2 RNA isolation and quantitative PCR

RNA was prepared by means of RNeasy (Qiagen) in accordance to the manufacture protocol. One microgram of total RNA was reverse-transcribed and 1/20th cDNA mixture was used for quantitative PCR by means of the Lightcycler LC24™ (Idaho Technology). A 10 μl reaction mixture contained 1x reaction buffer with BSA (Idaho Technology), 2 mM $MgCl_2$, 200 μmol dNTP, 0.4 U InViTaq polymerase (InViTek), 0.2 μl of a 1:1000 dilution of SYBR-Green I (Molecular Probes) and 0.5 μmol of the specific primers, either IL-12 (Biosource Int.), IL-2 and IFN-γ (Stratagen) primer set or DP IV/E3 (5'-GATGAAGCGGCCGCTGATAGC) and DP IV/E7 (5'-GTGACCATGTGATCCACTGTG). Experiments were performed in triplicate and resulting data statistically analysed using One-way-Anova test.

## 3. RESULTS

The Lightcycler allows a real time quantification of mRNA contents. During the cycling, SYBR-Green I is incorporated into double stranded DNA and the resulting fluorescence is measured at the end of each elongation step. The data were plotted and the mRNA contents were quantified directly using different concentrations of a standard.

Figure 1 shows the time course of mRNA expression of IL-2 (panel A), IFN-γ (B), IL-12 (C) and DP IV (D) of PBMC after PWM stimulation. At all time points investigated the IL-2, IFN-γ and IL-12 mRNA expression was found to be induced in response to stimulation. A typical time course of induction was observed for every cytokine. The inhibition of DP IV activity by means of the specific inhibitor Lys[Z($NO_2$)]-thiazolidide dose-dependently decreased the stimulatory effect of PWM. The inhibitory effect of Lys[Z($NO_2$)]-thiazolidide on IL-12 mRNA expression was less pronounced (panel C) and, in the case of IL-2 mRNA, the mRNA expression was significantly decreased at 24 hours only (panel A).

By measuring the protein expression levels of these cytokines we found that protein levels do well correspond to their mRNA levels (data not shown). Their amounts were in accordance to our earlier published results (Reinhold 1997a).

Interestingly, not only the cytokine mRNA expression is downregulated by the inhibition of DP IV activity but also the mRNA expression of DP IV itself (Figure 1, panel D). Over the period of time an increase of DP IV mRNA expression could be observed. At 72 hours after stimulation the expression level of DP IV is on average $5.012*10^6$ copies/μg RNA, a value decreased to $2.394*10^6$ copies/μg RNA by treatment with the inhibitor at a concentration of $10^{-5}$ M.

## 4. CONCLUSIONS

The presented data are in line with earlier reported findings showing that the inhibition of dipeptidyl peptidase IV activity by specific inhibitors led to a reduction of cytokine protein expression. Here, we could clearly demonstrate a reduced mRNA expression of the cytokines investigated.

Surprisingly, DP IV mRNA expression itself was affected by inhibitor treatment.

These results, taken together, support our hypothesis (Kähne 1999) that inhibition of DP IV leads to enhanced expression of TGF-β1 (Reinhold 1997a, Reinhold 1997b), which then acts in an autocrine feedback resulting in cell cycle arrest in late G1 (Schön 1985) and as shown here in

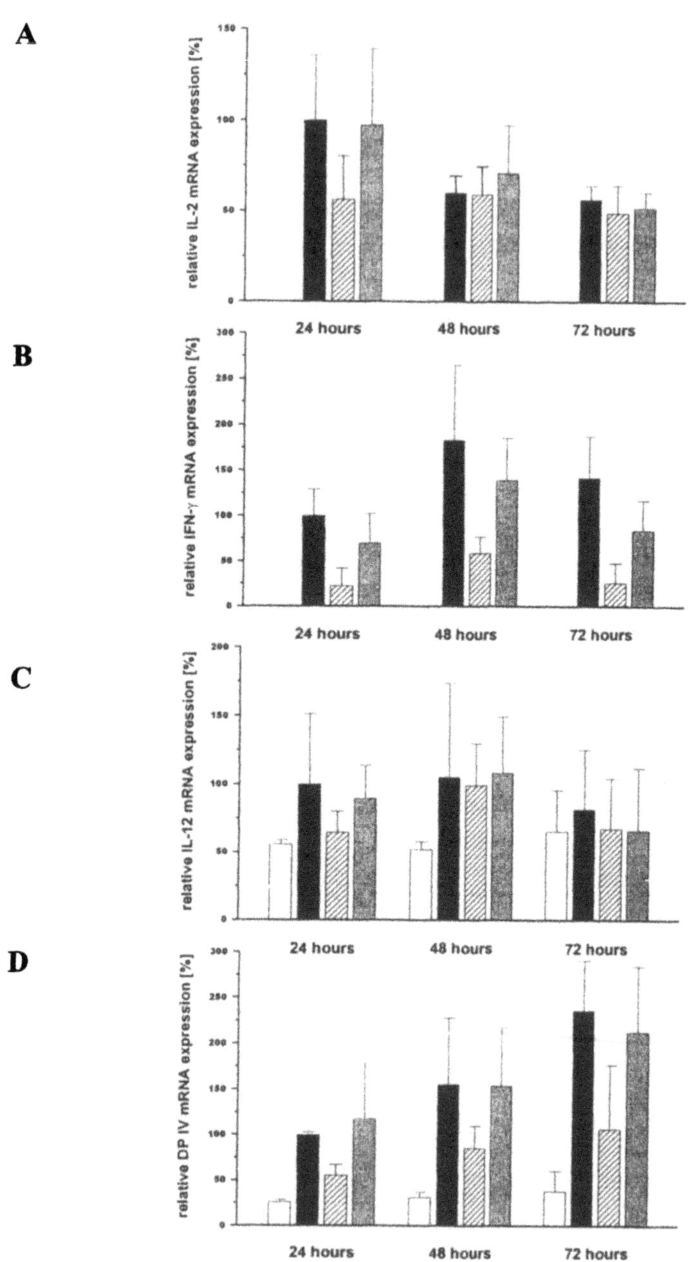


*Figure 1.* Time course of cytokine (panel A = IL-2, B = IFN-γ; C = IL-12) and DP IV mRNA (panel D) expression of PWM-stimulated PBMC after inhibition with different concentrations of DP IV inhibitor Lys[Z($NO_2$)]-thia-zolidide. white = control, black = stimulated with 2 μg/ml PWM (at 24 hours set as 100 %), hatched = stimulated with 2 μg/ml PWM and inhibitor $10^{-5}$ M, grey = stimulated with 2 μg/ml PWM and inhibitor $10^{-6}$ M

downregulation of the mRNA expression of IL-2, IL-12, IFN-γ and DP IV. Further examinations are under way to prove this hypothesis.

## ACKNOWLEDGMENTS

We are very grateful to R. H. Hädicke, K. Mnich, C. Wolf and K. Frank for their excellent technical assistance. The work was supported by a grant of the Deutsche Forschungsgemeinschaft (DFG), Germany (Sonderforschungsbereich 387).

## REFERENCES

Abbott, C.A., et al., 1994, Genomic organization, exact localization, and tissue expression of the human CD26 (dipeptidyl peptidase IV) gene. *Immunogenetics* **40:** 331-338.

Ansorge, S., et al., 1991, Membrane-bound peptidases of lymphocytes: functional implications. *Biomed.. Biochim. Acta* **50:** 799-807.

Boyum, A., 1968, Isolation of mononuclear cells and granulocytes from human blood. Isolation of monuclear cells by one centrifugation, and of granulocytes by combining centrifugation and sedimentation at 1 g. *Scand.. J. Clin. Lab. Invest. Suppl.* **97:** 77-89.

Bühling, F., et al., 1995, Functional role of CD26 on human B lymphocytes. *Immunol. Lett.* **45:** 47-51.

Bühling, F., et al., 1994, Expression and functional role of dipeptidyl peptidase IV (CD26) on human natural killer cells. *Nat. Immun.* **13:** 270-279.

Hegen, M., 1997, In *Leukocyte Typing VI* (T. Kishimoto, and H. Kikutani, eds), Garland Publishing Inc. pp. 478-481.

Julius, M.H., et al., 1973, A rapid method for the isolation of functional thymus-derived murine lymphocytes. *Eur. J. Immunol.* **3:** 645-649.

Kähne, T., et al., 1999, Dipeptidyl peptidase IV: A cell surface peptidase involved in regulating T cell growth. *Int. J. Mol. Med.* **4:** 3-15.

Rahfeld, J., et al., 1991, Extended investigation of the substrate specificity of dipeptidyl peptidase IV from pig kidney. *Biol. Chem. Hoppe Seyler* **372:** 313-318.

Reinhold, D., et al., 1997a, Inhibitors of dipeptidyl peptidase IV induce secretion of transforming growth factor-beta 1 in PWM-stimulated PBMC and T cells. *Immunology* **91:** 354-360.

Reinhold, D., et al., 1997b, Inhibitors of dipeptidyl peptidase IV (DP IV, CD26) induces secretion of transforming growth factor-*β*1 (TGF-*β*1) in stimulated mouse splenocytes and thymocytes. *Immunol. Lett.* **58:** 29-35.

Schön, E., et al., 1985, The dipeptidyl peptidase IV, a membrane enzyme involved in the proliferation of T lymphocytes. *Biomed. Biochem. Acta* **44:** 9-15.

Ulmer, A.J., et al., 1990, CD26 antigen is a surface dipeptidyl peptidase IV (DPPIV) as characterized by monoclonal antibodies clone TII-19-4-7 and 4EL1C7. *Scand. J. Immunol.* **31:** 429-435.

# DIPEPTIDYL PEPTIDASE IV IN INFLAMMATORY CNS DISEASE

Andreas Steinbrecher[1], Dirk Reinhold[2], Laura Quigley[1], Ameer Gado[1], Nancy Tresser[1], Leonid Izikson[1], Ilona Born[3], Jürgen Faust[3], Klaus Neubert[3], Roland Martin[1], Siegfried Ansorge[2] and Stefan Brocke[1]
*[1]National Institute of Neurological Disorders and Stroke, National Institutes of Health, Bethesda,MD, USA; [2]Institute of Experimental Internal Medicine, Department of Internal Medicine, Otto-von-Guericke University, Magdeburg, Germany; [3]Institute of Biochemistry, Department of Biochemistry and Biotechnology, Martin-Luther-University Halle-Wittenberg, Germany*

Key words Experimental autoimmune encephalomyelitis, dipeptidyl peptidase IV (DP IV, CD26), inhibitors, autoimmune disease, T cell

Abstract Current pathogenic concepts of inflammatory demyelinating disorders such as multiple sclerosis (MS) are based on the hypothesis that a T cell-mediated autoimmune response is involved in the disease process. One of the primary goals in the in the development of immunotherapies for autoimmune diseases has been to achieve inactivation of disease-inducing lymphocytes either by direct inhibition or suppression through regulatory cells and/or cytokines. The CD26 antigen is identical with the cell surface ectopeptidase dipeptidyl peptidase IV (DP IV, EC 3.4.14.5) which is involved in regulating T cell activation and growth. Activated T cells, including those specific for myelin antigens, express high levels of CD26/DP IV. In vitro, reversible DP IV inhibitors suppress T cell proliferation and pro-inflammatory cytokine production in response to myelin antigens. Further studies will evaluate the role of DP IV inhibition in T cell-mediated inflammatory disease of the central nervous system.

## 1. INTRODUCTION

Our current multi-step model for the initiation of T cell-mediated autoimmune inflammatory disease of the CNS includes peripheral activation of T cell specific for myelin antigens and Th1 differentiation (Fig 1).

Autoreactive T cells can be activated by variety of cross-reactive peptides, a mechanism which has been termed "molecular mimicry" (Fujinami and Oldstone, 1985). A very high level of cross-reactivity has now been recognized as an essential feature of the T cell receptor (Mason, 1998). These observations explain the fact that many pathogen-derived antigens with limited or no sequence identity to autoantigens are capable of stimulating autoreactive (Wucherpfennig and Strominger 1995, Hemmer *et al* 1997) and pathogenic T cells (Fujinami and Oldstone, 1985). Once activated, autoreactive T cells will cross the blood-brain barrier and respond to CNS antigens in situ. Naive $CD4^+$ T cells develop into either of two major subsets of Th helper cells (Ths) that produce distinct sets of cytokines (O'Garra *et al* 1997). Acquisition of a polarized cytokine phenotype by T cells is regulated by the strength of antigenic stimulation, the genetic background of the host, as well as by the cytokines present during priming. Thl/Th2 differentiation is further regulated by CC chemokines (Karpus, 1999). IL-12 is a major determinant directing Thl development and is produced by macrophages and dendritic cells (O'Garra *et al* 1997). In addition, IFN-γ is also involved in Thl differentiation (O'Garra *et al* 1997). In contrast, the addition of exogenous IL-4 to in vitro cultures is required for the differentiation of naïve $CD4^+$ T cells into Th2 cells. It is widely accepted that Thl cells, critical for cell-mediated immunity by their production of IL-2, IFN-γ, TNF-α and lymphotoxin are involved in the immunopathology of organ-specific autoimmune disease (O'Garra *et al* 1997). A role as regulators has been suggested for Th2 cells (O'Garra *et al* 1997) and cells producing TGF-β, recently characterized as Th3 and Tr (regulatory) $CD^{4+}$ T cells (Fig 1).

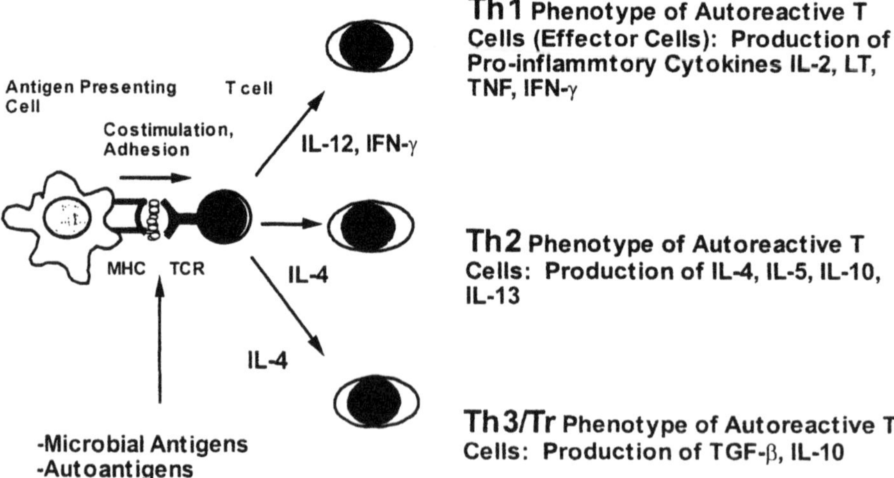


*Figure 1.* Peripheral activation and phenotypic differentiation of autoreactive T cells

The process of lesion formation in the CNS is further governed by a complex expression pattern of adhesion molecules and their ligands, cyto- and chemokines as well as enzymes by leukocytes, cerebrovascular endothelium and parenchymal cells of the CNS (Karpus and Ransohoff, 1998). Leukocyte accumulation and lesion formation in experimental autoimmune encephalomyelitis (EAE) critically depend on antigen-driven Th1-type T cell restimulation in the CNS compartment (Fig 2).

Activated T cells, including those specific for myelin antigens, express high levels of the dipeptidyl peptidase IV (DP IV, EC 3.4.14.5)/CD26. Surface expression of CD26 is upregulated after mitogenic, anti-CD3 or IL-2 stimulation of T cells, St. aureus protein stimulation of B cells and IL-2 stimulation of NK cells (Kähne *et al* 1999). Recent studies support the notion that DP IV may play an important role in the regulation of differentiation and growth of T lymphocytes and in CD3/T-cell receptor-mediated signal transduction (Morimoto and Schlossman, 1998, Von Bonin *et al* 1998, Kähne et al. 1999, De Meester *et al* 1999).

Using specific inhibitors of DP IV, it was demonstrated that suppression of the enzymatic activity of DP IV on leukocytes results in suppression of DNA synthesis and cytokine production (IL-2, IL-10, IL-12, IFN–$\gamma$) (Schön *et al* 1987, Reinhold *et al* 1993, Reinhold *et al* 1997a). Conversely, DP IV inhibitors stimulate the production of the immunoregulatory cytokine TGF-ß1 (Reinhold *et al* 1997a, Reinhold *et al* 1997b, Kähne *et al* 1999).

In a recent study, we demonstrated a protective effect of the two DP IV inhibitors Lys[Z($NO_2$)]-pyrrolidide and -thiazolidide on EAE (Steinbrecher *et al* submitted). Specific inhibitors of DP IV could interfere at several levels during CNS inflammation mediated by myelin-specific T cells. First, activation and Th1 differentiation of autoreactive T cells could be impaired by inhibition of CD26-dependent T cell growth (Fig 3). In addition, pathogenic T cells residing within inflammatory lesion can be targeted in situ with specific inhibitors of DP IV (Fig 4). Both direct effects including suppression of DNA synthesis and cytokine production as well as TGF-$\beta$-mediated regulation of Th1 immune responses could be responsible for the amelioration of inflammatory CNS disease (Fig 3, 4).

In the present study, we have examined the effect of the two DP IV inhibitors Lys[Z($NO_2$)]-thiazolidide and -pyrrolidide on the DNA synthesis of MBP-specific T cell clones generated from (PL X SJL)$F_1$ mice.

# 2. MATERIALS AND METHODS

## 2.1 Mice

Female (SJL X PL)$F_1$ mice were obtained from Jackson Laboratories (Bar Harbor, ME). Mice were 7 - 14 weeks of age when experiments were

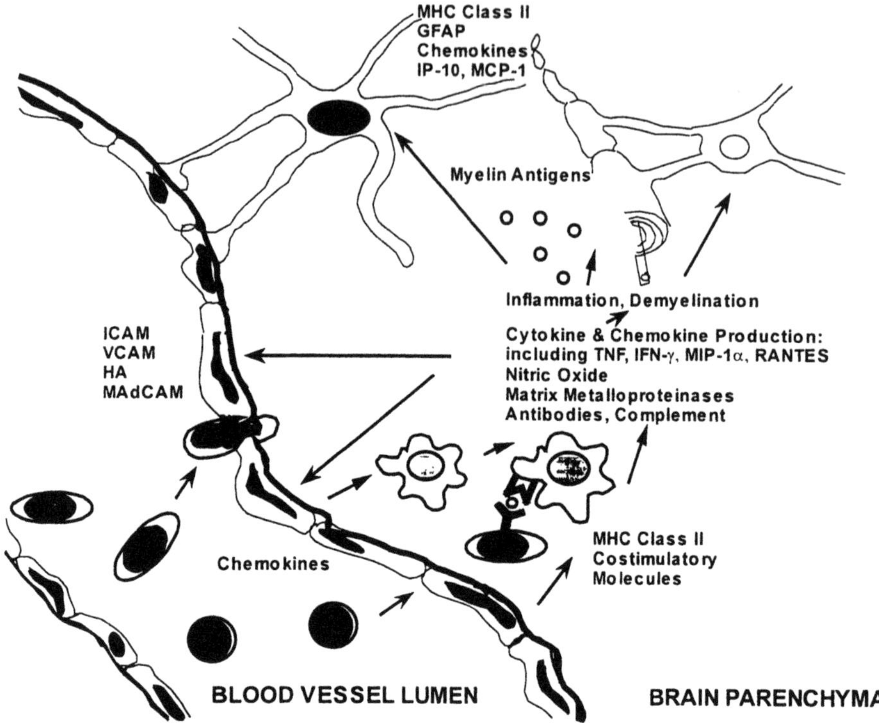


*Figure 2.* Cyto- and chemokine-regulated amplification of inflammation and lesion formation in the CNS during EAE

started. All procedures were carried out according to protocols approved by the ACUC of the National Institute of Neurological Disorders and Stroke.

## 2.2 Antigens

Myelin basic protein (MBP) peptide MBP(Ac1-11) was prepared by continuous flow solid phase synthesis according to the sequence for rat MBP (Ac-ASQKRPSQRHG) by the Protein and Nucleic Acid Facility, Beckman Center, Stanford University, Stanford, CA.

## 2.3 MBP-specific T cell clones

The T cell clones L1C1, L1C41 and L1C48 specific for MBP-(Ac1-11) were established from (PL X SJL)$F_1$ mice (Jackson Laboratories, Bar Harbor, ME) according to standard procedures (Brocke *et al* 1996).

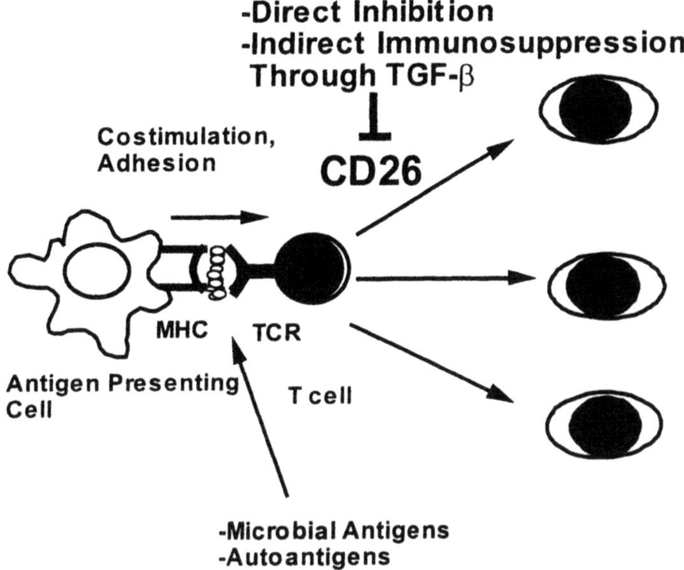


*Figure 3.* Possible effects of CD26-inhibition on activation and differentiation of autoreactive T cells

## 2.4 Proliferation assay

T cell clones were incubated in 96-well plates at $10^4$ T cells per well, with 10 μg/ml peptide presented on 2 x $10^5$ syngeneic irradiated spleen cells (3000 rad). To determine the effect of the inhibitors, varying concentrations were added at a fixed antigenic concentration. Wells without inhibitor or antigen, respectively, were used as controls. The culture medium was based on AIM-V (Life Technologies, Gaithersburg, MD), supplemented with 2 mM glutamine, 100 U/ml penicillin and 100 μg/ml streptomycin. T cell clones were cultured for 72 hours, the last 8 - 12 hours of which 1 μCi $^3$[H]thymidine per well was added (DuPont, Wilmington, DE). Incorporation of $^3$[H]thymidine was measured using a beta counter. Results are given as arithmetic means from cultures set up at least in triplicates.

## 2.5 Determination of DP IV activity

The enzymatic activity of DP IV was determined using 1.6 mM Gly-Pro-4-nitroanilide as substrate for DP IV (Schön *et al* 1984). The resulting 4-nitroaniline strongly absorbs at 392 nm. The enzymatic activity at 37 °C and pH 7.6 is expressed in pkat. All measurements with substrate and PBS-

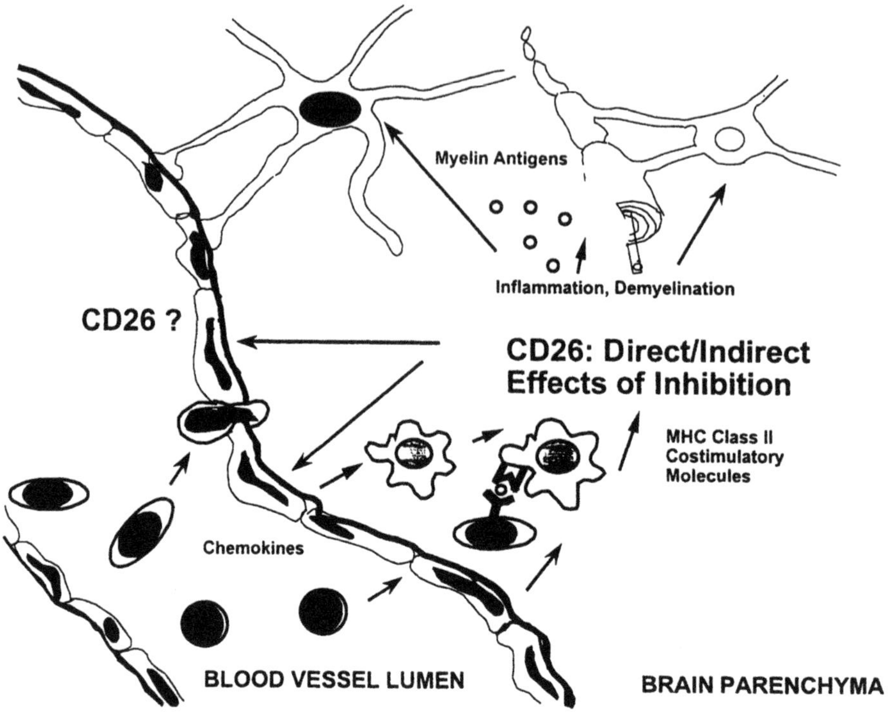


*Figure 4.* Possible effects of CD26-inhibition on inflammatory lesion formation by myelin antigen-specific T cells

controls were performed in duplicate. For the DP IV-activity assays in T cell clones 5 x $10^5$ viable T cells were used in the reaction mixture. T cells obtained immediately before and after 48 h of stimulation with 10 μg/ml MBP(Ac1-11) were compared.

# 3. RESULTS

## 3.1 DP IV enzymatic activity on murine MBP(Ac1-11)-specific T cell clones

Activated myelin-antigen reactive $CD4^+$ T cells are a requirement for the induction of adoptive transfer EAE. We therefore analyzed DP IV-activity and the effect of two synthetic inhibitors on the clonal level. Whereas L1C48 is an encephalitogenic Th1 $CD4^+$ T cell clone, L1C1 and L1C41 (both Th2) are non-encephalitogenic clones. All clones displayed DP IV-activity after

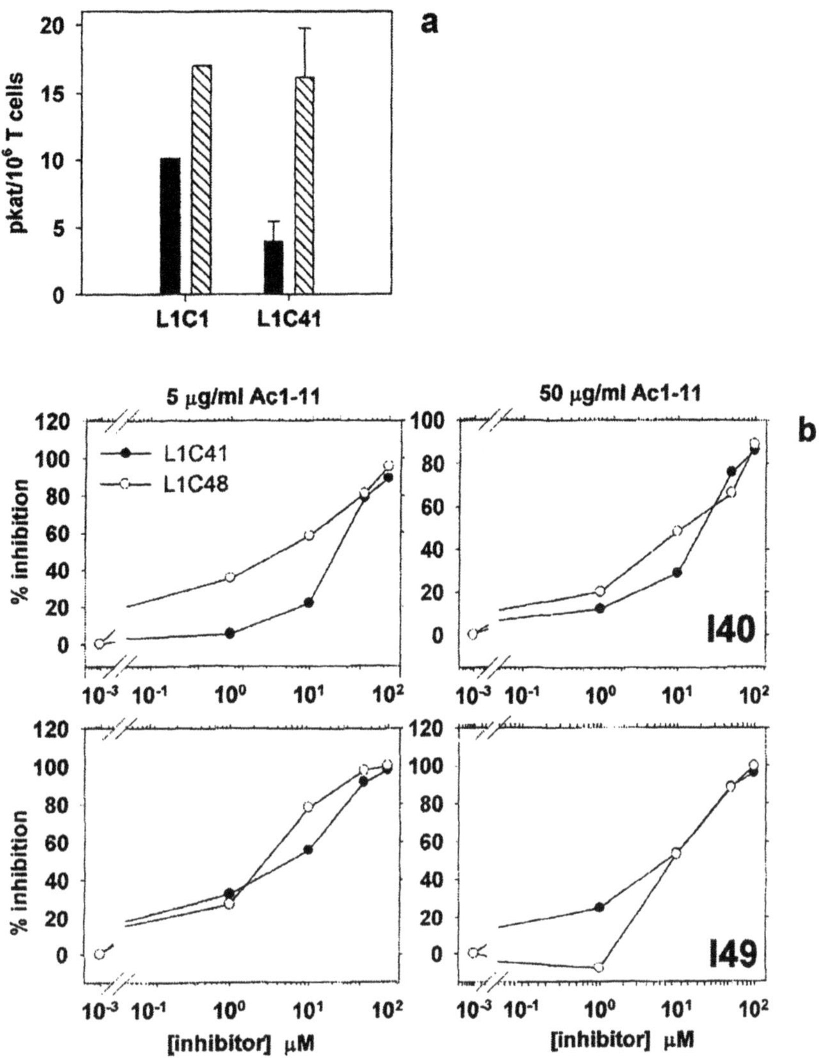


*Figure 5.* DP IV-activity of murine MBP-specific T cell clones and antiproliferative effect of DP IV-inhibition. (A) To determine DP IV-activity 5 x $10^5$ viable T cells from clones L1C1 and L1C41 were assayed immediately before (black bars) and after 48 h of stimulation with peptide (hatched bars). The results from duplicate measures are given as mean + SD. (B) The proliferation of clones L1C41 (filled circles) and L1C48 (open circles) to 5 µg/ml (left panel) and 50 µg/ml (right panel) of MBP(Ac1-11) and various concentrations of Lys[Z(NO2)]-pyrrolidide (I40) (upper panel) and Lys[Z(NO2)]-thiazolidide (I49) (lower panel) are shown. SIs without inhibitor: 103 and 123 (L1C41) and 31 and 22.4 (L1C48) for 5 and 50 µg/ml peptide, respectively. All cultures were performed in triplicates (SD of cpm in general < 15 %).

antigenic stimulation. When we compared the DP IV-activity between the resting and activating states of clones L1C1 and L1C41 a clear upregulation upon antigenic stimulation was detectable (Fig. 5A).

### 3.2 Lys[Z($NO_2$)]-pyrrolidide and thiazolidide inhibit DNA synthesis of MBP(Ac1-11)-specific T cell clones

The peptide-induced proliferation of all T cell clones examined (L1C1, and L1C41) was inhibited by Lys[Z($NO_2$)]-pyrrolidide (I40) and Lys[Z($NO_2$)]-thiazolidide (I49) in a dose-dependent manner (Fig 5B).

## 4. CONCLUSIONS

The synthetic CD26 inhibitors Lys[Z($NO_2$)]-pyrrolidide and -thiazolidide efficiently suppress antigen-induced activation of murine MBP-specific T cell clones. This observation together with preliminary data indicating a protective effect of CD26 inhibitors during EAE in vivo identifies DPIV/CD26 as a possible drug target in inflammatory autoimmune disease.

## ACKNOWLEDGMENTS

A. Steinbrecher was a postdoctoral fellow (Ste 813/1-1) of the Deutsche Forschungsgemeinschaft. D. Reinhold, I. Born, J. Faust, K. Neubert and S. Ansorge were supported by the Deutsche Forschungsgemeinschaft (DFG), SFB 387.

## REFERENCES

Brocke, S., *et al.*, 1996, Isolation and characterization of autoreactive T cells in experimental autoimmune encephalomyelitis of the mouse. *Methods* **9:** 458-462.

De Meester, I., *et al.*, 1999, CD26, let it cut or cut it down. *Immunol. Today* **20:** 367-375.

Fujinami, R.S., and Oldstone, M.B., 1985, Amino acid homology between the encephalitogenic site of myelin basic protein and virus: mechanism for autoimmunity. *Science* **230:** 1043-1045.

Hemmer, B., *et al.*, 1997, Identification of high potency microbial and self ligands for a human autoreactive class II-restricted T cell clone. *J. Exp. Med.* **185:** 1651-1659.

Kähne, T., *et al.*, 1999, Dipeptidyl peptidase IV: A cell surface peptidase involved in regulating T cell growth. (Review). *Int. J. Mol. Med.* **4:** 3-15.

Karpus, W. J., 1999, Chemokine regulation of inflammatory-mediated nervous system diseases. *J. Neurovirol.* **5:** 1-2.

Karpus, W.J., and Ransohoff, R.M., 1998, Chemokine regulation of experimental autoimmune encephalomyelitis: temporal and spatial expression patterns govern disease pathogenesis. *J. Immunol.* **161:** 2667-2671.

Mason, D., 1998, A very high level of crossreactivity is an essential feature of the T -cell receptor. *Immunol. Today* **19:** 395-404.

Morimoto, C., and Schlossman, S.F., 1998, The structure and function of CD26 in the T-cell immune response. *Immunol. Rev.* **161:** 55-70.

O'Garra, A., L. Steinman, and Gijbels, K., 1997, $CD4^+$ T-cell subsets in autoimmunity. *Curr. Opin. Immunol.* **9:** 872-883.

Reinhold, D., *et al.*,(1993, Dipeptidyl peptidase IV (CD26) on human lymphocytes. Synthetic inhibitors of and antibodies against dipeptidyl peptidase IV suppress the proliferation of pokeweed mitogen-stimulated peripheral blood mononuclear cells and IL-2 and IL-6 production. *Immunobiol.* **188:** 403-414.

Reinhold, D., *et al.*, 1997a, Inhibitors of dipeptidyl peptidase IV induce secretion of transforming growth factor-ß1 in PWM-stimulated PBMC and T cells. *Immunology* **91:** 354-360.

Reinhold, D., *et al.*, 1997b, Inhibitors of dipeptidyl peptidase IV (DP IV, CD26) induces secretion of transforming growth factor-ß1 (TGF-ß1) in stimulated mouse splenocytes and thymocytes. *Immunol. Lett.* **58:** 29-35.

Schön, E., *et al.*, 1984, Dipeptidyl peptidase IV of human lymphocytes. Evidence for specific hydrolysis of glycylprolin in T-lymphocytes. *Biochem. J.* **223:** 255-258.

Schön, E., *et al.*, 1987, The role of dipeptidyl peptidase IV in human T lymphocyte activation. Inhibitors and antibodies against dipeptidyl peptidase IV suppress lymphocyte proliferation and immunoglobulin synthesis in vitro. *Eur. J. Immunol.* **17:** 1821-1826.

Von Bonin, A., *et al.*, 1998, Dipeptidyl-peptidase IV/CD26 on T cells: analysis of an alternative T-cell activation pathway. *Immunol. Rev.* **161:** 43-53.

Wucherpfennig, K.W., and Strominger, J.L., 1995, Molecular mimicry in T cell-mediated autoimmunity: viral peptides activate human T cell clones specific for myelin basic protein. *Cell* **80:** 695-705.

# DIPEPTIDYL PEPTIDASE IV (CD26): ROLE IN T CELL ACTIVATION AND AUTOIMMUNE DISEASE

Dirk Reinhold[1], Bernhard Hemmer[2], Bruno Gran[3], Andreas Steinbrecher[3], Stefan Brocke[3], Thilo Kähne[1], Sabine Wrenger[1], Ilona Born[4], Jürgen Faust[4], Klaus Neubert[4], Roland Martin[3], and Siegfried Ansorge[1]

*[1]Institute of Experimental Internal Medicine, Department of Internal Medicine, Otto-von-Guericke-University, Magdeburg, Germany; [2]Clinical Neuroimmunology group, University of Marburg, Germany; [3]Neuroimmunology Branch, NINDS, National Institutes of Health, Bethesda, USA; [4]Institute of Biochemistry, Department of Biochemistry and Biotechnology, Martin-Luther-University Halle-Wittenberg, Germany*

Key words dipeptidyl peptidase IV (DP IV, CD26), inhibitors, autoimmunity, T lymphocyte, myelin basic protein , multiple sclerosis

Abstract The ectoenzyme dipeptidyl peptidase IV (DP IV; EC 3.4.14.5; CD26) has been shown to play a crucial role in T cell activation. In the present study, we show by flow cytometry and by enzymatic DP IV assay that myelin basic protein (MBP)-specific, $CD4^+$ T cell clones (TCC) derived from patients with multiple sclerosis (MS) express high levels of DP IV/CD26. The enzymatic activity of resting TCC was found to be three to fourfold higher than on resting peripheral blood T cells and close to that of T cells 48 hours after PHA stimulation. The DP IV inhibitors Lys[Z($NO_2$)]-thiazolidide and Lys[Z($NO_2$)]-pyrrolidide suppress in a dose-dependent manner DNA synthesis and IFN-γ, IL-4, and TNF-α production of the antigen-stimulated TCC. These data suggest that CD26 plays a role in regulating activation of autoreactive TCC. Further *in vivo* investigations will clarify, whether the inhibition of the enzymatic activity of DP IV could be a useful tool for therapeutic interventions in MS and/or other autoimmune diseases.

## 1. INTRODUCTION

The dipeptidyl peptidase IV (DP IV, EC 3.4.14.5) is a transmembrane type II glycoprotein, which is present on most mammalian cells (Fleischer *et al* 1994). DP IV is a serine peptidase that catalyses the release of N-terminal dipeptides from oligo- and polypeptides preferentially with proline, hydroxyproline, and with less efficiency alanine at the penultimate position (De Meester *et al* 1999, Kähne *et al* 1999). At the 4th Workshop on Leukocytes Differentiation Antigens a number of monoclonal antibodies recognising DP IV were subsumed under the term CD26 (Ulmer *et al* 1990).

DP IV is an activation marker for different immune cells. Surface expression of CD26 is upregulated after mitogenic, anti-CD3, or IL-2 stimulation of T cells, St. aureus protein stimulation of B cells, and IL-2 stimulation of NK cells (Kähne *et al* 1999). Data from several groups have provided evidence that DP IV plays a central role in the regulation of differentiation and growth of T lymphocytes and in CD3/T-cell receptor (TcR)-mediated signal transduction (Morimoto *et al* 1998; Von Bonin *et al* 1998, Kähne *et al* 1999, De Meester *et al* 1999).

Using specific inhibitors of DP IV, it was demonstrated that suppression of the enzymatic activity of DP IV on peripheral blood mononuclear cells (PBMC) and T cells results in suppression of DNA synthesis and cytokine production (IL-2, IL-10, IL-12, IFN-$\gamma$) (Schön *et al* 1987, Reinhold *et al* 1993, Reinhold *et al* 1997a). On the other hand, DP IV inhibitors stimulate the production of the immunosuppressive cytokine TGF-ß1 (Reinhold *et al* 1997a, Reinhold *et al* 1997b, Kähne *et al* 1999).

In the present study, we examined the effect of the two DP IV inhibitors Lys[Z($NO_2$)]-thiazolidide and -pyrrolidide on the DNA synthesis and cytokine production of MBP(87-99)-specific T cell clones generated from patients with MS. MBP-specific TCC generated from MS patients were used since reactivity of T cells to MBP in MS is well characterised, and MS may be considered as prototypic T cell-mediated autoimmune disease.

# 2. MATERIALS AND METHODS

## 2.1 MBP-specific TCC

MBP-specific TCC were grown in complete medium (CM) composed of Iscove's modified Dulbecco's medium with 2 mM L-glutamin, 50 µg/ml gentamycin, 100 U/ml penicillin, and 100 µg/ml streptomycin (Whittaker Bioproducts, Gaithersburg, MD, USA).

## 2.2 Immunofluorescence CD26-staining of cells

Indirect immunofluorescence staining of MBP-specific TCC was performed using the monoclonal anti-DP IV (CD26) antibody EF5/A3 (IgG1)

(Reinhold *et al* 1997c). Labelled cells were analysed by flow cytometry (FACScan, Becton Dickinson).

## 2.3 Enzymatic assay

Enzymatic activity of DP IV was determined using 1.6 mM Gly-Pro-4-nitroanilide and 4 x $10^4$ cells in the reaction mixture (Schön *et al* 1984).

## 2.4 Proliferation assay

Cells from TCC were rested for 8 days. 1 x $10^4$ cells were seeded in U-bottom 96-well plates with 5 x $10^4$ irradiated (3000 rad) PBMC and different concentrations of the MBP(87-99) peptide and the DP IV inhibitors Lys[Z($NO_2$)]-thiazolidide and -pyrrolidide. Cells were cultured for 72 h at 37 °C. During the last 16 h of culture, 0.5 μCi [$^3$H]thymidine was added to each well. Cells were then harvested and incorporated radioactivity was measured by scintillation counting.

## 2.5 Cytokine assay

For cytokine assays, 2 x $10^5$ T cells were seeded in 24-well plates together with 1 x $10^6$ irradiated autologous PBL in 1 ml CM in the presence or absence of 100 μg/ml of the MBP(87-99) peptide and the DP IV inhibitors Lys[Z($NO_2$)]-thiazolidide and -pyrrolidide ($10^{-5}$ M). Supernatants were collected after 48 hours and stored at -70 °C until the assays were performed. IFN-γ, IL-4, and TNF-α were measured using commercially available ELISA kits (Genzyme, Cambridge, MA, USA).

# 3. RESULTS

## 3.1 CD26 antigen expression and DP IV enzymatic activity on MBP-specific TCC of patients with MS

Six TCC generated from MS patients specific for the immunodominant epitope MBP(87-99) in the context of MS-associated HLA-DR molecules (DR2, DR4, DR6) were used in this study. TCC were rested for 8 days after restimulation and viable cells were analysed for enzymatic DP IV activity. Using Gly-Pro-4-nitroanilide as a substrate, high DP IV activity was detected on viable cells of all six TCC. The enzymatic activity of these resting TCC was found to be much higher (three to fourfold) than on resting peripheral

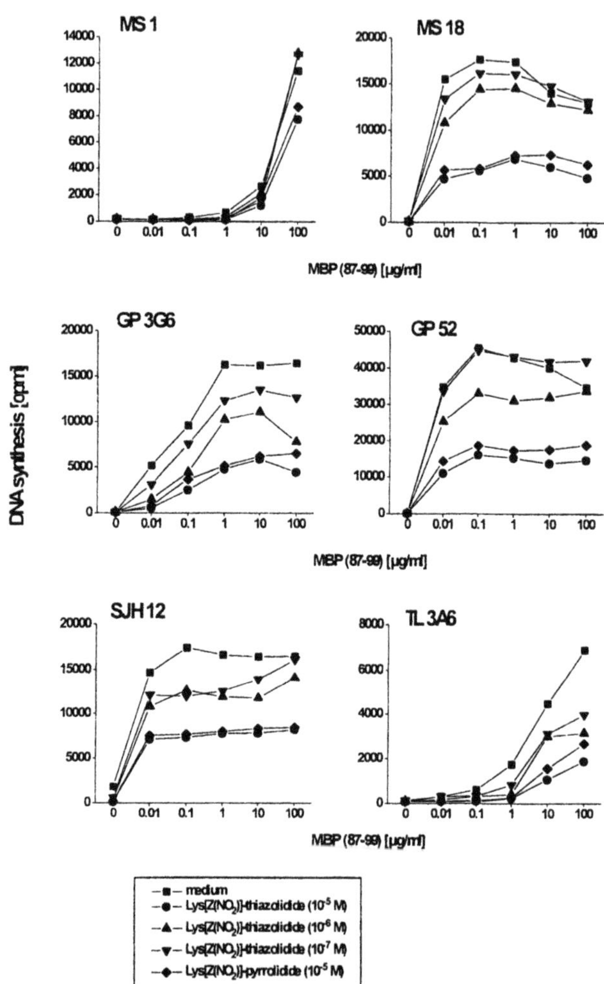


*Figure 1.* Influence of DP IV inhibitors on DNA synthesis of MBP-specific TCC generated from patients with MS. $[^3H]$dThd incorporation is indicated in cpm. Results are expressed as mean of three different experiments.

blood T cells of healthy probands (6 ± 2 pkat/$10^6$cells) and close to that of T cells 48 hours after PHA stimulation (23 ± 6 pkat/$10^6$ cells).These data were supported by flow cytometric studies obtained with the monoclonal anti-CD26 antibody EF5/A3. 95 % of cells from all six TCC expressed the CD26 antigen on the cell surface (data not shown).

## 3.2 Lys[Z($NO_2$)]-thiazolidide and -pyrrolidide inhibit DNA synthesis as well as cytokine production (IFN-γ, IL-4, TNF-α) of MBP-specific TCC

Previously, we reported that inhibitors of DP IV reduce the DNA synthesis of PWM-stimulated PBMC and purified T cells (Reinhold *et al* 1997a, Kähne *et al* 1999). To study the influence of these inhibitors on DNA synthesis of MBP-specific TCC, cells were rested for 8 days and stimulated with different concentrations of the MBP(87-99) peptide in the absence and presence of different concentrations of the synthetic DP IV inhibitors.

As shown in Figure 1, the inhibitors suppressed the DNA synthesis of all six TCC in a dose-dependent manner. To exclude possible cytotoxic effects of the DP IV inhibitors, we measured the viability of cell cultures by trypan blue staining. The viability was not impaired by the inhibitors in any of the experiments (data not shown).

To address the question whether the suppressive effect of these synthetic DP IV inhibitors on DNA synthesis of TCC correlates with a decrease in production and secretion of cytokines, we determined the concentrations of IFN-γ, IL-4, and TNF-α in supernatants of MBP(87-99)-stimulated TCC in presence and absence of Lys[Z($NO_2$)]-thiazolidide and Lys[Z($NO_2$)]-pyrrolidide ($10^{-5}$ M). The concentrations of cytokines released after 48 h were measured by enzyme immunoassays. The production of all three cytokines was significantly suppressed by both inhibitors (data not shown).

# 4. CONCLUSIONS

In conclusion, we demonstrate that synthetic inhibitors of DP IV/CD26 can suppress the activation of CD4+ TCC specific for MBP(87-99). The mechanism underlying the effect of DP IV inhibitors on T lymphocyte proliferation and cytokine production remains to be elucidated.

The inhibition of the enzymatic activity of DP IV and suppression of DNA synthesis, resp. by DP IV inhibitors could be a useful approach for therapeutic interventions in MS and other autoimmune diseases.

## ACKNOWLEDGMENTS

The work was supported by the Deutsche Forschungsgemeinschaft (DFG), SFB 387.

## REFERENCES


De Meester, I. et al., 1999, CD26, let it cut or cut it down. *Immunol. Today* **20:** 367-375.

Fleischer, B., 1994, CD26: a surface protease involved in T-cell activation. *Immunol. Today* **15:** 180-184.

Kähne, T. et al., 1999, Dipeptidyl peptidase IV: A cell surface peptidase involved in regulating T cell growth. *Int. J. Mol. Med.* **4:** 3-15.

Morimoto, C., and Schlossman, S.F., 1998, The structure and function of CD26 in the T-cell immune response. *Immunol. Rev.* **161:** 55-70.

Reinhold, D. et al., 1993, Dipeptidyl peptidase IV (CD26) on human lymphocytes. Synthetic inhibitors of and antibodies against dipeptidyl peptidase IV suppress the proliferation of pokeweed mitogen-stimulated peripheral blood mononuclear cells and IL-2 and IL-6 production. *Immunobiol.* **188:** 403-414.

Reinhold, D. et al., 1997a, Inhibitors of dipeptidyl peptidase IV induce secretion of transforming growth factor-ß1 in PWM-stimulated PBMC and T cells. *Immunology* **91:** 354-360.

Reinhold, D. et al., 1997b, Inhibitors of dipeptidyl peptidase IV (DP IV, CD26) induces secretion of transforming growth factor-ß1 (TGF-ß1) in stimulated mouse splenocytes and thymocytes. *Immunol. Lett.* **58:** 29-35.

Reinhold, D. et al., 1997c, The effect of anti-CD26 antibodies on DNA synthesis and cytokine production (IL-2, IL-10 and IFN-gamma) depends on enzymatic activity of DP IV CD26. *Adv. Exp. Med. Biol.* **421:** 149-155.

Schön, E. et al., 1984, Dipeptidyl peptidase IV of human lymphocytes. Evidence for specific hydrolysis of glycylprolin in T-lymphocytes. *Biochem. J.* **223:** 255-258.

Schön, E. et al., 1987, The role of dipeptidyl peptidase IV in human T lymphocyte activation. Inhibitors and antibodies against dipeptidyl peptidase IV suppress lymphocyte proliferation and immunoglobulin synthesis in vitro. *Eur. J. Immunol.* **17:** 1821-1826.

Ulmer, A.J. et al., 1990, CD26 antigen is a surface dipeptidyl peptidase IV (DPPIV) as characterised by monoclonal antibodies clone TII-19-4-7 and 4EL1C1. *Scand. J. Immunol.* **31:** 429-435.

Von Bonin, A. et al., 1998, Dipeptidyl peptidase IV/CD26 on T cells: analysis of an alternative T-cell activation pathway. *Immunol. Rev.* **161:** 43-53.

# EFFECTS OF NONAPEPTIDES DERIVED FROM THE N-TERMINAL STRUCTURE OF HUMAN IMMUNODEFICIENCY VIRUS-1 (HIV-1) TAT ON SUPPRESSION OF CD26-DEPENDENT T CELL GROWTH

Sabine Wrenger[1], Dirk Reinhold[1], Jürgen Faust[2], Carmen Mrestani-Klaus[2], Wolfgang Brandt[2], Annett Fengler[2], Klaus Neubert[2] and Siegfried Ansorge[1]

[1]*Institute of Experimental Internal Medicine, Department of Internal Medicine, Otto-von-Guericke-University, D-39120 Magdeburg, Germany;* [2]*Institute of Biochemistry, Department of Biochemistry and Biotechnology, Martin-Luther-University Halle-Wittenberg, D-06120 Halle, Germany*

Key words dipeptidyl peptidase IV; DP IV; CD26; HIV-1 Tat; peptidergic inhibitors; immunosuppression

Abstract The human immunodeficiency virus-1 (HIV-1) transactivator Tat occurs extracellularly and exerts immunosuppressive effects. Interestingly, Tat inhibits dipeptidyl peptidase IV (DP IV) activity of the T cell activation marker CD26. The short N-terminal nonapeptide Tat(1-9), MDPVDPNIE, also inhibits DP IV activity and suppresses DNA synthesis of tetanus toxoid-stimulated peripheral blood mononuclear cells (PBMC). Here, we present the influence of amino acid exchanges in the first three positions of Tat(1-9). For instance, the replacement of $D^2$ of Tat(1-9) by G or K generated peptides, which inhibit DP IV-catalyzed IL-2(1-12) cleavage nearly threefold stronger. Similar effects were observed on the suppression of DNA synthesis of Tetanus toxoid-stimulated PBMC. This correlation suggests that Tat(1-9)-deduced peptides mediate antiproliferative effects at least in part via specific DP IV interactions and supports the hypothesis that CD26 plays a key role in the regulation of lymphocyte growth.

## 1. INTRODUCTION

Dipeptidyl peptidase IV (DP IV, EC 3.4.14.5) catalyses the hydrolysis of N-terminal dipeptides from oligopeptides with protonated N-terminus if the penultimate amino acid is proline or alanine. DP IV is a transmembrane type II glycoprotein, which is identical with the activation marker CD26 of T and B lymphocytes and NK cells. Data from several groups have shown a key role of DP IV in the regulation of differentiation and growth of lymphocytes. Specific inhibitors of DP IV suppress mitogen- and alloantigen-induced T cell proliferation, B cell differentiation, immunoglobulin secretion and modulate cytokine production (Kähne *et al* 1999).

Evidence exists that the human immunodeficiency virus-1 (HIV-1) transactivator Tat occurs extracellularly and exerts immunosuppressive effects on non HIV-1-infected T cells (Rubartelli *et al* 1998). Native Tat protein inhibits the DP IV activity of the T cell activation marker CD26 with comparable potency as does Tat(1-9) (MDPVDPNIE) in a tenfold higher concentration (Wrenger *et al* 1996). Moreover, Tat(1-86) distinctly suppresses DNA synthesis of Tetanus toxoid- as well as pokeweed mitogen-stimulated peripheral blood mononuclear cells (PBMC) and pokeweed mitogen-stimulated T cells in the same concentration range as the synthetic, highly specific DP IV inhibitor Lys[Z($NO_2$)]-thiazolidide ($IC_{50} = 2.7 \pm 0.3\ \mu M$). Tat(1-9) exerts a comparable effect on DNA synthesis in a 20-fold higher concentration (Wrenger *et al* 1997). In this study, we investigated the influence of amino acid exchanges in the first three positions of the Tat(1-9) sequence on inhibition of DP IV activity and DNA synthesis of Tetanus toxoid-stimulated PBMC.

# 2. MATERIALS AND METHODS

## 2.1 DP IV and peptides

Human kidney DP IV was kindly provided by Dr. Thilo Kähne. IL-2(1-12) substrate peptide and Tat(1-9) peptides were synthesized by solid-phase peptide synthesis with Fmoc technique (Wrenger *et al* 1996). The synthetic peptides were purified by reversed-phase HPLC and analysed by mass spectrometry.

## 2.2 DP IV-catalyzed hydrolysis of IL-2(1-12)

After preincubation (30 min, 37 °C) of 550 fkat DP IV with the Tat(1-9) peptides the enzymatic reaction was started by addition of IL-2(1-12) substrate. Samples were incubated for 30 min at 37 °C. Thereafter, the reaction was stopped by addition of 30 mM phosphoric acid. Degradation of IL-2(1-12) was measured by capillary electrophoresis using Biofocus 3000 system (Reinhold *et al* 1996).

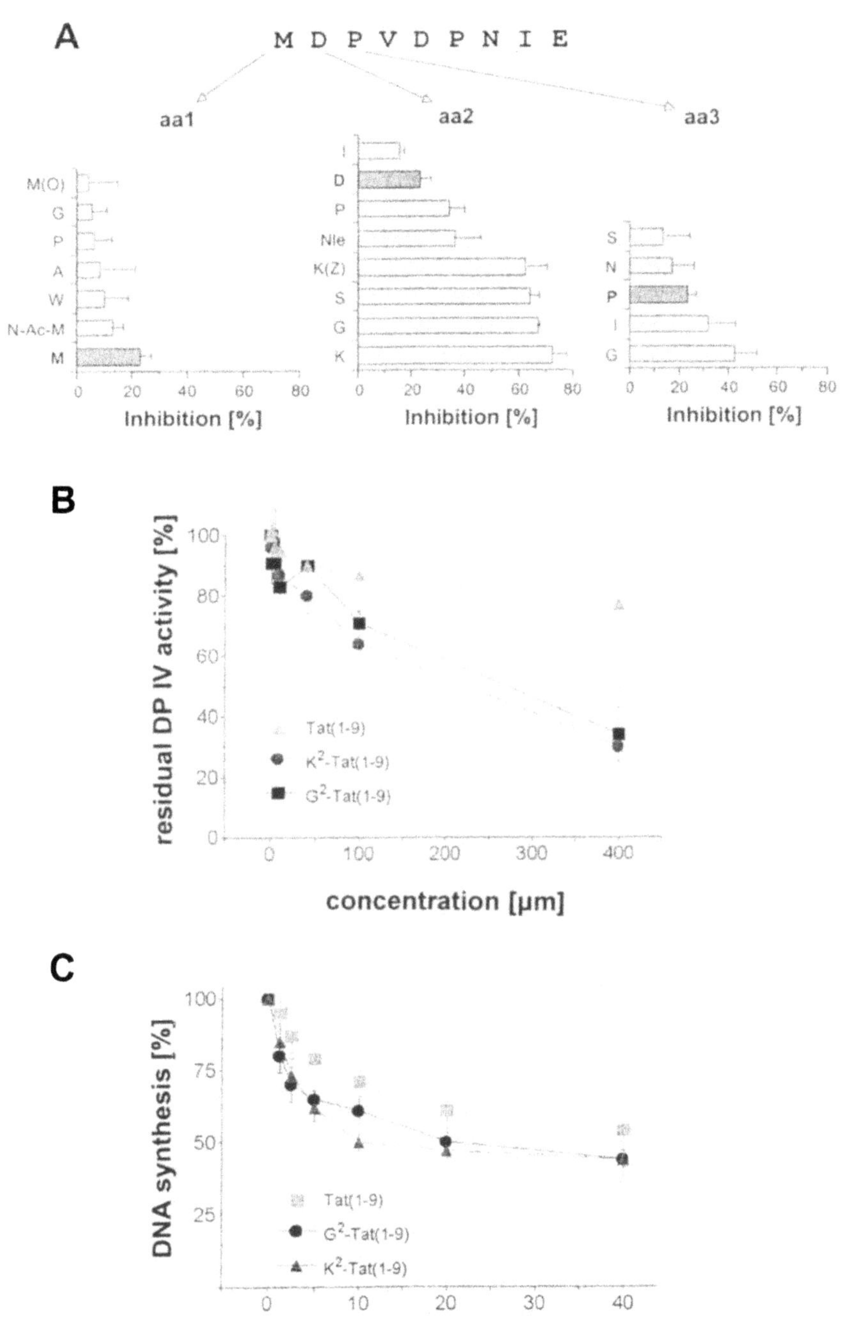


*Fig. 1.* Inhibition of DP IV-catalyzed IL-2 (1-12) degradation (A, B) and suppression of DNA synthesis in Tetanus-toxoid-stimulated PBMC (c) (K(Z) = $N^{\varepsilon}$-benzyloxycarbonyl-L-lysin, Ac-M = $N^{\alpha}$-acetyl-L-methionine).

## 2.3 Preparation of PBMC and proliferation assay

PBMC were prepared from heparinized blood of healthy donors and stimulated in serum-free CG medium with Tetanus toxoid in the presence of effectors in the concentrations indicated. After 6 days, cultures were pulsed for an additional 16 h with $^{3}$H-methyl-thymidine. Cells were harvested onto glass fibre filters and the incorporated radioactivity was measured by scintillation counting (Wrenger *et al* 1997).

# 3. RESULTS

N-Acetylation of the N-terminal amino group or oxidation of the N-terminal L-methionine strongly reduce inhibitory activity of the nonapeptide (Fig 1). None of the exchanges of the first amino acid performed generated peptides, which inhibited IL-2(1-12) cleavage by human DP IV stronger than Tat(1-9). However, L-Aspartic acid at the second position is not optimal for DP IV inhibition. Interestingly, the exchange of $D^2$ by a series of hydrophobic as well as hydrophilic amino acids leads to analogues, which are capable of suppressing the DP IV enzymatic activity up to threefold higher. $K^2$-, $G^2$- and $S^2$-Tat(1-9) inhibited nearly 70 % compared to 23 % for Tat(1-9) (Fig 1). Both peptides, $G^2$-Tat(1-9) and $K^2$-Tat(1-9) also suppress DNA synthesis of Tetanus toxoid-stimulated PBMC stronger than Tat(1-9) (Fig 1). The exchange of L-Proline at amino acid position 3 in Tat(1-9) (Fig 1) has only minor effects on the DP IV-catalysed IL-2(1-12) degradation.

# 4. CONCLUSION

Using synthetic, highly specific inhibitors for DP IV, we and other groups demonstrated that DP IV plays an important role in the activation and proliferation of lymphocytes. Evidence exists that the mechanisms for these effects are much more complex than first suggested. Recently, DP IV has been shown to process a number of chemokines including RANTES (regulating on activation normal T cell expressed and secreted) and SDF-1 (stromal cell-derived factor-1) generating naturally occuring truncated peptides with a significantly altered receptor specificity and thus biological activity (De Meester *et al* 1999). But the peptidase activity is probably not the only important feature of DP IV/CD26. Experiments with synthetic DP IV inhibitors clearly demonstrated that DP IV actively contributes as an accessory protein to the signalling of the T cell receptor/CD3 complex (Kähne *et al* 1999).

In former work, we have shown that the short nonapeptide Tat(1-9) of the HIV-1 transactivator Tat behaves as a competitive inhibitor and interacts directly with the active site of DP IV/CD26 (Wrenger *et al* 1997). Interestingly, a number of other natural peptides containing the N-terminal Xaa-Xaa-Pro motif, e. g. xenopsin-related peptide II, gastrin-releasing peptide(14-27), peptide YY(3-36), also inhibited DP IV activity (Wrenger *et al* 1996).

Here, we show that oxidation or N-acetylation of the N-terminal Met strongly reduces inhibition of DP IV activity. From the amino acid exchanges done, the following led to the Tat(1-9) analogues with the highest capability of inhibiting DP IV enzymatic activity: M at position 1, K, G, S at position 2 and G, I, P at position 3. Interestingly, the two peptides $G^2$-Tat(1-9) and $K^2$-Tat(1-9), which inhibited the DP IV activity nearly threefold stronger than Tat(1-9), also suppressed DNA synthesis of Tetanus-toxoid-stimulated PBMC stronger than Tat(1-9). This correlation suggests that the antiproliferative effects of these nonapeptides could be mediated at least in part by interaction and subsequent inhibition of DP IV.

In current investigations Tat(1-9) analogues with further amino acid exchanges are used to construct a consensus sequence for putative endogenous peptidergic DP IV inhibitors. Such effectors could play an important role in the regulation of lymphocyte proliferation by binding to DP IV and influencing its activity.

## ACKNOWLEDGMENTS

We thank T. Kähne for kindly providing us human DP IV and K. Mnich and A. Giese for their excellent technical assistance. This work was supported by the Deutsche Forschungsgemeinschaft (DFG), SFB 387.

## REFERENCES

De Meester, *et al.*, 1999, CD26, let it cut or cut it down. *Immunol. Today* **20,**:367-375.

Kähne, T., *et al.*, 1999, Dipeptidyl peptidase IV: A cell surface peptidase involved in regulating T cell growth. *Int. J. Mol. Med.* **4:** 3-15.

Reinhold, D., *et al.*, 1996, CD26 mediates the action of HIV-1 Tat protein on DNA synthesis and cytokine production in U937 cells. *Immunobiology* **195:** 119-128.

Rubartelli, A., *et al.*, 1998, HIV-I Tat: a polypeptide for all seasons. *Immunol. Today* **19:** 543-545.

Wrenger, S., *et al.*, 1996, The N-terminal X-X-Pro sequence of the HIV-1 Tat protein is important for the inhibition of dipeptidyl peptidase IV (DP IV/CD26) and the suppression of mitogen-induced proliferation of human T cells. *FEBS Lett.* **383:** 145-149.

Wrenger, S., *et al.*, 1997, The N-terminal structure of HIV-1 Tat is required for suppression of CD26-dependent T cell growth. *J. Biol. Chem.* **272:** 30283-30288.

# DNA SYNTHESIS IN CULTURED HUMAN KERATINOCYTES AND HACAT KERATINOCYTES IS REDUCED BY SPECIFIC INHIBITION OF DIPEPTIDYL PEPTIDASE IV (CD26) ENZYMATIC ACTIVITY

Robert Vetter[1], Dirk Reinhold[2], Frank Bühling[2], Uwe Lendeckel[2], Ilona Born[3], Jürgen Faust[3], Klaus Neubert[3], Siegfried Ansorge[2] and Harald Gollnick[1]

*[1]Dept. of Dermatology and Venereology, [2]Institute of Experimental Internal Medicine, Otto-von-Guericke University, D-39120 Germany, [3]Institute of Biochemistry, Dept. of Biochemistry and Biotechnology, Martin-Luther-University, Halle-Wittenberg, D-06120 Halle, Germany*

Key words CD26, keratinocytes, DP IV inhibitors, proliferation

Abstract The ectopeptidase dipeptidyl peptidase IV (DP IV, CD26, EC 3.4.14.5) is present on most mammalian cells. Using specific inhibitors of DP IV, it has been shown that this enzyme is involved in the regulation of DNA synthesis and in production of various cytokines in lymphocytes. The aim of the present work was to investigate the expression of DP IV/CD26 on human keratinocytes and to answer the question, whether the proliferation (DNA synthesis) of human keratinocytes is influenced by inhibition of the enzymatic activity of DP IV. Using flow cytometry, RT-PCR, and specific enzymatic activity assays, expression of DP IV-mRNA and CD26 antigen were shown on primary keratinocyte strains and on the HaCaT keratinocyte cell line. The synthetic DP IV inhibitors Lys[Z($NO_2$)]-thiazolidide and -pyrrolidide suppress the DNA synthesis of these cells in a dose-dependent manner. These data demonstrate that CD26 is also involved in the regulation of DNA synthesis of keratinocytes and that the enzymatic activity is required for mediating these effects.

## 1. INTRODUCTION

The dipeptidyl peptidase IV is a transmembrane type II glycoprotein that is present on most mammalian cells. Using specific inhibitors of DP IV, it has been shown that DP IV is involved in the regulation of DNA synthesis and in production of various cytokines in T lymphocytes (Reinhold *et al* 1997a, Kähne *et al* 1999).

On fibroblasts and hepatocytes, DP IV can serve as an adhesion molecule (Bauvois *et al* 1988). In animal experiments, DP IV inhibitors were capable of enhancing granulation tissue formation (Kohl *et al* 1991). Using immunohistochemical techniques, Moehrle et al. described a considerable DP IV expression in human basal cell carcinoma and Novelli et al. found a significant expression of DP IV on keratinocytes in various benign and malignant inflammatory and lymphocyte associated skin diseases in conjunction with an intraepidermal T cell infiltrate. In normal skin, no expression of CD26 on keratinocytes could be detected (Moehrle *et al* 1995, Novelli *et al* 1996).

The aim of the present study was to investigate how and to what extent the DNA synthesis of human keratinocytes and HaCaT cells is influenced by inhibition of the DP IV enzymatic activity of these cells *in vitro*.

## 2. MATERIALS AND METHODS

### 2.1 Cells

Commercially available human foreskin keratinocytes (NHEK, Clonetics) from adult and neonatal donors, primary human split-skin keratinocytes (RUM, ROL) from patients undergoing routine skin transplantation as well as the HaCaT keratinocyte cell line (generous gift of Prof. N.E. Fusenig, German Cancer Research Center, Heidelberg) were cultured in serum free Keratinocyte Growth Medium (KGM, Promocell).

### 2.2 RT-PCR, enzymatic assays, flow cytometry, proliferation assay

For RT-PCR, total RNA was isolated, reverse transcribed and PCR amplified using primers specific for DP IV. The enzymatic activity of DP IV was studied using Gly-Pro-4-nitroanilide as a substrate (Schön *et al* 1984). Indirect immunofluorescence staining of keratinocytes was performed using the monoclonal anti-CD26 antibody EF5/A3 (Reinhold *et al* 1997b). Labelled cells were analyzed by flow cytometry. For the proliferation assay, keratinocytes were incubated in KGM in the presence of different concentrations of the DP IV inhibitors Lys[Z($NO_2$)]-thiazolidide and Lys[Z($NO_2$)]-pyrrolidide. After 56 hours the cultures were pulsed with $^3$H-methyl-thymidine. Cells were harvested onto glass fibre filters and the incorporated radioactivity was measured by scintillation counting.

## 3. RESULTS

High DP IV activity was detected on viable HaCaT cells and normal keratinocyte cell preparations. These data were supported by RT-PCR and flow cytometric studies. HaCaT cells and all keratinocyte preparations showed considerable expression of DP IV-mRNA and of CD26 antigen on the cell surface.

*Table 1.* Expression of DP IV-mRNA, CD26 antigen, and enzymatic DP IV activity on human keratinocytes and HaCaT cells

| | RT-PCR (DP IV-mRNA) | CD26 antigen (%) | DP IV activity (pkat/$10^6$ cells) |
|---|---|---|---|
| NHEK Ad | + | 36 ± 6 | 61 ± 13 |
| NHEK neo | + | 25 ± 4 | 48 ± 10 |
| RUM | + | 34 ± 3 | 39 ± 5 |
| ROL | + | 33 ± 3 | 84 ± 7 |
| HaCaT | + | 21 ± 4 | 30 ± 5 |

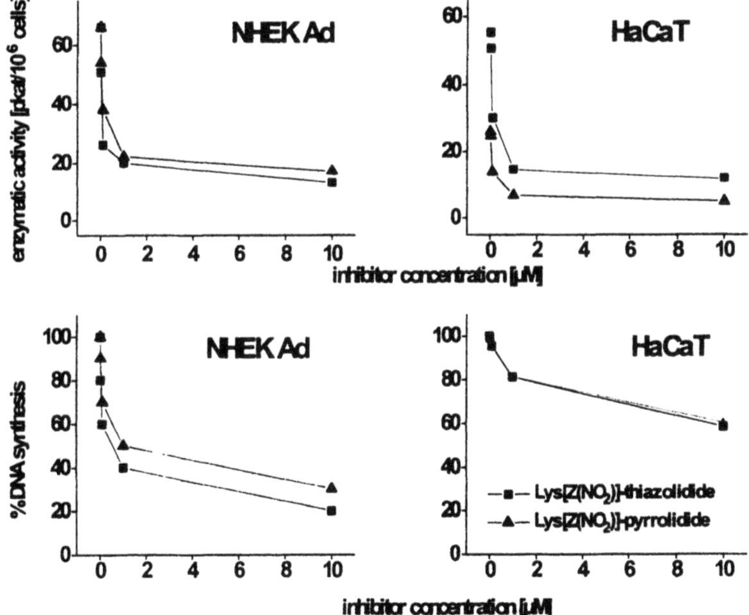


*Figure 1.* Influence of DP IV inhibitors on DP IV activity [pkat/$10^6$ cells] and DNA synthesis of human NHEK Ad keratinocytes and HaCaT cells

The DP IV inhibitors suppressed the enzymatic activity of these cell populations in a dose-dependent fashion. Starting from a concentration of 0.01 μM a visible reduction in enzymatic activity occurred. A concentration of 10 μM Lys[Z($NO_2$)]-thiazolidide the DP IV specific enzymatic activity of these keratinocytes was found to be significantly suppressed. The standard deviation of triplicate experiments did not exceed 15 % (Fig 1).

Moreover, the DNA synthesis of HaCaT cells and human keratinocytes was strongly suppressed by the DP IV inhibitors Lys[Z($NO_2$)]-thiazolidide and Lys[Z($NO_2$)]-pyrrolidide. We found a 10 – 30 % inhibition of normal keratinocyte (NHEK Ad) DNA synthesis at an inhibitor concentration of 0.5 μM and a 70 - 90 % inhibition at 10 μM. On HaCaT cells, a maximum suppression of DNA synthesis (45 %) were measured with the DP IV inhibitors at a concentration of 10 μM. The standard deviation of triplicate experiments did not exceed 18 % (Fig 1). Cell viability was not influenced by the inhibitors (data not shown).

## 4. CONCLUSION

Here, it was shown for the first time that the DP IV/CD26 is expressed not only on cultured human foreskin keratinocytes, neonatal keratinocytes, split-skin keratinocytes, as recently demonstrated (Reinhold *et al* 1998), but also on HaCaT keratinocytes.

Under basal cell culture conditions, the enzymatic activity of human keratinocytes and HaCaT cells was found to be much higher (3 to 6 fold) than on resting peripheral blood immune cells and close to mitogen-stimulated peripheral blood mononuclear cells (Kähne *et al* 1999). Thus, in our experiments considerable expression of CD26 and functional activity could be detected in absence of potential sources of cytokines like T cells. As cultured keratinocytes are in a highly proliferative and activated state resp., partially due to special media ingredients, i.e. low $Ca^{2+}$-concentration (0.15 mM), our observations correspond well to Novelli's, who concluded observations correspond well to Novelli's, who concluded that the CD26 molecule serves as a keratinocyte activation antigen (Novelli *et al* 1996).

The synthetic DP IV inhibitors Lys[Z($NO_2$)]-thiazolidide and -pyrrolidide directly suppress DNA synthesis and specific DP IV enzymatic activity of normal human keratinocytes (Reinhold *et al* 1998) and HaCaT keratinocytes in a dose-dependent manner. The less effective reduction of DNA synthesis of HaCaT cells occurs probably due to the accelerated proliferation cycle of these keratinocytes.

In summary, our findings show that CD26 plays an important role in the regulation of DNA synthesis of normal human keratinocytes as well as in keratinocytes of the HaCaT keratinocyte cell line, and that the specific enzymatic activity is required for mediating these effects. Thus, DP IV/CD26 could play a role in hyperproliferative skin diseases. Further investigations will clarify, whether the inhibition of the enzymatic activity of DP IV could be a useful tool for therapeutic interventions in these diseases.

# ACKNOWLEDGMENTS

We thank K. Mnich for their excellent technical assistance. This work was supported by the Deutsche Forschungsgemeinschaft (DFG), SFB 387, grants A4 and A5, and by the Kultusministerium Sachsen-Anhalt, grant 1816A/0084.

# REFERENCES

Bauvois, B., 1988, A collagen-binding glycoprotein on the surface of mouse fibroblasts is identified as dipeptidyl peptidase IV. *Biochem. J.* **252:** 723-731.

Boukamp, P., *et al.*, 1988, Normal keratinization in a spontaneously immortalized aneuploid human keratinocyte cell line. *J. Cell. Biol.* **106:** 761-771.

Kähne, T., *et al.*, 1999, Dipeptidyl peptidase IV: A cell surface peptidase involved in regulating T cell growth. *Int. J. Mol. Med.* **4:** 3-15.

Kohl, A., *et al.*, 1991, The role of dipeptidylpeptidase IV positive T cells in wound healing and angiogenesis. *Agents Actions* **32:** 125-127.

Moehrle, M.C., *et al.*, 1995, Aminopeptidase M and dipeptidyl peptidase IV activity in epithelial skin tumors: a histochemical study. *J. Cutan. Pathol.* **22:** 241-247.

Novelli, M., *et al.*, 1996, Keratinocytes express dipeptidyl-peptidase IV (CD26) in benign and malignant skin diseases. *Br. J. Dermatol.* **134:** 1052-1056.

Reinhold D., *et al.*, 1997a, Inhibitors of dipeptidyl peptidase IV induce secretion of transforming growth factor-beta 1 in PWM-stimulated PBMC and T cells. *Immunology* **91:** 354-60.

Reinhold, D., *et al.*, 1997b, The effect of anti-CD26 antibodies on DNA synthesis and cytokine production (IL-2, IL-10 and IFN-gamma) depends on enzymatic activity of DP IV CD26. *Adv. Exp. Med. Biol.* **421:** 149-155.

Reinhold, D., *et al.*, 1998, Dipeptidyl peptidase IV (DP IV, CD26) is involved in regulation of DNA synthesis in human keratinocytes. *FEBS Lett.* **428:** 100-104.

Schön E., *et al.*, 1984, Dipeptidyl peptidase IV of human lymphocytes. Evidence for specific hydrolysis of glycylprolin in T-lymphocytes. *Biochem. J.* **223:** 255-258.

# ATTRACTIN: A CUB-FAMILY PROTEASE INVOLVED IN T CELL-MONOCYTE/MACROPHAGE INTERACTIONS.

Jonathan S. Duke-Cohan, Wen Tang and Stuart F. Schlossman
*Tumor Immunology, Department of Cancer Immunology and AIDS, Dana-Farber Cancer Institute and Department of Medicine, Harvard Medical School, Boston, MA, U.S.A.*

Key words Attractin, CD26, Dipeptidyl peptidase IV, mahogany, T cells, chemokines, cytokines

Abstract Attractin is a rapidly upregulated membrane-associated molecule on activated T cells. It is a member of the CUB family of extracellular guidance and development proteins, sharing with them a protease activity similar to that of Dipeptidyl peptidase IV (DPPIV/CD26). Most remarkably, and in sharp contrast to CD26, it is released from the T cell and is presumed to be a major source of a soluble serum-circulating attractin. Genomic sequencing reveals that the soluble form is not a proteolytic product of the membrane form, but is in fact the result of alternative splicing. Recent results prove that the loss of murine membrane attractin results in the *mahogany* mutation with severe repercussions upon skin pigmentation and control of energy metabolism. In each of these latter instances, there is a strong likelihood that attractin is moderating the interaction of cytokines with their respective receptors. We propose that attractin is performing a similar function in the immune system through capture and proteolytic modification of the N-terminals of several cytokines and chemokines. This regulatory activity allows cells to interact and form immunoregulatory clusters and subsequently aids in downregulating chemokine/cytokine activity once a response has been initiated. These two properties are likely to be affected by the balance of membrane-expressed to soluble attractin.

## 1. INTRODUCTION

The continuous recirculation of T lymphocytes is a normal ongoing process consisting of the specific interaction of phenotypically distinct populations of T cells with the vascular endothelium leading to their extravasation and re-entry into the circulatory system. This process is markedly enhanced during antigenic challenge and provides for the rapid mobilisation of specialised T cells following the less specific response of the innate arm of the immune system. The extravasation of these lymphoid cells is primarily mediated by chemokines present on the endothelial surface (Lider *et al* 1995). These chemokines are released from damaged tissue (Moulin 1995), from monocytes (Cavaillon 1994) and granulocytes (Cassatella *et al* 1997) that participated in the primary non-specific rapid response and from T cells that have already arrived on the scene. After the initial selectin-mediated rolling on the endothelium followed by anchoring through integrins, the T cell then faces the quandary of migrating between the endothelial cells through the extracellular matrix (ECM) to the site of inflammation (Lider *et al* 1995). To aid in this process, there is an upregulation of the expression of a number of proteases by activated T cells including Dipeptidyl peptidase IV (CD26) (Masuyama *et al* 1992), gelatinases (Pender *et al* 1997) and heparanases (Fridman 1987).

The case of CD26 is intriguing: it is most certainly a marker of extravasating activated T cells (Masuyama *et al* 1992) but it is difficult to envisage how it would be involved in migration through the extracellular matrix. Recent clues have been provided by reports that CD26 is able to cleave the chemokine RANTES thus regulating its role in attracting cells to inflammatory sites (Oravecz *et al* 1997, Proost *et al* 1998, Iwata *et al* 1999a). Other chemoattractant proteins such as monocyte chemoattractant protein-2 (MCP-2) (Van Coillie *et al* 1998) and macrophage-derived chemokine (MDC) (Proost *et al.* 1999) both express a N-terminal penultimate proline and are potential substrates for CD26. Further supporting the notion that the proteolytic activity of CD26 is indirect and may be upon surrounding peptides rather than ECM degradation are the observations that aggressive melanoma cells lose CD26 and become less aggressive upon expression of transfected CD26 while $CD26^+$-T cell lymphomas are noticeably more aggressive (Iwata *et al* 1999b). The matter is further complicated by the observation that re-expression of CD26 in the melanoma cell line results in re-expression of the serine protease Fibroblast-Activation Protein-α (FAP-α), apparently chaperoned by formation of a heterodimer with CD26 (Wesley *et al* 1999). Since FAP-α is likely a product of the same gene as the gelatinase Seprase (initially identified in melanoma, Monsky *et al* 1994, Goldstein *et al* 1997) it is equally likely that some of the effects of CD26 are consequential to the surface activities of FAP-α and Seprase. The functional contribution of the DPPIV activity of CD26 to cell migration is thus still questionable (Iwata *et al* 1999b) and despite the presence of a putative collagen-interacting domain (Loster *et al* 1995), there is little else about CD26 that would identify it as being involved in cell adhesion or guidance.

## 2. ATTRACTIN IS A DPPIV BUT IS NOT CD26

Several years ago, we found that a significant proportion of the DPPIV activity of normal human serum could be accounted for by a protein of approximately 175 kDa - considerably larger than CD26 (Duke-Cohan *et al* 1995). In some preparations, a small amount of 105 - 115 kDa material could be identified of which amounts correlated with that reported by others (Schrader *et al* 1979, Shibuya-Saruta *et al* 1996). There was always the possibility that the large molecule was a carrier for CD26: initial evidence supported this proposal since polyclonal antibody prepared against the large molecule (initially termed DPPT-L but now named attractin) showed a strong activity against CD26. After removal of the CD26 activity using recombinant CD26 affinity columns, immunoprecipitation with the residual antibody could deplete almost completely the DPPIV activity of the serum attractin without having any effect whatsoever on the enzyme activity of recombinant CD26, and the little activity which remained could be completely removed with anti-CD26 antibodies. This suggested that there were similar epitopes shared by CD26 and attractin (Duke-Cohan *et al* 1995). Furthermore, attractin could be labelled with $^3$H-DFP with no band labelling at 105 - 115 kDa implying an active serine protease site independent of CD26. The specific activity of attractin was 10 – 50 % that of CD26, substrate specificity appeared similar but inhibitor specificity was not identical (Duke-Cohan *et al* 1996). In particular, serum attractin appeared more sensitive to EDTA suggesting a calcium dependency. Sequencing of tryptic N-terminal peptides revealed similarity with other serine proteinases but not with CD26. The issue was resolved by the cloning of attractin and expression of the recombinant protein (Duke-Cohan *et al* 1998). Attractin shares no sequence similarity with CD26, has a chymotrypsin-like catalytic motif and is able to hydrolyse artificial DPPIV substrates. Further below other structural motifs of attractin will be presented which characterize attractin as an extracellular glycoprotein with lectin and cytokine-binding potential.

## 3. ATTRACTIN IS A T CELL ACTIVATION ANTIGEN

The availability of the anti-attractin antibody without activity against CD26 provided us with a means to identify the cellular source of serum attractin. Attractin was expressed on many cell types with a distribution remarkably similar to that of CD26 with particularly high expression on kidney, pancreatic and hepatic cell lines. Unlike CD26, attractin is barely expressed on resting T cells, but within 24 - 48 hr of activation through CD3/TcR, attractin is expressed at high levels (Duke-Cohan *et al* 1996). Activation with PMA/ionomycin does not induce expression. Early levels of attractin expression are much higher than those of CD26 and unlike CD26, which maintains an increase in expression, the levels of attractin start to fall about 3 days after activation. To examine whether this was shedding, proteolytic release or internalisation, we purified attractin from the culture supernatants and showed it to be similar to soluble serum attractin with regard to size, enzyme activity and glycosylation. Examination of resting and activated T cells supported the proposal that attractin was localised to large vesicles containing an electron dense core that upon activation moved to the plasma membrane and released their contents (Duke-Cohan *et al* 1998). By analysis of DPPIV activity and attractin after subcellular fractionation, it appears that attractin is preformed in resting T cells and activation induces mobilisation and release – this would account for the rapid expression upon activation.

## 4. STRUCTURAL CHARACTERISATION OF ATTRACTIN

Attractin mRNA is found as two molecular weight species of 4.5 kb and 8 - 9 kb (Duke-Cohan *et al* 1998). Both bands were expressed in diverse tissues but hematopoietic expression seemed to be tissue specific with the higher band being preferentially expressed in thymus and the lower band in peripheral blood lymphocytes while both species were expressed in mixed tissues such as fetal liver or spleen. Both species would be large enough to code for the predicted open reading frame of 1198 amino acids and given the short 3' UTR, it seemed likely that the smaller form was coding for attractin in lymphocytes. We have identified a murine form of attractin which exhibits a C-terminal transmembrane domain, showing greater than 93 % identity at the amino acid level (Gunn *et al* 1999). Positional cloning of the murine form allowed us to use the murine genomic sequence to identify corresponding human sequence. It was immediately apparent that the soluble human attractin was the result of an alternatively-spliced exon (with a termination codon and weak polyadenylation site) upstream of further exons coding for a transmembrane domain and cytoplasmic tail. We cannot yet state for certain that the early membrane expression of attractin on activated T cells followed by release of a soluble form is the result of a switch in transcription of the membrane to soluble form but if this proves to be the case, the only other recorded instance is in B lymphocytes which switch between polyadenylation sites to generate soluble immunoglobulin rather than the membrane form upon differentiation into plasma cells (Takagaki *et al* 1996).

There appears to be a 5' alternatively-spliced version (attractin-2) which has a 74 amino acid insertion, similar to the murine membrane form (Genbank AF106861). The importance of this insertion seems to be its use as a glycosylation targeting motif and it may contain a cleavage site (supported by N-terminal sequencing). Cleavage at this site is likely to disrupt enzymic activity and may account for the wide variation we see in DPPIV activity from preparation to preparation. Subsequent to identification of the murine attractin, it became apparent that several deposited cDNA sequences matched up exactly with the putative membrane form of attractin. Furthermore, probes based on the 3' region of murine membrane attractin were able to hybridise to the 8 - 9 kb human mRNA but not to the 4.5 kB mRNA proving conclusively that the high molecular weight mRNA coded for the membrane form of attractin. Mapping of attractin places it firmly in the region of 20p13 with approximately 30 exons spanning a 1 Mb region. No pathologies except for the neurodegenerative disorder Hallervorden-Spatz Syndrome map to this region (Taylor *et al* 1996). The sequence of attractin is not only highly conserved between mouse and man but extends all the way down to the nematode that expresses a similar protein (F33C8.1).

The coding sequence directs synthesis of a protein containing several EGF motifs, a CUB domain, a $\gamma/\beta$ common cytokine receptor binding motif, a full C-type lectin (Selectin) domain and two laminin-type EGF domains in addition to the N-terminal putative serine protease catalytic site (Fig 1). Attractin is heavily N-glycosylated with no apparent O-glycosylation and minimal sialylation.

The EGF domains are regions heavily constrained by 3 disulphide bonds and are characteristic not only of extracellular proteins but also of proteins involved in cellular interactions – not a single EGF domain is expressed by any protein in unicellular organisms such as yeast (Chervitz *et al* 1998, Davis 1990). The CUB domain (termed after the three prototypic members C1r/s, U-EGF and BMP-1) (Bork *et al* 1993) is of considerable interest. Members of the CUB family usually also express EGF domains and have a serine protease or metalloproteinase catalytic activity. The $\gamma/\beta$ cytokine receptor signature of attractin is not functional alone but may well be functional in the context of the complementary ligand binding chain. It has already been demonstrated that interaction of the $\gamma/\beta$ chain requires a primary interaction with companion component of the cytokine receptor heterodimer for ligand binding to occur (Hage *et al* 1998).

***Genomic organisation of 3' exons of attractin***

*Soluble and membrane attractin arise from alternate splicing of the same gene*

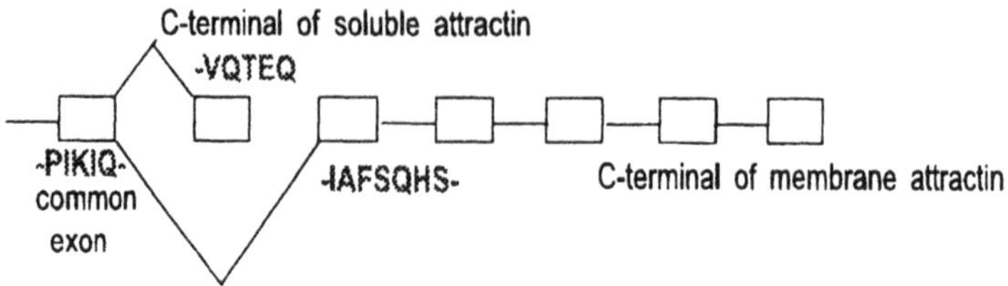


***Domain structure of soluble and membrane attractin***

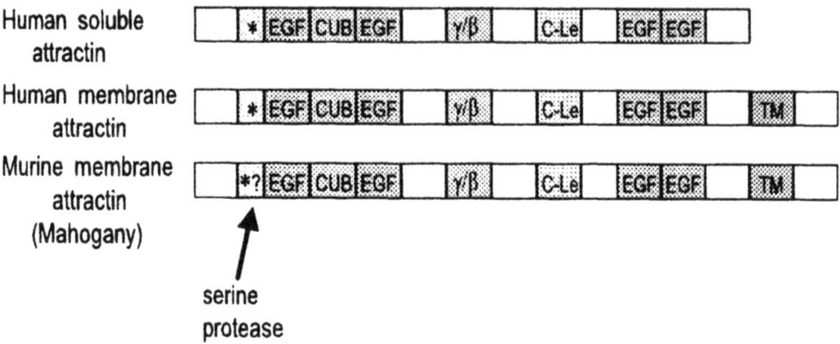


*Figure 1.* Genomic and protein organisation of attractin

We are in the process of determining the sugar specificity of the C-type lectin domain but two likely possibilities exist for its function – first as a means of membrane attractin mediating immune cell interactions and secondly, in its soluble form, as a means of initiating complement lysis after binding to repeat sugar arrays on microbial surfaces. In this latter respect, attractin shares a high degree of structural similarity with serum MASP-1, a Mannose-binding lectin-Associated Serine Protease-1) (Sato *et al* 1994) that is also related to C1r, expresses a CUB domain, a C-type lectin domain and interacts with the collectin Mannose-binding lectin that can interact with collectin receptors on the surface of phagocytes. Through its serine proteinase activity, MASP can also activate C4 and C2 of the classical complement pathway. This suggests an involvement of MASP-1 with primary exposure to pathogen and a similar role seems likely for attractin. The importance of attractin in this respect is enhanced further by its high circulating serum concentration, 10 – 30 μg/ml, comparable to that of the C1 components of complement and considerably higher than that of MASP at about 1-2 μg/ml. Despite one report that suggests CD26 as a proteolytic product of membrane CD26 is circulating at around 4-5 μg/ml (Iwaki-Egawa *et al* 1998), our results suggest a circulating concentration of 1 μg/ml or considerably less.

# 5. FUNCTIONAL ACTIVITY OF ATTRACTIN

In our early studies on the functional activity of purified attractin, we compared costimulatory activity with that of soluble CD26. Crosslinking of CD26 in the context of anti-CD3 stimulation had already been shown to enhance proliferative responses (Dang *et al* 1990). Soluble recombinant CD26 (rsCD26) also seemed to be costimulatory and this could have been mediated either by modification of signal transduction (Tanaka *et al* 1994), regulation of extracellular adenosine through the CD26-adenosine deaminase interaction (Dong *et al* 1997) or DPPIV processing of bioactive peptides. Attractin seemed as potent if not stronger than rsCD26 at mediating costimulation to recall antigen-driven proliferation (Duke-Cohan *et al* 1995, Duke-Cohan *et al* 1996). Neither reagent could induce any proliferative activity alone suggesting that the costimulation truly was a facilitating effect upon the CD3/TcR rather than a concomitant and independent activation, as provided through CD28, for example. Of note, both CD26 and attractin inhibit recall antigen-driven proliferation at concentrations of 5 μg/ml or greater.

We had already shown that attractin does not interact with adenosine deaminase (Duke-Cohan *et al* 1995) and no domain or motif of attractin peptide sequence suggested any association with known signal transducing mechanisms (Nagle *et al* 1999). Accordingly, the lectin, cytokine and proteolytic domains remain of interest and in the context of the EGF domains suggested that cell contact may account in part for the costimulatory activity. We found that addition of attractin to purified peripheral blood T cells had no noticeable effect upon proliferation or gross morphology. Similarly, addition of attractin to monocytes isolated by adherence had no noticeable effect. Over a 48 hr period, mixing of T cells and monocytes in serum-free medium also produced no remarkable effects. In the presence of attractin, however, the appearance of mixed cultures of monocytes and T cells underwent significant alteration over a 48 hr period (Duke-Cohan *et al* 1998). The monocytes elongated, expressed higher levels of CD14, MHC Class II and significantly upregulated CD80 (B7-1) – all hallmarks of preparing for antigen presentation. The smaller $CD3^+$ T cells moved towards the elongated adherent monocytes and clustered on top of them. The processes of the adherent monocytes continued to elongate forming a reticular network. It must be stressed that this network is not due to fibroblast contamination of either the T cells or the monocytes –

incubation of either population alone with attractin induces no effect. Furthermore, no proliferation occurs as a consequence of this interaction. We conclude that attractin is aiding in the formation of immunoregulatory clusters either by directly mediating an interaction between monocytes and T cells, or else by inducing or regulating the levels of chemotactic cytokines that induce T cells to move towards the adherent cells. It is also important to note that rsCD26 does not induce the clustering upon addition to similar cultures suggesting that the DPPIV activity is either unnecessary or else that it needs to be in the context of attractin rather than that of CD26.

Is there any evidence for this proposal? The reported cleavage of the chemokine RANTES by CD26 adds an important substrate to the roster of natural substrates available for hydrolysis by DPPIV (Oravecz *et al* 1997, Proost *et al* 1998, Iwata *et al* 1999a). We have found that attractin is also able to hydrolyse RANTES. Using Matrix-Assisted Laser Desorption-Ionisation-Time-of-Flight (MALDI-TOF) analysis, we find that attractin cleanly cleaves the N-terminal Ser-Pro of mature RANTES removing 184 daltons corresponding to the average monoisotopic mass of 87.07 and 97.12 for serine and proline, respectively (Fig 2).

Confirming the specificity of the digestion, attractin did not cleave a synthetic "cleaved" RANTES(3 - 68). Likewise, CD26 cleaved the Ser-Pro dipeptide from the N-terminal of RANTES. Surprisingly, however, once the RANTES was cleaved the CD26 was able to cleave a further 250 dalton dipeptide corresponding to the removal of Tyr-Ser at positions 3 and 4. Likewise, the same Tyr-Ser dipeptide was cleaved from the synthetic RANTES(3 - 68) by CD26. This raises questions as to the true specificity of CD26 and whether the potential to digest long natural bioactive peptides can be mimicked by the cleavage of short synthetic dipeptide-chromophore substrates which may not adapt to similar conformations. This has already been hinted at in previous studies that found N-terminal peptides of some cytokines could be hydrolysed by CD26 but that the full length cytokines were resistant to DPPIV activity (Ansorge *et al* 1995). More recent evidence suggesting that CD26 may remove sequential dipeptides from MDC (where the second peptide does not contain a proline) imply that CD26 may have a slightly more open specificity than previously thought (Proost *et al* 1999).

## 6. DISCUSSION

Prior to incorporating the results above into a more general picture of attractin function, a small digression is required to describe in detail the role of attractin in the mouse.

We found that attractin in its membrane isoform is the orthologue of the protein produced by the *mahogany* gene (Gunn *et al* 1999). The *Mahogany*

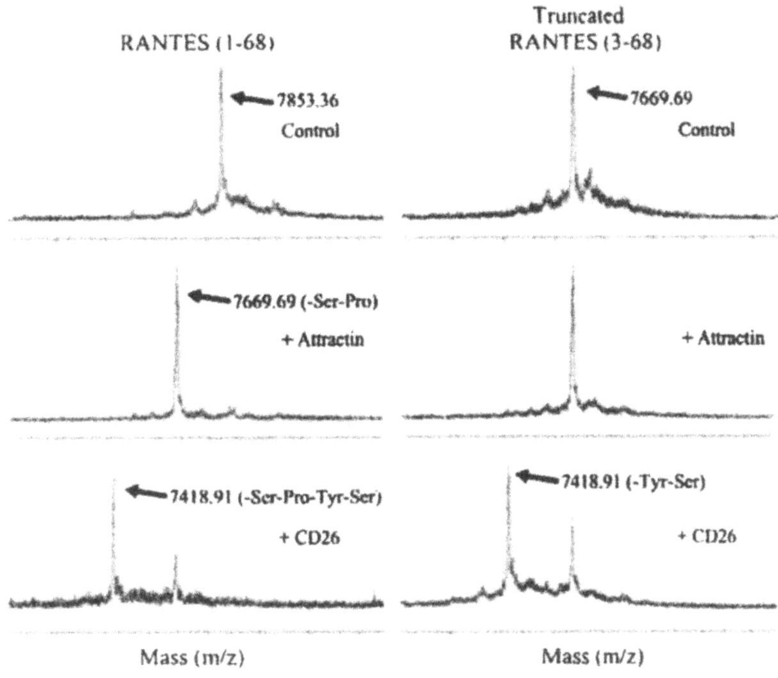


*Figure 2.* Digestion of RANTES by attractin and CD26/DPPIV

mouse arose as a natural mutation on an *Agouti* background (Silvers 1979). The subapical yellow band that characterises the hair of the *Agouti* mouse is absent in the *Mahogany* mouse. The cloning of the agouti protein and its biochemical characterisation showed that it is an antagonist of melanocyte-stimulating hormone (α-MSH) and upon binding to the melanocortin receptor (Mc1r) the production of yellow phaeomelanin rather than black eumelanin is induced. In the $mg^{3J}/mg^{3J}$ variant of the *Mahogany* mice, there is a complete absence of attractin mRNA despite normal levels of agouti protein. Genetic studies reveal that the action of attractin is downstream of the agouti protein but upstream of Mc1r (Miller *et al* 1997) from which one can conlude that attractin is either regulating agouti interaction with the Mc1r receptor or else modifying downstream signalling from the receptor. More recent studies suggest that although the C-terminal of agouti is required for its interaction, the amino terminal is required for maximal effect (Ollmann *et al* 1999) and perhaps, not surprisingly, the mature agouti is a potential candidate for hydrolysis by DPPIV.

A further abnormality in Mahogany mice consequent to the loss of functional attractin is the response to a high-fat diet (Nagle *et al* 1999). In contrast to normal littermates, the *Mahogany* mice do not gain weight. It is open to question whether this is due to a higher metabolic rate or aberrations in adipogenesis but the action appears to be related to the high expression of membrane attractin in the hypothalamus and control of energy metabolism at this point. Recent exciting results point to the Glucagon-like peptides 1 and 2 (GLP-1,2) being substrates for DPPIV activity and this may provide the link between energy metabolism and attractin (Mentlein *et al* 1993, Pauly *et al* 1996).

The role of attractin in pigmentation and energy metabolism in the mouse probably depends on the expression of membrane attractin since the 9kb form appears to be dominant in these tissues (Nagle *et al* 1999). The regulated expression of membrane to soluble attractin in T lymphocytes may reflect the mobility of these cells and their substrates, as opposed to the fixed tissues of the dermis and the hypothalamus. This, however, leads us to propose that attractin itself is not a particularly specific molecule. It's specificity is provided by the cells and cytokines in the environment in which it is found. A consequence of this then is not only that attractin plays a role in the initial interaction of T cells and monocytes but also leads to a reappraisal of the role of soluble circulating attractin. As described in the Introduction, chemokines are presented on the endothelial surface, acting as flags to attract passing activated T cells. Such a mechanism requires a clear gradient on the endothelium and further requires that chemokines are inactivated before being released into the circulation so as to preserve the local specific expression. Is it possible that circulating DPPIV activity, either as attractin or as the proteolytic fragment of membrane CD26, performs this inactivating role?

## ACKNOWLEDGMENTS

This work has benefited from collaboration with Drs Greg Barsh, Teresa Gunn and Lin He (Stanford University).

## REFERENCES

Ansorge, S., Buhling, F., Hoffman, T., Kahne, T., Neubert, K., Reinhold, D., (1995) DPIV/CD26 on human lymphocytes: functional roles in cell growth and cytokine regulation. In: Dipeptidyl Peptidase IV (CD26) in metabolism and the immune response, edited by Bernhard Fleischer. RG Landes, Austin, Texas.

Bork, P., and Beckmann, G.J., 1993, The CUB domain. A widespread module in developmentally regulated proteins. *J. Mol. Biol.* **231:** 539.

Cassatella, M.A., Gasperini, S., and Russo, M.P., 1997, Cytokine expression and release by neutrophils. *Ann. NY Acad. Sci.* **832:** 233.

Cavaillon, J.M., 1994, Cytokines and macrophages. *Biomed. Pharmacother.* **48:** 445.

Chervitz, S.A., Aravind, L., Sherlock, G., Ball, C.A., Koonin, E.V., Dwight, S.S., Harris, M.A., Dolinski, K., Mohr, S., Smith, T., Weng, S., Cherry, J.M., and Botstein, D., 1998, Comparison of the complete protein sets of worm and yeast: orthology and divergence. *Science* **282:** 2022.

Dang, N.H., Torimoto, Y., Deusch, K., Schlossman, S.F., and Morimoto, C., 1990, Comitogenic effect of solid phase immobilized anti-1F7 on human CD4 T cell activation via CD3 and CD2 pathways. *J. Immunol.* **144:** 4092.

Davis, C.G., 1990, The many faces of epidermal growth factor repeats. *New Biol.* **2:** 410.

Dong, R.P., Tachibana, K., Hegen, M., Munakata, Y., Cho, D., Schlossman, S.F., and Morimoto, C., 1997, Determination of adenosine deaminase binding domain on CD26 and its immunoregulatory effect on T cell activation. *J. Immunol.* **159:** 6070.

Duke-Cohan, J.S., Gu, J., McLaughlin, D.F., Xu, Y., Freeman, G.J., and Schlossman, S.F., 1998, Attractin (DPPT-L), a member of the CUB family of cell adhesion and guidance proteins, is secreted by activated human T lymphocytes and modulates immune cell interactions. *Proc. Natl. Acad. Sci. USA* **95:** 11336.

Duke-Cohan, J.S., Morimoto, C., Rocker, J.A., and Schlossman, S.F., 1995, A novel form of dipeptidyl peptidase IV found in human serum: isolation, characterization and comparison with T lymphocyte membrane dipeptidylpeptidase IV (CD26). *J. Biol. Chem.* **270:** 14107.

Duke-Cohan, J.S., Morimoto, C., Rocker, J.A., and Schlossman, S.F., 1996, Serum high molecular weight dipeptidyl peptidase IV (CD26) is similar to a novel antigen DPPT-L released from activated T cells. *J. Immunol.* **156:** 1714.

Fridman, R., Lider, O., Naparstek, Y., Fuks, Z., Vlodavsky, I., and Cohen, I.R., 1987, Soluble antigen induces T lymphocytes to secrete an endoglycosidase that degrades the heparan sulfate moiety of subendothelial extracellular matrix. *J. Cell Physiol.* **130:** 85.

Goldstein, L.A., Ghersi, G., Pineiro-Sanchez, M.L., Salamone, M., Yeh, Y., Flessate, D., and Chen, W.T., 1997, Molecular cloning of seprase: a serine integral membrane protease from human melanoma. *Biochim. Biophys. Acta.* **1361:** 11.

Gunn, T.M., Miller, K.A., He, L., Hyman, R.W., Davis, R.W., Azarani, A., Schlossman, S.F., Duke-Cohan, J.S., and Barsh, G.S., 1999, The mouse mahogany locus encodes a transmembrane form of human attractin. *Nature* **398:** 152.

Hage,T,, Reinemer, P., and Sebald, W., 1998, Crystals of a 1:1 complex between human interleukin-4 and the extracellular domain of its receptor alpha chain. *Eur. J. Biochem.* **258:** 831.

Iwaki-Egawa, S., Watanabe, Y., Kikuya, Y., and Fujimoto, Y., 1998, Dipeptidyl peptidase IV from human serum: purification, characterization, and N-terminal amino acid sequence. *J. Biochem.* **124:** 428.

Iwata, S., and Morimoto, C., 1999b, CD26/dipeptidyl peptidase IV in context. The different roles of a multifunctional ectoenzyme in malignant transformation. *J. Exp. Med.* **190:** 301.

Iwata, S., Yamaguchi, N., Munakata, Y., Ikushima, H., Lee, J.F., Hosono, O., Schlossman, S.F., and Morimoto, C., 1999a, CD26/dipeptidyl peptidase IV differentially regulates the chemotaxis of T cells and monocytes toward RANTES: possible mechanism for the switch from innate to acquired immune response. *Int. Immunol.* **11:** 417.

Lider, O., Hershkoviz, R., Kachalsky, S.G., 1995, Interactions of T lymphocytes, inflammatory mediators, and the extracellular matrix. *Crit. Rev. Immunol.* **15:** 271.

Loster, K., Zeilinger, K., Schuppan, D., and Reutter, W., 1995, The cysteine-rich region of dipeptidyl peptidase IV (CD 26) is the collagen-binding site. *Biochem. Biophys. Res. Commun.* **217:** 341.

Masuyama, J., Berman, J.S., Cruickshank, W.W., Morimoto, C., and Center, D.M., 1992,

Evidence for recent as well as long term activation of T cells migrating through endothelial cell monolayers in vitro. *J. Immunol.* **148:** 1367.

Mentlein, R., Gallwitz, B., Schmidt, W.E., (1993) Dipeptidyl-peptidase IV hydrolyses gastric inhibitory polypeptide, glucagon-like peptide-1(7-36)amide, peptide histidine methionine and is responsible for their degradation in human serum. Eur J Biochem **214:** 829.

Miller, K.A., Gunn, T.M., Carrasquillo, M.M., Lamoreux, M.L., Galbraith, D.B., Barsh, G.S., (1997) Genetic studies of the mouse mutations mahogany and mahoganoid. Genetics **146:** 1407.

Monsky, W.L., Lin, C.Y., Aoyama, A., Kelly, T., Akiyama, S.K., Mueller, S.C., Chen, W.T., 1994, A potential marker protease of invasiveness, seprase, is localized on invadopodia of human malignant melanoma cells. *Cancer Res.* **54:** 5702.

Moulin, V., 1995, Growth factors in skin wound healing. *Eur. J. Cell Biol.* **68:** 1.

Nagle, D.L., McGrail, S.H., Vitale, J., Woolf, E.A., Dussault, Jr. B.J., DiRocco, L., Holmgren, L., Montagno, J., Bork, P., Huszar, D., Fairchild-Huntress, V., Ge, P., Keilty, J., Ebeling, C., Baldini, L., Gilchrist, J., Burn, P., Carlson, G.A., and Moore, K.J., 1999, The mahogany protein is a receptor involved in suppression of obesity. *Nature* **398:** 148.

Ollmann, M.M., Barsh, G.S., (1999) Down-regulation of melanocortin receptor signaling mediated by the amino terminus of Agouti protein in Xenopus melanophores. J Biol Chem **274:** 15837.

Oravecz, T., Pall, M., Roderiquez, G., Gorrell, M.D., Ditto, M., Nguyen, N.Y., Boykins, R., Unsworth, E., and Norcross, M.A., 1997, Regulation of the receptor specificity and function of the chemokine RANTES (regulated on activation, normal T cell expressed and secreted) by dipeptidyl peptidase IV (CD26)-mediated cleavage. *J. Exp. Med.* **186:** 1865.

Pauly, R.P., Rosche, F., Wermann, M., McIntosh, C.H., Pederson, R.A., Demuth, H.U., (1996) Investigation of glucose-dependent insulinotropic polypeptide-(1-42) and glucagon-like peptide-1-(7-36) degradation in vitro by dipeptidyl peptidase IV using matrix-assisted laser desorption/ionization-time of flight mass spectrometry. A novel kinetic approach. J. Biol. Chem. **271:** 23222.

Pender, S.L., Tickle, S.P., Docherty, A.J., Howie, D., Wathen, N.C., and MacDonald, T.T., 1997, A major role for matrix metalloproteinases in T cell injury in the gut. *J. Immunol.* **158:** 1582.

Proost, P., De Meester, I., Schols, D., Struyf, S., Lambeir, A.M., Wuyts, A., Opdenakker, G., De Clercq, E., Scharpe, S., and Van Damme, J., 1998, Amino-terminal truncation of chemokines by CD26/dipeptidyl-peptidase IV. Conversion of RANTES into a potent inhibitor of monocyte chemotaxis and HIV-1-infection. *J. Biol. Chem.* **273:** 7222.

Proost, P., Struyf, S., Schols, D., Opdenakker, G., Sozzani, S., Allavena, P., Mantovani, A., Augustyns, K., Bal, G., Haemers, A., Lambeir, A.M., Scharpe, S., Van Damme, J., and De Meester, I., 1999, Truncation of macrophage-derived chemokine by CD26/ dipeptidyl-peptidase IV beyond its predicted cleavage site affects chemotactic activity and CC chemokine receptor 4 interaction. *J. Biol. Chem.* **274:** 3988.

Sato, T., Endo, Y., Matsushita, M., and Fujita, T., 1994, Molecular characterization of a novel serine protease involved in activation of the complement system by mannose-binding protein. *Int. Immunol.* **6:** 665.

Schrader, W.P., Woodward, F.J., and Pollara, B., 1979, Purification of an adenosine deaminase complexing protein from human plasma. *J. Biol. Chem.* **254:** 11964.

Shibuya-Saruta, H., Kasahara, Y., and Hashimoto, Y., 1996, Human serum dipeptidyl peptidase IV (DPPIV) and its unique properties. *J. Clin. Lab. Anal.* **10:** 435.

Silvers, W.K. (1979) The coat colors of mice. Springer-verlag, New York.

Takagaki, Y., Seipelt, R.L., Peterson, M.L., and Manley, J.L., 1996, The polyadenylation factor CstF-64 regulates alternative processing of IgM heavy chain pre-mRNA during B cell differentiation. *Cell* **87:** 941.

Tanaka, T., Duke-Cohan, J.S., Kameoka, J., Yaron, A., Lee, I., Schlossman, S.F., and Morimoto, C., 1994, Enhancement of antigen-induced T cell proliferation by soluble CD26/dipeptidyl peptidase IV. *Proc. Natl. Acad. Sci. USA* **91:** 3082.

Taylor, T.D., Litt, M., Kramer, P., Pandolfo, M., Angelini, L., Nardocci, N., Davis, S., Pineda, M., Hattori, H., Flett, P.J., Cilio, M.R., Bertini, E., and Hayflick, S.J., 1996, Homozygosity mapping of Hallervorden-Spatz syndrome to chromosome 20p12.3-p13. *Nat. Genet.* **14:** 479.

Van Coillie, E., Proost, P., Van Aelst, I., Struyf, S., Polfliet, M., De Meester, I., Harvey, D.J., Van Damme, J., and Opdenakker, G., 1998, Functional comparison of two human monocyte chemotactic protein-2 isoforms, role of the amino-terminal pyroglutamic acid and processing by CD26/dipeptidyl peptidase IV. *Biochemistry* **37:** 12672.

Wesley, U.V., Albino, A.P., Tiwari, S., and Houghton, A.N., 1999, A role for dipeptidyl peptidase IV in suppressing the malignant phenotype of melanocytic cells. *J. Exp. Med.* **190:** 311.

# ANALOGS OF GLUCOSE-DEPENDENT INSULINOTROPIC POLYPEPTIDE WITH INCREASED DIPEPTIDYL PEPTIDASE IV RESISTANCE

Kerstin Kühn-Wache°, Susanne Manhart°, Torsten Hoffmann°, Simon A. Hinke*, R.Gelling*, Raymond A. Pederson*, Christopher H.S. McIntosh*, Hans-Ullrich Demuth°

*°probiodrug GmbH, Weinbergweg 22 (Biocenter), 06120 Halle (Saale), Germany, *Department of Physiology, University of British Columbia, Vancouver, BC., Canada V6T 1Z3*

Key words dipeptidyl peptidase IV, glucose-dependent insulinotropic peptide

## 1. INTRODUCTION

The incretin GIP (glucose-dependent insulinotropic polypeptide), a 42 amino acid peptide, is released from the K-cells of the small intestine into the blood in response to oral nutrient ingestion. GIP inhibits the secretion of gastric acid and promotes the release of insulin from pancreatic islet cells (Brown *et al* 1970, Creutzfeldt 1979). It has been shown that GIP together with glucagon-like peptide-$1_{7\text{-}36}$ (tGLP-1) is sufficient for the full incretin effect of the entero-insular axis (Fehmann *et al* 1989). GIP and the related hormone, tGLP-1, have been considered to be involved in the pathogenesis of type II (non-insulin dependent) diabetes mellitus. The physiological actions of the incretins, and especially of GLP-1, are not only manifested by enhanced insulin secretion but also by inhibition of gastric emptying (Nauck *et al* 1997) and suppression of glucagon release (Gutniak *et al* 1992, Gutniak *et al* 1994, Nauck *et al* 1996, Nauck *et al* 1993) and may result in an improved glucose tolerance. Additionally, GIP is an important regulator of adipocyte function and may contribute to progression of obesity in man (McIntosh *et al* 1999).

*Cellular Peptidases in Immune Functions and Diseases 2*
Edited by Langner and Ansorge, Kluwer Academic/Plenum Publishers, 2000 

In serum both incretins, GIP and tGLP-1, are degraded by dipeptidyl peptidase IV (DPIV). The resulting short biological half-life (~2 min *in vivo*) limits the therapeutic use of GIP and tGLP-1 (Mentlein *et al* 1993, Kieffer *et al* 1995, Pauly *et al* 1996). In the case of tGLP-1 several studies have been directed to obtain biological active tGLP-1 analogs with improved DPIV-resistance (Deacon *et al* 1998, Siegel *et al* 1999). For GIP, systematic studies have been missing until now. Recently, a study has demonstrated that $Tyr^1$-glucitol-GIP displays DPIV-resistance and enhanced bioactivity (O'Harte *et al* 1999).

Simultaneously, there is a continual interest in defining the receptor binding region of GIP by syntheses of biologically active GIP-fragments. The GIP receptor, which is a member of the G-protein-coupled receptors (Gallwitz *et al* 1993, Amirano *et al* 1994), has a high specificity only for GIP and does not bind other peptides of the glucagon-family. For this reason GLP-1/GIP chimeric peptides also showed nearly no affinity to the GIP receptor (Gallwitz *et al* 1996). Hence it was concluded that the entire $GIP_{1\text{-}30}$ molecule is important for recognition. This was confirmed by Gelling et al (Gelling *et al* 1997) who showed that $GIP_{6\text{-}30}$-amide ($GIP_{6\text{-}30a}$) contains the high affinity binding region of GIP but exhibits antagonist activity like other N-terminal truncated forms. Both the DPIV degradation to $GIP_{3\text{-}42}$ and the reversing of the order of the first two amino acids in $Ala^1$-$Tyr^2$-GIP resulted in peptides with reduced affinity for the receptor in competition binding studies (in preparation) also emphasizing the importance of the correct N-terminus of GIP for recognition and activation.

Therefore in the current study N- and C-terminal truncated fragments as well as various GIP analogs with a reduced peptide bond or alterations of the amino acids close to the DPIV specific cleavage site were synthesized with the aim of improved DPIV-resistance and a prolonged half-time.

## 2. MATERIALS AND METHODS

### *Solid-phase Synthesis of Peptides*

The GIP analogs were synthesized with an automated synthesizer SYMPHONY (RAININ) using a modified Fmoc-protocol. Cycles were modified by using double couplings from the $15^{th}$ amino acid from the C-terminus of the peptide with five fold excess of Fmoc-amino acids and coupling reagent. The peptide couplings were performed by TBTU/NMM-activation using a 0.23 mmol substituted NovaSyn TGR-resin or the corresponding preloaded Wang-resin at 25 μmol scale. The cleavage from the resin was carried out by a cleavage-cocktail consisting of 94.5 % TFA, 2.5 % water, 2.5 % EDT and 1 % TIS.

Analytical and preparative HPLC were performed by using different gradients on the LiChrograph HPLC system of Merck-Hitachi. The gradients were made up from two solvents: (A) 0.1 % TFA in $H_2O$ and (B) 0.1 % TFA in acetonitrile. Analytical HPLC were performed under the following conditions: solvents were run (1 ml/min) through a 125 - 4 Nucleosil RP18-column, over a gradient from 5 % - 50 % B over 15 min and then up to 95 % B until 20 min, with UV detection ($\lambda$ = 220 nm). Purification of the peptides was carried out by preparative HPLC on either a 250 - 20 Nucleosil 100 RP8-column or a 250 - 10 LiChrospher 300 RP18-column (flow rate 6 ml/min, 220 nm) under various conditions depending on peptide chain length. For the identification of the peptide analogs, laser desorption mass spectrometry was employed using the HP G2025 MALDI-TOF system of Hewlett-Packard.

Tyr-Ala$\psi$($CH_2NH$)-$GIP_{3\text{-}30a}$ was synthesized by on resin fragment coupling of 2 equivalents of Fmoc-Tyr(tBu)$\psi$(CH2NH)-Glu(tBu)-Gly-OH by TBTU/DIPEA activation and double coupling over 4 hours. The corresponding $GIP_{5\text{-}30}$ fragments was synthesized as described above. The synthesis of the fully protected tetrapeptide Tyr-Ala$\psi$($CH_2NH$)-Glu(tBu)-Gly-OH was carried out on the acid sensitive Sasrin resin in a 0.7 mmol scale by Fmoc-strategy as described above using a half-automated peptide synthesizer Labortec (BACHEM). The protected tetrapeptide was cleaved from the resin by 1 % TFA. The reduced peptide bond was incorporated via reductive alkylation of the N-terminal deprotected peptide on the sasrin resin with Fmoc-alaninal (Meyer *et al* 1995).

## *Determination of DPIV Resistance by MALDI-TOF Mass Spectrometry*

The hydrolysis of peptide analogues by purified kidney DPIV was studied as described previously (Pauly *et al* 1996). In brief, peptides were incubated in 0.04 M Tris buffer pH 7.6 and DPIV for up to 24 h. Samples were removed from the incubation mixture and prepared for MALDI-TOF mass spectrometry, as described (Pauly *et al* 1996).

## *In Vitro Studies*

Chinese hamster ovary (CHO-K1) cells stably expressing the rat pancreatic islet (wild type) GIP receptor (wtGIP-R1 cells) were prepared as described previously (Gelling *et al* 1997, Wheeler *et al* 1995). Cells were cultured in DMEM/F12, supplemented with 10 % newborn calf serum, 50 units/ml penicillin G, and 50 μg/ml streptomycin (Culture media and antibiotics from Gibco BRL, Life Technologies). Cells were grown in 75 $cm^2$ flasks until 80 - 90 % confluent, when they were split and seeded onto 24 well plates at a density of 50,000 cells/well. Experiments were carried out 48 h later.

Binding studies using $^{125}$I-labeled $spGIP_{1-42}$, purified by high performance liquid chromatography (HPLC), were performed essentially as described previously (Wheeler *et al* 1995). wtGIP-R1 Cells (1 - 5 x $10^5$/well) were washed twice at 4 °C in binding buffer (BB), consisting of DMEM/F12 (GIBCO), 15 mM HEPES, 0.1 % bovine serum albumin (BSA), 1 % Trasylol (aprotinin; Bayer), pH 7.4. They were incubated for 12 - 16 h at 4 °C with $^{125}$I-spGIP (50,000 cpm) in the presence or absence of unlabeled $GIP_{1-42}$ or analogue. Following incubation, cells were washed twice with ice cold buffer, solubilized with 0.1 M NaOH (1 ml), and transferred to culture tubes for counting of cell-associated radioactivity. Nonspecific binding was defined as that measured in the presence of 1 μM $GIP_{1-42}$ or $GIP_{1-30}$, and specific binding expressed as percent of binding in the absence of competitor (percent B/Bo).

Wild type GIP-R1 cells were cultured for 48 h, washed in BB at 37 °C, and preincubated for 1 h prior to a 30 min stimulation period with test agents in the presence of 0.5 mM IBMX (Research Biochemicals Intl., Natick, MA) (Gelling *et al* 1997, Wheeler *et al* 1995). In inhibition experiments, cells were incubated with GIP analogues for 15 min prior to a 30 min stimulation with 1 nM $shGIP_{1-42}$. Cells were extracted with 70 % ethanol and cAMP levels measured by radioimmunoassay (Biomedical Technologies, Stoughton, MA) (19,21). Data are expressed as fmol/1000 cells or % maximal $GIP_{1-42}$–stimulated cAMP production (inhibition experiments).

## 3. RESULTS

The synthesized GIP fragments and analogs together with their MALDI-TOF data and DPIV resistance data are compiled in table 1. The resistance of the peptides against DPIV degradation was evaluated by MALDI-TOF mass spectrometry. Usually, DPIV inactivates GIP by removing the N-terminal dipeptide Tyr-Ala.

Since $GIP_{1-30a}$ showed nearly the same binding affinity and ability of stimulating adenylyl cyclase as the full length peptide (table 2), it was our aim to design analogs of this GIP fragment with improved DPIV resistance. It was demonstrated by MALDI-

TOF-MS that the substitution of amino acids in the cleavage position by D-Ala$^2$, NMeGlu$^3$, Pro$^3$ or the introduction of a reduced peptide leads to resistance against DPIV degradation for up to 24 hours in $GIP_{1-30}$ analogs as well as in the corresponding $GIP_{1-6}$ analogs. Analogs with Val-, Gly-, Ser-substitution for Ala$^2$ or D-Glu-substitution for Glu$^3$ showed reduced hydrolysis rates by DPIV. Gly$^2$-$GIP_{1-6}$ seems to be an exception because no degradation by DPIV was detectable.

For selected peptides *in vitro* data of their receptor binding affinity and their ability of stimulation adenylyl cyclase in CHO-K1 cells transfected with the GIP receptor are presented (table 2). All peptides were able to bind to the transfected GIP-receptor albeit with a small reduction of binding affinity (3- to 4-fold increased $IC_{50}$ in comparison to $GIP_{1-30}$, Fig 1).

Substitution of D-Glu for Glu$^3$ and D-Ala for Ala$^2$ resulted in peptides with only small reduction in ability to stimulate adenylyl cyclase whereas the Val$^2$-and Gly$^2$-analogs showed a significant reduction in efficacy. Interestingly, the introduction of the reduced peptide bond resulted in a dramatic deterioration of cAMP production (Fig 2). This confirms the importance of the integrity of the N-terminus of GIP.

None of the C-terminal truncated fragments tested ($GIP_{1-6}$, $GIP_{1-7}$, $GIP_{1-13}$, $GIP_{1-15}$, $GIP_{1-15a}$) were able to bind and activate the cloned rat GIP receptor (data not shown). Truncation of GIP from the N-terminus resulted in peptides able to bind the GIP receptor (Fig 3), but most act as antagonists (manuscript in preparation).

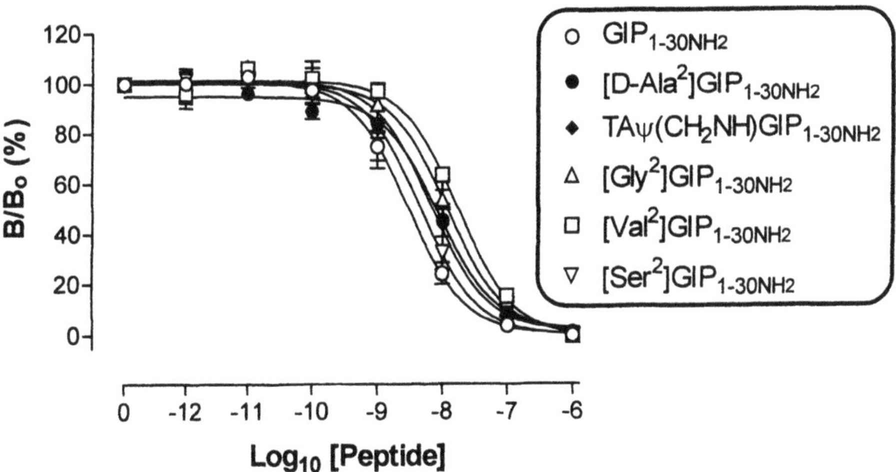


*Figure 1.* Displacement of $^{125}I$-$GIP_{1-42}$ binding by N-terminal modified $GIP_{1-30}$ analogs on the cloned rat pancreatic GIP receptor in transfected CHO-K1 cells

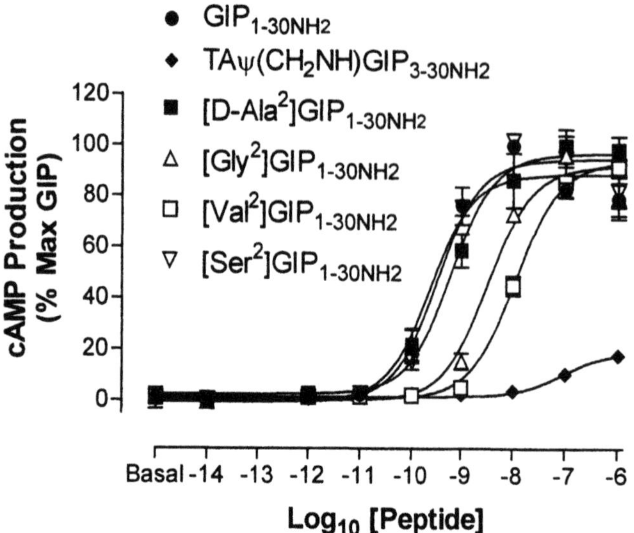


*Figure 2.* Stimulation of cyclic AMP in CHO-K1 cells transfected with the pancreatic GIP receptor by N-terminal modified $GIP_{1-30}$ analogs

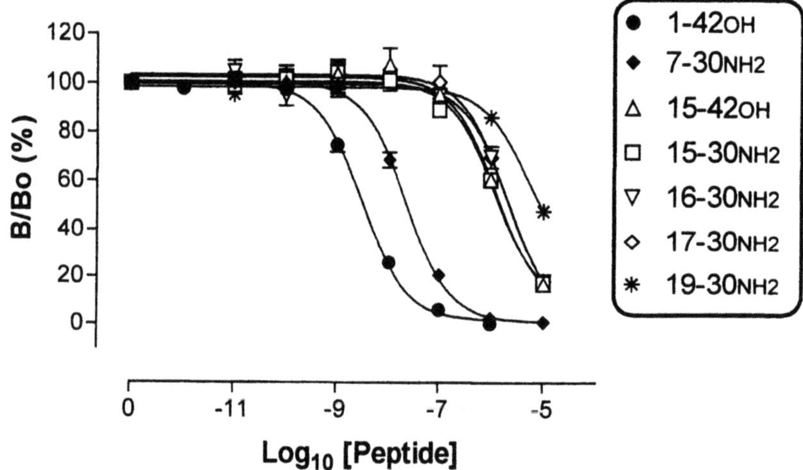


*Figure 3.* Competitive binding curves of synthetic GIP fragments versus $^{125}I$-$GIP_{1-42}$ on the cloned rat pancreatic GIP receptor in transfected CHO-K1 cells.

## 4. CONCLUSIONS

Fully and partially DPIV-resistant analogs of $GIP_{1-30}$ could be synthesized. The introduction of D-amino acids in P1- and P1'-position resulted in a slight reduction in binding and bioactivity. The examined C-terminal truncated fragments (with exception of the $GIP_{1-30}$ fragment) showed no binding affinity, whereas the antagonistic N-terminal truncated fragments were able

*Table 1.* N-terminal sequences, masses and DPIV-resistance of synthetic GIP analogs

| GIP-analog | N-terminal sequence | Mass (M) calculated | MALDI $M + H^+$ | half life after incubation with DP IV |
|---|---|---|---|---|
| $GIP_{1\text{-}42a}$ | Tyr-Ala-Glu-Gly.... | 4983.64 | 4983.9 | Not determined |
| $GIP_{1\text{-}30a}$ | Tyr-Ala-Glu-Gly.... | 3552.02 | 3553.3 | <15 min[a] |
| $GIP_{3\text{-}42a}$ | Glu-Gly... | 4749.38 | 4751.4 | Not determined |
| D-$Ala^2$-$GIP_{1\text{-}30a}$ | Tyr-D-Ala-Glu-Gly... | 3552.02 | 3553.8 | stable |
| N-$MeGlu^3$-$GIP_{1\text{-}30a}$ | Tyr-Ala-MeGlu-Gly.... | 3565.07 | 3566.1 | stable |
| D-$Glu^3$-$GIP_{1\text{-}30}$ | Tyr-Ala-D-Glu-Gly.... | 3551.07 | 3553.0 | 40.3 ± 4.8 |
| $Pro^3$-$GIP_{1\text{-}30}$ | Tyr-Ala-Pro-Gly.... | 3519.07 | 3522.9 | stable |
| $Ser^2$-$GIP_{1\text{-}30a}$ | Tyr-Ser-Glu-Gly.... | 3567.07 | 3568.0 | 137.1± 12.3 |
| $Val^2$-$GIP_{1\text{-}30a}$ | Tyr-Val-Glu-Gly.... | 3579.12 | 3580.7 | 298.3 ± 92.2 |
| $Gly^2$-$GIP_{1\text{-}30a}$ | Tyr-Gly-Glu-Gly.... | 3537.04 | 3539.1 | 150.5 ± 27.3 |
| YAψ(CH2NH)-$GIP_{3\text{-}30a}$ | Tyr-Alaψ(CH2NH)-Glu-Gly... | 3537.07 | 3539.0 | stable |
| $GIP_{1\text{-}6a}$ | Tyr-Ala-Glu-Gly.... | 685.74 | 686.9 | > 7.5 min |
| D-$Ala^2$-$GIP_{1\text{-}6a}$ | Tyr-D-Ala-Glu-Gly.... | 685.74 | 686.7 | stable |
| $Gly^2$-$GIP_{1\text{-}6a}$ | Tyr-Gly-Glu-Gly.... | 671.71 | 672.0 | Not detectable[b] |
| $Ser^2$-$GIP_{1\text{-}6a}$ | Tyr-Ser-Glu-Gly.... | 701.74 | 702.0 | 79.0± 12.2 |
| $Pro^2$-$GIP_{1\text{-}6a}$ | Tyr-Pro-Glu-Gly.... | 711.78 | 712.7 | > 7.5 min |
| $Val^2$-$GIP_{1\text{-}6a}$ | Tyr-Val-Glu-Gly.... | 713.79 | 715.2 | Not detectable |
| $Pro_3$-$GIP_{1\text{-}6a}$ | Tyr-Ala-Pro-Gly.... | 653.78 | 655.0 | stable |
| $GIP_{1\text{-}13}$ | Tyr-Ala-Glu-Gly.... | 1435.57 | 1435.6 | 11.5 ± 2.5 |
| $GIP_{1\text{-}15}$ | Tyr-Ala-Glu-Gly.... | 1681.85 | 1682.6 | 35.0 ± 5.2 |
| $GIP_{15\text{-}30a}$ | Asp-Lys-Ile-Arg.... | 2001.34 | 2003.3 | Not determined |
| $GIP_{17\text{-}30a}$ | Ile-Arg-Gln-Gln | 1758.07 | 1761.1 | Not determined |
| $GIP_{19\text{-}30a}$ | Gln-Gln-Asp-Phe | 1488.72 | 1489.8 | Not determined |
| $GIP_{7\text{-}30a}$ | Ile-Ser-Asp-Tyr | 2882.31 | 2886.9 | 130.1 ± 10.6 |

[a] After 15 min 92% of $GIP_{1\text{-}30}$ are hydrolyzed
[b] After 1500 min only 25 % of $G^2GIP_{1\text{-}30}$ are degraded

*Table 2.* Summary of competitive binding ($IC_{50}$), cAMP response ($EC_{50}$ and percent of maximal GIP-induced cAMP production) of GIP analogs

| GIP analog | $IC_{50}$ (nM) | $EC_{50}$ (nM) | maximal cAMP (%) |
|---|---|---|---|
| $GIP_{1\text{-}42a}$ | 3.56 ± 0.81 | 0.25 ± 0.07 | 100 |
| $GIP_{3\text{-}42a}$ | 58.42 ± 18.76 | No response | No response |
| $GIP_{1\text{-}30a}$ | 3.01 ± 0.69 | 0.195 ± 0.067 | 100 |
| D-$Ala^2$-$GIP_{1\text{-}42a}$ | 11.52 ± 1.08 | 1.78 ± 0.86 | 102.2 ± .03 |
| D-$Ala^2$-$GIP_{1\text{-}30a}$ | 10.26 ± 2.76 | 0.68 ± 0.21 | 94.5 ± 3.7 |
| D-$Glu^3$-$GIP_{1\text{-}30}$ | 3.84 ± 0.55 | 0.47 ± 0.13 | 103.5 ± 7.08 |
| $Ser^2$-$GIP_{1\text{-}30a}$ | 4.59 ± 0.34 | 0.336 ± 0.034 | 102.2 ± 1.7 |
| $Val^2$-$GIP_{1\text{-}30a}$ | 17.4 ± 1.8 | 11.23 ± 1.28 | 91.4 ± 3.6 |
| $Gly^2$-$GIP_{1\text{-}30a}$ | 11.8 ± 1.6 | 3.18 ± 0.8 | 96.2 ± 2.5 |
| YAΨ($CH_2NH$)-$GIP_{3\text{-}30a}$ | 12.7 ± 1.9 | 89.6 ± 15.9 | 17.6 ± 2 |

to bind to transfected rat GIP receptor. These results emphasize the hypothesis of an existing one-receptor-two-interaction-sites-model which was shown for peptides of the GRF-family.

Concerning the potential use of GIP analogs in the treatment of type II diabetes mellitus, these results offer the possibility to synthesize analogs with reasonable half-life times and physiologically relevant binding affinities and bioactivity.

## REFERENCES

Amiranoff, B., Vauclin-Jacques, N., Laburthe, M., 1984, Functional GIP receptors in a hamster pancreatic beta cell line, In 111: specific binding and biological effects. *Biochem. Biophys. Res. Commun.* **123:** 671-676.

Brown, J.C., Mutt, V., and Pederson, R.A., 1970, Further purification of a polypeptide demonstrating enterogastrone activity. *J. Physiol.* **209:** 57-64.

Creutzfeldt, W., 1979, The incretin concept today. *Diabetologia* **16:** 75-85.

Deacon, C.F., Knudsen, L.B., Madsen, K., et al., 1998, Dipeptidyl peptidase IV resistant analogues of glucagon-like peptide-1 which have extended metabolic stability and improved biological activity. *Diabetologia* **41:** 271-278.

Fehmann, H.C., Goke, B., Goke, R., et al., 1989, Synergistic stimulatory effect of glucagon-like peptide-1 (7-36) amide and glucose-dependent insulin-releasing polypeptide on the endocrine rat pancreas. *FEBS Lett.* **252:** 109-112.

Gallwitz, B., Witt, M., Folsch, U.R., et al., 1993, Binding specificity and signal transduction of receptors for glucagon-like peptide-1(7-36)amide and gastric inhibitory polypeptide on RINm5F insulinoma cells. *J. Mol. Endocrinol.* **10:** 259-268.

Gallwitz, B., Witt, M., Morys-Wortmann, C., et al., 1996, GLP-1/GIP chimeric peptides define the structural requirements for specific ligand-receptor interaction of GLP-1. *Regul. Pept.* **63:** 17-22.

Gelling, R.W., Coy, D.H., Pederson, R.A., et al., 1997, GIP(6-30amide) contains the high affinity binding region of GIP and is a potent inhibitor of GIP1-42 action in vitro. *Regul. Pept.* **69:** 151-154.

Gutniak, M., Orskov, C., Holst, J.J., et al., 1992, Antidiabetogenic effect of glucagon-like peptide-1 (7-36)amide in normal subjects and patients with diabetes mellitus. *N. Engl. J. Med.* **326:** 1316.

Gutniak, M.K., Linde, B., Holst, J J., et al., 1994, Subcutaneous injection of the incretin hormone glucagon-like peptide 1 abolishes postprandial glycemia in NIDDM. *Diabetes Care* **17:** 1039-1044.

Kieffer, T.J., McIntosh, C.H., Pederson, R.A., 1995, Degradation of glucose-dependent insulinotropic polypeptide and truncated glucagon-like peptide 1 in vitro and in vivo by dipeptidyl peptidase IV. *Endocrinology* **136:** 3585-3596.

McIntosh, C.H., Bremsak, I., Lynn, F.C., et al., 1999, Glucose-dependent insulinotropic polypeptide stimulation of lipolysis in differentiated 3T3-L1 cells: wortmannin-sensitive inhibition by insulin. *Endocrinology* **140:** 398.

Mentlein, R., Gallwitz, B., Schmidt, W.E., 1993, Dipeptidyl-peptidase IV hydrolyses gastric inhibitory polypeptide, glucagon-like peptide-1(7-36)amide, peptide histidine methionine and is responsible for their degradation in human serum. *Eur .J. Biochem.* **214:** 829-835.

Meyer, J.P., Davis, P., Lee, K.B., et al., 1995, Synthesis using a Fmoc-based strategy and biological activities of some reduced peptide bond pseudopeptide analogues of dynorphin A1. *J. Med. Chem.* **38:** 3462-3468

Nauck, M.A., Kleine, N., Orskov, C., et al., 1993, Normalization of fasting hyperglycaemia by exogenous glucagon-like peptide 1 (7-36 amide) in type 2 (non-insulin-dependent) diabetic patients. *Diabetologia* **36:** 741-744.

Nauck, M.A., Niedereichholz, U., Ettler, R., et al., 1997, Glucagon-like peptide 1 inhibition of gastric emptying outweighs its insulinotropic effects in healthy humans. *Am. J. Physiol.* **273:** E981-E988.

Nauck, M.A., Wollschlager, D., Werner, J., et al., 1996, Effects of subcutaneous glucagon-like peptide 1 (GLP-1 (7-36 amide)) in patients with NIDDM. *Diabetologia* **39:** 1546.

O'Harte, F.P., Mooney, M.H., Flatt, P.R., 1999, $NH_2$-terminally modified gastric inhibitory polypeptide exhibits amino-peptidase resistance and enhanced antihyperglycemic activity. *Diabetes* **48:** 758-765.

Pauly, R.P., Rosche, F., Wermann, M., McIntosh, C.H.S., Pederson, R.A., and Demuth, H.U., 1996, Investigation of glucose-dependent insulinotropic polypeptide-(1- 42) and glucagon-like peptide-1-(7-36) degradation in vitro by dipeptidyl peptidase IV using matrix-assisted laser desorption/ionization time of flight mass spectrometry - A novel kinetic approach. *J. Biol. Chem.* **271:** 23222-23229.

Siegel, E.G., Gallwitz, B., Scharf, G., et al., 1999, Biological activity of GLP-1-analogues with N-terminal modifications. *Regul. Pept.* **79:** 93-102.

Wheeler, M.B., Gelling, R.W., McIntosh, C.H., et al., 1995, Functional expression of the rat pancreatic islet glucose-dependent insulinotropic polypeptide receptor: ligand binding and intracellular signaling properties. *Endocrinology* **136:** 4629-4639.

# DIPEPTIDYL PEPTIDASE IV (DPP IV, CD26) IN PATIENTS WITH MENTAL EATING DISORDERS

Martin Hildebrandt, Matthias Rose, Christine Mayr, Petra Arck, Cora Schüler*, Werner Reutter**, Abdulgabar Salama, Burghard F. Klapp
*Department of Internal Medicine, Charité Campus Virchow-Klinikum, Humboldt-Universität zu Berlin, Germany, Institut für Anatomie, Med. Akademie Carl Gustav Carus der Technischen Universität Dresden, and **Institut für Molekularbiologie und Biochemie, Freie Universität Berlin, Germany*

Key words DPP IV/CD26 – Anorexia nervosa - Bulimia

Abstract

The notion that patients with eating disorders maintain a functional immunosurveillance in spite of severe malnutrition has attracted researchers for years. Dipeptidyl Peptidase IV (DPP IV), a serine protease with broad tissue distribution and known activity in serum, operates in the cascade of immune responses. Membrane-bound DPP IV expressed on lymphocytes, also known as the leukocyte antigen CD26, is considered to participate in T cell activation. We hypothesized that the activity of DPP IV in serum and expression of CD26 in lymphocytes may be altered in patients with eating disorders. Serum DPP IV activity and the number of CD26 (DPPIV)-positive peripheral blood lymphocytes were measured in 44 patients (anorexia nervosa (AN): n = 21, bulimia (B): n = 23) in four consecutive weekly analyses. The analysis of CD26-positive cells included the characterization of CD26-bright and CD26-dim positive subsets. Additionally, the expression of CD25 (IL-2 Receptor $\alpha$ chain) was evaluated to estimate the degree of T cell activation. The same analyses were carried out in healthy female volunteers (HC, n = 20).CD26-positive cells were reduced in patients as compared to healthy controls (mean 40.2 % (AN) and 41.1 % (B) vs. 47.4 % (HC), $p < 0.01$), while the DPP IV activity in serum was elevated (mean 108.4 U/l (AN) and 91.1 U/l (B) vs. 80.3 U/l (HC), $p < 0.01$). The potential implications of changes in DPP IV expression and serum activity on – and beyond - immune function are discussed.

# 1. INTRODUCTION

The clinical notion that patients with mental eating disorders, in spite of severe malnutrition, remain immunocompetent has been addressed by researchers in various studies (Pomeroy *et al* 1992). The results obtained showed many abnormalities including a reduction in complement components (Marcos *et al* 1993, Pomeroy *et al* 1997, Wyatt *et al* 1982), serum IgG (Barbouche *et al* 1993) and white blood cells (Rieger *et al* 1978). Changes in T cell subpopulation composition (Allende *et al* 1998, Fink *et al* 1996) and impaired cellular immune function (Allende *et al* 1998, Cason *et al* 1986) have also been described. Dipeptidyl peptidase IV (DPP IV, CD26), a serine-type protease which preferentially cleaves N-terminal dipeptides from polypeptides containing proline or alanine as the penultimate amino acid, has been shown to modulate immune responses (Reutter *et al* 1995). By virtue of its enzymatic activity, DPP IV is capable of expanding a T cell proliferative response in vitro (Tanaka *et al* 1994). Furthermore, potent cytokines are among the substrates for DPP IV described to date, suggesting a complex immunomodulatory role for DPP IV activity in serum. Crosslinking of CD26, i.e. membrane-bound DPP IV, with either CD2 or CD3 induces T cell activation and IL-2 production in vitro (Tanaka *et al* 1994). The capacity of CD26 to bind to ADA further adds to the importance of this membrane antigen.

Changes of DPP IV expression or serum activity appear to occur in several clinical and experimental situations of altered immune function. For example, a decrease in DPP IV serum activity has been reported in cyclosporin A-immunosuppressed patients after kidney transplantation (Scharpé *et al* 1990). In contrast, a sharp increase in DPP IV serum activity has been observed in rats immediately before rejection of cardiac allografts, whereas inhibition of DPP IV abrogated acute rejection (Korom *et al* 1997). Based on these experimental and clinical findings, we examined the DPP IV serum activity and the numbers of lymphocytes expressing CD26, i.e. membrane-bound DPP IV, in patients with hyporectic eating disorders, i.e. anorexia nervosa and bulimia.

# 2. PATIENTS AND METHODS

## 2.1 Patients

44 patients with eating disorders admitted to the Department of Internal Medicine, Charité Campus Virchow Klinikum, were enrolled in the study.

All patients gave their written informed consent. Among the patients included, twenty-one patients were diagnosed with severe anorexia nervosa and twenty-three patients with bulimia according to the International Classification of Diseases (WHO ICD-10). The anorectic patients had a history of eating disorders for 0.5 to 15 years (mean: 5.0 ± 4,2 y) and the bulimic patients of 0.5 to 20 years (mean: 6,7 ± 5,5 y). The observation period ranged between 3 and 23 weeks (mean: 5.6 weeks ± 3,4 weeks).

Fasting blood samples were drawn weekly by venepuncture. The controls, twenty female volunteers, were subjected to weekly venepunctures over a period of four consecutive weekly analyses. The body mass index in patients with eating disorders and controls was determined as follows: 14,6 ± 1,8 $kg/m^2$ (SD) in patients with severe anorexia nervosa, 18,9 ± 3,6 $kg/m^2$ (SD) in patients with bulimia, and 21,9 ± 2,3 $kg/m^2$ (SD) in healthy controls.

## 2.2 Laboratory analyses

Differential blood counts were determined using an automated blood count system (Technicon H3, Bayer, Germany). Quantitation of albumin and immunoglobulin fractions in serum was performed by standard techniques. Mononuclear cells were isolated from whole blood samples by density centrifugation using a Ficoll-Hypaque gradient (Seromed, Berlin, Germany). Immunophenotyping was performed using a FACScan flow cytometer (Becton-Dickinson, Mountain View, CA, U.S.A.). The lymphocyte population was gated by size/complexity criteria. Antibodies used for flow cytometry were obtained from Pharmingen (Heidelberg, Germany) and Coulter-Immunotech (Krefeld, Germany) and included the following: anti-CD2 (clone 39C1.5), anti-CD3 (clone UCHT1), anti-CD25 (clone B1.49.9), anti-CD26 (clone BA5).

Determination of DPP IV activity: 20 µl of serum were incubated with 10 µl of 2 mM Gly-Pro p-nitroanilide (Sigma Chemie GmbH, Deisenhofen, Germany) in 170 µl 0,1 M TRIS-HCl pH 8,0 for 30 min. The reaction was stopped by addition of 800 µl sodium acetate buffer (1M, pH 4,5). The DPP IV activity was deduced from the increase of extinction at 405 nm.

Statistical analyses were performed a) as a cross-sectional analysis of all data obtained from the first blood samples obtained from patients and controls, b) by comparison of the all data obtained in the first and last analyses of the observation period. All calculations were performed using SPSS 7.5 computer software. Among the tests applied were the following: Student´s t-Test, one-factor variance analysis (Oneway) with post-hoc comparison (Scheffé) and multifactorial analyses (ANOVA).

## 3. RESULTS

At admission to hospital (Table 1), DPP IV serum activity was elevated in patients with anorexia as compared to patients with bulimia and to healthy controls. Furthermore, reduced percentage numbers of CD2-/CD26 (DPP IV)-positive cells were found in patients with anorexia and bulimia. Expression of CD25 in CD2-positive and in CD26-positive cells was similar in all groups, as were the percentage numbers of CD2-positive cells. The percentage numbers of CD3-positive cells were decreased in patients with anorexia nervosa alone, but not to the extent seen in CD2-/CD26-positive cells.

*Table 1.* Results of analyses performed in blood samples of patients with anorexia nervosa

| | Healthy controls (HC, n = 20) | Anorexia nervosa (AN, n = 11) | Bulimia (B, n = 23) |
|---|---|---|---|
| Leukocytes/ul | 7210 ± 2395 | 5119 ± 1876 * | 6427 ± 1933 |
| Lymphocytes/µl | 2067 ± 695 | 1534 ± 825 * | 2284 ± 748 |
| % CD2+ | 80.8 ± 4.9 | 76.4 ± 11.6 | 75.4 ± 11.8 |
| % CD3+ | 61.2 ± 7.1 | 56.3 ± 8.5 | 57.9 ± 13.4 |
| % CD2+ CD25+ | 8.8 ± 3.9 | 10.2 ± 6.9 | 8.7 ± 4.7 |
| % CD2+ CD26+ | 47.4 ± 7.3 | 40.2 ± 6.9 ** | 40.9 ± 7.3 ** |
| % CD26+CD25+ | 5.3 ± 2.9 | 5.9 ± 4.3 | 4.6 ± 2.6 |
| CD2+/µl | 1679 ± 616 | 1155 ± 622 * | 1702 ± 587 |
| CD3+/µl | 1267 ± 484 | 842 ± 435 * | 1303 ± 506 |
| CD2+CD26/µl | 952 ± 229 | 589 ± 288 ** | 944 ± 359 |
| DPP IV (mU/ml) | 80.3 ± 17.1 | 108,4 ± 28,4 ** | 91.1 ± 18.7 |
| IgG (mg/dl) | 1120 ± 226 | 884 ± 259 * | 1026 ±226 |

(AN), bulimia (B) and of healthy controls (mean and standard deviation). Absolute numbers of CD2-, CD3- and CD2-/CD26-positive cells were calculated on the basis of lymphocyte numbers and the percentage numbers of cell populations as determined by flow cytometry. The asterisks label F-values of statistical significance (**p < 0.01 *p < 0.05).

In patients with bulimia, the numbers of CD2-positive, CD3-positive and of CD2-/CD26-double positive cells per µl were not different from the numbers observed in healthy controls. In patients with anorexia nervosa, however, the numbers of CD2-positive cells ($p < 0,05$), of CD3-positive cells ($p < 0,05$) and of CD2-/CD26-double positive cells per µl ($p < 0,01$) were decreased as compared to healthy controls. Additionally, reduced leukocyte numbers were noted in anorectic patients. An analysis of all data obtained at admission revealed that serum IgG and the leukocyte counts correlated with the body mass index, whereas DPP IV activity in serum showed an inverse correlation with the BMI ($p = 0.015$).

In the course of four consecutive weekly analyses, no changes could be observed in any of the parameters analyzed in patients with bulimia, with the exception of an increase of the relative proportion of CD2-/CD26-positive cells in week 4. In patients with anorexia nervosa, however, the initially decreased leukocyte counts and serum IgG levels increased, paralleled by an increase in the body mass index (14.6 kg/m$^2$ vs. 15.2 kg/m$^2$ ± SD) and a slight decrease of the DPP IV activity in serum (108.3 U/l vs. 97,9 U/l), yet without reaching statistical significance. The other parameters evaluated showed no remarkable changes.

The analysis of subsets of lymphocytes coexpressing CD26 and CD4 or CD8 (Fig 1) in patients with anorexia nervosa (n = 10) allowed the assessment of changes in lymphocytes brightly or weakly positive for CD26.

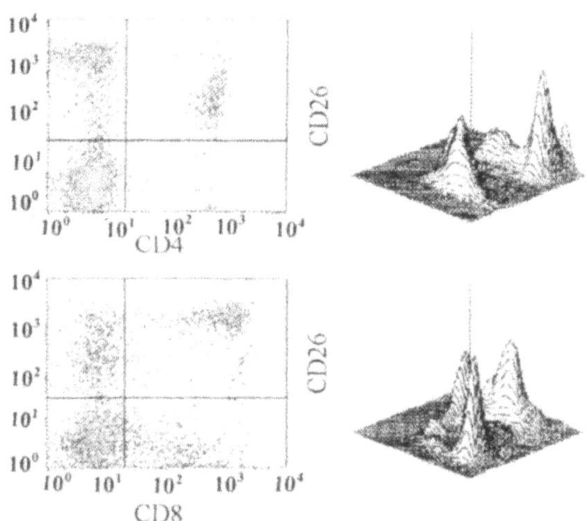


*Figure 1.* Flow cytometric analysis of cells coexpressing CD26 and CD4 and cells coexpressing CD26 and CD8. Differential gating of cells with bright or dim expression of CD26 was performed to determine the subsets.

*Table 2.* Subpopulations of cells coexpressing CD26 and CD4 or CD8.

| | CD26+ | CD26$^{bright}$ | CD4+ | | CD8+ | |
|---|---|---|---|---|---|---|
| | | | CD26$^{dim}$ | CD26$^{bright}$ | CD26$^{dim}$ | CD26$^{bright}$ |
| Anorexia (n = 10) | 36 ± 14 ** | 11 ± 7 | 12.6 ± 7 ** | 4.7 ± 2.6 | 8 ± 3.4 | 5 ± 4.6 |
| Controls (n = 18) | 49 ± 9 | 8.5 ± 4 | 24.1 ± 11 | 4.6 ± 2.6 | 8 ± 5.3 | 2.3 ± 1.8 |

Cells with bright or dim expression of CD26 were determined by differential gating. The numbers represent the percentage of all gated cells ± standard deviation. The asterisks label F-values of statistical significance (**$p < 0.01$ *$p < 0.05$).

The results were compared to similar analyses in healthy controls (n = 15). Cells brightly positive for CD26 were mostly CD8 positive, whereas cells with dim expression of CD26 coexpressed mostly CD4. Patients with Anorexia nervosa showed a selective decrease in the percentage of cells with dim expression of CD26 and coexpression of CD4, whereas the percentage of cells that were brightly positive for CD26 was normal or elevated (Tab 2).

## 4. DISCUSSION

Our findings shed new light on the immune function in patients with hyporectic eating disorders with regard to the notion that these patients, in spite of severe malnutrition, obviously remain immunocompetent. DPP IV activity in serum has been suggested to be a useful index of immune function (Maes *et al* 1993). Since recombinant soluble DPP IV had been shown to enhance antigen-induced T cell proliferation in vitro (Tanaka *et al* 1994), it had been concluded that, in subjects with low levels of CD26-positive cells and decreased levels of DPP IV serum activity, immune responses could be improved considerably by addition of exogenous soluble circulating DPP IV (Duke-Cohan *et al* 1996). If the decreased numbers of CD26-positive cells were indeed to be interpreted as a certain degree of immunosuppression, the elevated levels of DPP IV serum activity may be regarded as an attempt to compensate for the loss of CD26-positive cells, allowing an expansion of T cell proliferative responses as introduced above. Interestingly, this is not the case in HIV-infected patients where DPP activity in serum was shown to be normal (Vanham *et al* 1993). In addition, the analysis of subsets of CD26-positive cells revealed a selective decrease of CD4+/CD26(dim) cells, whereas other subsets, especially the CD26(bright)-positive cells presumed to resemble a memory cell population, remained unchanged. These changes are clearly different from those observed in HIV-positive individuals.

It should be pointed out that, in patients with hyporectic eating disorders, the implications of an increased activity of DPP IV in serum may go beyond a supportive role in the immune response. With regard to nutritional aspects, a vast amount of data points at a functionally important role of serum DPP IV in the inactivation of potent insulinotropic and peptides as GIP, GLP-1 and GLP-2, suggesting an involvement of DPP IV in blood glucose control, food intake and satiety. At present, a presumptive role of DPP IV as a crossing point in the network of bioactive peptide hormones and different regulatory circuits remains a speculative, yet intriguing, idea.

In conclusion, we report increased levels of DPP IV activity and reduced numbers of CD26 (DPPIV)-positive T lymphocytes in the peripheral blood of patients with anorexia nervosa, while the percentage of cells with bright expression of CD26 and presumed memory function remained unchanged. The changes observed may highlight attempts to maintain T cell function in these patients. In addition, the changes of DPP IV expression and serum activity may be important with regard to the metabolism of potent bioactive peptides presumed to be involved in the pathophysiology of mental eating disorders.

## ACKNOWLEDGMENTS

This work was supported by the Sonnenfeld-Stiftung, Berlin.

## REFERENCES

Allende, L.M., Corell, A., Manzanares, J., Madruga, D., Marcos, A., Madrono, A., Lopez-Goyanes, A., Garcia-Perez, M.A., Moreno, J.M., Rodrigo, M., Sanz, F., Arnaiz-Villena, A., 1998, Immunodeficiency associated with anorexia nervosa is secondary and improves after refeeding. *Immunology* **94:** 543-551.

Barbouche, M.R., Levy-Soussan, P., Corcos, M., Poirier, M.F., Bourdel, M.C., Jeammet, P., Avrameas, S., 1993, Anorexia and lower vulnerability to infections. *Am. J. Psychol.* **150:** 169-170.

Cason, J., Ainley, C., Wolstencroft, R.A., Norton, K.R.W., Thompson, R.P.H., 1986, Cell-mediated immunity in anorexia nervosa. *Clin. Exp. Immunol.* **64:** 370-375.

Duke-Cohan, J.S., Morimoto, C., Rocker, J.A., Schlossman, S.F., 1996, Serum high molecular weight dipeptidyl peptidase IV (CD26) is similar to a novel antigen DPPT-L released from activated T cells. *J. Immunol.* **156:** 1714-1721.

Fink, S., Eckert, E., Mitchell, J., Crosby, R., Pomeroy, C., 1986, T lymphocyte subsets in patients with abnormal body weight: Longitudinal studies in anorexia nervosa and obesity. *Int. J. Eat. Disord.* **20:** 295-305.

Korom, S., De Meester, I., Stadlbauer, T., Chandraker, A., Schaub, M., Sayegh, M., Belyaev, A., Haemers, A., Scharpé, S., Kupiec-Weglinski, J., 1997, Inhibition of CD26/Dipeptidyl Peptidase IV activity in vivo prolongs cardiac allograft survival in rat recipients. *Transplantation* **63:** 1495-1500.

Maes, M., Meltzer, H.Y., Scharpé, S., Bosmans, E., Suy, E., De Meester, I., Calabrese, J., Cosyns, P., 1993, Relationships between lower plasma L-tryptophan levels and immune-inflammatory variables in depression. *Psychiatry Res.* **49:** 151-165.

Marcos, A., Varela, P., Santacruz, I., Munoz-Velez, A., Morande, G., 1993, Nutritional status and immunocompetence in eating disorders. *Eur. J. Clin. Nutr.* **47:** 787-793.

Pomeroy, C., Mitchell, J.E., Eckert, E., 1992. Risk of infection and immune function in anorexia nervosa. *Int. J. Eat. Disord.* **12:** 47-55.

Pomeroy, C., Mitchell, J.E., Eckert, E., Raymond, N., Crosby, R., Dalmasso, A.P., 1997, Effect of body weight and caloric restriction on serum complement proteins, including factor D/adipsin: studies in anorexia nervosa and obesity. *Clin. Exp. Immunol.* **108:** 507-515.

Reutter, W., Baum, O., Löster, K., Fan, H., Bork, J.P., Bernt, K., Hanski, C., Tauber, R., 1995, Dipeptidyl Peptidase IV (CD26) *in Metabolism and Immune Response* ( B. Fleischer ed.) Springer Verlag, Heidelberg, pp. 55-73.

Rieger, W., Brady, J.P., Weisberg, E., 1978,Hematologic changes in anorexia nervosa. *Am. J. Psychol.* **135:** 984-985.

Scharpé, S., De Meester, I., Vanhoof, G., Hendriks, D., Uyttenbroek, W., Ntakarutimana, V., Deckx, R., 1990), Serum dipeptidyl peptidase IV activity in transplant recipients. *Clin. Chem.* **36:** 984

Tanaka, T., Duke-Cohan, J.S., Kameoka, J., Yaron, A., Lee, I., Schlossman, S.F., Morimoto, C., 1994, Enhancement of antigen-induced T-cell proliferation by soluble CD26/dipeptidyl peptidase IV. *Proc. Natl. Acad. Sci. USA* **91:** 3082-3086.

Vanham, G., Kestens, L., De Meester, I., Vingerhoets, J., Penne, G., Vanhoof, G., Scharpé, S., Heyligen, H., Bosmans, E., Ceuppens, J.L., 1993, Decreased expression of the memory marker CD26 on both CD4+ and CD8+ T lymphocytes of HIV-infected subjects. *J. Acquir. Immune Defic. Syndr.* **6:** 749-757.

Wyatt, R.J., Farrell, M., Berry, P.L., Forristal, J., Maloney, M.J., West, C.D., 1982, Reduced alternative complement pathway control protein levels in anorexia nervosa: Response to parenteral alimentation. *Am. J. Clin. Nutr.* **35:** 973-980.

# THE MEMBRANE-BOUND ECTOPEPTIDASE CPM AS A MARKER OF MACROPHAGE MATURATION *IN VITRO* AND *IN VIVO*

Michael Rehli, Stefan W. Krause, Reinhard Andreesen
*Department of Hematology and Oncology, University of Regensburg, 93042 Regensburg, Germany.*

Key words macrophage, differentiation, ectopeptidase

Abstract During terminal maturation of human blood monocytes into macrophages, a multitude of phenotypic and functional changes occurs: cells increase in size, they enhance their capacity for phagocytosis and tumor cytotoxicity but decrease their ability for T-lymphocyte stimulation. The pattern of secreted cytokines is shifted as is the profile of surface antigens. We recently identified carboxypeptidase M (CPM) as a macrophage maturation-associated antigen detected by mAb MAX.1/MAX.11. CPM, a phosphoinositol-linked ectopeptidase, is able to process a multitude of different substrates, among them immunologically important peptides like bradykinin, anaphylatoxins and enkephalins. It was previously shown to be expressed in placenta, lung, and kidney. CPM as detected by MAX.1/11 shows a strong expression on monocyte-derived macrophages in vitro and on macrophages *in vivo* accompanying T-lymphocyte activation like during allogeneic transplant rejection or allergic alveolitis. In contrast, its expression is suppressed on macrophages by some types of tumor cells. CPM expression seems to correlate with macrophage cytotoxic functions. However, the biological importance of CPM expression in human macrophages in vivo is difficult to predict. A wide range of biologically active peptides are cleaved by CPM, and the relevance of CPM peptide processing during an immune reaction is only poorly understood. The generation and analysis of CPM-deficient animals might improve our understanding of CPM function. Therefore we cloned a cDNA for the murine homologue of CPM. However, expression of mCPM was undetectable in murine primary macrophages and macrophage cell-lines, suggesting that CPM expression and function is not conserved between human and mouse macrophages.

# 1. INTRODUCTION

Monocytes, macrophages and their bone-marrow derived precursors represent different stages of differentiation in a functionally diverse hematopoietic cell lineage which is designated as „mononuclear phagocyte system“ (MPS) (van Furth 1989). The mature endpoints of this lineage include the heterogeneous group of tissue macrophages, multinucleated bone-resorbing osteoclasts and antigen-presenting dendritic cells (Udagawa, *et al* 1990, Akagawa *et al* 1996). The mature cell types develop from circulating blood monocytes upon migration through vascular endothelium. Their individual differentiation pathway is controlled by the surrounding microenvironment and local tissue-specific factors (van Furth 1989).

Cells of the MPS have a prominent role in host defence against microbial pathogens and malignant cells. Important functions include the secretion of cytokines, enzymes, oxygen radicals and other soluble products, and their capacity for phagocytosis and to mediate cellular cytotoxicity. Furthermore, they form an important bridge between innate and specific immunity as potent antigen-presenting cells (Steinman 1991).

In vitro, monocytes have been differentiated into macrophages, dendritic cells and osteoclast-like cells (Akagawa *et al* 1996, (Andreesen *et al* 1983a). The following review will mainly focus on „classical“ macrophage differentiation (as observed in the human system) and discuss carboxypeptidase M (CPM) as a marker for this differentiation process.

# 2. MACROPHAGE DIFFERENTIATION

Upon leaving the vasculature, blood monocytes encounter numerous extracellular stimuli like cytokines and cell-cell or cell-matrix interactions, which are tissue-specific and influence further differentiation (van Furth 1989). Macrophages of different tissues have variable functional properties and show a variable expression of surface markers (Rutherford *et al* 1993). The situation *in vivo* is very complex and access to tissue macrophage (at least in the human system) is limited. Therefore many studies utilise a model of monocyte to macrophage differentiation *in vitro*. When blood monocyte are cultured in the presence of human serum for about seven days, they adhere and increase in size, mostly resembling exsudate-type macrophages (Andreesen *et al* 1983a, Musson *et al* 1980).

In parallel with the changing cell-morphology, functional properties change: e.g. macrophages have an increased capacity for phagocytosis and for production of tumour necrosis factor (TNF), but have lost their ability to produce interleukin 1 (IL-1) (Scheibenbogen *et al* 1991). A list of differences between monocytes and macrophages is listed in Table 1. No cytokines have to be added for the induction of this process, but endogenous secretion of macrophage colony-stimulating factor is important (Brugger *et al* 1991). 1,25$(OH)_2$ Vitamin $D_3$ can partially substitute for human serum in the induction of macrophage maturation (Kreutz *et al* 1990).

*Table 1.* Properties of monocytes and monocyte-derived macrophages

| | | MO | MAC |
|---|---|---|---|
| Tumour cytotoxicity | (non-specific) | + | +++ |
| | (antibody-dependent) | - | +++ |
| Phagocytosis | | ++ | +++ |
| Adhesion to plastic | | + | +++ |
| Antigen presentation | (primed T-cells) | ++ | ++ |
| | (unprimed T-cells) | + | - |
| Production of lysozyme | | + | ++ |
| Cytokine production: | IL-1 | +++ | - |
| | IL-6 | +++ | + |
| | TNF | + | +++ |
| | GM-CSF[a] | + | ++ |
| Surface antigens: | CD11c | ++ | +++ |
| | CD14 | +++ | +++ |
| | CD33 | ++ | ++ |
| | HLA-DR | ++ | +++ |
| | HLA-DQ | +[b] | ++ |
| | CD16 | (+)[c] | ++ |
| | CD51 | - | + |
| | CD71 | - | ++[a] |
| | CPM (MAX.1, MAX.11) | (+)[d] | +++ |
| | MAX.3 (gp63) | - | ++ |

[a] granulocyte-macrophage colony-stimulating factor
[b] variable expression between cells of one population
[c] expressed on a small subset of blood monocyte
[d] homogeneous weak expression

Blood monocytes are positive for some broadly expressed myeloid antigens like CD33 and aminopeptidase N (CD13). Monocytes and most monocyte derived cells are positive for the CR4 a-chain (CD11c), Fcγ receptor II (CD64), and express the LPS-receptor CD14. Accordingly, human CD14 positive monocyte and macrophage are extremely sensitive to stimulation by bacterial endotoxin (lipopolysaccharide, LPS). HLA-class II molecules which are important for T-helper cell stimulation are expressed to some degree on monocytes and macrophages, however the strongest expression is found on monocyte-derived dendritic cells.

A number of surface antigens is markedly upregulated during the differentiation of human monocytes into macrophages, reflecting their changing properties and functions (Andreesen *et al* 1989). Expression patterns of several maturation-associated antigens are listed in Table 1. For example, the low affinity IgG receptor (CD16) is expressed only on a small subgroup of blood monocyte, but is detected on most monocyte-derived macrophage (Andreesen *et al* 1989, Ziegler-Heitbrock *et al* 1993). A series of antibodies (MAX) from our own laboratory detected previously undefined antigens that were differentially regulated during monocyte to macrophage differentiation (Andreesen *et al* 1986).

## 3. MONOCLONAL ANTIBODIES MAX.1 & 11 DETECT CARBOXYPEPTIDASE M

The monoclonal antibodies of the MAX series were produced by immunising mice with human monocyte-derived macrophages. The resultant hybridomas were screened for the production of antibodies that detect antigens present on macrophages but not or to a lower degree on blood monocytes. Amongst these antibodies, MAX.1 and MAX.11 both detect a glycophosphoinositol (GPI) linked antigen that is strongly expressed on the cell surface of macrophages but only weakly on monocytes. We did not detect a corresponding antigen on other hematopoietic cells (Andreesen *et al* 1986). A biochemical characterisation revealed that both antibodies precipitate a glycoprotein of about 62 - 65 kDa from macrophage membranes, which decreased to 46 kDa after deglycosylation with N-glycosidase F. Sequential purification and precipitation with both antibodies proved that MAX.1 and MAX.11 detect the same antigen (Rehli *et al* 1995).

$NH_2$-teminal amino acid sequencing of the immunopurified protein revealed its identity with the ectoenzyme carboxypeptidase M (CPM) (Rehli *et al* 1995), a known GPI-linked 62 kDa glycoprotein (Deddish *et al* 1990, Skidgel *et al* 1989, Tan *et al* 1989). Membrane-bound basic carboxypeptidase activity was then measured using a synthetic substrate (dns-Ala-Arg). The high activity found on macrophage membranes was almost completely precipitated with the antibodies MAX.1 and MAX.11 (Rehli *et al* 1995). We further showed that membrane-bound carboxypeptidase activity markedly increases during monocyte to macrophage differentiation in parallel with its mRNA and surface expression of MAX.1/11 antigen (Rehli *et al* 1995): CPM is a member of the group of basic carboxypeptidases that cleave C-terminal arginine or lysine residues from peptides or proteins (Skidgel 1988). It was initially purified from human placenta and was cloned in 1989 (Tan *et al* 1989). CPM is attached to cell membranes by phosphoinositol linkage and was the first membrane-bound carboxypeptidase to be described. Similar to other carboxypeptidases, a zinc ion resides at its active site. Like carboxypeptidase N, another member of the enzyme family which is synthesised in the liver and released into the blood, its activity has a neutral pH optimum and CPM (in contrast to CPN) cleaves argininc faster than lysine (Deddish *et al* 1990, Skidgel 1988). Natural substrates for CPM (and other basic carboxypeptidases) include a vast list of biologically active peptides, which are listed in Table 2.

CPM expression has been detected in several tissues. Its activity was originally purified from placenta, the strongest CPM expressing tissue (Skidgel *et al* 1989). More recent studies describe its expression on human ovarian follicles and corpora lutea of menstrual cycle and early pregnancy (Yoshioka *et al* 1998). CPM expression was also detected in lung (mainly on type I alveolar cells) and its enzymatic activity was described to be enhanced in infectious diseases (Nagae *et al* 1993). In kidney CPM expression is mainly associated with the glomerular mesangium (Andreesen *et al* 1988), in brain it colocalises with glial structures (Nagae *et al* 1992). Recently, another antiserum was described, which also detects CPM expression on activated T-cells and B-cells in germinal centres (de Saint-Vis *et al* 1995).

# 4. CARBOXYPEPTIDASE M IN HUMAN MONOCYTES AND MACROPHAGES

## 4.1 CPM expression and activity on in vitro differentiated macrophages

Our studies using MAX.1 and MAX.11 antibodies mainly focused on their reactivities on cells of hematopoietic origin. In contrast to the low expression on monocytes, CPM expression is strongly upregulated during macrophage differentiation in vitro (Andreesen *et al* 1990a). The upregulation of this differentiation marker parallels a normal serum-induced monocyte to macrophage maturation, no further stimulus is required to induce CPM expression. This is in contrast to inflammatory markers (e.g. TNF-α or IL-6) that are produced by macrophages only after stimulation, e.g. by LPS. Expression of CPM by monocyte or macrophage is neither upregulated nor downregulated by short term stimulation of monocyte or macrophage. However, if blood monocyte are stimulated by LPS, interferon-γ or other stimuli at the beginning of long-term cultures, normal macrophage maturation is disturbed: the morphological and functional changes that are usually observed during macrophage differentiation are suppressed. Accordingly, upregulation of CPM is blocked or greatly reduced (Brugger et al. 1991). If monocyte cultures are treated with IL-4 and granulocyte-macrophage colony-stimulating factor, a condition favouring the differentiation of dendritic cells

*Table 2.* Natural substrates of basic carboxypeptidases

| Substrate | C-terminal sequence | Substrate | C-terminal sequence |
|---|---|---|---|
| Bradykinine | -Pro-Phe-Arg | Erythropoietine | -Gly-Asp-Arg |
| Anaphylatoxine C3a | -Leu-Ala-Arg | Dynorphine A | -Lys-Leu-Lys |
| Anaphylatoxine C4a | -Leu-Gln-Arg | EGF[a] | -Glu-Leu-Arg |
| Anaphylatoxine C5a | -Leu-Gly-Arg | EGF-like protein 1 | -Asp-Gly-Lys |
| Enkephaline hexapeptide a | -Phe-Met-Arg | EGF-like protein 2 | -Asp-Gly-Lys |
| Enkephaline hexapeptide b | -Phe-Met-Lys | EGF-like protein 3 | -Asp-Arg-Lys |
| Enkephaline hexapeptide c | -Phe-Leu-Arg | EGF-like protein 4 | -Asp-Gly-Lys |
| Enkephaline hexapeptide d | -Phe-Leu-Lys | EGF-like protein 5 | -Leu-Ala-Arg |
| α-Neoendorphine | -Tyr-Pro-Lys | EGF-like protein 6 | -Cys-Gln-Arg |
| Fibrinopeptide 6A | -Pro-Ala-Lys | EGF-like protein 7 | -Leu-Asp-Arg |
| Fibrinopeptide 6D | -Glu-Trp-Lys | Amphireguline | -Gly-Glu-Lys |
| Fibrinopeptide A | -Gly-Val-Arg | HGF (α-chain)[b] | -Gln-Leu-Arg |
| Fibrinopeptide B | -Ser-Ala-Arg | MSP[c] | -Glu-Thr-Lys |

[a] epidermal growth factor
[b] hepatocyte growth factor
[c] macrophage stimulating protein

rather than macrophages (Peters *et al* 1996), CPM is only slightly upregulated (Grassi *et al* 1998).

Subnormal differentiation of macrophage in vitro can be observed in patients with aplastic anemia: monocytes of the majority of patients with this disease showed an abnormal cell surface antigen pattern after in vitro differentiation. Among the antigens with subnormal expression in these patients, CPM was most severely affected (Andreesen *et al* 1989). Monocytes of patients with HIV infection did not show the normal upregulation of CPM during macrophage maturation *in vitro* (Andreesen et al. 1990b).

CPM is absent or barely detectable on most myelomonocytic (e.g. U937, HL-60, or MonoMac-6) cell lines, which correlates with a rather immature myeloid or monocytic character of these cells. CPM is expressed on the myeloid cell line THP-1 that carries more features of mature macrophage and is additionally upregulated after stimulation of these cells with 1,25$(OH)_2$ Vitamin $D_3$ (Rehli *et al* 1995).

## 4.2 CPM expression in tissue macrophages *in vivo*

CPM expression on naturally occurring exsudate-type or resident tissue macrophages appears low in comparison to in vitro monocyte-derived macrophages, but like monocytes, these cells can be induced to express CPM at higher levels after in vitro culture (Andreesen *et al* 1988). In earlier studies, reaction of CPM-antibody MAX.1 was described as negative on different types of macrophage *in vivo*. Some reactivity, however, was detected with MAX.11 (Andreesen *et al* 1988). In recent experiments, using a more sensitive immunohistochemistry technique, positive staining of macrophages was found in most tissues with either antibody (unpublished observations).

*Table 3.* Expression of CPM on different types of myelomonocytic cells

| | | |
|---|---|---|
| Myeloid precursors (bone marrow) | | - |
| Blood MO | | (+) |
| Tissue MAC in situ: | normal lymph node, spleen | + |
| | rejection of transplanted renal allografts | +++ |
| Alveolar MAC: | healthy donors | + |
| | sarcoidosis, alveolitis | +++ |
| Peritoneal MAC | | (+) |
| *in vitro* MO derived MAC: | | |
| | healthy donors | +++ |
| | healthy donors, stimulated cultures (LPS, IFNγ) | (+) |
| | aplastic anemia | (+)[a] |
| | HIV infection | +[a] |
| | inside tumor spheroids of low/intermediate malignancy | +++ |
| | inside spheroids of highly malignant tumours | (+) |
| *in vitro* MO-derived dendritic cells | | + |

[a] variable expression in different patients

Strong expression was found on macrophages in two different pathological situations: In tissue samples collected from renal grafts during transplant rejections, a portion of the invading leukocytic infiltrate consisted of macrophage strongly expressing CPM (Andreesen *et al* 1987). Reactivity correlated with the severity of rejection and was suppressed by cyclosporin A treatment. Strong upregulation of CPM was also seen on alveolar macrophages in samples from patients with sarcoidosis or extrinsic allergic alveolitis (Andreesen *et al* 1988). In both situations, a T-lymphocyte mediated immune reaction can be assumed which seems to induce pronounced CPM expression on macrophages. Table 3 summarises the observed expression pattern of CPM in macrophages in vivo and in vitro.

## 4.3 CPM expression in tumour-associated macrophages

*In vivo*, macrophages are present in many tumours and may constitute a prominent part of the inflammatory infiltrate. Macrophages encountering tumour cells can exert nonspecific and antibody-mediated effector functions. Both properties increase during the differentiation from blood monocytes into mature macrophages. Tumour cytotoxicity of macrophages has been shown *in vitro* (Andreesen *et al* 1983b) and there is also evidence for its biological importance *in vivo* (Maass *et al* 1995). Unfortunately the interaction with macrophages can also be beneficial for tumour cells. Infiltrating macrophages sometimes are "deactivated", (e.g. production of oxygen radicals is reduced) and may support tissue invasion and neo-vascularization (Mantovani *et al* 1992).

We examined direct interactions between tumour cells and macrophages in a three-dimensional in vitro system of multicellular spheroids - an artificial model for a small tumour or a micrometastasis *in vivo* (Sutherland 1988). Three bladder tumour cell lines corresponding different degrees of malignancy were used to investigate their influence on macrophage differentiation and function (Konur *et al* 1996). In coculture experiments, spheroids of all three tumour cell lines were infiltrated by freshly isolated monocytes CPM expression was normally induced on differentiating macrophages inside the better differentiated and less malignant cell line RT4 and normal urothelial cells. However, the expression of CPM and other maturation-associated antigens, was inhibited inside spheroids derived from the highly malignant bladder carcinoma cell line J82 (Konur *et al* 1996). Macrophages differentiating inside J82 spheroids keep a rather "monocytic" cytokine secretion profile producing little TNF, much IL-6 and some IL-1. They also have a reduced capability of tumour cytotoxicity (Konur *et al* 1996). In this system malignancy of the tumour cell is inversely correlated with CPM expression on tumour associated macrophages.

## 5. BIOLOGICAL IMPLICATIONS

Cells of the MPS produce a wide variety of enzymes, either secreted or membrane-bound. In addition to CPM, these include a range peptidases (deamidase, dipeptidyl peptidase IV, prolylcarboxypeptidase, angiotensin-converting enzyme, leucine aminopeptidase, carboxypeptidase E, carboxy-peptidase D) which can be expressed depending on the type of macrophage and its stage of differentiation and/or activation.

In humans, CPM is an enzyme that is, on a low level, broadly expressed on most myelomoncytic cells, on monocyte-derived dendritic cells (and on many other types of cells) but is specifically and strongly upregulated during terminal maturation of macrophage. To understand possible functions of CPM on mature macrophages we need to consider its substrates *in vivo*. Many of the known substrates of CPM are peptides engaged in cell-cell signaling like bradykinin, anaphylatoxins and others. Cleavage of the terminal arginine or lysine may lead to activation, inactivation, or a change of biological activity, depending on the substrate under consideration.

The two kinins, bradykinin and kallidin (lys-bradykinin), are immunologically important substrates which are very efficiently metabolised into des-arg-kinins by CPM. Bradykinin (and kallidin) are released from precursor proteins (kininogens) by proteases called kallikreins (Bhoola *et al* 1992). Kallikrein activation in the blood with subsequent release of bradykinin is coupled to contact activation of the coagulation cascade. Kinins are also released in different tissues by tissue kallikreins where they support inflammatory processes by vasodilatation, formation of edema and enhancing lymphocyte motility and lead to pain sensation. Since both kininogen and kallikrein were described to be present in neutrophils, release of bradykinin directly in inflamed tissues by infiltrating neutrophils can be assumed (Bhoola *et al* 1992). Cleavage of the C-terminal arginine residue from bradykinin through CPM or carboxypeptidase N alters its binding specificity for the two known bradykinin receptors (BK1 and BK2). Vasodilatation and -permeabilisation are mainly induced by the binding of the complete bradykinin nonapeptide to BK2 receptors, whereas des-arg-bradykinin has a much lower affinity for the BK2 receptor and preferentially binds to BK1 receptors. Macrophages themselves have BK1 receptors on their surface and can respond to bradykinin, e.g. by secretion of IL-1 and TNF (Bockmann *et al* 1995, Tiffany *et al*1989). It is possible that a shift of bradykinin effects from BK2 to BK1 receptors is important in tumour immunology and metastasis. While BK2 receptor mediated vasodilatation might enhance metastasis formation, BK1 mediated stimulation of lymphocytes and macrophage might enhance anti-tumour responses. Interestingly, enhanced levels of kinins were found in ascites fluids of patients with peritoneal carcinosis (Matsumura *et al* 1990). This phenomenon could be connected to the suppression of CPM expression on macrophages by tumour cells as described above.

Anaphylatoxins represent another family of biologically important CPM substrates. Anaphylatoxin C5a, C4a and C3a are released from complement factors C5, C4 and C3 during complement activation. C5a is the most potent of these peptides, leading to the contraction of smooth muscle cells and bronchospasms. C5a induces histamine release, is chemotactic for lymphocytes and myeloid cells and induces the secretion of IL-8 and other cytokines from monocytes/macrophages (Ember *et al* 1994, Wetsel 1995). Most of the biological effects of anaphylatoxins are abolished by the cleavage of the C-terminal arginine. Inactivation of all three circulating anaphylatoxins in the blood stream is one of the important tasks of carboxypeptidase N. In inflamed tissues membrane-bound CPM on macrophages could carry out this function in a regulated manner.

Enkephalins are the last group of substrates with important functions in the immune system to be mentioned here. These opioid peptides together with other neuroendocrine hormones are probably responsible for the link between the neuroendocrine and the immune system. Morphine receptors were shown to be present on immune cells by a number of groups and correspondingly, direct effects of opioids on these cells like modulation of macrophage $H_2O_2$ release or lymphocyte proliferative response were shown (Radulovic *et al* 1995, Radulovic *et al* 1996, Roy *et al* 1996). Like for bradykinin, effects of enkephalins are mediated through different types of receptors and the receptor affinity of enkephalins can be modulated by the presence or absence of their C-terminal arginine. Enkephalin hexapeptides seem to preferentially bind to κ-opioid receptors, whereas des-arg/lys pentapeptides preferentially bind to δ-opioid receptors (Magnan *et al* 1982). A differential expression of these receptors on immune cells was shown (Tsuruta *et al* 1993) and might be of relevance, but the effect of opioid hexa- and pentapeptides on these cells has not been investigated extensively yet.

## 6. CPM AND MURINE MACROPHAGES

Gene targeting approaches in animal models, particularly in mice, are widely used to investigate the function of individual proteins. However, a murine homologue of CPM has not been described and there is no information available about its expression pattern in murine tissues. Only one murine EST-clone is currently present in public databases (Acc.-no. AA589379), that is partially homologous to the sequence of human CPM, but also contains intronic sequences. We generated primers corresponding to the homologous part of the sequence and used a nested 3'-RACE-PCR approach to amplify a 2.3 kb fragment of the murine homologue of CPM from bone marrow-derived mouse macrophages. However expression of murine CPM in primary mouse macrophages (bone marrow-derived and thioglycolate-elicited peritoneal macrophages), several myeloid and non-myeloid cell lines was below the detection limit in Northern Analysis (unpublished observations). There seems to be a significant difference in CPM biology between mice and humans and mice are probably not a useful animal model to study the function of CPM in macrophages.

## 7. CONCLUSIONS

The two monoclonal antibodies MAX.1 and MAX.11 were originally defined as differentiation markers and are now linked to carboxypeptidase M as a clearly defined molecule. A future challenge will be to define the function of CPM in human macrophages. It is difficult to predict biological effects of CPM on macrophages for immune mediated processes. A large group of substrates can be processed by the enzyme and our knowledge about the importance of these substrates in inflamed and normal tissues is still uncomplete. A striking feature is the very strong expression of CPM in renal transplant rejections, a process induced by activated T-cells. Sufficient evidence exists that macrophage are important effector cells in other T-cell mediated events. Biologically active peptides are involved in many of these immune reactions and a fast metabolism may be necessary to balance their effects. Macrophages carrying CPM on their surface may therefore play an important, yet to be defined role in normal or pathological turnover of these messenger molecules.

## ACKNOWLEDGMENTS

This work was supported by the Deutsche Forschungsgemeinschaft.

## REFERENCES

Akagawa, K.S., Takasuka, N., Nozaki, Y., Komuro, I., Azuma, M., Ueda, M., Naito, M., and Takahashi, K., 1996, *Blood* **88**: 4029-4039.

Andreesen, R., Boyce, N.W., and Atkins, R.C., 1987, *Transplant. Proc.* **19**: 2885-2888.

Andreesen, R., Bross, K.J., Osterholz, J., and Emmerich, F., 1986, *Blood* **67**: 1257-1264.

Andreesen, R., Brugger, W., Kunze, R., Stille, W., and von Briesen, H., 1990b, *Res. Virol.* **141**: 217-224.

Andreesen, R., Brugger, W., Scheibenbogen, C., Kreutz, M., Leser, H.G., Rehm, A., and Lohr, G.W., 1990a, *J. Leukoc. Biol.* **47**: 490-497.

Andreesen, R., Brugger, W., Thomssen, C., Rehm, A., Speck, B., and Lohr, G.W., 1989, *Blood* **74**: 2150-2156.

Andreesen, R., Gadd, S., Costabel, U., Leser, H.G., Speth, V., Cesnik, B., and Atkins, R.C., 1988, *Cell. Tissue Res.* **253**: 271-279.

Andreesen, R., Osterholz, J., Bross, K.J., Schulz, A., Luckenbach, G.A., and Lohr, G.W., 1983b, *Cancer Res.* **43**: 5931-5936.

Andreesen, R. Picht, J., and Lohr, G.W., 1983a, *J. Immunol. Methods* **56**: 295-304.

Bhoola, K.D., Figueroa, C.D., and Worthy, K.,1992, *Pharmacol. Rev.* **44**: 1-80.

Bockmann, S., and Paegelow, I., 1995, *Eur. J. Pharmacol.* **291**: 159-165.

Brugger, W., Kreutz, M., and Andreesen, R., 1991, *J. Leukoc. Biol.* **49**: 483-488.

Brugger, W., Reinhardt, D., Galanos, C., and Andreesen, R., 1991, *Int. Immunol.* **3**: 221-227.

Deddish, P.A., Skidgel, R. A., Kriho, V.B., Li, X.Y., Becker, R.P., and Erdos, E.G., 1990, *J. Biol. Chem.* **265**: 15083-15089.

Ember, J.A., Sanderson, S.D., Hugli, T.E., and Morgan, E.L., 1994, *Am. J. Pathol.* **144**: 393-403.

Furth, R. van, 1989, *Curr. Top. Pathol.* **79**: 125-150.

Grassi, F., Dezutter-Dambuyant, C., McIlroy, D., Jacquet, C., Yoneda, K., Imamura, S., Boumsell, L., Schmitt, D., Autran, B., Debre, P., and Hosmalin, A., 1998, *J. Leukoc. Biol.* **64**: 484-493.Kreutz, M., and Andreesen, R., 1990, *Blood* **76**: 2457-2461.

Konur, A., Kreutz, M., Knuchel, R., Krause, S.W., and Andreesen, R., 1996, *Int. J. Cancer* **66**: 645-652.

Konur, A., Kreutz, M., Knuchel, R., Krause, S.W., and Andreesen, R., 1998, *Int. J. Cancer* **78:** 648-653.

Maass, G., Schmidt, W., Berger, M., Schlicher, F., Koszik, F., Schneeberger, A., Stingl, G., Birnstiel, M.L., and Schweighoffer, T., 1995, *Proc. Natl. Acad. Sci.* USA **92:** 5540-5544.

Magnan, J., Paterson, S.J., and Kosterlitz, H.W., 1982, *Life Sci.* **31:** 1359-1361.

Mantovani, A., Bottazzi, B., Colotta, F., Sozzani, S., and Ruco, L., 1992, *Immunol. Today* **13:** 265-270.

Matsumura, Y., Maeda, H., and Kato, H., 1990, *Agents Actions* **29:** 172-180.

Nagae, A., Abe, M., Becker, R.P., Deddish, P.A., Skidgel, R.A., and Erdos, E.G., 1993, *Am. J. Respir. Cell. Mol. Biol.* **9:** 221-229.

Nagae, A., Deddisch, P.A., Becker, R.P., Anderson, C.H., Abe, M., Tan, F., Skidgel, R.A., and Erdos, E.G., 1992, *J. Neurochem.* **59:** 2201-2212.

Peters, J.H., Gieseler, R., Thiele, B., and Steinbach, F., 1996, *Immunol. Today* **17:** 273-278.

Radulovic, J., Dimitrijevic, M. Laban, O., Stanojevic, S., Vasiljevic, T., Kovacevic-Javanovic, V., and Markovic, B.M., 1995, *Peptides* **16:** 1209-1213.

Radulovic, J., Mancev, Z., Stanojevic, S., Vasiljevic, T., Kovacevic-Jovanovic, V., and Pesic, G., 1996, *J. Neuroimmunol.* **65:** 155-161.

Rehli, M., Krause, S.W., Kreutz, M., and Andreesen, R., 1995, *J. Biol. Chem.* **270:** 15644-15649.

Roy, S., Sedqi, M., Ramakrishnan, S., Barke, R.A., and Loh, H.H., 1996, *Cell. Immunol.* **169:** 271-277.

Rutherford, M.S., Witsell, A., and Schook, L.B., 1993, *J. Leukoc. Biol.* **53:** 602-618.

Saint-Vis, B. de, Cupillard, L., Pandrau-Garcia, D., Ho, S., Renard, N., Grouard, G., Duvert, V., Thomas, X., Galizzi, J.P., and Banchereau, J., 1995, *Blood* **86:** 1098-1105.

Scheibenbogen, C., and Andreesen, R., 1991, *J. Leukoc. Biol.* **50:** 35-42.

Skidgel, R.A., Davis, R.M., and Tan, F., 1989, *J. Biol. Chem.* **264:** 2236-2241.

Skidgel, R.A., 1988, *Trends Pharmacol. Sci.* **9:** 299-304.

Steinmann, R.M., 1991, *Annu. Rev. Immunol.* **9:** 271-296.

Sutherland, R.M., 1988, *Science* **240:** 177-184.

Tan, F. Chan, S.J., Steiner, D.F., Schilling, J.W., and Skidgel, R.A., 1989, *J. Biol. Chem.* **264:** 13165-13170.

Tiffany, C.W., and Burch, R.M., 1989, *FEBS Lett.* **247:** 189-192.

Tsuruta, T., Yamamoto, T., Matsubara, S. Nagasawa, S., Tanase, S., Tanaka, J., Takagi, K., and Kambara, T., 1993, *Am. J. Pathol.* **142:** 1848-1857.

Udagawa, N., Takahashi, N., Akatsu, T., Tanaka, H., Sasaki, T., Nishihara, T., Koga, T., Martin, T.J., and Suda, T., 1990, *Proc. Natl. Acad. Sci. USA* **87:** 7260-7264.

Wetsel, R.A., 1995, *Curr. Opin. Immunol.* **7:** 48-53.

Yoshioka, S., Fujiwara, H., Yamada, S., Nakayama, T., Higuchi, T., Inoue, T., Mori, T., and Maeda, M., 1998, *Mol. Hum. Reprod.* **4:** 709-717.

Ziegler-Heitbrock, H.W., Fingerle, G., Strobel, M., Schraut, W., Selter, F., Schutt, C., Passlick, B., and Pforte, A., 1993, *Eur. J. Immunol.* **23:** 2053-2058.

# MATRIX METALLOPROTEINASES (MMP-8, -13, AND -14) INTERACT WITH THE CLOTTING SYSTEM AND DEGRADE FIBRINOGEN AND FACTOR XII (HAGEMANN FACTOR)

[1]Harald Tschesche, Andrea Lichte, Oliver Hiller, André Oberpichler, [2]Frank H. Büttner, Eckart Bartnik
*[1]University Bielefeld, Faculty of Chemistry, Department of Biochemistry, Bielefeld, Germany;*
*[2]Biochemical Research, Hoechst Marion Roussel Deutschland GmbH, Frankfurt/M., Germany*

Key words Fibrinogen, factor XII (Hagemann factor), matrix metalloproteinases, clotting, degradation

## 1. INTRODUCTION

Matrix metalloproteinases (MMPs, matrixins) form a family of structurally and functionally related zinc-containing endopeptidases. Altogether they are able to degrade most of the constituents of the extracellular matrix such as basement membrane and interstitial collagens, proteoglycans, fibronectin and laminin. Thus, they are implicated in connective tissue remodelling processes associated with embryonic development, pregnancy, growth and wound repair (Woessner 1991). The deleterious potential of the MMPs is normally controlled by the endogenous and specific tissue inhibitors of metalloproteinases (TIMPs) or the more general, nonspecific $\alpha_2$-macroglobulin (Nagase and Woessner 1999). Disturbance of the well balanced equilibrium of MMPs and TIMPs results in pathologic situations such as rheumatoid and osteoarthritis, atherosclerosis, tumour growth, metastasis and fibrosis (Nagase *et al* 1997, Johnson *et al* 1998, Yong *et al* 1998, Coussens and Werb 1996, Chembers and Matrisian 1997). Seventeen different human MMPs have so far been identified and several have been cloned. They all share significant sequence homologies especially within the zinc binding region and exhibit a common multidomain organization. According to their structural and functional properties the members of the MMP family are subdivided into five groups: (i) the collagenases (MMP-1, -8, and -13), (ii) the gelatinases A and B (MMP-2 and 9), (iii) the stromelysins 1, 2 and 3 (MMP-3, -10, and -11), (iv) a more heterogenous subgroup with matrilysin (MMP-7), macrophage metallo elastase (MMP-12), enamelysin (MMP-20), and the unnamed MMP-19, and finally (v) the membrane-type MMPs -14 to -17 (MT-MMPs-1 to -4). As a family of evolutionary related proteins their zinc-containing catalytic domains share a common structural folding pattern with similar enzymatic mechanisms, but different specificities due to differences in subsites evolved from divergent evolution. The TIMP family currently includes four members (TIMPs-1 to -4), also homologous proteins, each containing six disulfide bridged loops. The N-terminal three loops generate the inhibitory activity while the C-terminal parts can develop binding properties to the hemopexin-like domains of the progelatinases e. g. TIMP-1 for proMMP-9 and TIMP-2 for proMMP-2 (Strongin and Bannikov *et al* 1995, Strongin *et al* 1993). The TIMPs also seem to have other functions such as growth factor-like and antiangiogenic activity (Gomez *et al* 1997, Cawston 1998). All TIMPs form tight 1 : 1 stoichiometric complexes with the catalytic site of MMPs and do not discriminate much between the various MMPs except for TIMP-1 and MT1-MMP or MT2-MMP where the interaction is rather weak (Sato *et al* 1996, Zucker *et al* 1998).

The intrinsic pathway of the clotting cascade is initiated by activation of the Hagemann factor to yield active factor XIIa. After several steps of further zymogen activations it terminates in the proteolytic conversion of prothrombin into thrombin, which cleaves the propeptides A and B from fibrinogen to allow fibrin monomer polymerization (Rock and Wells 1997).

## 2. DEGRADATION OF FIBRINOGEN BY CDMMP-14

Human fibrinogen was incubated with the catalytic domain of MMP-14 (cdMT1-MMP [Ile 114-Ile 318]) in a molar ratio of 10:1 for up to 3 h at body temperature (37 °C). The fragments obtained after certain time intervals were separated according to their molecular weights in a SDS-polyacrylamide gel electrophoresis (PAGE). A successive degradation with time becomes obvious. The fragment mixture was separated by HPLC and individual fragments were isolated and subjected to automated Edman degradation for identification of the cleavage sites. Thus, six major fragments with $M_r$ of 45, 38, 35, 15, 13 and 6 kDa detected in the SDS-polyacrylamide gel (PAGE) could be identified. Their amino-terminal sequences revealed three major cleavages in the fibrinogen α-chain and two major cleavages in the fibrinogen (γ-chain. The α-chain was cleaved between residues Ser 104-Leu 105, Asp 116-Phe 117, and Lys 432-Leu 433. The α-chain seemed to be subject to more rapid cleavage than the β-chain and the (γ-chain (Fig 1). Therefore, larger fragments and distinct cleavage sites have so far not been identified from the fibrinogen β-chain. The (γ-chain is cleaved between Gln 91-Leu 92 and Met 104-Ile 105.

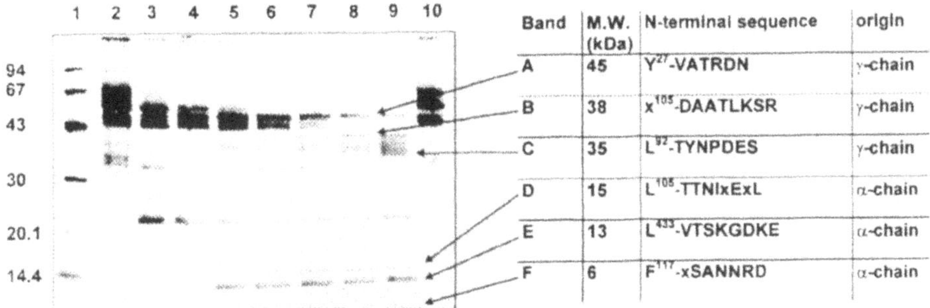


| Band | M.W. (kDa) | N-terminal sequence | origin |
|---|---|---|---|
| A | 45 | $Y^{27}$-VATRDN | γ-chain |
| B | 38 | $x^{105}$-DAATLKSR | γ-chain |
| C | 35 | $L^{92}$-TYNPDES | γ-chain |
| D | 15 | $L^{105}$-TTNIxExL | α-chain |
| E | 13 | $L^{433}$-VTSKGDKE | α-chain |
| F | 6 | $F^{117}$-xSANNRD | α-chain |

*Figure 1.* Time course of fibrinogen digestion by cdMT1-MMP.
At the indicated time intervals aliquots were taken and the digestion was stopped by adding EDTA and denaturing buffer. Samples were then subjected to SDS-PAGE in 10 % gels. Fibrinogen was digested by cdMT1-MMP in a 1 : 10 enzyme to substrate molar ratio for 3 h at 37 °C. Samples were taken at 0, 1 min, 5 min, 15 min, 30 min, 1 h, 2 h, and 3 h. Lane 1 : low molecular weight standard, lane 2: fibrinogen at time 0, lanes 3 - 9: samples taken at indicated time intervals, lane 10: fibrinogen after 3 h of incubation with buffer alone.

The cleavage within the α-chain between Lys 432 and Leu 433 detaches the C-terminal part of the amino acid sequence, because this part is no longer covalently bound, e.g. by disulfide bridging (Fig 2). This part contains the terminal RGD sequence, residues 591 to 593, responsible for cell attachment mediated by β-integrin receptors (Cheresh and Berliner 1989). The two additional cleavages in the α-chain at residues 104 and 116 remove a second RGD sequence within a dodeka fragment also not covalently attached to the rest of the molecule. The cleavage within the (γ-chain at Met104 might further lead to removal of the thrombocyte aggregation site. These cleavages finally lead to a situation similar to the physiological degradation of fibrinogen by plasmin. It has previously been demonstrated that removal of the C-terminal RGD-sequences from the α-chain and removal of the thrombocyte aggregation site from the C-terminus of the (γ-chain by plasmin prevents the $\alpha_v\beta_3$-integrin mediated cell adhesion of M 21 melanoma cells to fibrinogen (Felding-Habermann *et al* 1993).

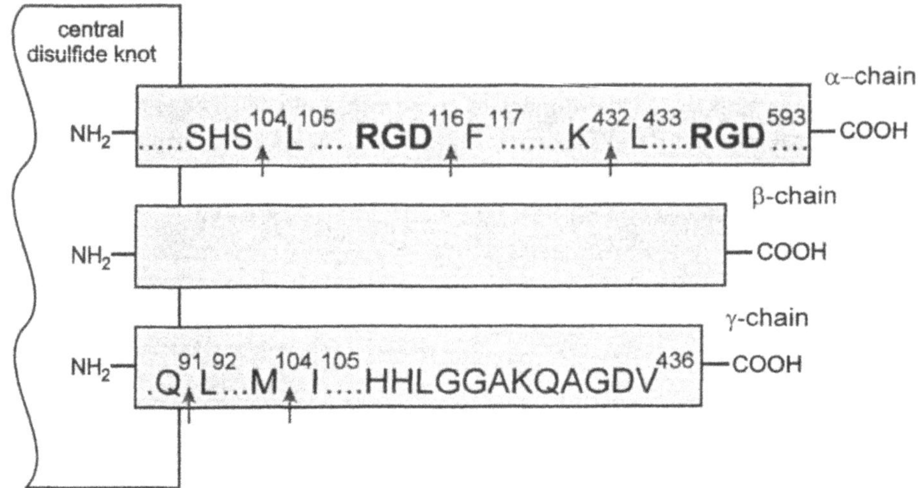


*Figure 2.* Simplified model of the fibrinogen molecule and its cleavage sites for cdMT1-MMP.

Only one half of the symmetrical molecule is shown. The locations of the putative adhesive sites identified for αIIbβ3 integrins are indicated in bold letters. These include RGD both at the NH2-terminus and the COOH-terminus of the α-chain, and the dodecapeptide sequence HHLGGAKQAGDV at the COOH-terminus of the γ-chain, respectively. The cleavage sites for cdMT1-MMP are indicated by arrows.

## 3. DEGRADATION OF FIBRINOGEN BY CDMMP-13

Fibrinogen was also rapidly degraded by the catalytic domain of human collagenase 3 (cdMMP-13 [Tyr 85-Gly 248]) at an enzyme to substrate ratio of 1 : 10 at body temperature (37 °C). The time course of the cleavage of the α-, β-, and (γ-chain within minutes is shown in Figure 1. Six fragments with $M_r$ of 45, 35, 32, 28, 20, 16, and 14 from all three fibrinogen chains were isolated by HPLC-separation and purified by SDS-polyacrylamide gel electrophoresis (PAGE) and subjected to N-terminal sequence determination (Fig 3). Fragment G indicated a proteolytic cleavage in the α-chain at the peptide bond Glu 441-Leu 442. This fragmentation at a site similar to the cleavage observed by MMP-14 also releases the C-terminal RGD cell attachment site from fibrinogen, i.e. residues 591 to 593. The impairment of the coagulability of fibrinogen by MMP-13 was even more dramatic as with MMP-14, see below, point 6.

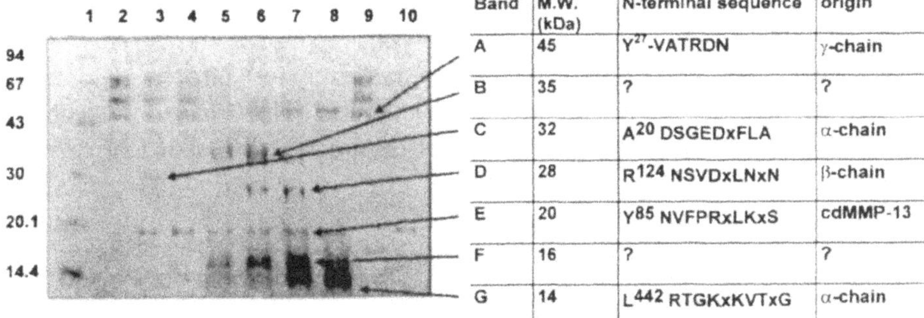


| Band | M.W. (kDa) | N-terminal sequence | origin |
|---|---|---|---|
| A | 45 | $Y^{27}$-VATRDN | γ-chain |
| B | 35 | ? | ? |
| C | 32 | $A^{20}$ DSGEDxFLA | α-chain |
| D | 28 | $R^{124}$ NSVDxLNxN | β-chain |
| E | 20 | $Y^{85}$ NVFPRxLKxS | cdMMP-13 |
| F | 16 | ? | ? |
| G | 14 | $L^{442}$ RTGKxKVTxG | α-chain |

*Figure 3.* Time course of fibrinogen digestion by cdMMP-13
At the indicated time intervals aliquots were taken and the digestion was stopped by adding EDTA and denaturing buffer. Samples were separated by SDS-PAGE in 10 % gels. Fibrinogen was digested by cdMMP-13 in a 1:10 enzyme to substrate molar ratio for about 2 h at 37 °C. Samples were taken at 0, 4 min, 8 min, 16 min, 32 min, 64, and 128 min. Lane 1: low molecular weight standard, lanes 2-8: samples taken at indicated time intervals, lane 9: fibrinogen after 2 h of incubation with buffer alone, lane 10: cdMMP-13 after 2 h of incubation alone.

# 4. DEGRADATION OF FIBRINOGEN BY CDMMP-8

The catalytic domain of neutrophil collagenase, cdMMP-8 [Met 80-Gly 242], also rapidly cleaved the α-chain of human plasma fibrinogen into several fragments. Obviously, the β- and (γ-chain are cleaved at a slower rate (Fig 4).

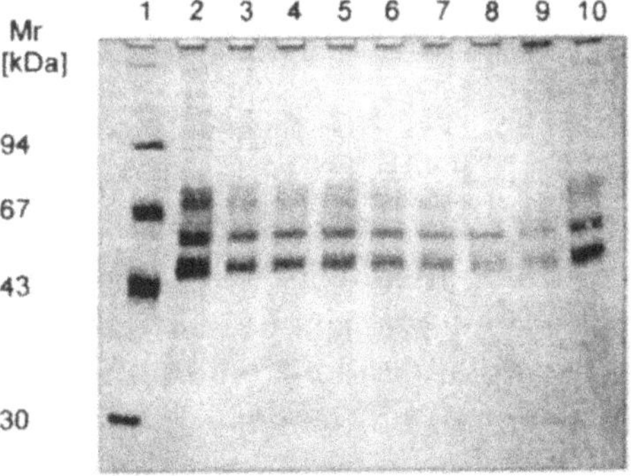


*Figure 4.* Time course of fibrinogen digestion by cdMMP-8.
At the indicated time intervals aliquots were taken and the digestion was stopped by adding EDTA and denaturing buffer. Samples were denatured and then electrophoresed on 10 % SDS gels. Fibrinogen was digested by cdMMP-8 in a 1 : 50 enzyme to substrate molar ratio for 3 h at 37 °C. Samples were taken at 0, 1 min, 5 min, 15 min, 30 min, 1 h, 2 h, and 3 h. Lane 1: low molecular weight standard, lanes 2 - 9: samples taken at indicated time intervals, lane 10: fibrinogen after 3 h of incubation with buffer alone (control).

About a dozen of fragments were separated by SDS-polyacrylamide gel electrophoresis and blotted onto PVDF-membrane (polyvinylidene difluoride). Individual bands were cut out and placed into an automated sequencer for N-terminal sequence determination. Three of the five bands with $M_r$ 10, 12, and 13 kDa revealed the N-terminus with Leu 442 (Fig 5).

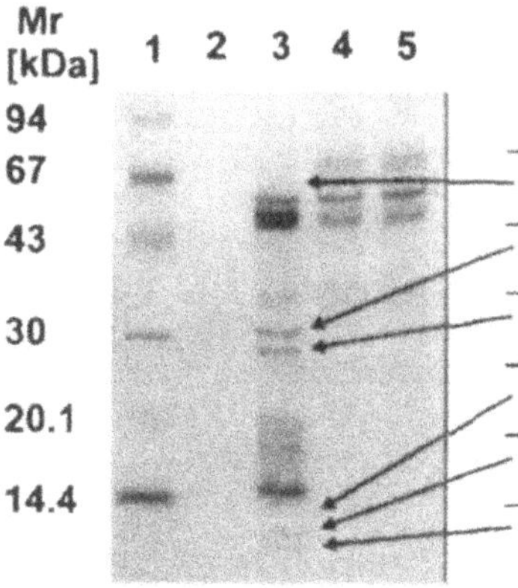

Mr [kDa] 1 2 3 4 5
94
67
43
30
20.1
14.4

| Band | M. W. [kDa] | N-terminal sequence | origin |
|---|---|---|---|
| A | 68 | $A^{20}$-D-S-G-E-G-D- | α-chain |
| B | 32 | $A^{20}$-D-S-G-E-G-D- | α-chain |
| C | 30 | $A^{20}$-D-S-G-E-G-D- | α-chain |
| D | 13 | $L^{442}$-R-T-G-K-E-K-V | α-chain |
| E | 12 | $L^{442}$-R-T-G-K-E-K-V | α-chain |
| F | 10 | $L^{442}$-R-T-G-K-E-K-V | α-chain |

*Figure 5.* Degradation of fibrinogen by cdMMP-8.
Fibrinogen was incubated with cdMMP8 in a 1 : 50 enzyme to substrate molar ratio for 2 h at 37 °C. Lane 1: low molecular weight standard, lane 2: cdMMP-8 after two hours of incubation alone, lane 3: incubation of fibrinogen with cdMMP-8 for 2 h at 37°, lane 4: fibrinogen at time 0, lane 5: fibrinogen after 2 h incubation with buffer alone.

This indicated the same cleavage site as was identified with MMP-13. Thus, leucocyte collagenase also removes the C-terminal RGD cell attachment site from the fibrinogen α-chain, residues 591 to 593.

## 5. IMPAIRMENT OF COAGULABILITY BY MMPS

In a coagulator (Amelung Modul 17, K10, Amelung GmbH, Lemgo) the amount of functional fibrinogen can be determined. The coagulation is induced with a huge excess of thrombin with the effect that the coagulation time depends only on the amount of functional fibrinogen. The specific time period from the application of active thrombin to the fibrinogen solution is measured until coagulation takes place. A calibration curve was established with different concentrations of fibrinogen that was used to determine the residual amount of functional fibrinogen after various time intervals (3, 5, 8, 20, and 40 min.) of incubation with the MMPs. Immediately after removal of the individual time aliquots the respective MMP was inhibited by addition of batimastat. The coagulability decreases rapidly with time after addition of MMP-8, -13, or -14 to the fibrinogen solution (Fig 6). After a few minutes the amount of functional fibrinogen drops to about 50 % of the original value and then levels off to about 25 %. This behavior is shown by all three MMPs with a slightly more drastic effect of MMP-13 (Fig 6).

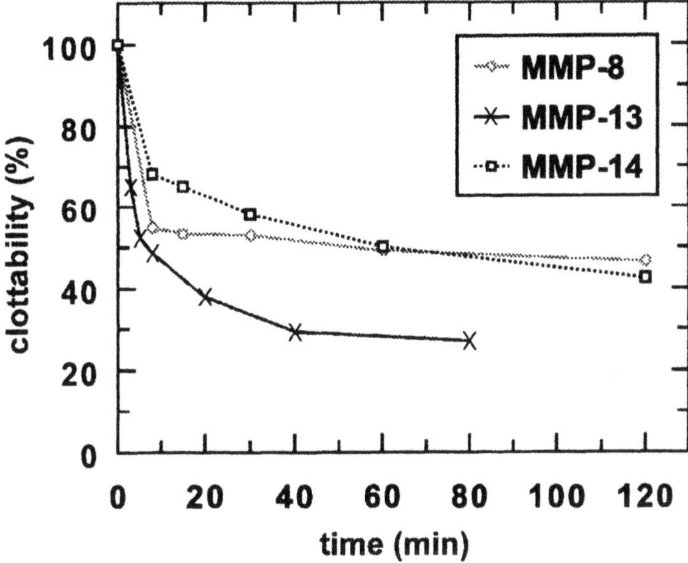


*Figure 6.* Effect of cdMT1-MMP on the thrombin-induced clotting of fibrinogen. Coagulability was expressed as the percentage of clottable protein after incubation with MMPs in a time-dependent manner. The clottability of untreated fibrinogen (control, time point 0 min) was set to 100 %. Values are the mean of three determinations and the mean standard deviations are always < 10 %.

## 6. DEGRADATION OF FACTOR XII (HAGEMANN-FACTOR) BY MMP-13

Hagemann factor (factor XII) is a glycoprotein of about 80 kDa present as a proenzyme in the circulation in human blood plasma (Fujikawa and McMullen 1983, McMullen and Fujikawa 1985). After activation by a single cleavage in the clotting cascade the active factor XIIa finally leads to the conversion of prothrombin to active thrombin. The degradation with time of factor XII by MMP-13 at an enzyme to substrate ratio of 1:30 was investigated at 37 °C. The proteolytic cleavages resulted in a breakdown of factor XII (80 kDa) into several fragments in minutes as revealed by the SDS-polyacrylamide gel (PAGE) electrophoresis (Fig 7). After 4 to 8 minutes two new fragments with $M_r$ 45 and 30 kDa (A and C) were formed that were subject to further degradation. Fragments B and C disappeared almost completely after two hours of incubation, but other fragments with $M_r$ 27 and 22 kDa (D and E) and after two hours fragment B with 43 kDa appeared. The bands of the SDS-PAGE were blotted onto a PVDF-membrane and subjected to automated Edman degradation.

The N-terminal sequences of the bands A to D were determined (Fig 7) and revealed the two major cleavage sites in the polypeptide chain of factor XII. Fragment B with N-terminal Val 94 indicated a cleavage prior to the EGF-like domain I, while fragments C and D with N-terminal Leu 377 indicated a cleavage at the beginning of the catalytic domain of factor XIIa.

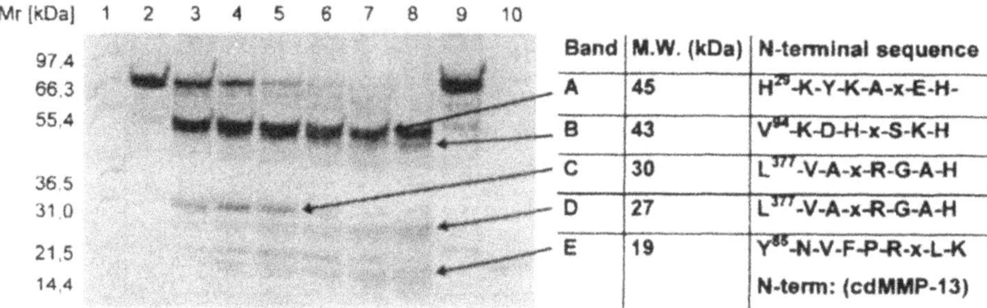


| Band | M.W. (kDa) | N-terminal sequence |
|---|---|---|
| A | 45 | $H^{29}$-K-Y-K-A-x-E-H- |
| B | 43 | $V^{94}$-K-D-H-x-S-K-H |
| C | 30 | $L^{377}$-V-A-x-R-G-A-H |
| D | 27 | $L^{377}$-V-A-x-R-G-A-H |
| E | 19 | $Y^{85}$-N-V-F-P-R-x-L-K |
| | | N-term: (cdMMP-13) |

*Figure 7.* Time course of factor XII digestion by cdMMP-13.
At the indicated time intervals aliquots were taken and the digestion was stopped by adding EDTA and denaturing buffer. Samples were then electrophoresed on 10 % SDS gels. Factor XII was digested by cdMMP-13 in a 1 : 30 enzyme to substrate molar ratio for 3 h at 37 °C. Samples were taken at 0, 4 min, 8 min, 16 min, 32 min, 64 min, and 128 min. Lane 1 : low molecular weight standard, lanes 2 - 8: samples taken at indicated time intervals, lane 9: factor XII after 3 h of incubation with buffer alone, and lane 10: active cdMMP-13 (control).

## 7. DEGRADATION OF FACTOR XII BY MMP-14

A comparable degradation as with MMP-13 of factor XII by the catalytic domain of MMP-14 (cd MT1-MMP) could be observed (Fig 8) after incubation at body termperature (37 °C) for up to 2 hours at an enzyme to substrate ratio of 1 : 15. Fragments of $M_r$ 45, 33, 30, and 12 kDa were generated that were separated by SDS-PAGE, blotted onto PVDF-membrane, cut out and subjected to automated Edman degradation. The N-terminus of the $M_r$ 30 kDa fragment started with Leu 377 and indicated the same cleavage as observed with cdMMP-13. The $M_r$ 33 kDa fragment revealed an N-terminus with Leu 351 (Fig 8) which is obviously an earlier fragment that is processed further to the $M_r$ 30 kDa degradation product. The active factor XIIa produced by plasma kallikrein, however, exhibits Asn 354 which contains three residues less than the 33 kDa fragment from MMP-14 cleavage.

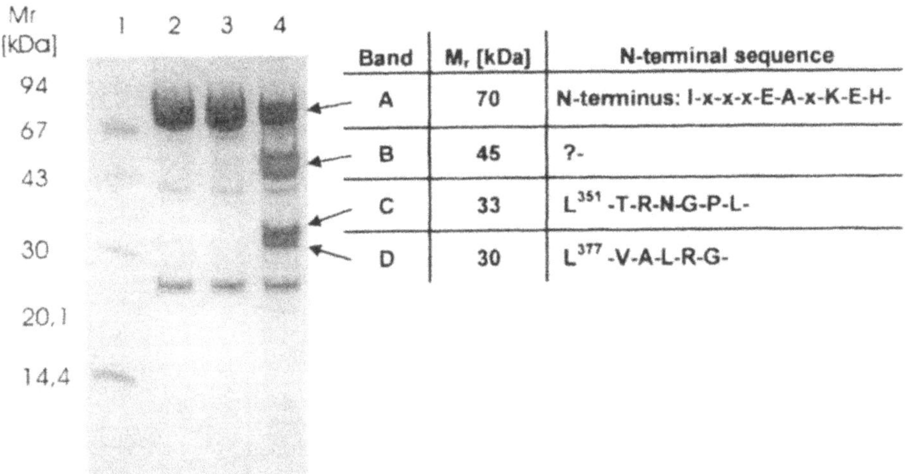


| Band | $M_r$ [kDa] | N-terminal sequence |
|---|---|---|
| A | 70 | N-terminus: I-x-x-x-E-A-x-K-E-H- |
| B | 45 | ?- |
| C | 33 | $L^{351}$-T-R-N-G-P-L- |
| D | 30 | $L^{377}$-V-A-L-R-G- |

*Figure 8.* Degradation of factor XII by cdMMP-14.
Factor XII was incubated with cdMMP-14 in a 1:15 enzyme to substrate molar ratio for 2 h at 37 °C. Lane 1: low molecular weight standard, lane 2: Factor XII alone at time 0, lane 3: Factor XII alone after 2 h at 37°, lane 4: Factor XII after 2 h of incubation with cdMT-1MMP at 37 °C.

The cleavage of the Gly 379-Leu 377 peptide bond generating the $M_r$ 30 kDa fragment could perhaps produce a factor XIIa like fragment similar to the β-factor XIIa generated by plasma kallikrein (Fig 9). However, no intermediate factor XIIa activity could be observed against the specific chromogenic substrate S-2302 (Chromogenix, Sweden). Neither could any activity be observed from the $M_r$ 33 kDa fragment generated from the cleavage of the Ser 350-Leu 351 peptide bond. This fragment is similar to the first part of the β-factor XIIa, residues at Asp 354-Leu 363. The catalytic site is present in the sequence domain from residues 373 to 615. It is interesting to note that the fragment, residue 377 to 615, generated by either MMP-13 or MMP-14 does not exhibit any activity though it contains only three N-terminal residues less than the active catalytic domain with the N-terminal amino acid sequence Val 374-Gly-Gly-Leu 377-Val-Ala-Leu.

It can be assumed that the N-terminus of active factor XIIa requires the terminal, positive charge of Val 374 for formation of the activating salt bridge as demonstrated in other active serine proteinases, e.g. Ile-Val-terminus in trypsin (Bode *et al* 1978, Huber and Bode 1978).

The cleavage specificity of the MMPs obviously varies considerably in different enzymes. No cleavages could be observed in factor XII with MT2-MMP, MMP-8, or MMP-9. However, there are similar cleavages of MMP-13 and MMP-14 (cdMT1-MMP). As has been shown previously (Li *et al* 1998), the catalytic domain of MMP-14 (MT1-MMP) is shed from the

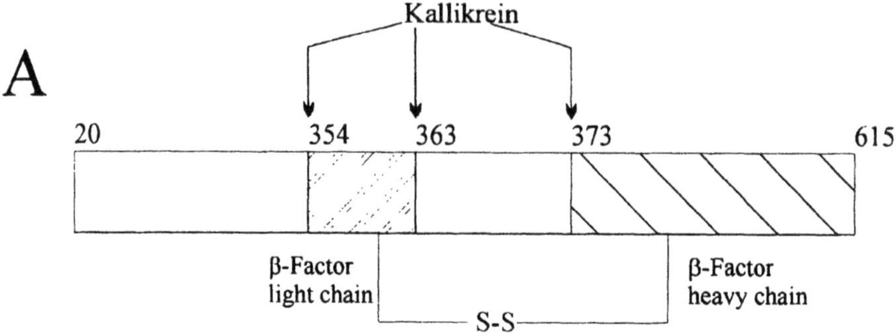


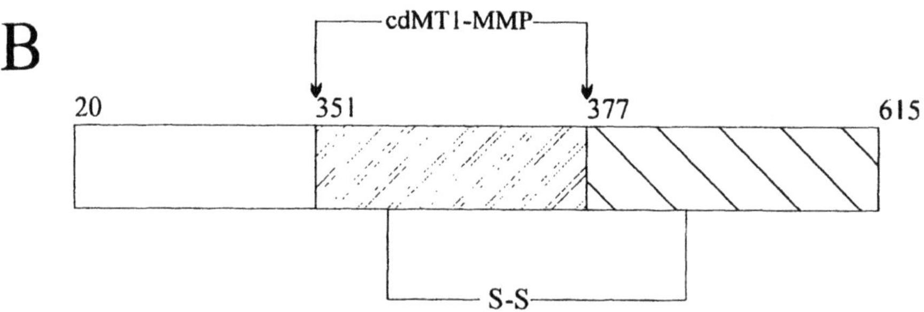


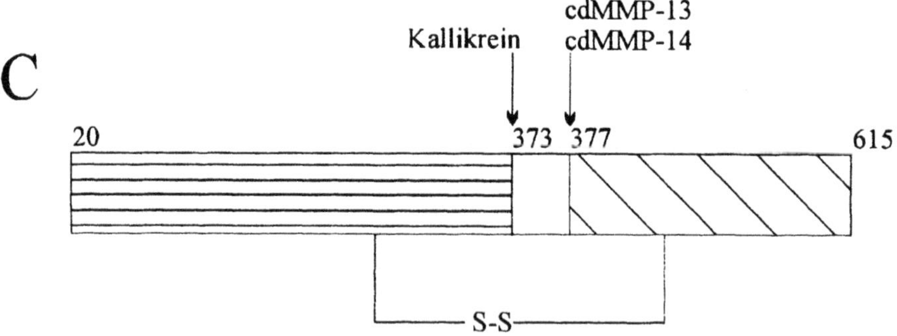


*Figure 9.* Comparison of factor XII cleavage sites.
The similarity of the factor XII degradation induced by both kallikrein and MMPs is displayed. Cleavage by kallikrein leads to activation of factor XII, whereas degradation by MMPs leads to a loss of catalytic potential.

membrane and can be detected as a soluble enzyme in plasma. Further studies will elucidate the pathophysiological relevance of the interaction of MMPs with the clotting system in vivo.

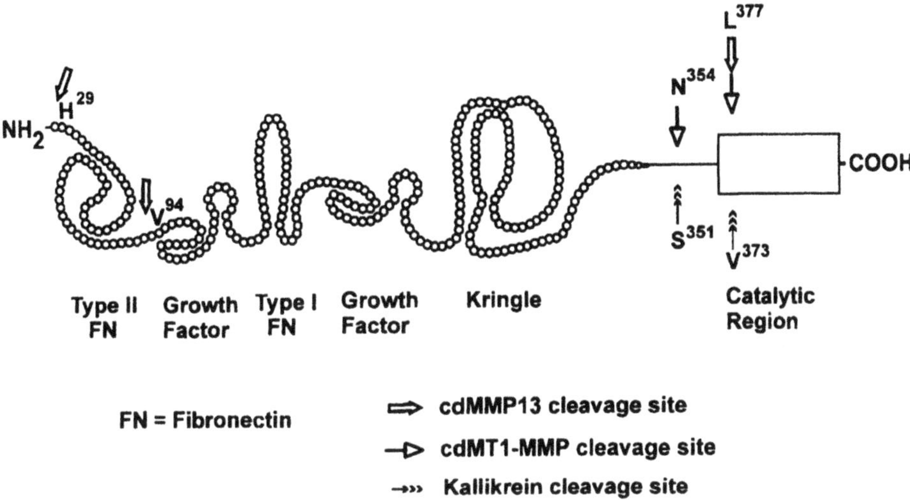


*Figure 10.* Schematic representation of factor XII and its cleavage sites. The protein domain structure and the cleavage sites for MMP-13 and -14 are displayed. Sites cleaved by kallikrein during initiation of blood coagulation are displayed for comparison.

# ACKNOWLEDGMENTS

This work was supported by the Deutsche Forschungsgemeinschaft, Bonn, SFB 549, project A05.

# REFERENCES


Bode, W., Schwager, P., and Huber, R., 1978. The Transition of Bovine Trypsinogen to a Trypsin-like State upon Strong Ligand Binding. *J. Mol. Biol.* **118**: 99-112.

Cawston, T., 1998, Matrix Metalloproteinases and TIMPs: Properties and Implications for the Rheumatic Diseases. *Mol. Med. Today* **4**: 130-137.

Chembers, A.F., and Matrisian L.M., 1997, Changing Views of the Role of Matrix Metalloproteinases in Metastasis. *J. Natl. Cancer Inst.* **89**: 1260-1270.

Cheresh, D.A., and Berliner, S.A., 1989, Recognition of Distinct Adhesive Sites on Fibrinogen by Related Integrins or Platelets and Endothelial Cells. *Cell* **58**: 945-953.

Coussens, L.M., and Werb Z., 1996, Matrix Metalloproteinases and the Development of Cancer. *Chem. Biol.* **3**: 895-904.

Felding-Habermann, B., Ruggeri, Z.M., and Cheresh, D.A., 1993, Distinct Biological Consequences of Integrin $\alpha_v\beta_3$ Mediated Melanoma Cell Adhesion to Fibrinogen and its Plasmic Fragments. *J. Biol. Chem.* **267**: 5070-5077.

Fujikawa, K., and McMullen, B A., 1983, Amino Acid Sequence of Human β-Factor XIIa. *J. Biol. Chem.* **258**: 10924-10933

Gomez, D.E., Alonso, D.F., Yoshiji, H., and Thorgeirsson U.P., 1997, Tissue Inhibitors of Metalloproteinases: Structure, Regulation and Biological Functions. *Eur. J. Cell Biol.* **74**: 111-122

Huber, R., and Bode, W., 1978, Structural Basis of the Activation and Action of Trypsin. *Accounts of Chemical Research* **11**: 114-122.

Johnson L.L., Dyer R., and Hupe D.J.; 1998; Matrix Metalloproteinases. *Curr. Opin, Chem. Biol.* **2**: 466-471.

Li, H., Bauzon, D.E., Xu, X., Tschesche H., Cao, J., and Sang, Q.A., 1998, Immunological Characterization of Cell-Surface and Soluble Forms of Membrane Type 1 Matrix Metalloproteinase in Human Breast Cancer Cells and in Fibroblasts. *Molecular Carcinogenesis* **22**: 84-94.

McMullen, B.A., and Fujikawa. K., 1985, Amino Acid Sequence of the Heavy Chain of Human ∀-Factor XIIa (Activated Hagemann Factor). *J. Biol. Chem.* **260**: 5328-5338.

Nagase. H., Das, S.K., Dey, S.K., Fowlkes, J.L., Huang, W., and Brew, K. 1997. In: *Inhibitors of Metalloproteinases in Development and Disease* (S. P. Hawkes, D. R. Edwards and Khokha R., eds), Harwood, Lausanne.

Nagase, H., and Woessner, J.F. Jr., 1999, Matrix Metalloproteinases. *J. Biol. Chem.* **274**: 21491-21494.

Rock, G., and Wells, P., 1997, New Concepts in Coagulation. *Critical Reviews in Clinical Laboratory. Sciences* **34**: 475-501.

Sato, H., Kinoshita T., Takino, T., Nakayama K., and Seiki M., 1996, Activation of a Recombinant Membrane Type 1-Matrix Metalloproteinase (MT1-MMP) by Furin and its Interaction with Tissue Inhibitor of Metalloproteinases (TIMP)-2. *FEBS Lett.* **393**: 101-104.

Strongin, A.Y., Collier, I.E., Bannikov, U., Marmer, B L., Grant, G.A., and Goldberg, G.I., 1995, Mechanism of Cell Surface Activation of 72 kDa Type IV Collagenase. *J. Biol. Chem.* **270**: 5331-5338.

Strongin, A.Y., Marmer, B.L., Grant, G.A., and Goldberg, G.I., 1993, Plasma Membrane-Dependent Activation of the 72-kDa Type IV Collagenase is Prevented by Complex Formation with TIMP-2. *J. Biol. Chem.* **268**: 14033-14039.

Woessner, J.F. Jr., 1991, Matrix Metalloproteinases and their Inhibitors in Connective Tissue Remodeling, *FASEB J.* **5**: 2145-2155.

Yong V.W., Krekoski C.A., Forsyth P.A., Bell R., and Edwards D.R., 1998 Matrix Metalloproteinases and Diseases of the CNS. *Trends Neurosci* **21**: 75-80.

Zucker, S., Drews, M., Conner, C., Foda, H. D., DeClerck, Y.A., Langley, K.E., et al. 1998, Tissue Inhibitor of Metalloproteinase-2 (TIMP-2) Binds to the Catalytic Domain of the Cell Surface Receptor, Membrane Type 1-Matrix Metalloproteinase 1 (MT1-MMP). *J. Biol. Chem.* **273**: 1216-1222.

# THE NEPRILYSIN FAMILY IN HEALTH AND DISEASE

Anthony J. Turner, Carolyn D. Brown, Julie A. Carson and Kay Barnes
*School of Biochemistry and Molecular Biology, University of Leeds, Leeds LS2 9JT, U.K.*

Key words neprilysin, endothelin, peptidase, converting enzyme, phosphoramidon.

Abstract The mammalian neprilysin (NEP) family comprises at least seven members: NEP itself, Kell blood group antigen (KELL), the endothelin-converting enzymes (ECE-1 and ECE-2), the enzyme PEX, associated with X-linked hypophosphataemia, "X-converting enzyme" (XCE) a CNS-expressed orphan peptidase and a soluble, secreted endopeptidase (SEP). These zinc metallopeptidases are all type II integral membrane proteins. Where identified, these enzymes have roles in the processing or metabolism of regulatory peptides and therefore represent potential therapeutic targets. A distinct feature of ECE-1 species is their existence as distinct isoforms differing in their N-terminal cytoplasmic tails. These tails play a role in enzyme targeting and turnover with di-leucine and tyrosine-based motifs affecting localization. Additional anchorage of these enzymes can also occur through palmitoylation. Bacterial homologues of the neprilysin family exist, for example the products of the *pepO* genes from *L. lactis* and *S. parasanguis*, and a recently described gene product of *P. gingivalis* which is an ECE-1 homologue that can catalyse the conversion of big endothelin to endothelin. A genomics based approach to understanding the functions of this proteinase family is aided by the completion of the *C. elegans* and *Drosophila* genomes, both of which encode multiple copies of NEP-like enzymes.

## 1. INTRODUCTION

Neprilysin (NEP) is the prototype of a family of zinc metalloproteinases that are involved in the processing or metabolism of regulatory peptides such as the opioids, tachykinins, endothelins, natriuretic peptides, and others. Thus, this family of enzymes are potential therapeutic targets in cardiovascular disease, pain and inflammatory disorders. The dramatic down-regulation of NEP reported in cases of prostate cancer may also implicate these enzymes in tumour development (Papandreou *et al* 1998).

Hence there has been substantial interest in the molecular and cellular characterization of neprilysins and the development of selective and potent inhibitors. NEP itself is a 90-100 kDa membrane glycoprotein which is widely distributed but especially abundant in the renal brush border. In its catalytic activity, NEP shares some similarity with the bacterial zinc metalloproteinase thermolysin particularly in its substrate specificity, hydrolysing oligopeptides preferentially on the amino side of hydrophobic residues, and in its sensitivity to the inhibitor phosphoramidon (PR). Molecular cloning of NEP (Devault *et al* 1987, Malfroy *et al* 1987) revealed it to be a type II integral membrane protein of approx 700 residues with a short cytoplasmic domain, a transmembrane hydrophobic region and with the bulk of the protein forming an extracellular catalytic domain with what are now recognised as typical characteristics of a "gluzincin" of the M13 peptidase family (Barrett *et al* 1998).

The first identified physiological function for NEP was in the brain where its location as a synaptic enzyme allowed it to terminate the actions of many peptide neurotransmitters, especially the enkephalins and substance P (Malfroy *et al* 1978, Matsas *et al* 1983, Barnes *et al* 1993). Its identification as the elusive "enkephalinase" led to the development of the potent inhibitor, thiorphan (Roques *et al* 1980) which has subsequently been valuable for distinguishing between NEP and other members of the family. NEP has now been shown to have equally important roles in the periphery, for example in terminating the action of the vasodilator, atrial natriuretic peptide (ANP) (Kenny and Stephenson, 1988) and in the metabolism of chemotactic and pro-inflammatory peptides (Shipp *et al* 1991). NEP is also identical with the common acute lymphoblastic leukaemia antigen (CALLA, CD10) (Letarte *et al* 1988) which is mainly associated with precursors of B lymphocytes and is also expressed on a number of malignant cell types.The structure and function of NEP has been extensively reviewed in recent years (Roques *et al* 1993, Turner, 1998) and so this chapter will focus on the roles of other family members: the endothelin converting enzymes (ECE-1 and ECE-2), the erythrocyte surface protein KELL, the PEX protein involved in the regulation of bone metabolism, and finally the recently described peptidases, designated X-converting enzyme (XCE) and soluble secreted endopeptidase (SEP). Table 1 summarises some of the properties of all these proteins. The application of functional genomics to understanding the roles of this important family of proteins will also be highlighted.

## 2. ENDOTHELIN CONVERTING ENZYMES

The biosynthesis of the potent, vasoactive endothelin peptides (ET-1, -2 and -3) involves a unique processing step in which an inactive intermediate

*Table 1.* Comparison of mammalian neprilysin family members.

| Enzyme | GenBank/EBI Accession no. | Subunit $M_r$ (kDa) | Preferred substrates |
|---|---|---|---|
| NEP | A30431 | 90-100 | SP, ANP, ET-1, enkephalins |
| ECE-1 | D49471 | 120-130 | big ETs, BK |
| ECE-2 | U27341 (bovine) | 130 | big ETs, (other?) |
| KELL | M64934 | 93 | big ET-3 |
| PEX | U81970 | 100 | PTH-related peptides |
| XCE | Y16187/Y16188 | 95 | unknown |
| SEP | AF157105 (mouse) | 110 | big ET-1, ET-1, SP, ANP, BK |

The accession numbers are for the human cDNA unless otherwise stated. Note that ECE-1 exists in several isoforms. ANP, atrial natriuretic peptide; BK, bradykinin; ET, endothelin;PTH, parathyroid hormone; SP, substance P.

termed big endothelin is hydrolysed at a unique site by ECE to generate the biologically active product (see Turner and Tanzawa, 1997 for review). Although NEP itself can catalyse the conversion process, it further degrades ET-1 ruling it out as a functional ECE. The cloning of the first physiologically relevant ECE (now termed ECE-1) was reported by several groups in 1994 and the enzyme was revealed as a homologue of NEP (37 % identity in the rat) conserving the known catalytic residues. Subsequent site-directed mutagenesis studies revealed that ECE-1 exhibits a similar catalytic mechanism but a distinct substrate binding mechanism compared with NEP (Shimada *et al* 1996). The most significant structural difference between NEP and ECE-1 is that ECE-1 exists as a disulphide-linked homodimer. Site-directed mutagenesis has established that it is residue $C^{412}$ (in rat ECE-1) that is responsible for dimerisation (Shimada *et al* 1996). The equivalent residue in NEP is glutamate ($E^{403}$). However, if this is changed to cysteine, NEP too becomes a covalent dimer indicating considerable structural similarity between the two enzymes (Hoang *et al* 1997). A further difference between ECE-1 and NEP is in their sensitivity to inhibitors, phosphoramidon and thiorphan inhibiting NEP at nanomolar concentrations whereas ECE-1 is inhibited only at micromolar levels of PR and is relatively insensitive to thiorphan. The accumulated comparative data on this family of enzymes has allowed molecular modelling of the active sites of both ECE-1 and NEP, largely based on their catalytic similarity with the bacterial enzyme, thermolysin (Sansom *et al* 1998, Tiraboschi *et al* 1999). This should aid greatly in the future design of more potent and selective inhibitors that may have therapeutic value. Although ECE-1 and ETs have mainly been studied in relation to regulation of the cardiovascular system, gene knock-out studies in mice indicate other critical biological functions, especially in the regulation of embryogenesis, by affecting development of neural crest-derived cells (Yanagisawa *et al* 1998). A loss-of-function mutation in the ECE-1 gene in a human patient was recently shown to result in a severe phenotype exhibiting Hirschsprung Disease (a congenital disorder characterized by an absence of enteric ganglia along the bowel), as well as cardiac defects and autonomic dysfunction (Hofstra *et al* 1999). The mutation was shown to result in the change of an arginine residue close to the active site ($R^{742}$) into a cysteine which abolished enzyme activity.

## 3. THE SUBSTRATE SPECIFICITY OF ECE-1

NEP has a very broad specificity for peptide substrates, hydrolysing on the amino side of hydrophobic residues. In contrast, the dogma concerning ECE-1 was that it was a highly specific peptidase reflecting its presumed unique role in converting big ET to ET. Bearing in mind the broad similarities between NEP and ECE-1, we reasoned that other substrates may exist for ECE-1 and set about examining its specificity in more detail. We used recombinant rat ECE-1 expressed in CHO cells as enzyme source and were able to show that bradykinin was efficiently hydrolysed with a $k_{cat}/K_m$ comparable to that for the conversion of big ET (Hoang and Turner, 1997). The hydrolysis of bradykinin was solely at the $Pro^7$-$Phe^8$ bond, producing bradykinin-(1-7) and bradykinin-(8-9). The bond cleaved, (Pro-Phe), is very distinct from that cleaved in the big ETs (-Trp-Val/Ile-) which indicates that ECE-1 can behave as a peptidyl dipeptidase (similar to angiotensin converting enzyme), as well as an endopeptidase. Subsequently, we and others have been able to show that ECE-1 can hydrolyse the tachykinin peptide, substance P, with hydrolysis occurring at similar sites to those previously described for NEP (Carson, J., Barnes, K. and Turner, A.J., unpublished, Johnson *et al* 1999). Johnson et al (1999) have extended the study of the specificity of ECE-1 and have identified a range of susceptible peptide substrates, which include neurotensin, bradykinin, angiotensin I and insulin B chain. It is clear, therefore, that new roles may emerge for ECE-1, and not only in cardiovascular physiology and embryology.

## 4. LOCALIZATION AND ISOFORMS OF ECE-1

ECE-1 is most abundant in endothelial cells from which it was first isolated but it is also present in a wide variety of other cells, for example on smooth muscle cells, where it can be found in association with α-actin filaments, and on neurons and glia in the brain (Barnes and Turner 1997, 1999). Whereas NEP is exclusively a cell-surface ectoenzyme, both an intracellular and surface localization have been reported for ECE-1. This heterogeneous distribution is, in part, explained by the existence of multiple isoforms of ECE-1 arising from the use of alternative promoters of a single gene located on chromosome 1 (Valdenaire *et al* 1995). To date, four isoforms (termed a, b, c, d) have been reported (Valdenaire *et al* 1999a). The nomenclature for the isoforms has proved confusing since the original two isoforms decribed in the rat were referred to as α and β (see e.g. Turner and Murphy 1996), which are now known to be equivalent to human isoforms c and a. Studies on the stable expression of individual ECE-1 isoforms in CHO cells indicate distinct subcellular localizations: thus ECE-1b appears to be exclusively intracellular (Azarani *et al* 1998) whereas a, c and d are mainly found at the cell-surface (Schweizer *et al* 1997, Valdenaire *et al* 1999a, b). Various targeting motifs have been implicated in ECE-1 isoform localization by us (Barnes *et al* 1998, Turner *et al* 1998) and recent data has confirmed the importance of dileucine-based motifs in the extreme N-terminal region of the protein in this process (Cailler *et al* 1999, Valdenaire *et al* 1999b). The functional significance of the isoforms is unclear since they appear to display identical catalytic properties but we have suggested that intracellular ECE-1, located in the Golgi and vesicle compartments, may be involved in processing of big ET whereas cell-surface ECE-1 may have additional roles in metabolism of other regulatory peptides (Hoang and Turner 1997). The development of isoform-specific antibodies against ECE-1 has greatly aided in understanding the cell and subcellular localizations of the isoforms (Brown *et al* 1998) and should be valuable in investigating changes in isoform expression in human diseases, such as atherosclerosis and cancer.

A further modification of ECE-1 is that all isoforms appear to be palmitoylated on a conserved cysteine residue on the cytoplasmic side of the membrane (Schweizer *et al* 1999a). The role of palmitoylation is unclear since it does not appear to affect the localisation or activity of the enzyme. While palmitoylation has been shown, for some proteins, to cause them to locate to specialized regions of the plasma membrane (caveolae), we failed to show co-localization of ECE-1 with caveolin-1, a marker for caveolae (Barnes *et al* 1996). Schweizer et al (1999a), subsequently, were also unable to detect palmitoylated isoforms of ECE-1 in caveolar fractions. Palmitoylation may stabilise the enzyme against retrieval from the cell-surface and subsequent degradation, since some forms of ECE-1 are short-lived (Emoto *et al* 1999) and we have shown that the enzyme can be recycled, possibly under the control of a tyrosine-based retrieval motif (Barnes *et al* 1998, Turner *et al* 1998). Another unexpected observation concerned with the subcellular localization of ECE-1 is the effect of metalloproteinase inhibitors, especially PR. When endothelial cells are cultured in the presence of PR, there is a remarkable relocation of the enzyme from the cell-surface to intracellular compartments, which is combined with an increase in total enzyme activity. We have suggested that this is due to the inhibition of a PR-sensitive, intracellular metalloproteinase involved in the degradation of ECE-1 (Barnes *et al* 1996).

## 5. OTHER MEMBERS OF THE NEPRILYSIN FAMILY

A second gene encoding an ECE-like protein (termed ECE-2) has been described, although relatively little is known about its localization and specificity (Emoto and Yanagisawa 1995). Although it can convert big endothelins to the active endothelins *in vitro*, there is no evidence that it plays any significant role in ET production *in vivo*, especially as it is predominantly expressed in brain and is of low abundance in endothelial and smooth muscle cells. ECE-2 shows 59 % amino acid sequence identity with ECE-1 but, unlike ECE-1, it has an acidic pH optimum and is virtually inactive at neutral pH. It was originally reported that ECE-2 is much more sensitive to inhibition by PR than ECE-1 but this probably just reflects the pH at which the assays were performed since the sensitivity of ECE-1 to PR inhibition increases markedly at more acidic pH values (Ahn *et al* 1998). The acidic pH optimum of ECE-2 implies localization in a secretory compartment and a role in protein processing but its physiological substrates remain an enigma. A more detailed investigation of its substrate specificity is essential.

The KELL blood group protein has 32 % to 36 % amino acid identity with NEP and ECE-1 with the similarities especially in its extracellular catalytic domain. It conserves 10 extracellular cysteine residues that are preserved in all members of this family. Kell proteins differ from other members of the family in that they are covalently linked to a protein termed XK that spans the membrane 10 times (Russo *et al* 1998). The absence of XK, which occurs in McLeod patients, is correlated with acanthocytic red blood cells and a late-onset form of nerve and muscle disorders, but its specific cellular functions are unknown. Although the Kell protein conserves all the essential catalytic residues, no enzymic activity had been ascribed to all the essential catalytic residues, no enzymic activity had been ascribed to the protein until recently. Lee et al (1999) have now shown that Kell can convert big ET-3 to ET-3 and its activity is inhibited by PR, although even more weakly than is ECE-1. Conversion of big ET-1 and big ET-2 was much less efficient. ET-3 is important in the development of the enteric nervous system and the migration of neural crest-derived cells (Baynash *et al* 1994), yet individuals lacking Kell are perfectly healthy suggesting that Kell is redundant in terms of ET-3 biosynthesis and that the protein probably has other substrates and other biological functions, both in erythroid and non-erythroid tissues.

The *PEX* gene encodes a membrane-bound metallopeptidase homologous with NEP and was identified from studies of patients with X-linked hypophosphataemic rickets (HYP Consortium, 1995). The substrate of PEX protein is presumed to be an unidentified peptide hormone (termed "phosphatonin") that regulates renal tubular phosphate handling. The full length human PEX cDNA has recently been cloned and expressed, and shown to be located as an ectoenzyme at the plasma membrane (Lipman *et al* 1998). Recombinant PEX was shown to hydrolyse parathyroid-hormone derived peptides although it is unlikely that these are the physiological substrates as they do not have the properties of the sought-after phosphatonin. A more detailed assessment of PEX activity has not been reported although it failed to hydrolyse an enkephalin analogue. A survey of PEX mRNA expression by *in situ* hybridization revealed its presence in osteoblasts and odontoblasts implicating the enzyme in the development of bones and teeth (Ruchon *et al* 1998). Two other novel members of the neprilysin family have recently been revealed. X-converting enzyme (XCE), so-called because it is an orphan peptidase for which no substrate has yet been identified, was cloned from human caudate nucleus and spinal cord cDNA libraries and was shown to be preferentially expressed in the central nervous system (Valdenaire *et al* 1999c). Disruption of the XCE gene in mice by homologous recombination resulted in neonatal lethality with the mice dying of respiratory failure (Schweizer *et al* 1999b). Thus, XCE is an essential gene, which may play a critical role in the control of respiration. A neprilysin relative, described as soluble secreted endopeptidase (SEP), has been cloned from ECE-1$^{-/-}$ embryo cDNA in a search for an alternative endothelin-converting enzyme (Ikeda *et al* 1999). Alternative splicing resulted in the generation of two isoforms, one membrane-bound and another secreted into the culture medium This shedding probably occurred via the action of a membrane protein secretase, in a mechanism similar to that causing the shedding of angiotensin converting enzyme. Characterization of the secreted form revealed it to have broad substrate specificity, hydrolysing, for example, big ET-1, ET-1, atrial natriuretic peptide, bradykinin and enzyme more closely resembled NEP than other members of the family. Whether it plays any physiological role in the metabolism of vasoactive peptides remains to be established.

## 6. GENOMICS-BASED APPROACHES TO PEPTIDASE DISCOVERY

A number of bacterial homologues of NEP have been described of which the closest similarity is seen in the product of the *pepO* gene of *Lactococcus lactis* which has an overall identity of 27.1 % with human NEP but does not resemble thermolysin in sequence. Endopeptidases from *Mycobacterium leprae* and *Streptococcus parasanguis* are also closely related to NEP, the latter enzyme hydrolysing methionine-enkephalin (Froeliger *et al* 1999) . A recently described endopeptidase from *Porphyromonas gingivalis* resembles ECE-1 more closely than NEP and is able to convert big ET-1 to ET-1, can hydrolyse bradykinin and is sensitive to inhibition by PR but not by thiorphan (Awano *et al* 1999). The *P. gingivalis* enzyme had a putative signal sequence in the *N*-terminal region but no membrane-spanning domains suggesting it may be secreted. The *L. lactis Pep O* gene product, on the other hand, lacked a signal sequence and was shown to be localized on the inner side of the cell membrane. The functions of these bacterial peptidases are unknown but the secreted *P. gingivalis* ECE-like enzyme may have pathological consequences. *P. gingivalis* has been found to invade aortic and heart endothelial cells *in vitro* and infection with this organism has been implicated in the development of atherosclerosis and coronary artery disease. Whether this is as a consequence of the metabolism of vasoactive and mitogenic peptides remains to be elucidated.

The availability of the complete genome sequences of the nematode *C. elegans* and, shortly, of *Drosophila* has allowed the full spectrum of NEP-like activities in a eukaryotic organism to be studied. Thus, there appear to be over 20 NEP/ECE-like genes in *C. elegans* and 7 have been identified in approximately half of the *Drosophila* genome. Thus, it is highly likely that there are other members of the mammalian NEP family to discover. Identifying substrates and physiological roles for new peptidases is often problematic and studies on simpler organisms such as *C. elegans* and *Drosophila* may provide useful insights. Here, expression patterns can be readily examined, as well as the effects of gene knock-outs and RNA interference studies.

## 7. CONCLUSIONS

For almost twenty years NEP was a unique mammalian metallopeptidase, typified by a broad substrate specificity, wide tissue distribution and location as an ectoenzyme at the cell-surface. It also appeared to serve diverse physiological roles, for example in terminating the actions of neuropeptide transmitters, vasoactive and inflammatory peptides. In the last few years the NEP family has become more diverse, implicated both in protein processing and inactivation, and in the regulation of bone metabolism, in respiration, reproduction, and other events. Although the NEP family would appear to provide useful therapeutic targets, no clearcut clinical application is yet available and development of inhibitors selective for individual members of the family is in its infancy. Finally, although modelling and mutagenesis has told us much about the active sites of NEP and ECE, a structural model for either of these enzymes is much in need.

# ACKNOWLEDGMENTS

We thank the British Heart Foundation and the National Heart Research Fund for financial support of this work.

# REFERENCES

Ahn, K., Herman, S.B., and Fahnoe, D.C., 1998, Soluble human endothelin-converting enzyme-1: Expression, purification, and demonstration of pronounced pH sensitivity. *Arch. Biochem. Biophys.* **359:** 258-268.

Awano, S., Ansai, T., Mochizuki, H., Tanzawa, K., Turner, A.J., and Takehara, T., 1999, Sequencing, expression and biochemical characterization of the *Porphyromonas gingivalis pepO* gene encoding a protein homologous to human endothelin converting enzyme. *FEBS Lett.* in press.

Azarani, A., Boileau, G., and Crine, P., 1998, Recombinant human endothelin-converting enzyme ECE-1b is located in an intracellular compartment when expressed in polarized Madin-Darby canine kidney cells. *Biochem. J.* **333:** 439-448.

Barnes, K., and Turner, A.J., 1997 The endothelin system and endothelin-converting enzyme in the brain: molecular and cellular studies. *Neurochem. Res.* **22:** 1033-1040.

Barnes, K., and Turner, A.J., 1999, Endothelin converting enzyme is located on α-actin filaments in smooth muscle cells. *Cardiovasc. Res.* **42:** 814-822.

Barnes, K., Turner, A.J., and Kenny, A.J., 1993, An immunoelectron microscopic study of pig substantia-nigra shows colocalization of endopeptidase-24.11 with substance P. *Neuroscience* **53:** 1073-1082.

Barnes, K., Shimada, K., Takahashi, M., Tanzawa, K., and Turner, A.J., 1996, Metallopeptidase inhibitors induce an up-regulation of endothelin-converting enzyme levels and its redistribution from the plasma membrane to an intracellular compartment. *J Cell Sci.* **109:** 919-928.

Barnes, K., Brown, C., and Turner, A.J., 1998, Endothelin-converting enzyme. Ultrastructural localization and its recycling from the cell surface. *Hypertension* **31:** 3-9.

Barrett, A.J., Rawlings, N.D., and Woessner, J.F. (eds.), 1998, *Handbook of Proteolytic Enzymes*, Academic Press, London and San Diego, pp. 1033-1037.

Baynash, A.G., Hosoda, K., Giaid, A., Richardson, J.A., Emoto, N., Hammer, R.E., and Yanagisawa, M., 1994, Interaction of endothelin-3 with endothelin-B receptor is essential for development of epidermal melanocytes and enteric neurons. *Cell* **79:** 1277-1285.

Brown, C., Barnes, K., and Turner, A.J., 1998, Anti-peptide antibodies specific to rat endothelin-converting enzyme-1 isoforms reveal isoform localisation and expression. *FEBS Lett.* **424:** 183-187.

Cailler, F., Zappulla, J.P., Boileau, G., and Crine, P., 1999, The N-terminal segment of endothelin-converting enzyme (ECE)-1b contains a di-leucine motif that can redirect neprilysin to an inracellular compartment in Madin-Darby canine kidney (MDCK) cells. *Biochem. J.* **341:** 119-126.

Devault, A., Lazure, C., Nault, C., Le Moual, H., Seidah, N.G., Chrétien, M., Kahn, P., Powell, J., Mallet, J., Beaumont, A., Roques, B.P., Crine, P., and Boileau, G., 1987, Amino acid sequence of rabbit kidney neutral endopeptidase-24.11 (enkephalinase) deduced from a complementary DNA. *EMBO J.* **6:** 1317-1322.

Emoto, N., and Yanagisawa, M., 1995, Endothelin-converting enzyme-2 is a membrane-bound, phosphoramidon-sensitive metalloprotease with acidic pH optimum. *J. Biol. Chem.* **270:** 15262-15268.

Emoto, N., Nurhantari, Y., Alimsardjono, H., Xie, J., Yamada, T., Yanagisawa, M., and Matsuo, M., 1999, Constitutive lysosomal targeting and degradation of bovine endothelin-converting enzyme-1a mediated by novel signals in its alternatively spliced cytoplasmic tail. *J. Biol. Chem.* **274:** 1509-1518.

Froeliger, E.H., Oetjen, J., Bond, J.P., and Fives-Taylor, P., 1999, *Streptococcus parasanguis pepO* encodes an endopeptidase with structure and activity similar to those of enzymes that modulate peptide receptor signaling in eukaryotic cells. *Infect. Immunity* **67:** 5206-5214.

Hoang, M.V., and Turner, A.J., 1997, Novel activity for endothelin-converting enzyme: hydrolysis of bradykinin. *Biochem. J.* **327:** 23-26.

Hoang, M.V., Sansom, C.E., and Turner, A.J., 1997, Mutagenesis of Glu$^{403}$ to Cys in rabbit neutral endopeptidase-24.11 (neprilysin) creates a disulphide-linked homodimer: analogy with endothelin converting enzyme. *Biochem. J.* **327:** 925-929.

Hofstra, R.M.W., Valdenaire, O., Arch, E., Osinga, J., Kroes, H., Löffler, B-M., Hamosh, A., Meijers, C., and Buys, C.H.C.M., 1999, A loss-of-function mutation in the endothelin-converting enzyme-1 (ECE-1) associated with Hirschsprung Disease, cardiac defects, and autonomic dysfunction. *Am. J. Hum. Genet.* **64:** 304-308.

HYP Consortium, 1995, A gene (*PEX*) with homologies to endopeptidases is mutated in patients with X-linked hypophosphatemic rickets. *Nature Genet.* **11:** 130-136.

Ikeda, K., Emoto, N., Raharjo, S.B., Nurhantari, Y., Saiki, K., Yokoyama, M., and Matsuo, M., 1999, Molecular identification and characterization of novel membrane-bound metalloprotease, the soluble secreted form of which hydrolyzes a variety of vasoactive peptides. *J. Biol. Chem.* **274:** 32469-32477.

Johnson, G.D., Stevenson, T., and Ahn, K., 1999, Hydrolysis of peptide hormones by endothelin-converting enzyme-1. A comparison with neprilysin. *J. Biol. Chem.* **274:** 4053-4058.

Kenny, A.J., and Stephenson, S.L., 1988, Role of endopeptidase-24.11 in the inactivation of atrial natriuretic peptide. *FEBS Lett.* **232:** 1-8.

Lee, S., Lin, M., Mele, A., Cao, Y., Farmar, J., Russo, D., and Redman, C., 1999, Proteolytic processing of big endothelin-3 by the Kell blood group protein. *Blood* **94:** 1440-1450.

Lipman, M.L., Panda, D., Bennett, H.P.J., Henderson, J.E., Shane, E., Shen, Y., Goltzman, D., and Karaplis, A.C., 1998, Cloning of human *PEX* cDNA. Expression, subcellular localization, and endopeptidase activity. *J. Biol. Chem.* **273:** 13729-13737.

Malfroy, B., Swerts, J.P., Guyon, A., and Roques, B.P., 1978, High-affinity enkephalin-degrading peptidase in brain is increased after morphine. *Nature* **276:** 523-526.

Malfroy, B., Schofield, P.R., Kuang, W.J., Seeburg, P.H., Mason, A.J., and Henzel, W.J., 1987, Molecular cloning and amino acid sequence of rat enkephalinase. *Biochem. Biophys. Res. Commun.* **144:** 59-66.

Matsas, R., Fulcher, I.S., Kenny, A.J. and Turner, A.J., 1983, Substance P., and [Leu]enkephalin are hydrolysed by an enzyme in pig caudate synaptic membranes that is identical with the endopeptidase of kidney microvilli. *Proc. Natl. Acad. Sci. USA* **80:** 3111-3115.

Papandreou, C.N., Usmani, B.A., Geng, Y., Bogenreider, T., Freeman, R., Wilk, S., Finstad, C.L., Reuter, V.E., Powell, C.T., Scheinberg, D., Magill, C., Scher, H.I., Albino, A.P., and Nanus, D.M., 1998, Neutral endopeptidase 24.11 loss in metstatic human prostate cancer contributes to androgen-independent progression. *Nature Med.* **4:** 50-57.

Roques, B.P., Fournié-Zaluski, M-C., Soroca, E., Lecomte, J-M., Malfroy, B., Llorens-Cortes, C., and Schwartz, J.-C., 1980, The enkephalinase inhibitor thiorphan shows anti-nociceptive activity in mice. *Nature* **288:** 286-288.

Roques, B.P., Noble, F., Daugé, V., Fournié-Zaluski, M-C., and Beaumont, A., 1993, Neutral endopeptidase 24.11. Structure, inhibition, and experimental and clinical pharmacology. *Pharmacol. Rev.* **45:** 87-146.

Ruchon, A.F., Marcinkiewicz, M., Siegfried, G., Tenehouse, H.S., DesGroseillers, L., Crine, P., and Boileau, G., 1998, Pex mRNA is localized in developing mouse osteoblasts and odontoblasts. *J. Histochem. Cytochem.* **46:** 459-468.

Russo, D., Redman, C., and Lee, S., 1998, Association of XK and Kell blood group proteins. *J. Biol. Chem.* **273:** 13950-13956.

Sansom, C.E., Hoang, M.V., and Turner, A.J., 1998, Molecular modelling and sire-directed mutagenesis of the active site of endothelin converting enzyme. *Protein Eng.* **11:** 1235-1241.

Schweizer, A., Valdenaire, O., Nelbock, P., Deuschle, U., Edwards, J.B.D.M., Stumpf, J.G., and Löffler, B.M., 1997, Human endothelin-converting enzyme (ECE-1): three isoforms with distinct subcellular locations. *Biochem. J.* **328:** 871-877.

Schweizer, A., Löffler, B-M., and Rohrer, J., 1999a, Palmitoylation of the three isoforms of human endothelin-converting enzyme-1. *Biochem. J.* **340:** 649-656.

Schweizer, A.,Valdenaire, O., Köster, A., Lang, Y., Schmitt, G., Lenz, B., Bluethmann, H., and Rohrer, J., 1999b, Neonatal lethality in mice deficient in XCE, a novel member of the endothelin-converting enzyme and neutral endopeptidase family. *J. Biol. Chem.* **274:** 20450-20456.

Shimada, K., Takahashi, M., Turner, A.J., and Tanzawa, K., 1996, Rat endothelin-converting enzyme-1 fprms a dimer through Cys(412) with a similar catalytic mechanism and a distinct substrate binding mechanism compared with neutral endopeptidase-24.11. *Biochem. J.* **315:** 863-867.

Shipp, M.A. Stefano, G.B., Switzer, S.N., Griffin, S.N., and Reinherz, E.L., 1991, CD10 (CALLA)/neutral endopeptidase 24.11 modulates inflammatory peptide-induced changes in neutrophil morphology, migration and adhesion proteins and is itself regulated by neutrophil activation. *Blood* **78:** 1834-1841.

Tiraboschi, G., Jullian, N., Thery, V., Antonczak, S., Fournie-Zaluski, M-C., and Roques, M.P., 1999, A three-dimensional construction of the active site (region 507-749) of human neutral endopeptidase (EC3.4.24.11). *Protein Eng.* **12:** 141-149.

Turner, A.J., 1998, Neprilysin. In: *Handbook of Proteolytic Enzymes.* (A.J. Barrett, N.D. Rawlings and J.F. Woessner, eds), Academic Press, London and San Diego, pp. 1080-1083.

Turner, A.J., 1998, Neprilysin. In:. *Handbook of Proteolytic Enzymes.* (A.J. Barrett, N.D. Rawlings and J.F. Woessner, eds), Academic Press, London and San Diego, pp. 1080-1083.

Turner, A.J., and Murphy, L.J., 1996, Molecular pharmacology of endothelin converting enzymes. *Biochem. Pharmacol.* **51:** 91-102.

Turner, A.J., and Tanzawa, K., 1997, Mammalian membrane metallopeptidases: NEP, ECE, KELL and PEX. *FASEB J.* **11:** 355-364.

Turner, A.J., Barnes, K., Schweizer, A., and Valdenaire, O., 1998, Isoforms of endothelin-converting enzyme: why and where? *Trends Pharmacol. Sci.* **19:** 483-486.

Valdenaire, O., Rohrbacher, E., and Mattei, M.G., 1995, Organisation of the gene encoding the human endothelin-converting enzyme (ECE-1). *J. Biol. Chem.* **270:** 29794-29798.

Valdenaire, O., Lepailleur-Enouf, D., Egidy, G., Thouard, A., Barret, A., Vranckx, R., Tougard, C., and Michel, J.-B., 1999a, A fourth isoform of endothelin-converting enzyme-1 is generated from an additional promoter. Molecular cloning and characterization. *Eur. J. Biochem.* **264:** 1-10.

Valdenaire, O., Barret, A., Schweizer, A., Rohrbacher, E., Mongiat, F., Pinet, F., Corvol, P., and Tougard, C., 1999b, Two di-leucine-based motifs account for the different subcellular localizations of the human endothelin-converting enzyme (ECE-1) isoforms. *J. Cell Sci.* **112:** 3115-3125.

Valdenaire, O., Richards, J.G., Faull, R.L.M., and Schweizer, A., 1999c, XCE, a new member of the endothelin-converting enzyme and neutral endopeptidase family, is preferentially expressed in the CNS. *Mol. Brain Res.* **64:** 211-221.

Yanagisawa, H., Yanagisawa, M., Kapur, R.P., Richardson, J.A., Williams, S.C., Clouthier, D.E., deWit, D., Emoto, N., and Hammer, R.E., 1998, Dual genetic pathways of endothelin-mediated intercellular signaling revealed by targeted disruption of endothelin converting enzyme-1 gene. *Development* **125:** 825-836.

# REVIEW: NOVEL CYSTEINE PROTEASES OF THE PAPAIN FAMILY

Frank Bühling[1], Annett Fengler[4], Wolfgang Brandt[4], Tobias Welte[3], Siegfried Ansorge[1], and Dorit K. Nägler[2]
*[1]Institue of Immunology, [2]Institute of Pathology, [3]Dept. of Cardiology, Angiology and Pneumology, Otto von Guericke University Magdeburg, [4]Institute of Biochemistry, Martin Luther University Halle, Germany*

Key words cysteine protease, cathepsin F, cathepsin K, cathepsin O, cathepsin V, cathepsin W, cathepsin X, cystatin

## 1. INTRODUCTION

The papain family of cysteine proteases, by far the largest among cysteine proteases, comprises enzymes from bacteria, plants, invertebrates, and vertebrates. It includes the lysosomal cathepsins B, C, H, L, and S, as well as the more recently described cathepsins F, K, O, V, W, and X. The term "cathepsin" stands for "lysosomal proteolytic enzyme", regardless of the enzyme class. Therefore, in addition to cysteine proteases, this term includes serine proteases (cathepsins A and G), and aspartic proteases (cathepsins D and E) as well. Sequence homologies of cathepsins B, C, H, L, and S indicate that these enzymes diverged early during eukaryotic evolution (Berti and Storer 1994). Some of the genes encoding novel enzymes, however, may be the result of relatively recent gene duplication events, as suggested by three gene pairs with common chromosomal localization and high sequence homologies (Fig 1). Mammalian cysteine proteases of the papain family have been implicated in general protein degradation and turnover within the endosomal/lysosomal system (Kirschke and Barrett 1987), as well as in

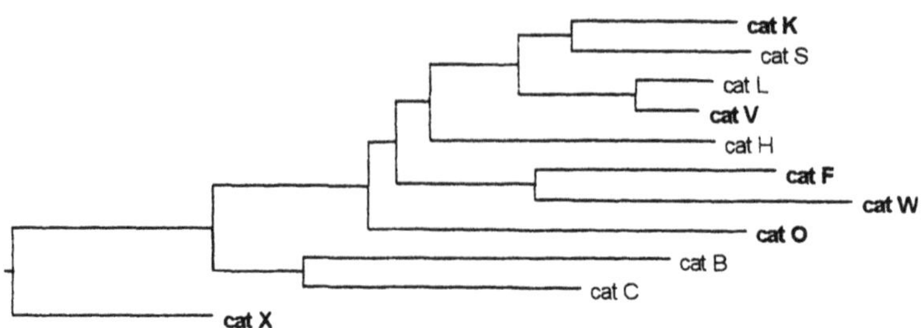


*Figure 1.* Phylogenetic tree of human cysteine proteases of the papain family based on the sequence alignment of the mature enzymes. The alignment and tree were produced using MegAlign (DNASTAR Inc.). Sequences are from SWISS-PROT (accession numbers CATB_HUMAN, CATC_HUMAN, CATK_HUMAN, CATH_HUMAN, CATK_HUMAN, CATS_HUMAN, CATV_HUMAN, CATW_HUMAN, CATO_HUMAN) and from Nägler *et al* 1998 and Nägler *et al* 1999a. Enzymes reviewed in this article are in bold type.

limited proteolysis events such as proenzyme activation (Samarel *et al* 1989), antigen processing (Riese *et al* 1996, Nakagawa *et al* 1998), and hormone maturation (Docherty *et al* 1982). Evidence has accumulated that cysteine proteases are secreted from cells in a variety of physiological and pathophysiological processes such as bone resorption (Inui *et al* 1997), muscular dystrophy (Katunuma and Kominami, 1987), arthritis (Mort *et al* 1984), tumor invasion and metastasis (Sloane *et al* 1981, Sloane and Honn, 1984), and Alzheimer's disease (Lemere *et al* 1995, Figueiredo *et al* 1999). Due to their involvement in several disease states, these enzymes are of clinical interest as targets for the development of inhibitors.

Lysosomal cysteine proteases are synthesized as inactive precursors and targeted to the endosomal/lysosomal compartment via the mannose 6-phosphate receptor pathway. Activation of the proenzyme usually occurs following cleavage and dissociation of the N-terminal segment. Proregions of varying lengths are found ranging from 38 amino acid residues for cathepsin X to as many as 251 residues for the recently described cathepsin F (Fig 2). In addition to being selective inhibitors of the parent enzyme (Fox *et al* 1992, Carmona *et al* 1996, Maubach *et al* 1997, Wiederanders, this book ch. 28), the proregions appear to be important for the correct folding of the newly synthesized polypeptide chain and for stabilizing the protein against denaturation due to neutral to slightly alkaline pH (Tao *et al* 1994, Mach *et al* 1994). Several three-dimensional structures of cysteine proteases are available and demonstrate that enzymes of the papain family share a common fold. Structural and functional aspects of mature enzymes and of cysteine protease precursors were reviewed recently (Storer and Ménard 1996, Cygler and Mort 1997, Turk *et al* 1997, Turk *et al* 1998, Groves *et al*

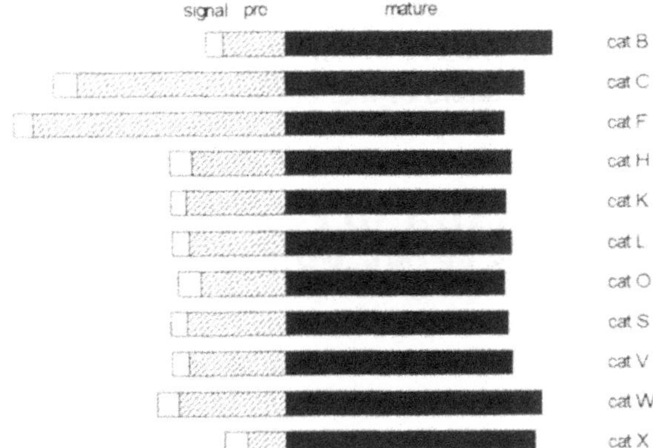


*Figure 2.* Structural organization of mammalian cysteine proteases of the papain family. Signal sequences are shown as white bars; proregions are depicted as hatched bars; mature enzyme are shown in black

1998). The main structural features of a papain-like two-domain catalytic platform are depicted in Fig 3 using a model of mature cathepsin K. The active site is located at the interface between the two domains. Most cysteine proteases of the papain family including cathepsin K, are "by design" endopeptidases. Exopeptidases, such as cathepsins B and H, are likely to have evolved from the endopeptidase template by addition of structural elements capable of interacting with the C- or N-terminus of a substrate (Turk *et al* 1998).

Advances in molecular biology and cell biology as well as improved methods for detection and localization of enzymatic activities have led to the discovery of cysteine proteases with restricted tissue distribution or with

*Figure 3.* Secondary structure representation of mature cathepsin K. Catalytic residues are highlighted as capped sticks.

unique properties. In this review, we summarize recent findings on cathepsins F, K, O, V, W, and X. Details on several of these enzymes can be found in the following chapters of this book.

# 2. STRUCTURAL AND FUNCTIONAL ASPECTS OF NOVEL CYSTEINE PROTEASES

## 2.1 Cathepsin F

The human cathepsin F cDNA was originally identified from a λgt10-skeletal muscle cDNA library. In contrast to all known human cysteine proteases of the papain family, the open reading frame of the cathepsin F cDNA apparently lacked a signal sequence. The protease was therefore believed to be targeted to the lysosomal compartment via an alternative pathway (Wang *et al* 1998). Recent work has shown that the initially described cDNA was truncated. Full-length cDNA clones of human cathepsin F revealed a typical signal sequence as well as an additional segment unique to cathepsin F (Nägler *et al* 1999a, Santamaria *et al* 1999). Cathepsin F is ubiquitously expressed in human tissues with higher expression levels in skeletal muscle, testis, heart, and brain.

The mature part of the cathepsin F sequence was found to share a significant degree of similarity with other members of the papain family, in particular with cathepsin W (Nägler *et al* 1999a, Wex *et al* 1999). The proregion, however, is unique with regard to its length and sequence. It is the longest proregion described so far for a cysteine protease of this family (251 amino acid residues, Fig 2) and contains an additional N-terminal segment which is predicted to share structural similarities with cysteine protease inhibitors of the cystatin superfamily (Nägler *et al* 1999a). The cystatin-like domain of cathepsin F contains some of the elements known to be important for inhibitory activity (Stubbs *et al* 1990). Functional studies are required to examine if this domain is an inhibitor of mature cathepsin F or any other cysteine protease. Alternatively, this additional domain in cathepsin F could have evolved to perform a different function.

Cathepsin F is the first hybrid cysteine protease precursor linking a cystatin-like domain to the protease. Interestingly, the cathepsin F gene has been mapped to chromosome 11 at position q13.1-3 (Santamaria *et al* 1999) in close proximity to the cathepsin W gene (Wex *et al* 1999) and the gene encoding cystatin M (Stenman *et al* 1997). Sequence homology to cathepsin W and to cystatins as well as a common chromosomal localization suggest that cathepsin F may be the product of both a gene duplication and a gene fusion event.

Full-length procathepsin F has not yet been expressed in a heterologous system. However, constructs including the mature enzyme and a fragment of the proregion or fusion constructs of the mature enzyme with glutathione S-transferase have been expressed in *P. pastoris* and in *E. coli*, respectively, and appear to hydrolyze small synthetic substrates (Wang *et al* 1998, Santamaria *et al* 1999). A homology model of human mature cathepsin F (PDB code: 1D5U; see also chapter 27 in this book), based on the crystal structures of cathepsins K and L, papain, and actinidin as well as kinetic data available to date indicate a relatively broad specificity, overlapping with cathepsins K, L, S and V (Wang *et al* 1998). The residue at position P2 of a substrate seems to be the most significant specificity determinant. Defining the physiological role of both the cystatin-like domain and the protease domain as well as monitoring post-translational proteolytic processing events will be important and challenging tasks.

## 2.2 Cathepsin K

Rabbit cathepsin K was first cloned in 1994 (Tezuka *et al* 1994) and named OC-2 because of its predominant expression in osteoclasts. In 1995 the human ortholog of this cysteine protease was described by Brömme (cathepsin O2, ) (Brömme and Okamoto 1995), Shi (cathepsin O) (Shi *et al* 1995) and Inaoka (cathepsin K) (Inaoka *et al* 1995). Similar to the rabbit enzyme, human cathepsin K is predominantly expressed in osteoclasts, being the second enzyme besides cathepsin S with a restricted expression pattern. However, in addition to osteoclasts, cathepsin K was recently shown to be highly expressed in multinucleated cells, for example in giant cell tumors (Littlewood-Evans *et al* 1997a), in osteoarthritic synovial tissues (Dodds *et al* 1999), or in epitheloid-like cells of granulomas during sarcoidosis (Bühling, unpublished). Cathepsin K was also detected in breast cancer cells (Littlewood-Evans *et al* 1997b), in synovial fibroblasts of patients with rheumatoid arthritis (Hummel *et al* 1998), in chordoma cells (Häckel personal communication) and in bronchial epithelial cells (Bühling *et al* 1999). Immunhistochemical investigations of the cathepsin K expression during the human embryonal development revealed a significant expression in epithelial cells of various organ systems (Häckel *et al* 1999). Interestingly, cathepsin K appears to be expressed earlier than cathepsin B or cathepsin L in bronchial epithelial cells during the embryonal lung development.

The amino acid sequence of mature cathepsin K shows a relatively high homology to cathepsins S and L (52 % and 46 %, respectively). The cathepsin K gene is clustered with the cathepsin S gene on chromosome

1q21, therefore suggesting a common origin of these genes. Similarly to cathepsin S, the cathepsin K promotor shows a number of binding sites for regulatory transcription factors indicating a strongly regulated expression (Gelb *et al* 1997).

The X-ray structures of both mature cathepsin K and procathepsin K have been determined recently (PDB code: 1MEM (McGrath *et al* 1997), PDB code: 7PCK (Sivaraman *et al* 1999); PDB code: 1BY8 (LaLonde *et al* 1998). There is no change in the overall conformation of the mature portion of the enzyme in procathepsin K, indicating that no significant rearrangement occurs following activation. Several structures of cathepsin K complexed with inhibitors are available that can be used to identify the residues in the substrate binding site which make contacts with the residues of a substrate (McGrath *et al* 1997, Zhao *et al* 1997, Yamashita *et al* 1997). As with other cysteine proteases, the residues in the S2 subsite (Tyr67, Met68, Ala134, Leu160, Ala163, Leu209) form a hydrophobic region suggesting a preference for hydrophobic residues in the P2 position of a substrate. However, the S2 subsite of cathepsin K appears to be spatially more restricted than that of cathepsins S or L, due to the orientation of Tyr67 and Leu209 side chains (Fengler and Brandt 1998).

Functional characterization of cathepsin K revealed several distinctive features of this enzyme. Regarding substrate specificity with small artificial substrates, Aibe and coworkers described a preference for proline in the P2-position and a more effective cleavage of Z-Gly-Pro-Arg-MCA than Z-Phe-Arg-MCA by rabbit cathepsin K (Aibe *et al* 1996). Z-Gly-Pro-Arg-MCA was not hydrolyzed by the cathepsins S and L and cleaved at about 10-fold lower rates by cathepsin B. Even more remarkable is the enzymatic activity of cathepsin K against connective tissue proteins such as collagen and elastin. Cathepsin K was characterized to be one of the most effective mammalian elastases. At pH 5.5, the elastinolytic activity exceeds that of pancreatic elastase (Chapman *et al* 1997). In addition, cathepsin K is capable of catalyzing the hydrolysis of type I, II and IV collagens (Kafienah *et al* 1998). Interestingly enough, the cleavage sites of cathepsin K in the collagen molecule are unique among mammalian proteases (Garnero *et al* 1998). Most of the collagenolytic cathepsins (L, S, B) generate collagen monomers by cleavage of the collagen triple helix telopeptide regions. The interstitial collagenases (MMP-1, MMP-2, MMP-8, MMP-13, neutrophil elastase) are characterized by a distinct cleavage site in the collagen molecule, resulting in N-terminal and C-terminal fragments of the collagen molecule. In contrast, cathepsin K cleaves collagen in the telopeptides and at multiple sites within the native triple helix generating fragments of various sizes. Therefore, cathepsin K is the only mammalian protease known to solubilize collagen without the action of other proteases.

The physiological relevance of the collagenolytic activity of cathepsin K is illustrated by the finding that the autosomal recessive bone disorder pycnodysostosis is based on an inactive form of cathepsin K (Gelb *et al* 1996). Cathepsin K knock out mice showed osteopetrosis (Saftig *et al* 1998 and this book, ch. 32). A decreased matrix-degrading activity of osteoclasts was observed in patients with pycnodysostosis as well as in cathepsin K knock-out mice. *In vitro* cathepsin K inhibitors are capable of inhibiting a significant part but not all of the bone resorbing activity of osteoclasts. Thus, evidence is now accumulating that cathepsin K is one of the major proteases responsible for bone resorption and specific inhibitors of cathepsin K are considered to be a usefull tool for the therapy of destructive bone diseases, for example osteoporosis (Smith and Abdelmeguid, 1999). The physiological role of cathepsin K in other cells is not known.

## 2.3 Cathepsin O

Cathepsin O was initially cloned from a breast cancer cell library (Velasco *et al* 1994). It is ubiquitously expressed, however the highest expression levels were found in human ovary, kidney, liver and placenta.

The cathepsin O cDNA encodes a preproenzyme of 321 amino acids. Sequence alignments revealed a sequence homology below 30 % with other papain-like cysteine proteases. The genomic organization of cathepsin O has been determined recently (Santamaría *et al* 1998a) and revealed that the number and distribution of exons and introns differs from those reported for other human cysteine proteases. The gene was mapped to chromosome 4q31-q32, a unique localization for cysteine proteases.

Preliminary investigations of the enzymatic activity of cathepsin O expressed in *E. coli* revealed that the enzyme can cleave the substrates Z-Phe-Arg-MCA and Z-Arg-Arg-MCA although at lower rates than cathepsin B.

## 2.4 Cathepsin V

Two different approaches were employed to identify and clone the gene for human cathepsin V. The first group performed an extensive search of the EST database, which led to the identification of a short mouse DNA fragment with high similarity to cathepsin L (Santamaría *et al* 1998b). A cDNA containing part of the EST clone was prepared and used as a probe to screen a human brain cDNA library. One of the isolated clones encoded an enzyme with high similarity to cathepsin L (78 % identity) and was therefore initially referred to as cathepsin L2. The second group identified cathepsin V among other genes highly expressed in corneal epithelium (Adachi *et al*

1998). The two groups screened several human tissues for the presence of the cathepsin V cDNA, they reached, however, different conclusions regarding the tissue distribution of cathepsin V. While the first group found high expression in thymus, testis, and several human tumors (Santamaría *et al* 1998b), the second group reports its unique expression in corneal epithelium (Adachi *et al* 1998). Interestingly, cathepsin V was mapped to the chromosomal region 9q22.2, a site adjacent to the cathepsin L locus (Brömme *et al* 1999), suggesting the evolution of both enzymes from a common ancestral cathepsin L-like precursor by gene duplication.

Recombinant procathepsin V was expressed in *Pichia pastoris* and shown to be autocatalytically activated at acidic pH (Brömme *et al* 1999). Kinetic data available indicate that the P2 substrate specificity of cathepsin V overlaps with that of cathepsins S and L. According to a homology model of cathepsin V, minor differences in the substrate specificity may be attributed to alterations of amino acid residues defining the S2 subsite. Leu182 and Met272 in cathepsin L are replaced by Phe182 and Leu275 in cathepsin V. The cathepsin V model revealed completely different electrostatic potential maps when compared with human cathepsin L (Brömme *et al* 1999). Surprisingly, the electrostatic potential of the human cathepsin V model structure resembled that of mouse cathepsin L, therefore suggesting that mouse cathepsin L may be more closely related to human cathepsin V than to human cathepsin L. Based on these findings and the high expression in thymus, it was suggested that cathepsin V may play a role in MHC class II antigen presentation.

## 2.5 Cathepsin W

Cathepsin W was identified through a search of the EST database and initially cloned from a human placental cDNA library. Detailed expression pattern analysis, however, revealed its predominant expression in lymphocytes, particularly in CD8-positive T-lymphocytes (Linnevers *et al* 1997). Further investigations indicated a strong expression of the cathepsin W mRNA in natural killer cells characterized by one of the surface markers CD16 or CD56 which partially overlap with the population of CD8-positive cells. Cathepsin W was also detected in large granular lymphocyte leukaemia cells and hematopoietic progenitor cells (Brown *et al* 1998). Coexpression of CD16 or CD8 and cathepsin W in single cells was demonstrated using polyclonal and monoclonal anti-cathepsin W antibodies (Bühling, unpublished). The subcellular localization of cathepsin W is unclear at the moment. In analogy to other cysteine proteases of the papain family, one would predict a lysosomal localization of this enzyme. However, overexpression of cathepsin W in Cos-7 and Hela cells led to a reticular staining pattern (Wex *et al* 1998). Preliminary experiments suggest that the expression of cathepsin W may be regulated by cytokines as IL-2 or PKC effectors as PMA (Bühling, unpublished).

Sequence alignments of cathepsin W with the related cathepsins revealed a relatively low amino acid sequence identity (lower than 30 %). The"ERFNIN"-motif found in the proregion of other cathepsin L-like proteases is only partially conserved in cathepsin W (Wex *et al* 1998 and this book, ch. 29). Amino acid sequence comparison of cathepsin W with another novel member of the papain family, cathepsin F, revealed that these two enzymes are more closely related to each other than to any other known cysteine protease (Wang *et al* 1998, Nägler *et al* 1999a, Wex *et al* 1999 and this book, ch. 29, Fig 1).

All of the highly conserved residues in papain-like cysteine proteases are present in cathepsin W. The two most significant differences in the primary structure of cathepsin W compared to other cathepsins are a 21 amino acid residue insertion between His164 and Asn204 of the active site and an 8 amino acid residue C-terminal extension (Linnevers *et al* 1997). A homology model of mature cathepsin W shows that the insertion loop between the active site histidine and asparagine residues is located on the surface of the enzyme remote from the active site (Fengler, unpublished). The function of this insertion is unknown so far. The second insertion at the C-terminus of cathepsin W is predicted to form a short helical structure. The S2 subsite of cathepsin W is more restricted in size and depth as compared to cathepsin K, based on the position of the side chains which form this subsite (Phe68, Val69, Thr136, Val162, Ser165, Phe230; Brandt, unpublished). The enzymatic activity of cathepsin W has not yet been characterized.

## 2.6 Cathepsin X

A cDNA clone isolated from bovine heart encoding a short fragment of a novel cysteine protease of the papain family was described many years ago and named cathepsin X (Gay and Walker, 1985). Recently, the human ortholog of cathepsin X has been cloned independently by two groups and referred to as cathepsin X (Nägler *et al* 1998) and cathepsin Z (Santamaria *et al* 1998c). The cathepsin X gene has been mapped by fluorescent *in situ* hybridization to chromosome 20q13, a locus that differs from other cysteine proteases of the papain family (Santamaria *et al* 1998c). The corresponding cDNA is ubiquitously expressed in human tissues and contains an open reading frame of 912 nucleotides encoding a predicted protein of 303 amino acids. All highly conserved regions in papain-like cysteine proteases including the catalytic residues, are present in cathepsin X. The mature part of cathepsin X is 26 – 32 % identical to human cathepsins B, C, H, K, L, O, S, V, and W.

Unlike other recently discovered members of the papain family of cysteine proteases, which are closely related to cathepsin L, cathepsin X possesses unique and interesting features: (1) a very short proregion (Fig 1) and (2) a three amino acid residue insertion in a highly conserved region between the glutamine of the putative oxyanion hole and the active site cysteine. In a homology model of the mature cathepsin X, this insertion was found to be located in the primed region of the binding cleft as part of a surface loop corresponding to residues His23 to Tyr27, and has been termed the "mini-loop" (Nägler *et al* 1999c). According to the model, which is described in more detail by Menard *et al* (this book, ch. 34), this distinctive structural feature might confer exopeptidase activity to the enzyme. Indeed, cathepsin X was found to display excellent carboxypeptidase activity against the substrate Abz-FRF($4NO_2$), with a $k_{cat}/K_M$ value of 1.23 X $10^5$ $M^{-1}s^{-1}$ at the optimal pH of 5.0. The bound carboxypeptidase substrate is predicted to establish a number of favorable contacts within the cathepsin X binding site, in particular with residues His23 and Tyr27 from the mini-loop. The presence of the mini-loop restricts the accessibility of endopeptidase and MCA substrates in the primed subsites of the protease and explains the extremely low activity of cathepsin X against the substrates Cbz-FR-MCA and Abz-AFRSAAQ-EDDnp (Nägler *et al* 1999c). While other exopeptidases, such as cathepsin B, are still able to act as endopeptidases, although at slower rates (Nägler *et al* 1999b), efficient hydrolysis by cathepsin X requires substrates with a free C-terminus. In addition, cathepsin X is not inhibited by the endogenous inhibitor cystatin C up to a concentration of 4 μ M of inhibitor. This raises the question of whether this enzyme is regulated *in vivo* by endogenous inhibitors. The marked structural and functional differences of cathepsin X relative to other cysteine proteases will be of great value both in discriminating between different members of this enzyme class and in designing specific inhibitors as research tools to investigate the physiological and potential pathological roles of this novel enzyme.

## 3. CONCLUSIONS

In analogy to other gene families, novel genes encoding cysteine proteases of the papain family have been discovered recently, and completion of the human genome project will almost certainly reveal additional sequences. In particular, tissue specific or developmentally regulated proteases such as the recently described cathepsin P (Sol-Church *et al* 1999, Tisljar *et al* 1999), have yet to be discovered. Since information available to date indicates relatively broad, overlapping specificities for the mature proteases, the increasing number of cysteine proteases raises the possibility of functional redundancy for this family of enzymes. However, differences in cellular distribution and intracellular localization can contribute to defining specific functional roles for these relatively similar proteases. The incorporation of additional domains with specific roles might also confer functional diversity. Cathepsin F is the first cysteine protease containing a cystatin-like domain in its precursor, a particularly interesting finding considering that many cystatins are reversible tight binding inhibitors of lysosomal cysteine proteases and are believed to regulate proteolytic activity *in vivo*.

Little is known so far about the nature of the protein targets of these novel enzymes. In this regard, it will be interesting to explore if the apparent functional redundancy described for the hydrolysis of small synthetic substrates also applies to limited proteolysis and/or degradation of protein substrates. The great challenge for the next few years will certainly consist in identifying natural substrates for individual cysteine proteases and gaining further insight into their protein substrate specificity. For most of these recently discovered enzymes, with the exception of cathepsin K, their physiological role and potential involvement in diseases are currently unknown. A better understanding of the specific physiological and pathophysiological function of a protease should prove useful in evaluating its potential value as a target for therapeutic interventions.

## REFERENCES

Adachi, W., *et al.*, 1998, Isolation and characterization of human cathepsin V: A major proteinase in corneal epithelium. *Invest. Ophthalmol. Vis. Sci.* **39:** 1789-1796.

Aibe, K., *et al.*, 1996, Substrate specificity of recombinant osteoclast-specific cathepsin K from rabbits. *Biol. Pharm. Bull.* **19:** 1026-1031.

Berti, P.J., and Storer, A.C., 1995, Alignment/phylogeny of the papain superfamily of cysteine proteases. *J. Mol. Biol.* **246:** 273-283.

Brömme, D., *et al.*,(1999, Human cathepsin V functional expression, tissue distribution, electrostatic surface potential, enzymatic characterization, and chromosomal localization. *Biochemistry* **38:** 2377-2385.

Brömme, D., and Okamoto, K., 1995, Human cathepsin O2, a novel cysteine protease highly expressed in osteoclastomas and ovary molecular cloning, sequencing and tissue distribution. *Biol. Chem. Hoppe Seyler* **376:** 379-384.

Brömme, D., *et al.*, 1996, Human cathepsin O2, a matrix protein-degrading cysteine protease expressed in osteoclasts. Functional expression of human cathepsin O2 in Spodoptera frugiperda and characterization of the enzyme. *J. Biol. Chem.* **271:** 2126-2132.

Brown, J., *et al.*, 1998, Lymphopain, a cytotoxic T and natural killer cell-associated cysteine proteinase. *Leukemia* **12:** 1771-1781.

Bühling, F., *et al.*, 1999, Expression of cathepsin K in lung epithelial cells. *Am. J. Cell. Mol. Biol.* **20:** 612-619.

Carmona, E., *et al.*, 1996, Potency and selectivity of the cathepsin L propeptide as an inhibitor of cysteine proteases. *Biochemistry* **35:** 8149-8157.

Chapman, H.A., *et al.*, 1997, Emerging roles for cysteine proteases in human biology. *Annu. Rev. Physiol.* **59:** 63-88.

Cygler, M., and Mort, J.S., 1997, Proregion structure of members of the papain superfamily. Mode of inhibiton of enzymatic activity. *Biochimie* **79:** 645-652.

Docherty, K., *et al.*, 1982, Conversion of proinsulin to insulin: involvement of a 31,500 molecular weight thiol protease. *Proc. Natl. Acad. Sci.* **79:** 4613-4617.

Dodds, R.A., *et al.*, 1999, Expression of cathepsin K messenger RNA in giant cells and their precursors in human osteoarthritic synovial tissues. *Arthritis Rheum.* **42:** 1588-1593.

Fengler, A., and Brandt, W., 1998, Three Dimensional Structures of the Cysteine Proteases Cathepsin K and S Deduced by Knowledge-Based Modelling and Active Site Characteristics. *Prot. Eng.* **11:** 1007-1013.

Figueiredo-Pereira, M.E., *et al.*, 1999, Distinct secretases, a cysteine protease and a serine protease, generate the C termini of amyloid beta-proteins Abeta1-40 and Abeta1-42, respectively. *J. Neurochem.* **72:** 1417-1422.

Fox, T., *et al.*, 1992, Potent slow-binding inhibition of cathepsin B by its propeptide. *Biochemistry* **31:** 12571-12576.

Garnero, P., *et al.*, 1998, The collagenolytic activity of cathepsin K is unique among mammalian proteinases. *J. Biol. Chem.* **273:** 32347-32352.

Gay, N. J., and Walker, J.E., 1985, Molecular cloning of a bovine cathepsin. *Biochem. J.* **225:** 707-712.

Gelb, B.D., *et al.*, 1996, Pycnodysostosis, a lysosomal disease caused by cathepsin K deficiency. *Science* **273:** 1236-1238.

Gelb, B.D., *et al.*, 1997, Structure and chromosomal assignment of the human cathepsin K gene. *Genomics* **41:** 258-262.

Groves, M.R., *et al.*, 1998, Structural basis for specificity of papain-like cysteine protease proregions toward their cognate enzymes *Proteins* **32:** 504-514.

Häckel, C., *et al.*, 1999, Expression of cathepsin K in the human embryo and fetus. *Dev. Dyn.* **216:** 89-95.

Hummel, K.M., *et al.*, 1998, Cysteine proteinase cathepsin K mRNA is expressed in synovium of patients with rheumatoid arthritis and is detected at sites of synovial bone destruction. *J. Rheumatol.* **25:** 1887-1894.

Inaoka, T., *et al.*, 1995, Molecular cloning of human cDNA for cathepsin K: novel cysteine proteinase predominantly expressed in bone. *Biochem. Biophys. Res. Commun.* **206:** 89-96.

Inui, T., *et al.*, 1997, Cathepsin K antisense oligodeoxynucleotide inhibits osteoclastic bone resorption. *J. Biol. Chem.* **272:** 8109-8112.

Kafienah, W., *et al.*, 1998, Human cathepsin K cleaves native type I and II collagens at the N-terminal end of the triple helix. *Biochem. J.* **331:** 727-732.

Katunuma, N., and Kominami, E., 1987, Abnormal expression of lysosomal cysteine proteinases in muscle wasting diseases, Rev. Physiol. Biochem. Pharmacol. **108:** 1-20.

Kirschke, H., and Barrett, A. J., 1987, In *Lysosomes: their role in protein breakdown* (H. Glaumann, and F. J. Ballard, eds.), Academic Press, London pp. 193-238.

Lemere, C.A., *et al.*, 1995, The lysosomal cysteine protease, cathepsin S, is increased in Alzheimer's disease and Down syndrome brain. An immunocytochemical study. *Am. J. Pathol.* **146:** 848-860.

LaLonde, J.M., *et al.*, 1998, Use of papain as a model for the structure-based design of cathepsin K inhibitors: Crystal structures of two papain-inhibitor complexes demonstrate binding to S'-subsites. *J. Med. Chem.* **41:** 4567-4576.

Linnevers, C., *et al.*, 1997, Human cathepsin W, a putative cysteine protease predominantly expressed in CD8+ T-lymphocytes. *FEBS Letters* **405:** 253-259.

Littlewood-Evans, A., *et al.*, 1997a, Localization of cathepsin K in human osteoclasts by in situ hybridization and immunohistochemistry. *Bone* **20:** 81-86.

Littlewood-Evans, A.J., *et al.*, 1997b, The osteoclast-associated protease cathepsin K is expressed in human breast carcinoma. *Cancer Res.* **57:** 5386-5390.

Mach, L., *et al.*, 1994, Noncovalent complexes between the lysosomal proteinase cathepsin B and its propeptide account for stable, extracellular, high molecular mass forms of the enzyme. *J. Biol. Chem.* **269:** 13036-13040.

Maubach, G., *et al.*, 1997, The inhibition of cathepsin S by its propeptide--specificity and mechanism of action. *Eur. J. Biochem.* **250:** 745-750.

McGrath, M.E., *et al.*, 1997, Crystal structure of human cathepsin K complexed with a potent inhibitor. *Nature Struct. Biol.* **4:** 105-109.

Mort, J.S., *et al.*, 1984, Extracellular presence of the lysosomal proteinase cathepsin B in rheumatoid synovium and its activity at neutral pH. *Arthr. Rheum.* **27:** 509-515.

Nägler, D. K., and Ménard, R., 1998, Human cathepsin X: A novel cysteine protease of the papain family with a very short proregion and unique insertions. *FEBS Lett.* **434:** 135-139.

Nägler, D.K., *et al.*, 1999a, Full-length cDNA of human cathepsin F predicts the presence of a cystatin domain at the N-terminus of the cysteine protease zymogen. *Biochem. Biophys. Res. Commun.* **257:** 313-318.

Nägler, D.K., *et al.*, 1999b, Interdependency of sequence and positional specificities for cysteine proteases of the papain family. *Biochemistry* **38:** 4868-4874.

Nägler, D.K., *et al.*, 1999c, Human cathepsin X: A cysteine protease with unique carboxypeptidase activity. *Biochemistry* **39:** 12648-12654.

Nakagawa, T., *et al.*, 1998, Cathepsin L: critical role in Ii degradation and CD4 T cell selection in the thymus. *Science* **280:** 450-453.

Riese, R.J., *et al.*, 1996, Essential role for cathepsin S in MHC class II-associated invariant chain processing and peptide loading. *Immunity* **4:** 357-366.

Saftig, P., *et al.*, 1998, Impaired osteoclastic bone resorption leads to osteopetrosis in cathepsin-K-deficient mice. *Proc. Natl. Acad. Sci.* **95:** 13453-13458.

Samarel, A.M., *et al.*, 1989, Effects of cysteine protease inhibitors on rabbit cathepsin D maturation. *Am. J. Physiol.* **257:** 1069-1079.

Santamaría, I., *et al.*, 1998a, Genomic structure and chromosomal localization of the human cathepsin O gene (CTSO). *Genomics* **53:** 231-234.

Santamaría, I., *et al.*, 1998b, Cathepsin L2, a novel human cysteine proteinase produced by breast and colorectal carcinomas. *Cancer Res.* **58:** 1624-1630.

Santamaría, I., *et al.*, 1998c, Cathepsin Z, a novel human cysteine proteinase with a short propeptide domain and a unique chromosomal location. *J. Biol. Chem.* **273:** 16816-16823.

Santamaria, I., *et al.*, 1999, Molecular cloning and structural and functional characterization of human cathepsin F, a new cysteine proteinase of the papain family with a long propeptide domain. *J. Biol. Chem.* **274:** 13800-13809.

Shi, G.P., *et al.*, 1995, Molecular cloning of human cathepsin O, a novel endoproteinase and homologue of rabbit OC2. *FEBS Lett.* **357:** 129-134.

Sivaraman, *et al.*, 1999, Crystal structure of wild-type human procathepsin K. *Protein Sci.* **8:** 283-290.

Sloane, B.F., *et al.*, 1981, Lysosomal cathepsin B: correlation with metastatic potential. *Science* **212:** 1151-1153.

Sloane, B.F., and Honn, K.V., 1984, Cysteine proteinases and metastasis. *Cancer Metastasis Rev.* **3:** 249-263.

Smith, W.W., and Abdelmeguid, S.S., 1999, Cathepsin K as a target for the treatment of osteoporosis. *Expert Opin. Ther. Patents* **9:** 683-694.

Sol-Church, K., *et al.*, 1999, Cathepsin P, a novel protease in mouse placenta. *Biochem. J.* **343:** 307-309.

Stenman, G., *et al.*, 1997, Assignment of a novel cysteine proteinase inhibitor (CST6) to 11q13 by fluorescence in situ hybridization Cytogenet. *Cell Genet.* **76:** 45-46.

Storer, A.C., and Ménard, R., 1996, Recent insights into cysteine protease specificity: Lessons for drug design. *Perspect. Drug Discovery Des.* **6:** 33-46.

Stubbs, M.T., *et al.*, 1990, The refined 2.4 A X-ray crystal structure of recombinant human stefin B in complex with the cysteine proteinase papain: a novel type of proteinase inhibitor interaction. *EMBO J.* **9:** 1939-1947.

Tao, K., *et al.*, 1994, The proregion of cathepsin L is required for proper folding, stability, and ER exit. *Arch. Biochem. Biophys.* **311:** 19-27.

Tezuka, K., *et al.*, 1994, Molecular cloning of a possible cysteine proteinase predominantly expressed in osteoclasts. *J. Biol. Chem.* **269:** 1106-1109.

Tisljar, K., *et al.*, 1999, Cathepsin J, a novel murine cysteine protease of the papain family with a placenta-restricted expression. *FEBS Lett.* **459:** 299-304.

Turk, B., *et al.*, 1997, Structural and functional aspects of papain-like cysteine proteinases and their protein inhibitors. *Biol. Chem.* **378:** 141-150.

Turk, D., *et al.*, 1998, Revised definition of substrate binding sites of papain-like cysteine proteases. *Biol. Chem.* **379:** 137-147.

Velasco, G., *et al.*, 1994, Human cathepsin O. Molecular cloning from a breast carcinoma, production of the active enzyme in Escherichia coli, and expression analysis in human tissues. *J. Biol. Chem.* **269:** 27136-27142.

Wang, B., *et al.*, 1998, Human cathepsin F. *J. Biol. Chem.* **273:** 32000-32008.

Wex, T., *et al.*, 1999, Human cathepsins F and W: A new subgroup of cathepsins. *Biochem. Biophys. Res. Commun.* **259:** 401-407.

Wex, T., *et al.*, 1998, Structure and chromosomal localization of the human cathepsin W-gene. *Biochem. Biophys. Res. Commun.* **248:** 255-261.

Yamashita, D.S., *et al.*, 1997, Structure and design of potent and selective cathepsin K inhibitors. *J. Am. Chem. Soc.* **119:** 11351-11352.

Zhao, B.G., *et al.*, 1997, Crystal structure of human osteoclast cathepsin K complex with E-64. *Nature Struct. Biol.* **4:** 109-111.

# DEVELOPMENT AND VALIDATION OF HOMOLOGY MODELS OF HUMAN CATHEPSINS K, S, H, AND F

Annett Fengler and Wolfgang Brandt
*Department of Biochemistry and Biotechnology, Institute of Biochemistry, Martin Luther University Halle-Wittenberg, Kurt-Mothes-Str. 3, 06120 Halle (Saale), Germany*

Key words cathepsins K, S, H, F, tertiary structure, homology modelling

Abstract Models of the tertiary structures of cathepsins K, S, H, and F were constructed by using homology protein modelling methods and refinements by interactive graphics and energy minimisation. The predicted structures yield information regarding their substrate binding sites and indicate the residues surrounding these sites. The ligand binding sites were characterised and compared with each other by means of calculated molecular electrostatic surface potentials. This will allow designing and development of new ligands specific for these cathepsins in future investigations.

## 1. INTRODUCTION

The lysosomal cathepsins are papain-like cysteine proteases that are responsible for normal cellular protein degradation. The cathepsins K, S, and H are implicated in a variety of physiological processes and their proteolytic activities may also be relevant to human diseases such as neoplasia, arthritis, emphysema, and pycnodysostosis (Kirschke *et al* 1995, Gelb *et al* 1996). Up to now, the precise functional role of cathepsin F could be not defined (Santamaria *et al* 1999). All these enzymes have closely related amino acid sequences and overall folding structures (Turk *et al* 1998). In spite of these similarities, there is a great divergence in their proteolytic specificity (Turk *et al* 1997). Therefore, it is of interest to investigate the role of the amino acid residues around the active site evoking such a divergence in their activities. For this purpose models of the tertiary structures of cathepsins K, S, H, and F were constructed by using homology protein modelling methods.

## 2. MATERIALS AND METHODS

X-ray structures of selected proteins with high homology to the proteins under investigation were extracted from the Brookhaven Protein Data Bank (PDB). Using the programs MODELLER and COMPOSER (Sali and Blundell 1993, Blundell *et al* 1988) homology based models of the cathepsins K, S, H, and F were generated. The refinement procedure included manual corrections of the most unfavourable van der Waals contacts, energy minimisation restricted to the side chain atoms, and minimisation extended to the whole molecule (TRIPOS Assoc. Inc.). The models were evaluated for self-consistency using PROCHECK and PROSAII (Laskowski *et al* 1993, Sippl 1993).The MOLCAD software was used for the computation of the electrostatic potentials (EP) on the molecular surface of the cathepsins K, L, S, H, and F (Heiden *et al* 1993).

## 3. RESULTS

The models of the cathepsins K, S, H, and F contain common features in their overall structure and folding. The superposition of the Cα atoms of the generated cathepsin models shows the high structural homology (Fig 1). The cathepsins K, H, and F have three disulfide bonds conserved among the cysteine proteases. The sequence of cathepsin S contains two additional cysteine residues which are involved in a fourth disulfide bond.

In contrast to the other cathepsins the protein structure of cathepsin H contains a mini-chain (EPQNCSAT) (Machleidt and Müller-Esterl 1986). This octapeptide is a part of its propeptide. Guncar et al. (1997) described the mini-chain binds into the active site cleft of cathepsin H in substrate direction. The position and the interaction behaviour of this octapeptide to the active site residues could be determined (Fig 2) (Fengler and Brandt 1999). The S2 and S3 subsites are occupied with the amino acid residues of the mini-chain and form hydrophobic interactions and hydrogen bonds.

The models passed all criteria implemented in PROCHECK. A more stringent test of the overall fold and the side chain packing of the model may be provided by PROSAII. Fig 3a shows two energy graphs calculated from the modelled structure of cathepsin K and its corresponding X-ray structure. We can observe a strong correlation of both energy courses (see also Fig 4a).

In Fig 3b the energy graphs of the models of cathepsins S and F in comparison with the X-ray structure of cathepsin L are plotted.

*Figure 1.* Cα plot of the models of cathepsins K, S, H, and F. The structures are superimposed.

*Figure 2.* The model of cathepsin H including the mini-chain (dark).

(a)

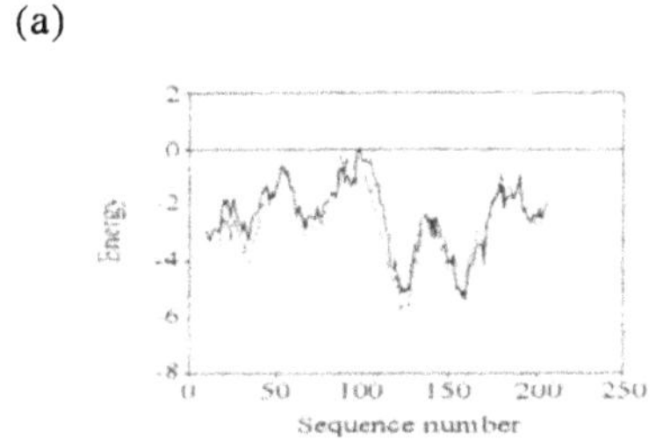


(b)

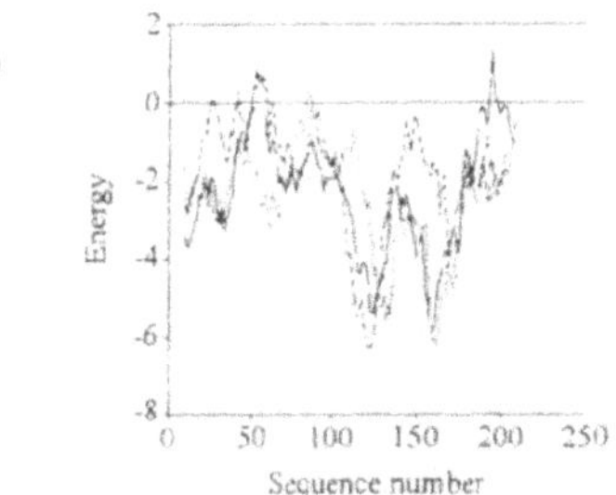


*Figure 3.* Energy graphs of the model of cathepsin K (solid line) in comparison with its X-ray structure (dotted line) (a) and of the models of cathepsins S (dotted line), F (dashed line), and the related X-ray structure of cathepsin L (solid line) (b).

An analysis of the EPs of cathepsins K, S, H, and F was performed in comparison with cathepsin L. Contrary to the high sequence homology and folding similarity these potentials are clearly different (Fig. 4b - f). The active site cleft of cathepsin K is weakly electronegative. The left-hand region is characterised by residues with a negative potential. The molecular surface of cathepsin S is more electropositive over extended regions in contrast to the other cathepsins. The active site-cleft of cathepsin H is partly occupied with the mini-chain. The visible surface of this cathepsin is neutral and similar to that of cathepsin K. The potential of the cleft of cathepsin F is electronegative or neutral. The potential of cathepsin L is more electronegative suggesting electrostatic interactions are most important for substrate docking to this enzyme.

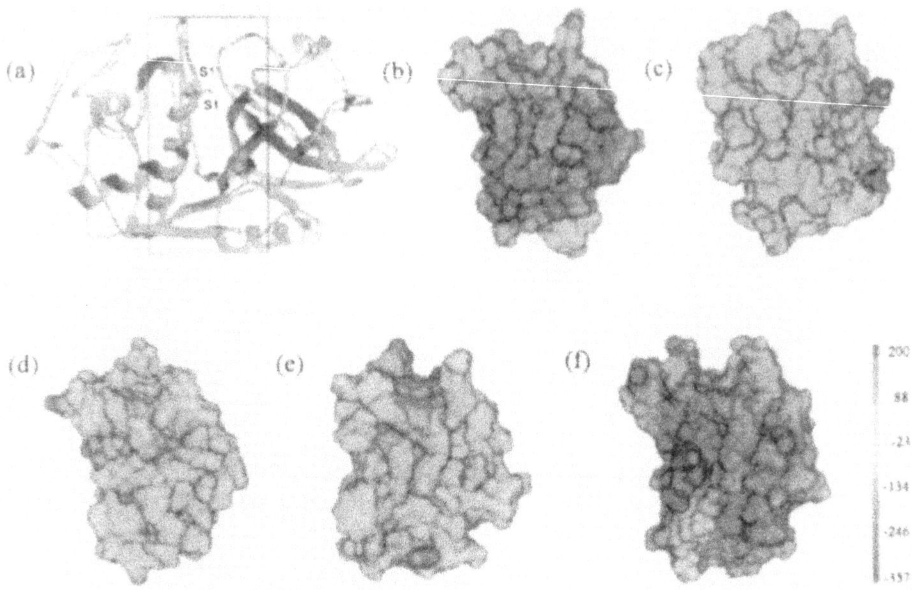


*Figure 4.* The ribbon plot of the model of cathepsin K (green) and its X-ray structure (orange) (a). The EPs of the active site clefts of the cathepsins K (b), S (c), H (d), F (e), and L (f). Electronegative potentials are coloured blue and electropositive potentials are coloured red. (The coloured Figures may be viewed at http://www.biochemtech.uni-halle.de/~mdqba/catfig.htm)

The different properties of the potentials could be explained by the residues in the cleft. Those residues that might be responsible for different active site characteristics of the cathepsins are listed in Table 1.

*Table 1.* S1´, S2, and S3 subsites forming residues in human cathepsins K, S, H, F, and L

| | Residue number* | Cat K | Cat S | Cat H | Cat F | Cat L |
|---|---|---|---|---|---|---|
| S1´ | 138 | A | A | V | A | A |
| | 144 | Q | F | M | Y | L |
| | 162 | N | N | N | D | D |
| S2 | 69 | Y | F | L | L | L |
| | 70 | M | M | P | P | M |
| | 135 | A | G | A | A | A |
| | 161 | L | V | V | I | M |
| | 164 | A | G | A | A | G |
| | 214 | L | F | C | M | A |
| S3 | 63 | D | K | Y | K | E |
| | 69 | Y | F | L | L | L |

* Residue numbers used are according to the cathepsin L numbering

# 4. CONCLUSION

Reasonable 3D structures for the cathepsins K, S, H, and F have been developed. The results of the verifications of the models in their geometry support the principal correctness of these protein models. The models of the protein structure and their corresponding molecular electrostatic potentials provide insights into the active site characteristics of these enzymes. These results will help to understand the substrate specificity of these cathepsins and will be the basis to design new specific ligands.

# ACKNOWLEDGMENT

We are grateful for research support from Deutsche Forschungsgemeinschaft (SFB 387, TP A8). Some work was performed by A. F. at sabbatical leave at the Mount Sinai School of Medicine in New York.

# REFERENCES

Blundell, T., *et al.*, 1988, Knowledge-based protein modelling and design. *Eur. J. Biochem.* **172:** 513-520.

Fengler, A., and Brandt, W., 1999, The tertiary structure of the human cathepsin H including the mini-chain, deduced by knowledge-based approaches and active site characteristics. *J. Mol. Mod.* **5:** 177-188.

Gelb, B.D., *et al.*, 1996, Pycnodysostosis, a lysosomal disease caused by cathepsin K deficiency. *Science* **273:** 1236-1238.

Guncar, G., *et al.*, 1997, Crystal structure of porcine cathepsin H determined at 2.1 Å resolution: location of the mini-chain C-terminal carboxyl group defines cathepsin H aminopeptidase function. *Structure* **6:** 51-61.

Heiden, W., *et al.*, 1993, Fast generation of molecular surfaces from 3D data fields with an enhanced 'marching cube' algorithm. *J. Comput. Chem.* **14:** 246-250.

Kirschke, H., *et al.*, 1995 In *Protein Profile, Vol. 2* (Sheterline, P., ed.), Academic Press, London, pp. 1587-1646.

Laskowski, R.A. *et al.*, 1993, PROCHECK: a program to check the stereochemical quality of protein structures. *J. Appl. Cryst.* **26:** 283-289.

Machleidt, W., and Müller-Esterl, W. 1986 In *Cystein Proteases and Their Inhibitors* (Turk, D., ed), Walter de Gruyter and Co., pp. 3-18.

Sali, A., and Blundell, T.L., 1993),Comparative protein modelling by satisfaction of spatial restraints. *J. Mol. Biol.* **234:** 779-815.

Santamaría, I., *et al.*, 1999, Molecular cloning and structural and functional characterization of human cathepsin F, a new cysteine proteinase of the papain family with a long propeptide domain. *J. Biol. Chem.* **274:** 13800-13809.

Sippl, M.J., 1993, Recognition of errors in three-dimensional structures of proteins. *Proteins* **17:** 355-362.

Turk, D., *et al.*, 1998, Revised definition of substrate binding sites of papain-like cysteine proteases. *Biol. Chem. Hoppe-Seyler* **379**: 137-147.

Turk, B., *et al.*, 1997, Structural and functional aspects of papain-like cysteine proteinases and their protein inhibitors. *Biol. Chem. Hoppe-Seyler* **378**: 141-150.

# THE FUNCTION OF PROPEPTIDE DOMAINS OF CYSTEINE PROTEINASES

Bernd Wiederanders
*Friedrich-Schiller-Universität Jena, Klinikum, Institut für Biochemie, Nonnenplan 2, D-07743 Jena*

Key words ERFNIN-motif, mutagenesis, procathepsin S, maturation, targeting, inhibitor function,

Abstract The papain-like cysteine proteinases can be divided into cathepsin L-like and cathepsin B-like enzymes because of the extended proregion of the former ones. We performed a series of mutations (alanine scan) in the prodomain of procathepsin S in order to elucidate the function of this extended domain in the L-like cathepsins. One of the most striking results was that the structural stability and the folding of procathepsin S were considerably dependent on an aromatic stack built by the residues Trp 28, Trp 31 and Trp 52. Replacement either of one or of all of these residues by alanine resulted in loss of transport, maturation and secretion of the mutated zymogen. Recombinant propeptides carrying the same mutations are not longer selective and powerful inhibitors of cathepsin S. Therefore we postulated an essential role of the propeptide for proper folding of the whole enzyme. This assumption was further proved by in vitro studies. We investigated the capability of recombinant cathepsin S propeptide to catalyze the renaturation of denatured mature cathepsin S. The experiments showed a 10-25fold faster renaturation rate in presence than in absence of the propeptide underlining its function as intramolecular chaperone.

## 1. INTRODUCTION

Some lysosomal cysteine peptidases occur ubiquitously in nearly all organs as the cathepsins B, L, H (Kirschke *et al* 1998) or F (Nägler *et al* 1999). Others can be found only in a restricted number of cell populations as cathepsin S in antigen presenting cells (Shi *et al* 1994) or cathepsin K in osteoclasts (Inaoka *et al* 1995). Enzymes of the first group can obviously replace each other in their physiological functions, enzymes of the latter group cannot. Mice whose genes coding for either of these enzymes have taught us this lesson: a cathepsin B knockout mouse shows no pathological phenotype (Deussing *et al* 1998), a cathepsin L knockout mouse resembles or is identical to the furless mice (Roth *et al* 1998), however, it survives without any disease symptoms. In contrast, the cathepsin K knockout mice show a phenotype resembling the human disease pyknodysostosis (Saftig *et al* 1998). Cathepsin S knockout mice exhibit defects in the immune system (Shi *et al* 1999).

In contrast to enzyme depletions there are much more frequently pathological situations in which an enhanced activity of cysteine peptidases can be observed e.g. in some forms of chronic inflammatory diseases (Cunnane *et al* 1999), cancer (Mignatti and Rifkin 1993) and osteoporosis (Bossard *et al* 1996). These few examples underline the need to study the regulation of expression and activity of cathepsins.

The cysteine peptidases consist of two catalytic domains forming the active site cleft built by a cysteine, a histidine and an asparagine residue. These amino acid residues are completely conserved in all cysteine peptidases of plant and animal origin (Kirschke *et al* 1998). The enzymes are more or less inactive after the synthesis because a third domain - the propeptide part - covers this cleft and makes it unaccessible to protein substrates (Coulombe *et al* 1996). The enzymes become active after cleavage of this domain. It is noticeable that this inhibitory domain contains also some highly conserved sequence parts whose function remained unknown for a long time (Karrer *et al* 1993). This is natural since the interest in the enzyme activity was much higher than the interest in the function of the propeptide domain. Recently, the propeptide part raises more interest because this part plays a regulatory role in cellular targeting, in correct folding and in activation of the mature enzyme (Fox *et al* 1992, Carmona *et al* 1996, Coulombe *et al* 1996, Völkel *et al* 1996, Maubach *et al* 1997). We studied these regulatory functions by introducing mutations in conserved amino acid sequences of the propeptide domain, following the targeting, the maturation and the inhibitory properties of the mutated enzymes and propeptides, respectively.

## 2. METHODS

Site directed mutagenesis: We used human procathepsin S as model enzyme and constructed mutants of procathepsin S and of the propeptide of procathepsin S by the ExSite™ protocol (Stratagene) using non-overlapping primer pairs. The following mutants were constructed (Fig 1):

1. E44A, R48A, W52A, N55A, V59A, N63A, E44A/R48A, W52A/N55A/V59A/N63A, ΔE44-N63, ΔE44-I51, ΔW52-N63 in order to study the function of the highly conserved ERFNIN-motif (Karrer *et al* 1993)

2. W28A, W31A in order to study the function of an aromatic core in the propeptide domain

3. N104Q in order to remove the only glycosylation site in procathepsin S.

Transfection and localization: Human embryonic kidney (HEK) cells which do not express cathepsin S were stabily transfected with the cDNA of the various mutants as well as with wild type procathepsin S-cDNA. The localization of the respective proteins was followed by immune histochemistry using polyclonal antibodies to human cathepsin S and to LAMP-1 and secondary fluorescent labelled antibodies.

Expression of rec human cathepsin S propeptide and inhibition studies: The expression vectors for human cathepsin S propeptide and mutants of this propeptide were constructed on the basis of the site directed mutagenesis experiments and using the vector pASK75. The propeptides were expressed in *E. coli* and solubilized from inclusion bodies according to Maubach *et al* (1997). Inhibition experiments were performed using mature human cathepsins S and cathepsin L from *Paramecium tetraulia* (Völkel *et al* 1996).

Denaturation/renaturation experiments: Denaturation was performed in 6 M GdnHCl/100 mM dithiothreitol. The composition of the renaturation buffer is indicated in the legend to Fig 6. Activity determination is described in the legend to Fig 5.

## 3. RESULTS AND DISCUSSION

### *The helical ERFNIN motif is essential for transport and maturation of cathepsin L type cysteine peptidases*

Karrer *et al* (1993) described for the first time a structural motif in a group of cysteine peptidase precursors called ERFNIN after the highly conserved amino acids found in this helix. The orientation of this domain in

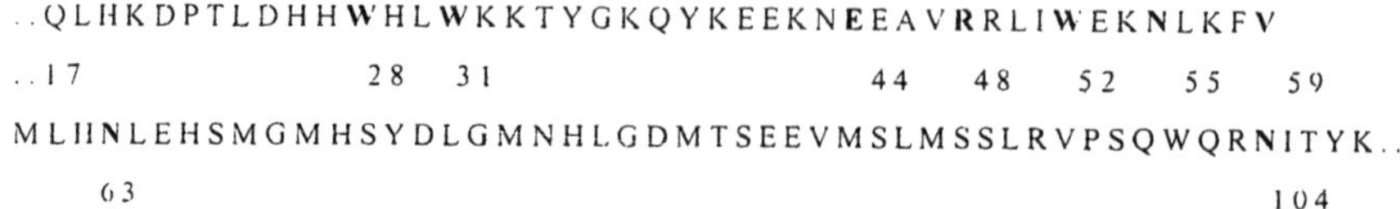


*Figure 1.* The position of the mutants in the propeptide region of human procathepsin S Partial primary sequence of human procathepsin S with the positions of the potentially mutated amino acids highlighted by bold letters.

the whole molecule has been elucidated in the 3-D-structure of procathepsin L. It is one of three helices (α1, α2, α3) in the propeptide part of cysteine peptidase precursors of the cathepsin L-type (Coulombe *et al* 1996). It is located on top of the two catalytic domains.

We constructed partial and complete deletion mutants lacking this motif and we transfected HEK 293 cells with the respective cDNAs. Cells expressing the proteins were labelled with [35]S-methionine and analyzed in pulse-chase experiments by immune precipitation of antigenic proteins. Fig 2 shows one representative result. The ΔW52-N63 mutant procathepsin S is expressed in HEK 293 cells as stable protein. However, it is not processed to mature cathepsin S as is the wild type procathepsin S. There was no cathepsin S activity measurable in the cells transfected with the mutant procathepsin S. Other deletions resulted in identical protein patterns (not shown). We conclude from these results that the mutant procathepsin exhibits a structure preventing the enzyme from further transport and processing.

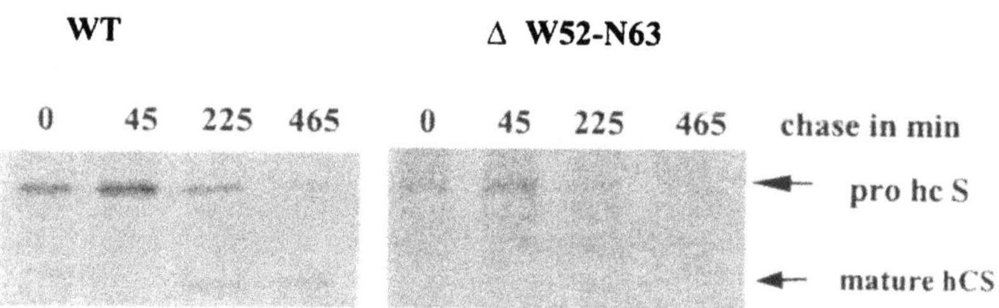


*Figure 2.* Cells expressing the deletion mutant ΔW52-N63 of cathepsin S do not show mature cathepsin S. HEK 293 cells (8 x $10^5$ cells) expressing wild type procathepsin S and the mutant ΔW52-N63, respectively, were labelled with [35S] methionine for 15 min. Chase times (min) are indicated. Cathepsin S was immunoprecipitated from cell extracts with a polyclonal rabbit antiserum to human procathepsin S. The immune complexes were removed by protein-A-sepharose CL-4B absorption, dissolved in sample buffer and subjected to SDS-PAGE. Dried gels were analyzed by phosphoimaging. Extracts of cells transfected with WT procathepsin S (left) and with the mutant ΔW52-N63 (right), respectively.

## *An aromatic stack built by the residues Trp 28, Trp 31 and Trp 52 is a structure stabilizing core of the propeptide domain*

In order to localize the structure stabilizing elements more precisely we constructed point mutations in procathepsin S replacing all six amino acids of the conserved ERFNIN motif by alanine. Again, we transfected HEK 293 cells with the cDNAs and analyzed the protein pattern in a similar fashion as we did before with the deletion mutants. The result is shown in Fig 3. The W52A mutant resulted in a non-processed procathepsin S, whereas the other alanine mutants ended up in mature processed cathepsin S.

Since it was known from the 3-D-structure of procathepsin L (Coulombe *et al* 1996) that this tryptophane residue is part of a core formed by two additional aromatic residues we simulated the procathepsin S structure by homology modelling using the homology module of the INSIGHT II (MSI) software package on the basis of the procathepsin L crystal structure data (pdb1CJL, Coulombe *et al* 1996). We found the aromatic core must also be present in procathepsin S and it must be formed by the two additional tryptophane residues W28 and W32. Therefore, we replaced these two tryptophane residues also by alanine and transfected the HEK 293 cells with the respective cDNAs. The results of the pulse-chase experiments are also represented in Fig 3: mutations in the aromatic stack resulted again in a complete loss of enzyme maturation. We conclude from these results, that the aromatic stack formed between the $\alpha$1- and the $\alpha$2-helices is an essential structure stabilizing element in the prodomains of cysteine peptidase precursors.

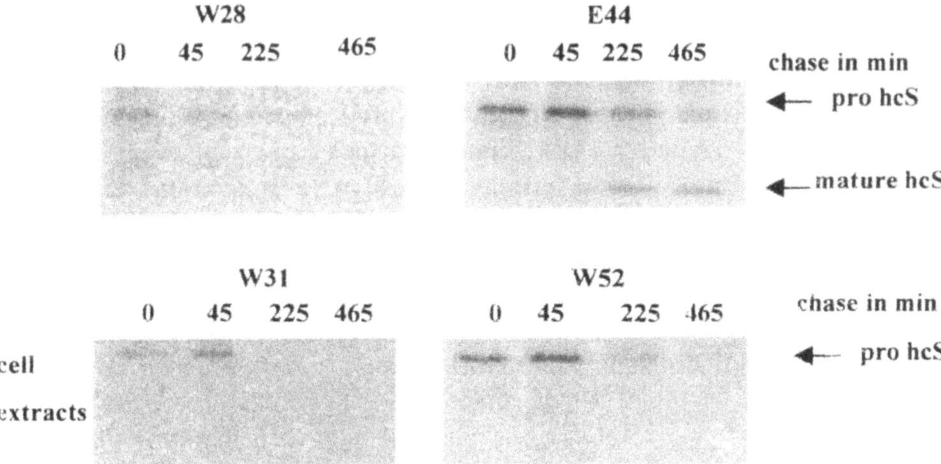


*Figure 3.* Cells expressing the mutants W28A, W31A and W52A do not show mature cathepsin S. The experiments were performed as in Figure 2

Immunohistochemistry of HEK cells transfected with either of the tryptophane mutants and with a mutant not influencing the aromatic stack (E44A), resp., documented the intracellular localization of the expressed proteins. Procathepsin S of the mutant E44A showed a lysosomal localization indicated by the colocalization with the constitutive lysosomal membrane glycoprotein LAMP-1. The mutant W52A did not show a lysosomal localization (Fig 4). These immunohistochemical data underline the significant role of the tryptophane stack for the precursor transport.

We followed also the transport of a procathepsin S mutant lacking the only glycosylation site (N104Q). No substantial maturation of this mutant has been observed, however, the non-glycosylated procathepsin bound to the plasma membrane of HEK 293 cells at 2 °C suggesting a specific interaction (Nissler *et al* 1998).

## *The inhibition of cathepsin S by propeptide variants of human procathepsin S suggests a correctly folded conformation of the propeptide*

The selective inhibitory potency of propeptide parts of cysteine peptidases towards their respective mature enzymes has been shown for some examples, whereas the inhibitory capacity of the propeptides to non-respective cysteine peptidases was only (Fox *et al* 1992, Carmona *et al* 1996, Völkel *et al* 1996, Maubach *et al* 1997 ).

We suggested the structure stabilizing effects of the aromatic stack in the prodomain should be reflected by the inhibitory properties of wild type and mutant propeptides. In order to test this hypothesis we expressed wild type and mutant propeptides of procathepsin S in *E.coli*. We performed kinetic experiments with all mutants and detected the mode of inhibition with cathepsin S as target enzyme. The wild type propeptide inhibited cathepsin S with Ki-values around 50 pM. Other cathepsins as well as papain are inhibited with much lower affinity. The type of inhibition was tight binding/slow reacting (Fig 5A). Similar results were obtained with the mutants E44A (Ki 0.3 nM) and R48A (Ki 1.3 nM).

However, the tryptophane mutants showed a different behaviour. The initial Ki values were three orders of magnitude higher (around 100 nM), and the kinetics revealed a two step mechanism

$$(E + I \Leftrightarrow EI \Leftrightarrow EI^*)$$

indicating that binding of the propeptide to the enzyme initiates a conformational shift in the propeptide towards better affinity (Fig 5B). We

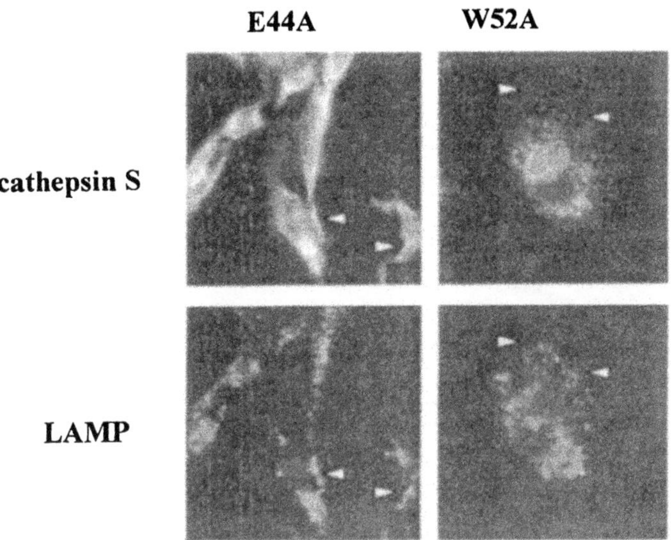


*Figure 4.* Cathepsin S with mutations in the tryptophane stack of the proregion does not co-localize with LAMP-1, a reporter protein of lysosomal structures. HEK 293 cells were transfected with cDNA coding for the mutant E44A and the mutant W52A, resp. LAMP-1 was recognized in transfected cells by the monoclonal antibody H4A3 (Developmental Hybridoma Bank, Dept. Biol. Sci., Univ. of Iowa), cathepsin S was recognized by a polyclonal rabbit antiserum. The secondary antibodies were labelled with different fluorescence dyes. Cells were monitored with a Zeiss Axioskop microscope equipped with an Atto Arc HBO 100 W mercury lamp and a 100 x /1.3 oil immersion objective lens. Cells were exposed for different time periods in order to get identical light intensities of the various fluorescence signals. Left: mutant E44A; right: mutant W52A. Staining for cathepsin S (upper panel) and for LAMP-1 (lower panel).

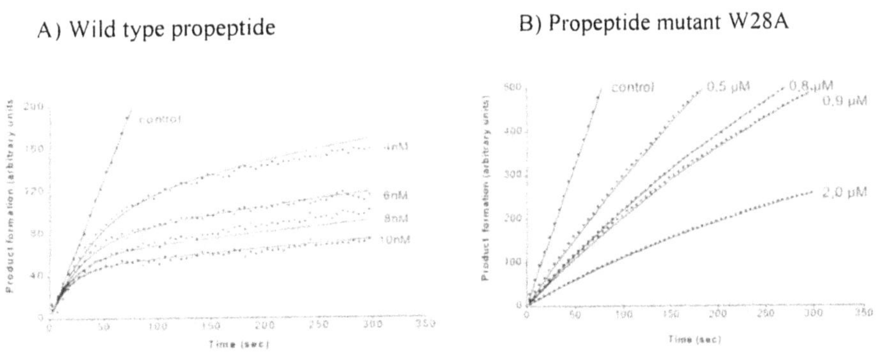


*Figure 5.* Inhibition of cathepsin S by wild type and mutant W28A propeptide of cathepsin S. Measurements were performed at 37 °C in 50 mM NaCl, 50 mM K phosphate, pH 6.5, 0.01% Triton X-100, 2mM EDTA, 2 mM DDT. [E] was 25 pM cathepsin S, [S] was 40 µM Z-Val-Val-Arg-AMC, [I] as indicated. The reaction was continuously followed in a LS50B fluorimeter (Perkin Elmer) at 360 nm $\lambda_{ex}$ and 460 nm $\lambda_{emm}$.

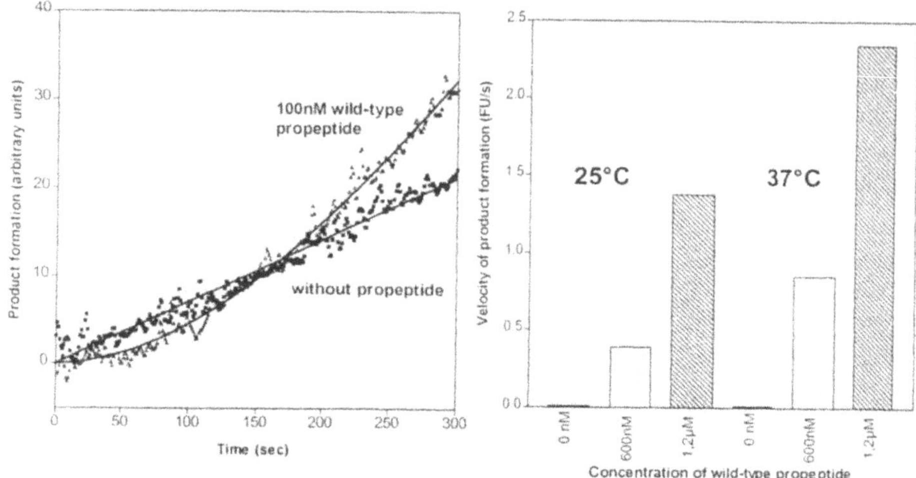


*Figure 6.* Renaturation rate of denatured cathepsin S. Left: Representative example of primary data acquisition. Denatured cathepsin S (80 nM) was transferred to renaturation buffer, 20 mM TRIS-HCl, 5 mM reduced / 1 mM oxidized glutathione, 2 mM Na-EDTA, 6 µM BSA, with and without wild type propeptide. Enzyme activity was measured after 1 : 100 dilution into the assay buffer specified in the legend to Figure 5. Right: Estimated velocities of uninhibited reaction as function of propeptide concentration and temperature. Cathepsin S concentration was 80 nM.

conclude from these results that the two parts of the molecule, the propeptide and the mature enzyme, influence each other in their conformation.

## *The propeptide domain of procathepsin S shows chaperone function*

If the latter suggestion is right, the propeptide part should - vice versa - influence the conformation of the mature part, too. A similar effect has been reported for propeptides of serine proteases on mature enzymes (Shinde *et al* 1993). In order to check the postulated chaperone effect we tried to renature denatured cathepsin S in the presence and in the absence of recombinant propeptide. The result is shown in Fig 6. Reactivation of denatured cathepsin S was accelerated in the presence of the propeptide by a factor of ≈20. Further experiments will reveal whether mature enzymes expressed in *E. coli* in inclusion bodies can be renatured in the presence of the respective propeptides.

## ACKNOWLEDGMENTS

The work was supported in part by the Deutsche Forschungsgemeinschaft (Wi 1102/1-4). The contribution to the experimental work of the following colleagues is gratefully acknowledged:Mark Fehn, Stefan Kreusch, Karl Nissler: Site directed mutagenesis experiments. Gunter Maubach, Ingrid Wenz: Expression and purificationof recombinant proteins. Sandra Pietschmann, Klaus Schilling,: Kinetic analyses, inhibition and renaturation experiments. Ekkehard Weber (Halle): Antibody production. Winfried Rommerskirch: Advice.

## REFERENCES

Bossard, M.J., Tomaszek, T.A., Thompson, S.K., Amegadzie, B.Y., Hanning, C.R., Jones, C., Kurdyla, J. T., McNulty, D. E., Drake, F. H., Gowen, M., and Levy, M. A., 1996, Proteolytic activity of human osteoclast cathepsin K. Expression, purification, activation, and substrate identification. *J. Biol. Chem.* **271:** 12517-12524.

Carmona, E., Dufour, E., Plouffe, C., Takebe, S., Mason, P., Mort, J.S., and Menard, R., 1996, Potency and selectivity of the cathepsin L propeptide as an inhibitor of cysteine proteases. *Biochemistry* **35:** 8149-8157.

Coulombe, R., Grochulski, P., Sivaraman, J., Menard, R., Mort, J.S., and Cygler, M., 1996, Structure of human procathepsin L reveals the molecular basis of inhibition by the prosegment. *EMBO J.* **15:** 5492-5503.

Cunnane, G., FitzGerald, O., Hummel, K.M., Gay, R.E., Gay, S., and Bresnihan, B., 1999, Collagenase, cathepsin B and cathepsin L gene expression in the synovial membrane of patients with early inflammatory arthritis. *Rheumatology-Oxford* **38:** 34-42.

Deussing, J., Roth, W., Saftig, P., Peters, C., Ploegh, H.L., and Villadangos, J.A., 1998, Cathepsins B and D are dispensable for major histocompatibility complex class II-mediated antigen presentation. *Proc. Natl. Acad. Sci. USA* **95:** 4516-4521.

Fox, T., de-Miguel, E., Mort, J.S., and Storer, A C., 1992, Potent slow-binding inhibition of cathepsin B by its propeptide. *Biochemistry* **31:** 12571-12576.

Inaoka, T., Bilbe, G., Ishibashi, O., Tezuka, K., Kumegawa, M., and Kokubo, T., 1995, Molecular cloning of human cDNA for cathepsin K: novel cysteine proteinase predominantly expressed in bone. *Biochem. Biophys. Res. Commun.* **206:** 89-96.

Karrer, K.M., Peiffer, S.L., and DiTomas, M. E., 1993, Two distinct gene subfamilies within the family of cysteine protease genes. *Proc. Natl. Acad. Sci. USA* **90:** 3063-3067.

Kirschke, H., Barrett, A J., and Rawlings, N.D. 1998 *Protein Profile. Lysosomal cysteine proteases*, Oxford University Press, Oxford/New York/Tokyo.

Maubach, G., Schilling, K., Rommerskirch, W., Wenz, I., Schultz, J.E., Weber, E., and Wiederanders, B., 1997, The inhibition of cathepsin S by its propeptide - specificity and mechanism of action. *Eur. J. Biochem.* **250:** 745-750 .

Mignatti, P., and Rifkin, D B., 1993, Biology and biochemistry of proteinases in tumor invasion. *Physiol. Rev.* **73:** 161-195.

Nägler, D.K., Sulea, T., and Menard, R., 1999, Full-length cDNA of human cathepsin F predicts the presence of a cystatin domain at the N-terminus of the cysteine protease zymogen. *Biochem. Biophys. Res. Comm.* **257:** 313-318.

Nissler, K., Kreusch, S., Rommerskirch, W., Strubel, W., Weber, E., and Wiederanders, B., 1998, Sorting of non-glycosylated procathepsin S in mammalian cells. *Biol. Chem.* **379:** 219-224.

Roth, W., Deussing, J., Saftig, P., Hafner, A., Schmidt, P., Schmahl, W., Rudensky, T., Nakagawa, T., Botchkarev, R., Paus, R., Villadangos, H., Ploegh, H., and Peters, C., 1998, Towards in vivo functions of lysosomal proteinases. *Biol. Chem.* **379:** Suppl. S120.

Saftig, P., Hunziker,E., Wehmeyer, O., Jones, S., Boyde, A., Rommerskirch, W., Moritz, J. D., Schu, P., and v. Figura, K., 1998, Impaired osteoclastic bone resorption leads to osteopororsis in cathepsin-K-deficient mice. *Proc. Natl. Acad. Sci. USA* **95:** 13453 – 13458.

Shi, G.P., Webb, A.C., Foster, K.E., Knoll, J.H., Lemere, C.A., Munger, J.S., Chapman, and H.A., 1994, Human cathepsin S: chromosomal localization, gene structure, and tissue distribution. *J. Biol. Chem.* **269:** 11530-11536.

Shi, G.P., Villadangos, J.A., Dranoff, G., Small, C., Gu, L., Haley, K.J., Riese, R., Ploegh, H.L., and Chapman, H. A., 1999, Cathepsin S required for normal MHC class II peptide loading and germinal center development. *Immunity* **10:** 197-206.

Shinde, U., Li, Y., Chatterjee, S., and Inouye, M., 1993, Folding pathway mediated by an intramolecular chaperone. *Proc. Natl. Acad. Sci. USA* **90:** 6924-6928.

Volkel, H., Kurz, U., Linder, J., Klumpp, S., Gnau, V., Jung, G., and Schultz, J.E., 1996, Cathepsin L is an intracellular and extracellular protease in Paramecium tetraurelia. Purification, cloning, sequencing and specific inhibition by its expressed propeptide. *Eur. J. Biochem.* **238:** 198-206.

# HUMAN CATHEPSINS W AND F FORM A NEW SUBGROUP OF CATHEPSINS THAT IS EVOLUTIONARY SEPARATED FROM THE CATHEPSIN B- AND L-LIKE CYSTEINE PROTEASES.

Thomas Wex[#], Brynn Levy, Heike Wex and Dieter Brömme[#]
*Department of Human Genetics, Mount Sinai School of Medicine, New York, NY 10029, USA*
*#Correspondence to Thomas Wex, Ph.D. or Dieter Brömme, Ph.D., Department of Human Genetics, Mount Sinai School of Medicine, Box 1498, Fifth Avenue at 100th Street, New York, NY 10029, USA, Fax 212- 849-2508, bronmd01@doc.mssm.edu, kabawex@aol.com*

## 1. INTRODUCTION

Thiol-dependent cathepsins have been characterized as cysteine proteases responsible for the terminal protein breakdown in lysosomes but also for specific cellular processes such as bone resorption, MHC-II class-mediated antigen presentation and activation of other zymogens like granzymes (Brömme *et al* 1999, Chapman *et al* 1997, Kirschke *et al* 1995). Based on the current classification, these proteases are members of the papain family C1 within the clan CA (Rawlings *et al* 1993). In recent years, the number of human cathepins in this family, whose complete cDNAs have been isolated, has increased to eleven [cathepsins B, C (dipeptidylpeptidase I), F, H, K, L, O, S, V (L2), W (lymphopain), X (Z), reviewed in Brömme *et al* 1999 and Santamaria *et al* 1999]. In addition to that classification, human cathepsins have been divided in two subgroups. Karrer and co-workers described a highly conserved interspersed amino acid motif ("ERW/FNIN") in the propeptide region of papain-like cysteine proteases (Karrer *et al* 1993). Members of the first subgroup, the "cathepsin L-like" proteases, contain the "ERW/FNIN“ motif in their proregions, whereas members of the second subgroup, the "cathepsin B-like" proteases do not contain this motif. Recently, two novel members of the papain superfamily, cathepsins W and

F, were identified and their cDNAs cloned (Santamaria *et al* 1999, Nagler *et al* 1999, Wang *et al* 1998). Amino acid sequence comparisons revealed that both cathepsins share a higher degree of homology to each other than to any other known human cathepsin (Wang *et al* 1998). In order to investigate the relatedness of human cathepsins F and W towards other human cathepsins, we characterized their genes and performed evolutionary analyses.

## 2. MATERIAL AND METHODS

### *Cloning and analysis of cathepsin W and F genes*

A human genomic library was screened with $\alpha$-$^{32}$P-labeled cathepsin F- and W-cDNA fragments, respectively. Two genomic clones for both cathepsins were identified, and used as a template for studying their genes. In general, for both cathepsin F and W genes, several genomic fragments identified by PCR or Southern analysis were subcloned as described previously (Wex *et al* 1998, Wex *et al* 1999a). Amplified PCR products and plasmid DNA containing subcloned DNA fragments were purified and sequenced on an Applied Biosystems Model 377 Automated Sequencer. The DNA fragments were sequenced using a series of primers based on the sequencing vector, exon and intron sequences.

### *Chromosomal localization of cathepsin W and F genes by FISH*

Plasmids containing either the cathepsin F or W gene were individually labeled by nick translation using Digoxigenin-11-dUTP and biotin-16-dUTP (Boehringer Mannheim), respectively. These labeled probes were used either for single-color or dual-color hybridization that was carried out and analyzed as previously described (Wex *et al* 1999a).

### *Amino acid sequence alignments and phylogenetic analyses*

Multiple sequence alignments of amino acid sequences of human cathepsins and related proteases from other organisms were created using the ClustalW program. Protein sequences were retrieved from the GenBank™ Database. Accession numbers for the human cathepsins were as follows: cathepsin K: S79895; cathepsin O: X77383; cathepsin C: U79415; cathepsin L: X12451; cathepsin H: X16832; cathepsin B: M14221; cathepsin S: M90696; cathepsin V: AF070448; cathepsin F: AF132894; cathepsin X: AF032906; cathepsin W (lymphopain): AF055903. Murine and rat sequences are as follows: cathepsin P: AF158182; cathepsin W: AF014941,

Cathepsin F: AI156216; cathepsin Q: AF187323. Other accession numbers from sequences subjected to the phylogenetic analysis are identified in Fig 4. Phylogenetic dendrograms were created and verified using the Puzzle 4.0 program package.

# 3. RESULTS AND DISCUSSION

## *Genomic organization and chromosomal localization of human cathepsins W and F*

In order to study the genomic organization of human cathepsins W and F, genomic BAC or PAC clones were identified and isolated from a human library (Genome Systems Inc.). For the cathepsin F gene, a 4 kb genomic PCR fragment and an overlapping 6 kb BamHI-fragment were subcloned and analyzed. Altogether, about 6.2 kb genomic DNA including the complete cathepsin F-cDNA as well as 1.1 kb of an upstream fragment were sequenced (GenBank™, accession number: AF132894). Sequence analysis revealed that the cathepsin F gene is composed of 13 exons that all conformed to the "GT/AG" rule (Shapiro *et al* 1987) (Fig 1).

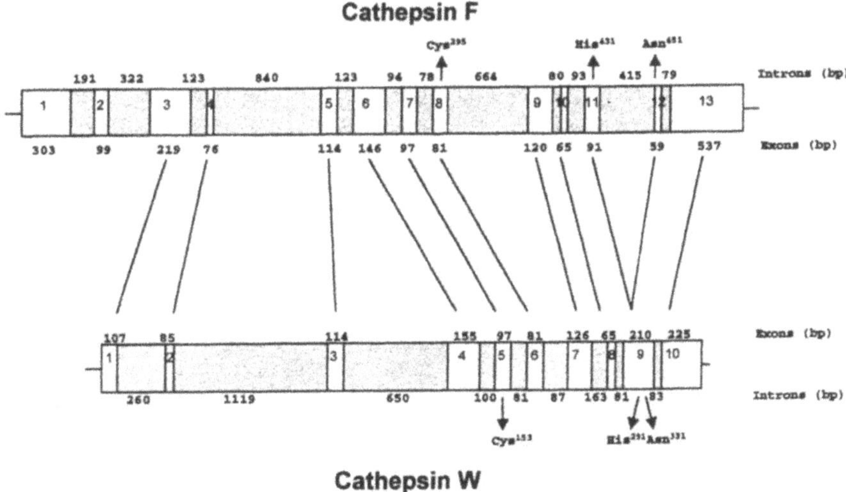


*Figure 1.* Schematic representation of the genomic structure of the human cathepsin F (5.1 kb) and W (3.8 kb) genes. White boxes represent exons and light gray boxes introns. The exon numbers and the conserved residues of the catalytic triad (Cys, His, Asn) identified in their corresponding exons are shown below or above the columns, resp. The lines between both genes indicate homologous exons between cathepsins W and F. The exon and intron sizes of cathepsins F and W are shown at the top or bottom of each column, respectively.

For the cathepsin W gene, several genomic DNA fragments were subcloned, and at all more than 9 kb of the cathepsin W gene were sequenced (partially released at GenBank™, accession number: AF 055903). The cathepsin W was found to be encoded in 10 exons whose splice sites conformed to the "GT/AG" rule as well (Fig 1). A putative 5'-untranslated exon was detected, but its exact location has not been determined (not shown). The comparison of genomic structures of cathepsins F and W with those of other cathepsin genes revealed that most cathepsin genes share a conserved gene structure in their 5'-region. Notably, splice sites #4, 5, and 6 (cathepsin F numbering) are conserved among several other cathepsins such as L, V, K, S, and O (Fig 2). The gene structure of cathepsin O revealed partial similarity to that of cathepsins W and F. In addition to the generally conserved splice sites 4, 5 and 6, the splice sites #8 and #9 were identical to those in cathepsins F and W. Futhermore, the splice site #11 that interrupts the region encoding for the putative active residues histidine and asparagine was found to be conserved in cathepsins F and O, but not in cathepsin W. However, the highest degree of conservation was identified between cathepsins F and W. Nine out of 12 splice sites (#3-10, and #12) of cathepsin F were identical with respect to their location and splice phase to those identified in cathepsin W (Fig 2). In addition, the cathepsin F gene contained two unique splice sites (#1, #2) that interrupted the very long proregion of cathepsin F within the "cystatin-like" domain identified by Nagler *et al* (1999) recently. Therefore, based on this fact, we compared the genomic organization of these two splice sites of the cathepsin F gene to those of cystatin-encoding genes (Wex *et al* 1999b). Interestingly, this study revealed striking similarities. The "cystatin like" domain in cathepsin F was encoded in three exons, a feature that has been described for most cystatin genes such as human cystatins C, D and SN. In addition, the two splice sites of cathepsin F (#1 and #2) were conserved regarding their location and splice phase with those of human cystatin C, D, SN genes and corresponding orthologues in mouse, rat and chicken. Based on this conservation, it seems to be likely that cathepsin F represents a fusion product between an ancestral cathepsin and cystatin gene.

In order to identify the chromosomal localization of both cathepsins W and F, *FISH* analyses were performed using fluorescence labeled probes either separately (mono-color analysis) or together (dual-color analysis). These studies revealed that both genes form a gene pair on the long arm of chromosome 11 at position q13.1-3 (data not shown). Since this gene pair is already the third cathepsin gene pair known [others are: CTSK and S at 1q21; CTSL and V at 9q21-22], gene duplication events have clearly contributed to the diversity of this protease class. Other cathepsin genes such as cathepsin B, C, H, X/Z, and O were mapped to unique places (CTSB 8p22-23.1; CTSC 11q14.1; CTSH, 15q24-25; CTSX/Z, 20q13; CTSO, 4q31-

32; reviewed in Wex *et al* 1999a). Based on the gene structures of cathepsins F, W, K, S, L, (Chauhan *et al* 1993, Gelb *et al* 1997, Santamaria *et al* 1998, Shi *et al* 1993, Wex *et al* 1998, Wex *et al* 1999b) and V (Itoh *et al* 1999) as well as on the degree of homology of their amino acid sequences, it can be concluded that all three gene pairs were formed by separate gene duplications. Since cathepsins F and W have a more complex gene structure than other cathepsins, it can be assumed that the ancestral gene of cathepsins F and W split before the divergence within the cathepsin L-like gene pairs occurred. Furthermore, based on the very high degree of homology between cathepsins L and V, it can be concluded that this gene pair at chromosome 9q21 resulted from the most current duplication event.

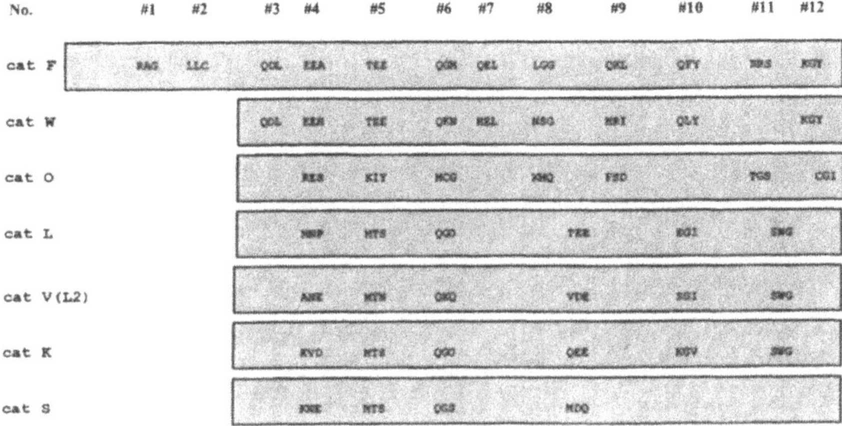


*Figure 2.* Comparison of human cathepsin genes. Columns represent the protein sequences aligned by the ClustalW program. The placements of splice sites within the amino acid sequence are represented by each tripeptide. Note that the diagrams neither reflect the sizes of the introns or the size of the genes. Furthermore, note that splice sites within untranslated regions described for several cathepsin genes are not shown. Data for the genomic organization of cathepsins W, F, K, S, L, V, and O were taken from literature.

## *The evolutionary relatedness of human cathepsins*

Comparisons of amino acid sequences among the cathepsins F, W, K, L, S, H, O and B revealed that both cathepsins F and W shared the highest degree of homology reaching a score at 58 % in their mature and overlapping propeptide sequences (Wang *et al* 1998). A comparison of propeptide sequences from several cathepsins revealed that the "ERW/FNIN" motif originally identified in human cathepsin L has been found in other human cathepsin L-related proteases such as cathepsins V, K, S as well (Fig 3). Interestingly, the novel mouse cathepsin P (Sol-Church, 1999) and the rat cathepsin Q (Sol-Church et al, 1999, accession number: AF187323 in the Genbank™ Database) whose human orthologues have not been identified as yet also contain the conserved "ERWNIN" motif. In contrast, cathepsins F and W share a conserved but partial different motif, the "ERFNAQ" motif that was also found in their mouse orthologues as well as slightly changed in two other cysteine proteases from invertebrates (Fig 3). An extended search in the Genbank™ Database revealed that "ERFNAQ"-related motifs are present in a variety of organisms including plants, invertebrates and mammals, suggesting that this motif has been conserved through evolution.

The evolutionary relatedness among human papain-like cathepsins was further studied using a phylogenetic analysis program. In a previous study focused on human papain-like proteases, we documented that cathepsins with adjacent chromosomal loci (cathepsins K and S on 1q21; cathepsins L and V on 9q22; cathepsins W and F on 11q13.1-3) represent separate dendrogram branches with a high significance (Wex *et al* 1999b). In order to extend this study, 33 different cysteine proteases from plants, invertebrates and mammals including all 11 human members were subjected to an

```
Consensus                    E  R   W N  I  N
                             +  +   + +  +  +
human cathepsin L           -EEGWRRAVWEKNMKMIELHNQEYRE
human cathepsin K           VDEISRRLIWEKNLKYISIHNLEASL
human cathepsin S           NEEAVRRLIWEKNLKFVMLHNLEHSM
human cathepsin V           -EEGWRRAVWEKNMKMIELHNGEYSQ
human cathepsin H           -EYHHRLQTFASNWRKINAHNNGNHT
mouse cathepsin P           KEEALRRAVWEENMRMIKLHNKENSL
mouse cathepsin Q           EEVLKRVVKWEENVKKIELHNRENSL

Consensus                    E  R   F N  A  Q
                             +  +   + +  +  +
human cathepsin F           EEARWRLSVFVNNMVRAQKIQALDRG
human cathepsin W           EEHAHRLDIFAHNLAQAQRLQEEDLG
mouse cathepsin W           AEYTRRLSIFAHNLAQAQRLQQEDLG
mouse cathepsin F           EEAQWRLTVFARNMIRAQKIQALDRG
Paragonium westermani       EDDQKRFAIFKDNLVRAQQYQTQEQG
Caenorhabditis elegans      REVLKRFRVFKKNAKVIRELQKNEQG
```

*Figure 3.* Comparison of conserved motifs in the proregions of human cathepsins. Amino acid sequences from several papain-like proteases (accession numbers are identified in the section "Material and Methods") were aligned using the ClustalW program, and the conserved motifs are highlighted and identified above the alignments as consensus sequence.

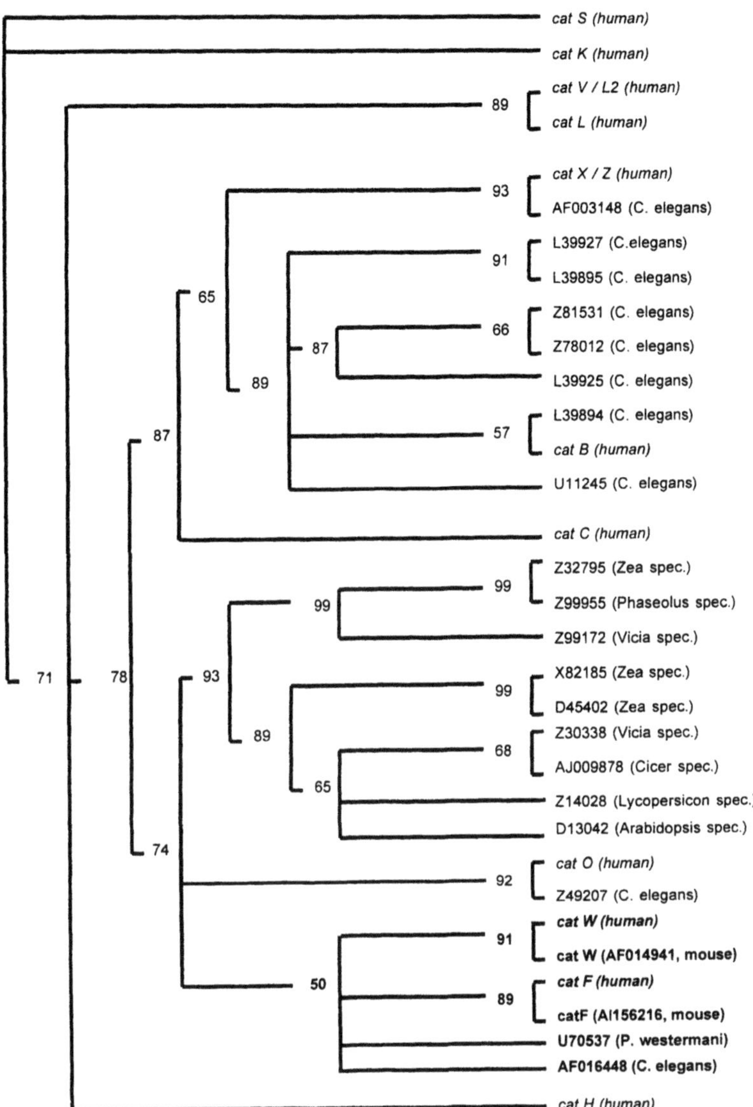


*Figure 4.* Evolutionary relatedness of papain-like proteases. Amino acid sequences of several cysteine proteases from plants, invertebrates and all 11 human cathepsins were subjected to an evolutionary analysis as described in the section "Material and Methods". The separate branch of "cathepsin F-like proteases" containing human cathepsins F and W, their mouse orthologues and related cysteine proteases from Caenorhabditis elegans and Paragonium westermani is highlighted with bold letters. The numbers at intersections and branches represent relative "support values" with a maximum at 100 indicating the significance of that particular branch or intersection.

evolutionary analysis using a the program package "PUZZLE". The computer analysis revealed known and novel aspects. First, it revealed that the three human gene pairs (cathepsins K/S; L/V and F/W) are found in separate branches of the tree, and that all four members of the "cathepsin L-like" proteases subgroup are directly linked to each other. Second, it confirmed the existence of a "cathepsin F-like protease" subgroup. In addition to human cathepsins F and W, their mouse orthologues and related proteases from two invertebrates (cathepsin F-related: Caeno elegans; cathepsin W-related: Paragonium westermani) were found in the same branch of the dendrogram (Fig 4). Third, the study revealed several orthologues from human cathepsins in *C. elegans*. Since the genome of the invertebrate *C. elegans* has been resolved, a comprehensive comparison between cysteine proteases of this organism and known human papain-like proteases was possible for the first time. As shown in Fig 4, *C. elegans* has one cathepsin O, X, and F orthologue and several cathepsin B-like cysteine proteases. The cathepsin F-like protease (AF016448) also contains a "cystatin-like" domain as described in the human cathepsin F (Nagler *et al* 1999). Therefore, it can be assumed that cathepsins O, X, F, and B represent evolutionary older proteases. Interestingly, no *C. elegans* orthologues were found for any of the human "cathepsin L-like" proteases such as cathepsin K, S, L and V suggesting that these proteases have been evolved after the last common ancestor between mammals and *C. elegans* has split. The genomic structure and degree of homology among human cathepsins as discussed above also support this hypothesis. Notably, *C. elegans* has seven different cathepsin B-like proteases, whereas only one human member, cathepsin B, has been identified as yet. Whether the variety of these cathepsin B-like proteases has been lost through evolution, or they have not been identified in humans so far will be finally revealed after completing the human genome project.

## 4. CONCLUSIONS

Based on the phylogenetic analysis, the presence of the "ERFNAQ" motif in the propeptides of both cathepsins as well as the highly conserved genomic organization and chromosomal localization of their genes, we concluded that cathepsins F and W are members of a novel subgroup of cathepsin proteases. According to the current nomenclature of the papain family that includes the cathepsin L-and B-like proteases, we proposed to name this novel third subgroup "cathepsin F-like" proteases.

Despite this assignment, both cathepsins F and W seem to be very different regarding their functions. Cathepsin W is exclusively expressed in natural killer cells and presumably in cytotoxic T cells suggesting a very specific function in the immune response. Based on northern analysis, cathepsin F was characterized as ubiquitously expressed protease suggesting a house-keeping function. Interestingly, cathepsin F contains a very long propeptide that encodes a "cystatin-like" domain, which might function as an endogenous cysteine protease inhibitor after its release from the cathepsin F zymogen as proposed by Nagler *et al* 1999. Future studies will shed light on the regulation and functional implications of human cathepsins F and W.

## ACKNOWLEDGMENTS

This work was supported by a fellowship (We/2170/2-1) of the "Deutsche Forschungsgemeinschaft", Germany and in part by the National Health Institutes research grants AR 39191 and AR 41331.

## REFERENCES

Brömme D., 1999, Cysteine Proteases as Therapeutic Targets. *Drug News Persp.* **2:** 73-82.

Chauhan, S.S., Popescu, N.C., Ray, D., Fleischmann, R., Gottesman, M.M., and Troen, B.R., 1993, Cloning, genomic organization, and chromosomal localization of human cathepsin L. *J. Biol. Chem.* **268:** 1039-1045.

Chapman, H.A., Riese, R.J., and Shi, G.P., 1997, Emerging roles for cysteine proteases in human biology. *Annu. Rev. Physiol.* **59:** 63-88.

Gelb, B.D., Shi, G.-P., Heller, M., Weremowicz, S., Morton, C., Desnick, R.J., and Chapman, H.A., 1997, Structure and chromosomal assignment of the human cathepsin K gene. *Genomics* **41:** 258-262.

Itoh R., Kawamoto S., Adachi W., Kinoshita S., and Okubo K., 1999, Genomic organization and chromosomal localization of the human cathepsin L2 gene. *DNA Res.* **6:** 137-140.

Karrer, K.M., Peiffer, S.L., and DiTomas, M.E., 1993, Two distinct gene subfamilies within the family of cysteine protease genes. *Proc. Natl. Acad. Sci. USA* **90:** 3063-3067.

Kirschke, H., Barrett, A.J., and Rawlings, N.D., 1995, Proteinases 1: lysosomal cysteine proteinases. *Protein Profile* **2:** 1581-1643.

Linnevers, C., Smeekens, S.P., and Brömme, D., 1997, Human cathepsin W, a putative cysteine protease predominantly expressed in CD8+ T-lymphocytes. *FEBS Lett.* **405:** 253-259.

Nagler, D.K., Sulea, T., and Menard, R., 1999, Full-length cDNA of human cathepsin F predicts the presence of a cystatin domain at the N-terminus of the cysteine protease zymogen. *Biochem. Biophys. Res. Commun.* **257:** 313-318.

Rawlings, N.D., and Barrett, A.J. (1993). Evolutionary families of peptidases. *Biochem J.* **15:** 290, 205-18.

Santamaria, I., Velasco, G., Pendas, A.M., Paz, A., and Lopez-Otin, C., 1999, Molecular cloning and structural and functional characterization of human cathepsin F, a new cysteine proteinase of the papain family with a long propeptide domain. *J.Biol.Chem.* **274:** 13800-13809.

Santamaria I., Pendas A.M., Velasco G., and Lopez-Otin C., 1998, Genomic structure and chromosomal localization of the human cathepsin O gene (CTSO). *Genomics* **53:** 231-234.

Shapiro, M.B., and Senapathy, P., 1987, RNA splice junctions of different classes of eukaryotes, sequence statistics and functional implications in gene expression. *Nucl. Acids Res.* **15:** 7155-7174.

Shi, G.P., Webb, A.C., Foster, K.E., Knoll, J.H., Lemere, C.A., Munger, J.S., and Chapman, H.A., 1994, Human cathepsin S: chromosomal localization, gene structure, and tissue distribution. *J. Biol. Chem.* **269:** 11530-11536.

Sol-Church K., Frenck J., Troeber D., and Mason R.W., 1999, Cathepsin P, a novel protease in mouse placenta.. *Biochem J.* **15:** 307-309.

Strimmer, K., and von Haeseler, A., 1997, Likelihood-mapping: a simple method to visualize phylogenetic content of a sequence alignment. *Proc. Natl. Acad. Sci. USA* **94:** 6815-6819.

Wang, B., Shi, G.P., Yao, P.M., Li, Z., Chapman, H.A., and Brömme, D., 1998, Human cathepsin F. Molecular cloning, functional expression, tissue localization, and enzymatic characterization. *J.Biol.Chem.* **273:** 32000-32008.

Wex, T., Levy, B., Smeekens, S.P., Ansorge, S., Desnick, R.J., and Brömme, D., 1998, Genomic structure, chromosomal localization, and expression of human cathepsin W. *Biochem. Biophys. Res. Commun.* **248:** 255-261.

Wex, T., Levy, B., Wex, H., and Brömme, D., 1999a, Human cathepsins F and W: A new subgroup of cathepsins. *Biochem. Biophys. Res. Commun.* **259:** 401-407.

Wex, Th., Wex, H., and Brömme, D., 1999b, The Human Cathepsin F Gene - A Fusion Product Between an Ancestral Cathepsin and Cystatin Gene. *Biol. Chem.* in press.

# CATHEPSIN K EXPRESSION IN HUMAN LUNG

Frank Bühling[1], Nadine Waldburg[1], Annegret Gerber, Carsten Häckel[4], Sabine Krüger[4], Dirk Reinhold[3], Dieter Brömme[5], Ekkehard Weber[6], Siegfried Ansorge and Tobias Welte[2]

*[1]Inst. of Immunology, [2]Dept. of Pneumology and Critical Care, [3] Inst. Exptl. Int. Med., [4] Inst. of Pathology, Otto-von-Guericke University Magdeburg, Magdeburg, Germany; [5]Dept. of Human Genetics, Mt. Sinai Medical School New York, U.S.A.; [6]Inst. of Physiological Chemistry, Martin-Luther-University Halle, Germany*

Key words cysteine proteinases, cathepsin K, cathepsin B, cathepsin L

Abstract Tissue remodeling is crucial in different lung diseases, in the embryonal development as well as in bronchial carcinoma. Cathepsins were proposed to be involved in the degradation of matrix proteins. Cathepsin K is one of the most potent matrix-degrading cysteine proteinases known as yet. The elastinolytic and collagenolytic activity of this papain-like protease is comparable with that of neutrophil elastase. We have investigated the cathepsin K expression in normal adult lung tissues, in embryonal lung tissue and in bronchial carcinoma. With help of specific anti-cathepsin K antibodies it could be shown that cathepsin K was expressed in bonchial epithelial cells. These data could be confirmed at mRNA level using a quantitative RT-PCR as well as by visualisation of the specific enzymatic activity in epithelial cell lines. During the embryonal development cathepsin K was expressed in the epithelial cells of the developing bronchi. The expression seemed to be upregulated in parallel with the development of the bronchial and alveolar lumen. In the later phase of lung development the cathepsin K expression was restricted to bronchial epithelial cells. Furthermore, using quantitative RT-PCR it could be shown that cathepsin K-mRNA was upregulated in lung tumor tissues in comparison to normal tissues from the same patients. These data suggest that cathepsin K may play an important role in matrix remodeling of the lung under physiological and pathological conditions.

# 1. INTRODUCTION

Cysteine proteinases (such as cathepsins B, L, S, H) are thought to be the major proteinases involved in intracellular protein degradation (Barrett and Kirschke, 1981). Several studies have shown the release of enzymatically active cathepsins from bronchial epithelial cells (Burnett *et al* 1995) and the involvement of these proteases in tissue degeneration and tumor progression (Reddy *et al* 1995, Werle *et al* 1997).

Cathepsin K is a member of the growing family of papain-like cysteine proteases. It shows high sequence homology to cathepsins S and L. This protease was shown to be expressed predominantly in osteoclasts (Brömme and Okamoto, 1995). A number of recent papers described the expression at other sites (Littlewood-Evans *et al* 1997, Sukhova *et al* 1998). The expression of cathepsin K in the lung was controversially discussed.

Cathepsin K seems to be the cysteine protease with the highest matrix degrading activity known as yet. It was shown that cathepsin K cleaves collagen type I, II and IV with higher efficiency than cathepsin L or cathepsin S. Furthermore, the catalytic mechanism of collagen cleavage by cathepsin K seems to be unique among the mammalian proteases (Garnero *et al* 1998). Its elastinolytic activity is higher when compared to cathepsin L and pancreatic elastase (Chapman *et al* 1997). The capability of cathepsin K to degrade components of the extracellular matrix suggests an involvement of this protease in the process of tissue destruction and remodelling also in the lung.

Using a combination of immunological, enzymatic and molecular biological methods this study will demonstrate that cathepsin K is expressed in the lung tissue and will show the cellular distribution of this enzyme.

# 2. MATERIALS AND METHODS

Lung tissue samples were obtained at surgery from patients with lung tumors. The samples were isolated from areas with and without morphological signs of neoplastic cells. Part of the tissue was formalin fixed and paraffin embedded for use in immunohistological studies or shock frozen in liquid nitrogen for use in PCR analyses. Primary cultures of lung fibroblasts were established as described by Becerril (Becerril *et al* 1999).

Recombinant cathepsin K was purified from the culture supernatant of *Pichia pastoris* transfected with human procathepsin K-cDNA. The polyclonal anti-cathepsin K-antiserum was raised in rabbits.

For RT-PCR experiments sense and antisense primers were designed using previously published cDNA sequences for the human Cathepsins K, B, and L.

DNA amplification was performed using LightCycler® (Idaho Technologies, Idaho Falls, ID, USA). SYBRGreen was used as dsDNA specific dye. PCR product specificity was verified by taking a melting curve and by agarose gel electrophoresis. Quantitative analysis was performed using the LightCycler software.

Paraffin embedded tissue blocks were cut, using a microtome, and 5 μm sections were placed on coated glass slides. The sections were dried overnight, dewaxed in xylene and rehydrated into 0.1 M Tris-buffer. Dewaxed paraffin sections were overlaid with either the rabbit cathepsin K-antiserum or with antiserum preincubated with saturating amounts of purified cathepsin K (negative control). A standard avidin/biotin-complex method (alkaline phosphatase) was performed according to the manufacturers protocol (Vectastain-Kit, Boehringer Ingelheim, Germany). New fuchsin was used as substrate. Sections were counterstained with hematoxylin and mounted in gelatine.

# 3. RESULTS

## 3.1 Normal lung tissues

Using a specific cathepsin K antiserum, immunoreactivity was tested in paraffin embedded lung tissues of 15 patients. Distinct immunoreactions were abundant in ciliated and non-ciliated bronchial epithelial cells (*Fig* 1 left). In contrast, only a weak immunoreactivity was found in alveolar epithelial cells or in alveolar macrophages (data not shown). The specificity of the immunoreaction was shown in control experiments where the antiserum was preincubated with purified cathepsin K - no staining was detectable in these slides. The staining specificity was further confirmed in western blots. The expression was shown at mRNA-level and using cytochemical approaches (Bühling *et al* 1999).

## 3.2 Embryonal lung tissues

The expression of cathepsin K in comparison to cathepsin B- and cathepsin L-expression was studied in lung tissues of embryos between 10th and 37th week of gestation. Cathepsin K was expressed in the developing bronchi in the 10. week of gestation (Fig 1 right). At this time no immunoreactivity for cathepsin L or cathepsin B was found. Later the expression of cathepsin K was higher in the epithelial cells of the larger bronchi than in the basal or intermediate epithelial cells of the small branching bronchi. Therefore, it was suggested that the expression of this protease was upregulated during differentiation of epithelial cells. In contrast to the expression pattern of cathepsin B and cathepsin L, an immunoreactivity of the cathepsin K antiserum was found only in epithelial cells.

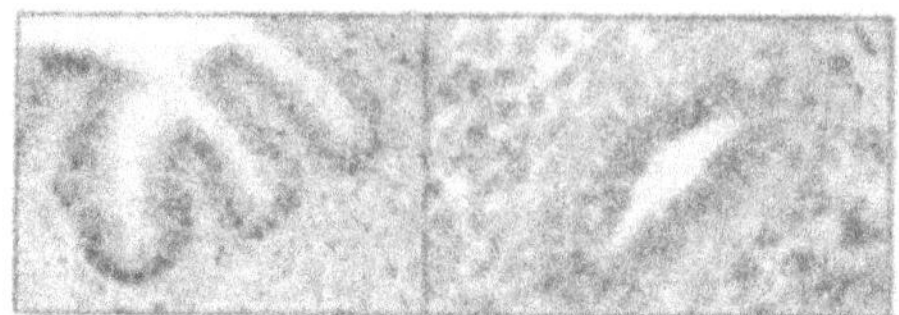

*Figure 1.* Cathepsin K expression in adult and embryonal bronchial epithelial cells.

## 3.3 Lung tumor tissues

Tumor cells represent undifferentiated cells of a defined origin. Lung carcinoma cells are an example of undifferentiated epithelial cells. The expression of cathepsin K in squamous cell carcinoma and adenocarcinoma of the lung was studied at mRNA and protein levels. Using quantitative RT-PCR analyses it was shown that lung tumor tissues contained significantly more cathepsin K-mRNA than the corresponding non-tumor tissues from the same patients. No significant difference was found for cathepsin B or cathepsin L (Fig 2). In correlation to the findings in the embryonal lung tissues immunohistochemical investigations revealed that the expression of cathepsin K in the undifferentiated tumor cells was lower than in the corresponding normal, differentiated epithelial cells (not documented).

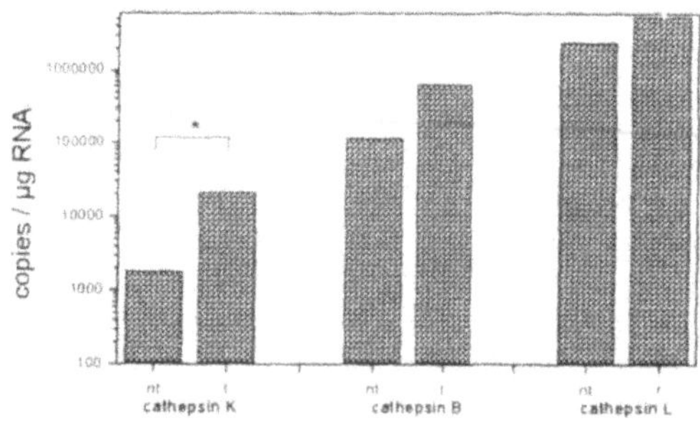


*Figure 2.* Cathepsin mRNA expression in lung tissues. nt – non-tumor tissue; t-tumor tissue

## 4. CONCLUSIONS

Throughout the human lifetime lung tissue undergoes continuos remodelling. During the embryonal development there is a rapid synthesis and degradation of connective tissue proteins as the developing bronchial tree and later the alveoli substitute the initial interstitial tissue. Later the lung is one of the organs with direct contact to the environment. Therefore a number of injuries and microbial infections lead to a destruction of lung tissue, provoking a further degradation and substitution of the injured matrix proteins. These processes of matrix remodeling directly depend on the presence and the activity of proteolytic enzymes.

The neoplastic growth of tumor cells is another example of matrix destruction and remodeling. A number of studies provided evidence that proteolytic enzymes play a key role in the process of tumor invasion and growth.

Cysteine proteases were shown to be expressed in tumor cells of different origin. The description of the highly collagenolytic and elastinolytic enzyme cathepsin K raised the question whether this protease could be involved in the processes of matrix remodeling in the lung.

Cathepsin K seems to be constitutively expressed in bronchial epithelial cells of the adult lung independent of whether they are ciliated and localized in the proximal bronchi or non-ciliated and localized in the distal bronchi. Interestingly, its expression occurs earlier during the embryonal development than the expression of cathepsin B and cathepsin L.

In embryonal lung tissues and in bronchial carcinoma it could be shown that undifferentiated epithelial cells expressed lower but readily detectable amounts of cathepsin K. Therefore it was hypothesized that the expression could be upregulated during differentiation. Further studies should investigate the functional consequences of the cathepsin K expression in the lung.

## REFERENCES


Barrett, A.J., and Kirschke, H., 1981, Cathepsin B, Cathepsin H, and cathepsin L. *Methods Enzymol.* **80:** 535-561.

Becerril, C., et al., 1999, Acidic fibroblast growth factor induces an antifibrogenic phenotype in human lung fibroblasts. *Am. J. Resp. Cell Molec. Biol.* **20:** 1020-1027.

Brömme, D., and Okamoto, K., 1995, Human cathepsin O2, a novel cysteine protease highly expressed in osteoclastomas and ovary molecular cloning, sequencing and tissue distribution. *Biol. Chem. Hoppe Seyler* **376:** 379-384.

Burnett, D., et al., 1995, Synthesis and secretion of procathepsin B and cystatin C by human bronchial epithelial cells in vitro: modulation of cathepsin B activity by neutrophil elastase. *Arch. Biochem. Biophys.* **317:** 305-310.

Bühling, F., et al., 1999, Expression of cathepsin K in lung epithelial cells. *Am. J. Cell. Mol. Biol.* **20:** 612-619.

Chapman, H.A. et al., 1997, Emerging roles for cysteine proteases in human biology. *Ann. Rev. Physiol.* 59, 63-88.

Garnero, P., et al., 1998, The collagenolytic activity of cathepsin K is unique among mammalian proteinases. *J. Biol.. Chem.* **273:** 32347-32352.

Littlewood-Evans, A.J., et al., 1997, The osteoclast-associated protease cathepsin K is expressed in human breast carcinoma. *Cancer Res.* **57:** 5386-5390.

Reddy, V.Y., et al., 1995, Pericellular mobilization of the tissue-destructive cysteine proteinases, cathepsins B, L, and S, by human monocyte-derived macrophages. *Proc. Natl. Acad. Sci.* **92:** 3849-3853.

Sukhova, G.K., et al., 1998, Expression of the elastolytic cathepsins S and K in human atheroma and regulation of their production in smooth muscle cells. *J. Clin. Invest.* **102:** 576-583.

Werle, B., et al., 1997, Cathepsin B fraction active at physiological pH of 7.5 is of prognostic significance in squamous cell carcinoma of human lung. *Br. J. Cancer* **75:** 1137-1143.

# EXPRESSION OF CATHEPSINS B AND L IN HUMAN LUNG EPITHELIAL CELLS IS REGULATED BY CYTOKINES

Annegret Gerber[1], Tobias Welte[2], Siegfried Ansorge[1] and Frank Bühling[1]
*[1]Institute of Immunology, [2]Department of Pneumology and Critical Care, Otto von Guericke University Magdeburg, Magdeburg, Germany*

Key words cathepsin, cytokine

Abstract The cathepsins B, L, and H are expressed ubiquitously and represent the major proportion of lysosomal enzymes. They are involved in bulk proteolysis in the lysosomes, processing of proteins and matrix degradation. Under pathological conditions the participation of cathepsins, especially their secreted forms, was observed in inflammation, tumor progression and metastasis. The enzymatic activity of cathepsins is regulated by posttranslational modification, localization, maturation, changes in pH, and their interaction with inhibitors. Regulation at the level of transcription is not well elucidated. The aim of this study was to investigate the effect of IL-1β, IL-6, IL-10, TGF-β1, and HGF on mRNA expression and protein level in human lung epithelial cell lines A-549 and BEAS-2B. IL-6 leads to a twofold increase in cathepsin L mRNA expression, whereas TGF-β1 decreases the amount of cathepsin L mRNA. At protein level, using enzyme immunoassay, it was shown that IL-6 induced increased amounts of cathepsin L but not cathepsin B. In contrast, after incubation of bronchial epithelial cells with TGF-β1 the cathepsin L concentration was decreased. In conclusion, gene expression of cathepsins B and L is variable. The cytokines IL-6 and TGF-β1 modulate cathepsin gene expression.

## 1. INTRODUCTION

Neoplastic cells often show upregulation, membrane association and secretion of cathepsin B. In lung carcinoma cathepsin B activity is increased in tumor as compared to associated lung parenchyma. A significant correlation was established between increased cathepsin B activity in tumor cells and shorter survival rates (Krepela 1990, Werle 1995, Ebert 1994). An association of tumor progression and the cathepsin L concentration was also found in lung cancer (Werle 1997). Thus, the understanding of the mechanism of the alterations in cathepsin B and cathepsin L in tumor cells and the surrounding normal lung parenchyma is of pathophysiological interest. Until now the role of cytokines in the regulation of cathepsin gene expression is not well elucidated.

## 2. MATERIALS AND METHODS

### *Cell lines*

The cell line A-549, an alveolar epithelial cell line, was obtained from DSMZ (Braunschweig, Germany). BEAS-2B, which was derived from virus-transformed bronchial epithelial cells, was a kindly gift from A. Gillesen (University Bonn, Germany). Cells were incubated for 48 h with cytokines in the following concentrations: 200 ng/ml IL-1β; 1000 U/ml IL-6; 5 ng/ml IL-10; 50 ng/ml HGF; 20 ng/ml TGF-β1.

### *RT-PCR*

For RT-PCR 1.5 μg of total RNA was reverse transcribed by First Strand DNA Synthesis kit (Amersham Pharmacia Biotech, Freiburg, Germany). The PCR reaction was optimized with the following primer pairs (5′→ 3′): cathepsin L sense CAG GCA GGT GAT GAA TGG CT, cathepsin L antisense CAG GCC TCC ATT ATC CTG AA, cathepsin B sense GCC TGC AAG CTT CGA TGC AC, cathepsin B antisense ATC ATC TCT CCG GTG ACG TGT, β-actin sense TGA CGG GGT CAC CCA CAC TGT GCC CAT CTA, β-actin antisense CTA GAA GCA TTT GCG GTG GAC GAT GGA GGG. DNA amplification was performed using LightCycler (Idaho technologies, Idaho Falls, ID, USA) according to the instructions of the manufacturer. SYBRGreen was used as dsDNA specific dye. Reactions were cycled 40 times in glass capillaries (95 °C denaturation for 0 s, 66 °C annealing for 3 s, and 72 °C extension). Extension time was 13 s, 18 s, and 20 s for cathepsin L and B and for β-actin, respectively. Product accumulation was monitored at the end of extension phase at a temperature 5 °C below the product melting temperature. PCR product specificity was verified by taking a melting curve and by agarose gel electrophoresis. Quantification data were analyzed using LightCycler software. Serial dilutions of plasmid DNA containing the cloned amplicons were used as standards. Results for cathepsin mRNA expression were calculated in relation to β-actin mRNA expression. Experiments were performed in quadruplicate and results are given as mean ± SEM.

### *ELISA*

For detection of cathepsin B and cathepsin L in cell lysates, commerially available ELISA kits were used (Krka d.d., Novo mesto, Slovenia). Protein concentration was determined using Fast Protein Assay Kit (Pierce, Rockford, IL, USA). Results are expressed as mean ± SEM of three independent experiments.

### *Statistical analysis*

The data were evaluated using the Student´s t test. Significance was established at $p < 0.05$.

## 3. RESULTS

Using quantitative RT-PCR we found a twofold increase in cathepsin L mRNA expression after incubation of cells with IL-6, whereas TGF-β1 decreased the amount of cathepsin L mRNA. This effect was more clearly pronounced in BEAS-2B than in A-549 cells. In contrast, cathepsin B mRNA expression was nearly unaffected by cytokines except for TGF-β1, which decreased cathepsin B mRNA in A-549 but not in BEAS-2B (Table 1). After incubation with IL-6, cathepsin B mRNA was slightly reduced. Incubation of cells with IL-1β, IL-10, and HGF showed no effects on cathepsin B and L mRNA expression.

*Table 1.* Effects of cytokines on relative mRNA concentration of cathepsin B and cathepsin L

| Cytokine | Relative amount of cathepsin L mRNA (control = 1.0) | | Relative amount of cathepsin B mRNA (control = 1.0) | |
|---|---|---|---|---|
| | A-549 | BEAS-2B | A-549 | BEAS-2B |
| IL-1β | 1.0 ± 0.2 | 1.2 ± 0.5 | 1.0 ± 0.3 | 1.2 ± 0.4 |
| IL-6 | 1.9 ± 0.4 * | 2.0 ± 0.5 * | 0.7 ± 0.1 | 0.7 ± 0.3 |
| IL-10 | 1.3 ± 0.3 | 1.0 ± 0.2 | 0.8 ± 0.2 | 0.9 ± 0.3 |
| HGF | 1.3 ± 0.3 | 1.1 ± 0.3 | 0.8 ± 0.2 | 0.7 ± 0.2 |
| TGF-β1 | 0.7 ± 0.1 * | 0.6 ± 0.2 * | 0.5 ± 0.2 * | 1.0 ± 0.1 |

Effect of IL-1β, IL-6, IL-10, HGF, and TGF-β1 on relative mRNA concentration of cathepsins B and L. Cathepsin mRNA concentration was determined in relation to β-actin mRNA level by quantitative RT-PCR using a microvolume fluorimeter with rapid temperature control and continuous fluorescence monitoring (n = 4, mean ± SEM). *$p < 0.05$.

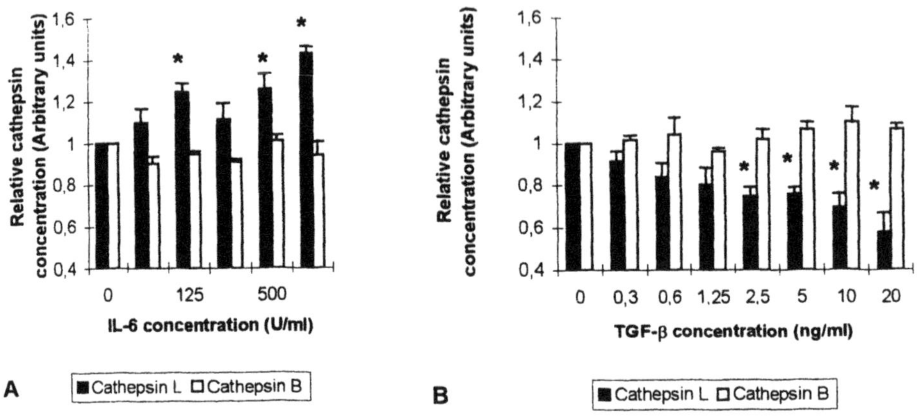


*Figure 1*. Effect of IL-6 (A) and TGF-β1 (B) on cathepsin L and cathepsin B protein concentration. A-549 cells were incubated for 48 h with cytokines in the indicated concentrations. The amount of cathepsin B and and cathepsin L in cell lysates was measured by enzyme immunoassay (n = 3, mean ± SEM). *p < 0.05.

Cathepsin B and L protein levels were measured in lysates of A-549 cells after incubation of cells with IL-6 (0 to 1000 U/ml) and TGF-β1 (0 to 20 ng/ml). Using enzyme immunoassay it was shown that IL-6 induced increased amounts of cathepsin L but not cathepsin B. The increase in cathepsin L protein after incubation with IL-6 was dose-dependent. Highest amount of cathepsin L was detected after incubation with 1000 U/ml IL-6 (Fig 1A).

Incubation with TGF-β1 clearly reduced the cathepsin L concentration in a dose-dependent manner. After incubation with 20 ng/ml TGF-β1 maximally decreased cathepsin L was found. Cathepsin B concentration was nearly unaffected by the incubation with TGF-β1 in the concentration range of 0 to 20 ng/ml (Fig 1B).

## 4. CONCLUSION

It has been reported that the 5′untranslated region of the cathepsin B and L genes does not contain a TATA-box and is GC rich, suggesting that cathepsins B and L are housekeeping genes (Chauhan 1993, Berquin 1995). Besides their housekeeping gene-type promotors considerable variability in mRNA expression has been found. Examination of the main promotor region has indicated possible control by several regulating factors including SP-1 binding elements.

In NIH3T3 cells the cathepsin L gene is activated by PDGF, EGF, PMA, and activated oncogens (Kane 1990). TGF-β1 was found to decrease the amount of cathepsin L mRNA. The suppressing role of TGF-β1 on cathepsin activity was also established in cultured kidney tubule cells (Ling 1995). Our results showed decreased cathepsin L mRNA expression and decreased protein concentration in human lung epithelial cells A-549 after incubation with TGF-β1. In contrast, cathepsin B mRNA level and protein concentration were nearly unchanged. Several studies have shown that IFN-γ modulates cathepsin expression. IFN-γ-induced elevation of cathepsin B led to an increase in cathepsin B activity in THP-1 cells (Li 1997). Investigation of the underlying mechanism revealed increased transcription and mRNA stability (Li 1998). In another study elevated cathepsin B expression at the mRNA level was found with differentiation of promonocytic U937 cell line induced by PMA and GM-CSF (Ward 1990). We investigated the effects of IL-1β, IL-6, IL-10, TGF-β1, and HGF on mRNA expression and protein levels of cathepsin B and L. We observed an inhibitory effect of TGF-β1 on cathepsin L, whereas IL-6 increased mRNA expression and protein concentration of cathepsin L in lung tumor cells.

The results suggest that the cytokines IL-6 and TGF-β1 are involved in regulation of cathepsin L gene expression at transcriptional and post-transcriptional levels. Cathepsin B expression was nearly unaffected by the investigated cytokines. Further studies are necessary to elucidate the mechanism regulating altered mRNA expression.

## ACKNOWLEDGMENTS

This work was supported by the Deutsche Forschungsgemeinschaft, SFB 387 (B7) and by the Deutsche Krebshilfe DKH 10-1355-Ge 1.

## REFERENCES

Berquin, I.M., *et al.*, 1995, Identification of two new exons and multiple transcription start points in the 5'-untranslated region of the human cathepsin-B-encoding gene. *Gene* **159:** 143-149.

Chauhan, S.S., *et al.*, 1993, Cloning, genomic organization, and chromosomal localization of human cathepsin L. *J. Biol. Chem.* **268:** 1039-1045.

Ebert, W., *et al.*, 1994, Prognostic value of increased lung tumor tissue cathepsin B. *Anticancer. Res.* **14:** 895-899.

Kane, S.E., and Gottesman, M.M., 1990, The role of cathepsin L in malignant transformation. *Semin. Cancer. Biol.* **1:** 127-136.

Krepela, E., *et al.*, 1990, Increased cathepsin B activity in human lung tumors. *Neoplasma* **37:** 61-70.

Li, Q., *et al.*, 1997, Endotoxin induces increased intracellular cathepsin B activity in THP-1 cells. *Immunopharmacol. Immunotoxicol.* **19:** 215-237.

Li, Q., *et al.*, 1998, Interferon-gamma induces cathepsin B expression in a human macrophage-like cell line by increasing both transcription and mRNA stability. *Int. J. Mol. Med.* **2:** 181-186.

Ling, H., *et al.*, 1995, Suppressing role of transforming growth factor- 1 on cathepsin activity in cultured kidney tubule cells. *Am. J. Physiol.* **269:** 911-917.

Ward, C.J., *et al.*, 1990, Changes in the expression of elastase and cathepsin B with differentiation of U937 promonocytes by GM-CSF. *Biochem. Biophys. Res. Commun.* **167:** 659-664.

Werle, B., *et al.*, 1995, Assessment of cathepsin L activity by use of the inhibitor CA-074 compared to cathepsin B activity in human lung tumor tissue. *Biol. Chem. Hoppe Seyler* **376:** 157-164.

Werle, B., *et al.*, 1997, In *Proteolysis in cell function* (V.K. Hopsu-Havu, M. Jarvinen, and H. Kirschke, eds), IOS Press, pp. 472-478.

# FUNCTIONS OF CATHEPSIN K IN BONE RESORPTION

*Lessons from cathepsin K deficient mice*

Paul Saftig[1*], Ernst Hunziker[2], Vincent Everts[3], Sheila Jones[4], Alan Boyde, Olaf Wehmeyer[1], Anke Suter[1], Kurt von Figura[1]

*[1]Zentrum Biochemie und Molekulare Zellbiologie, Abteilung Biochemie II, Universität Göttingen, Heinrich-Düker-Weg 12, 37073 Göttingen, Germany; [2]Universität Bern, M.E. Müller-Institut für Biomechanik, Murtenstrasse 35, CH-3010 Bern, Switzerland. [3]Dept Periodontology and Dept. Cell Biology and Histology, Academic Centre for Dentistry, Amsterdam; University of Amsterdam, Amsterdam, The Netherlands;[4]Department of Anatomy and Developmental Biology, University College London, Gower Street, London WC1E 6BT, UK. [*]To whom correspondence should be addressed at: Zentrum Biochemie und Molekulare Zellbiologie, Abteilung Biochemie II, Universität Göttingen, Heinrich-Düker-Weg 12, D-37073 Göttingen, FRG, Tel: ++49-551-395932, Fax: ++49-551-395979, e-mail: saftig@uni-bc2.gwdg.de*

Key words Key wordscathepsin K, knockout, osteopetrosis, osteoporosis, pycnodysostosis, bone resorption

Abstract Cathepsin K is a cysteine proteinase expressed predominantly in osteoclasts. Cathepsin K cleaves key bone matrix proteins and is believed to play an important role in degrading the organic phase of bone during bone resorption. Pycnodysostosis, an autosomal recessive osteosclerosing skeletal disorder has recently been shown to result from mutations in the cathepsin K gene. Cathepsin K deficient mice generated by targeted disruption of this proteinase phenocopy many aspects of pycnodysostosis. They display an osteopetrotic phenotype with excessive trabeculation of the bone-marrow space accompanied by an altered ultrastructural appearance of the cathepsin K deficient osteoclasts. These cells also demonstrate an impaired resorptive activity in vitro. In contrast to other forms of osteopetrosis, which are due to disrupted osteoclastogenesis, cathepsin K deficiency is associated with an inhibition of osteoclast activity. Taken together the phenotype of cathepsin K knockout mice underlines the importance of this proteinase in bone remodelling.

# 1. INTRODUCTION

Bone remodelling is a constant process that involves bone resorption and rebuilding (Erlebacher *et al* 1995). The resorption phase of this process is carried out by osteoclasts, which adhere to the surface of bone leading to the creation of an extracellular compartment which is isolated from general extracellular fluid. The resorption pit is maintained at an acidic pH by active secretion of protons via vacuolar adenosine triphosphatase through the ruffled border membrane. Morphological data suggest that the degradation of inorganic matrix precedes the degradation of organic matrix (Everts *et al* 1998). Active proteases released from osteoclasts into the resorption lacunae are known to be involved in matrix degradation. Cysteine proteinases have been suggested to play a vital role in bone matrix degradation (Delaisse *et al* 1984, Everts *et al* 1992).

Cathepsin K (EC 3.4.22.38), is a recently identified lysosomal cysteine proteinase. It has been identified by several groups, as well as having been cloned from human, rabbit, and murine osteoclast libraries (Tezuka et al 1994, Inaoka *et al* 1995, Li *et al* 1995, Drake *et al* 1996, Brömme *et al* 1995, Shi *et al* 1995, Velasco *et al* 1994). By means of *in situ*-hybridisation and immunolocalisation (Tezuka *et al* 1994, Li *et al* 1995, Drake *et al* 1996) it has been shown to be highly and selectively expressed within human osteoclasts.

Over-expressed *baculovirus* cathepsin K is known to degrade bone-matrix proteins, including type I and type II collagen, osteopontin and osteonectin at low pH (Bossard *et al* 1996, Kafienah *et al* 1998). The view that cathepsin K plays a major role in bone resorption is strengthened by the recent identification of cathepsin K gene mutations which are linked to pycnodysostosis, a hereditary bone disorder in which osteoclast function in bone resorption is defective (Gelb *et al* 1996). To date five separate mutations (nonsense, missense or stop codon mutations) in the cathepsin K gene have been identified in unrelated pedigrees in which pycnodysostosis occurs (Gelb *et al* 1996, Ho *et al* 1999). In two cases, it was shown that cathepsin K expression was virtually absent, while in the heterozygote parents expression was reduced by 50 – 80 % compared with controls (Ho *et al* 1999). These expression studies suggest that a significant reduction of enzyme activity is well tolerated and that only the total loss of activity leads to a clinical phenotype. This is especially interesting since cathepsin K antagonists are being advocated as potentially effective therapies for bone disorders such as osteoporosis (Thompson *et al* 1997). Pycnodysostosis is characterised by a variable clinical picture which includes short stature, open fontanelles, obtuse mandibular angles, partial or total aplasia of the terminal phalanxes, a predisposition to bone fractures and an increased radiographic density of the entire skeleton (Edelson *et al* 1992). In these patients osteoclasts have the ability to demineralise bone but lack the tools to digest bone matrix (Everts *et al* 1985).

## 2. CATHEPSIN K-DEFICIENCY AND OSTEOPETROSIS

To delineate the role of cathepsin K in bone resorption, mice with a targeted mutation of this proteinase were generated (Saftig *et al* 1998). Cathepsin K deficient mice are viable and fertile. They lack cathepsin K mRNA and protein expression. Despite the lack of this proteinase gross inspection of these mice did not reveal abnormalities. However, radiological examination of bones from these mice revealed pronounced osteosclerotic abnormalities. This phenotype appeared to become more pronounced with age. The osteopetrotic changes predominantly affect bone tissue which undergoes rapid remodelling during development such as in long bones and vertebrae, and especially in the medullary cavity of the mid-diaphysis. Detailed radiological, microradiographic and scanning electron microscopy analyses of the long bones (Fig 1) and vertebrae (Saftig *et al* 1998) revealed an unusually dense trabeculation of the bone-marrow spaces. Increased trabecular and cortical bone mass were most pronounced in regions of rapid longitudinal growth (e. g. the distal femur and proximal tibia).

Histological examination of long bones and vertebrae confirmed the microradiographic and scanning electron microscopy findings. Primary trabeculae which represent remnants of neonatal endochondral bone formation persisted within the epiphyses, metaphyses and diaphyses of all long bones. This is a characteristic feature of osteopetrosis (Shapiro *et al* 1980). The marrow cavities of all endochondral bones derived from cathepsin K-deficient mice contained abundant unresorbed primary spongiosa, which reduced the effective space therein by up to about 60 %. Vertebrae of cathepsin K-deficient mice also contained a higher-than-normal proportion of primary spongiosa. The overall structure of epiphyseal and vertebral growth plates appeared to be normal (Saftig *et al* 1998).

Although the heights of tibial growth plates were greater in cathepsin K-deficient (169 μm ± 8 μm) mice than in controls (141 μm ± 5μm) (Fig 2A), the daily longitudinal bone growth rates did not differ significantly between the two types of animals (Fig 2B). The numbers of TRAP-stained osteoclasts within the tibial metaphysis (Fig 2C) and cortex (Fig 2D) of cathepsin K-deficient mice appeared not to be significantly different from controls.

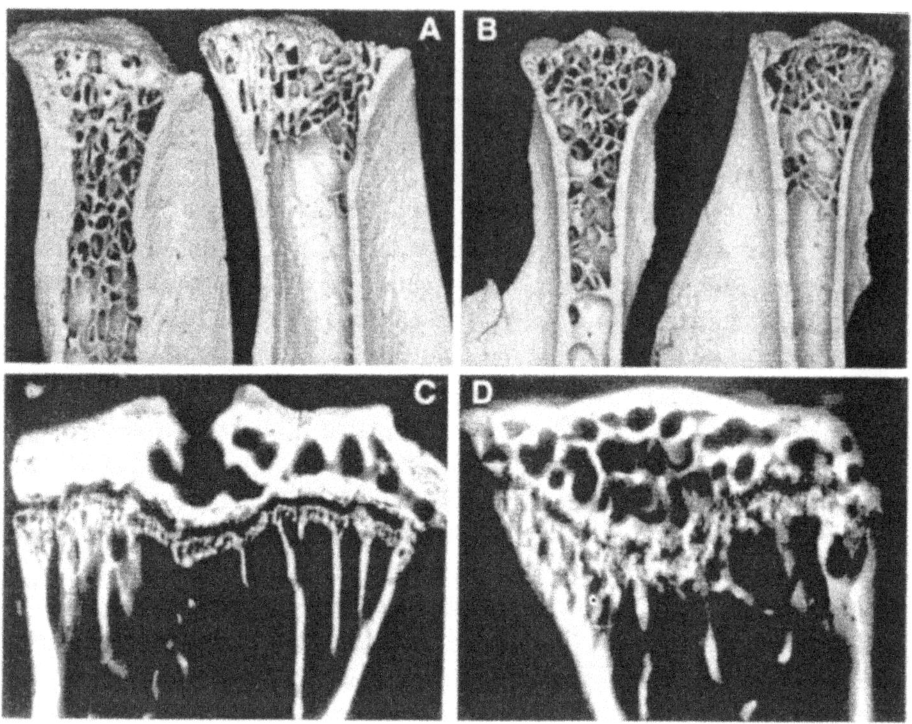


*Figure 1.* Unresorbed primary spongiosa in cathepsin K deficient bones.(A) Scanning electron microscopy (SEM) images of 3 month-old proximal tibiae (with epiphyses removed). *left*, cathepsin K -/-; *right*, control bone. (B) SEM images of humeri. *left*, cathepsin K -/-; *right*, control bone. Note the retention of cancellous bone in the shafts of cathepsin K-deficient (-/-) bones. (C) Microradiography of a control (+/+, 8 weeks old) proximal tibia. (D) Microradiography of a cathepsin K-deficient (-/-, 8 weeks old) proximal tibia.

## 3. OSTEOCLASTS WITHOUT CATHEPSIN K

Although the number of TRAP-stained osteoclasts was not altered in cathepsin K deficient mice (Fig 2C, D), large areas of demineralised bone matrix underlying the ruffled borders of such osteoclasts were frequently found. This suggests that bone demineralisation proceeds normally in cathepsin K deficiency, but that resorption of the bone matrix is impaired.

Cathepsin K deficient osteoclasts display a poorly defined resorptive surface with irregular conformation and extensions, open and detached adhesion sites, a broad fringe of demineralised matrix, no electron dense

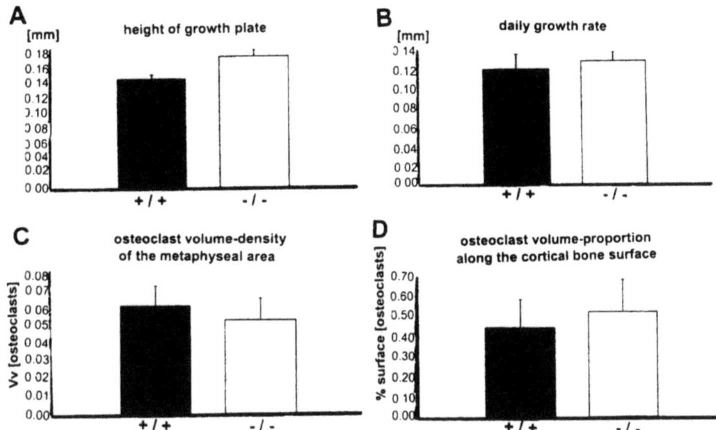


*Figure 2.* Growth plate, growth rate and osteoclast densities in cathepsin K deficient mice. (A) Heights of tibial growth plates are greater in cathepsin K-deficient mice than in controls. (B) Daily longitudinal bone growth rate (tibia of 4 weeks old mice measured after calcein injection: Saftig et al. 1998). (C) Volume-density of TRAP-positive osteoclasts within the tibial metaphysis (D) Surface-density of TRAP-positive osteoclasts within the tibial cortex.

particles in the lacunae, but a large number of collagen fibrils (Saftig *et al* 1998). The latter observation suggests that these collagen fibrils persist for prolonged periods in the resorption lacunae before being removed.

The impaired resorptive activity of cathepsin K deficient osteoclasts *in vivo* was confirmed using an *in vitro* osteoclast resorption assay which measures the activity of isolated osteoclasts after seeding on dentine slices. The resorption pits of cathepsin K deficient mice displayed an increased mean area and perimeter whereas the mean volume and depth (Fig 3B) of the pits were significantly reduced (Saftig *et al* 1998).The pits generated by cathepsin K deficient osteoclasts were very shallow and of irregular form (Fig 3D). The layer of unprocessed residual demineralised matrix was much thicker in cathepsin K -/- pits (Saftig *et al* 1998). Scanning electron microscopy demonstrated that inhibition of osteoclastic movement is not responsible for the excess bone. Migration tracks of osteoclasts can be clearly seen on the surface of cathepsin K deficient bone (Fig 3A).Despite the severe inhibition of bone resorption *in vitro* it appears that it is not totally inhibited *in vivo*. Looking at the bone surfaces there is still a considerable extent of resorption occurring in cathepsin K deficient mice. The osteoclasts may be working inefficiently, but they do work, perhaps over a longer time. In this context it is worth mentioning that it has also been suggested that

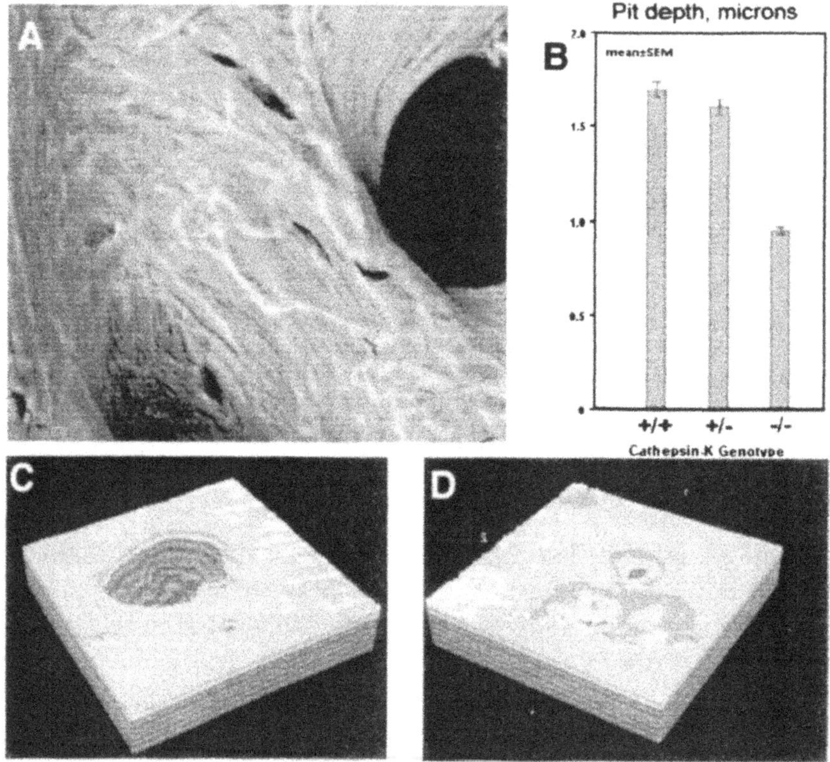


*Figure 3.* (A) Scanning electron microscopy demonstrating resorption pits in cathepsin K deficient bone: their elongation indicates the direction of osteoclastic migration (B) Mean depth of pits after resorption of dentine by mouse osteoclasts *in vitro*. (C ) Confocal reflection light microscope topographic map image of a pit area created by a control osteoclast (D) Confocal reflection light microscope topographic map image of a pit area created by a cathepsin K deficient osteoclast. Note that the control pit has steep sides and a max depth of 13 microns whereas the pit created by a cathepsin K deficient osteoclast is only 3 microns deep.

other cysteine proteinases such as cathepsin B, L, or S contribute to bone resorption, although it has been shown that these are expressed at much lower levels in osteoclasts than cathepsin K (Drake *et al* 1996). The role of other cysteine proteinases *in vivo* is currently under investigation using cathepsin B and cathepsin L deficient mice, as well as mice doubly deficient for both cathepsin K/cathepsin L and cathepsin K/cathepsin B. It appears that there are also subtle alterations in bone resorption in the other cysteine proteinase knockout mouse models (Everts and Saftig, unpublished data). By comparing long bones and calvariae of cathepsin K -/- mice, it appeared that in long bones adjacent to osteoclasts large areas were present of non-digested demineralised bone matrix adjacent to osteoclasts. In the calvaria, however, such areas proved to be much smaller. Morphometric analyses supported this observation and showed that the amount of non-digested bone matrix in long bones was four times higher (Everts 1999). It is possible that calvarial osteoclasts use other cysteine proteinases for digestion of bone matrix (Everts *et al* 1999).

## 4. CATHEPSIN K DEFICIENT MICE AS AN ANIMAL MODEL FOR HUMAN PYCNODYSOSTOSIS

Cathepsin K knockout mice display many of the phenotypic features of pycnodysostosis, including increased bone density and bone deformity. As in the human disease, bones that are rapidly remodelled during normal bone development and homeostasis are especially affected in cathepsin K knockout mice. Osteopetrosis is a characteristic feature of pycnodysostosis (Edelson *et al* 1992) and it may be responsible for the enhanced bone fragility and/or predisposition to bone fractures observed in these individuals (Meredith *et al* 1978). Preliminary data suggest that bones that form and grow in cathepsin K null conditions are of increased mass with appropriately increased strength (Pennypacker *et al* 1999). Cathepsin K mutations lead to the phenotype of short stature due to the reduced size of the long bones (Sedano *et al* 1968) and craniofacial anomalies. The suppressed growth and the craniofacial defects seen in pycnodysostosis patients were not observed in cathepsin K deficient mice indicating a somewhat altered clinical outcome or syndrome between humans and mice.

Ultrastructurally, osteoclasts derived from cathepsin K deficient mice display the same morphological abnormalities as those seen in pycnodysostosis, namely that they exhibit large areas of demineralised bone and large, irregular-shaped collagen containing vacuoles within the osteoclasts (Fig 4A - 4D) and a reduced *in vitro* resorption capacity. The

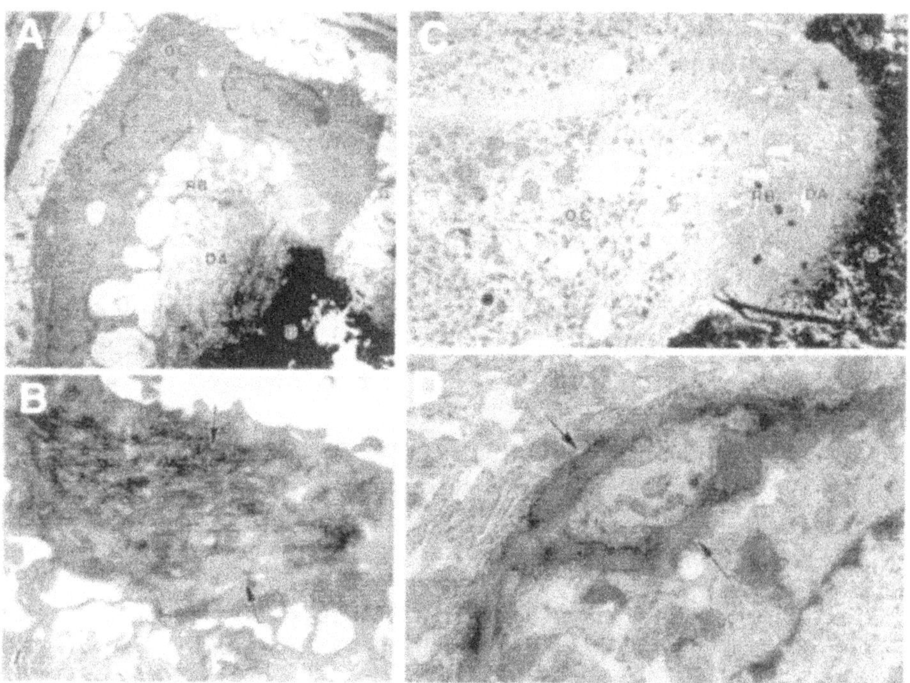


*Figure 4.* Similarities between the ultrastructure of osteoclasts from cathepsin K deficient mice (A,B) and pycnodysostosis patients (C,D).
A: Low magnification of an osteoclast (OC) of long bone of a 5 day old mouse deficient for cathepsin K. Adjacent to the ruffled border (RB) of the cell a large area of demineralised bone (DA) is apparent.
B: bone. X 2,100. B. High power electron micrograph of part of an osteoclast of a cathepsin K -/- mouse. Arrows indicate a vacuole containing internalised bone collagen. X 7,100.
C: Electron micrograph of an osteoclast (OC) attached to the bone surface. The micrograph is from a bone biopsy of a pycnodysostosis patient (Everts *et al* 1985). Note the large area of demineralised bone matrix (DA) adjacent to the ruffled border (RB). B: bone. X 2,100. D. Higher magnification of part of an osteoclast of a pycnodysostosis patient demonstrating internalised bone collagen (arrows) in a lysosomal vacuole. X 9,500.

cathepsin K knockout mice are an appropriate animal model for aspects of pycnodysostosis and might be useful to optimise therapeutic regimes, e. g. bone marrow transplantation and gene therapy for the treatment of this disorder.

## 5. CATHEPSIN K AND OSTEOPOROSIS

Cathepsin K could also play a pivotal role in the pathogenesis of post-menopausal osteoporosis. Oestrogen has been suggested to down regulate cathepsin K expression (Mano *et al* 1996) and its deficiency is thought to be causative for the development of post-menopausal osteoporosis. Conceptually, it therefore seems possible to use cathepsin K protease inhibitors as a more downstream target to inhibit increased osteoclastic activity in patients with osteoporosis due to glucocorticoid excess, oestrogen deficiency, rheumatoid arthritis or hypercalcemia of malignancy. Indeed, potent and selective active-site-spanning inhibitors have been developed for cathepsin K. These inhibitors act by mechanisms that involve tight binding intermediates which can result in irreversible and (or) reversible active-site modification. Such inhibitors have been shown to have antiresorptive activity both *in vitro* and *in vivo* (Thompson *et al* 1997).

## 6. ROLE OF CATHEPSIN K IN OTHER TISSUES

It has been reported that there is a restricted tissue distribution of cathepsin K in humans. There is still a debate as to whether human and mouse cathepsin K expression is comparable (Drake *et al* 1996, Inaoka *et al* 1995, Ratankoko *et al* 1996, Brömme *et al* 1995) and whether cathepsin K expression is limited to osteoclasts. There are also reports on the presence of cathepsin K mRNA in thyroid gland, lung epithelial cells (K. Brix, and F. Bühling unpublished data) as well as in heart, muscle, brain, spleen, and pancreas. Since these studies are mostly based on RNA expression patterns, extraskeletal cathepsin K expression needs to be confirmed using immunoblot and immunocytochemical approaches. In situ hybridisation experiments revealed a developmental stage-dependent pattern of cathepsin K expression in osteoclasts and preosteoclasts at sites of cartilage and bone remodelling in 15-17 day foetuses: no expression was found in any non-skeletal tissue at these time points (Dodds *et al* 1998). A more detailed morphological and ultrastructural analysis of cathepsin K deficient mice will help to unravel the possible role(s) of cathepsin K in cell types other than osteoclasts. Such an analysis is needed especially to evaluate possible side effects of the application of cathepsin K inhibitors for treating the excessive bone loss incurred in diseases such as rheumatoid arthritis and osteoporosis.

## 7. CONCLUSION

To investigate the putative crucial role of cathepsin K in bone resorption, we generated cathepsin K deficient mice by targeted inactivation of the cathepsin K gene (Saftig *et al* 1998). Cathepsin K-deficient mice develop osteopetrosis due to an impaired osteoclastic resorption of bone matrix. In contrast to other forms of osteopetrosis which are due to disrupted osteoclastogenesis (Yoshida *et al* 1990, Wang *et al* 1992), cathepsin K deficiency is associated with an inhibition of osteoclastic bone matrix resorption. Bone demineralisation does not appear to be affected by the protease deficiency suggesting that demineralisation and organic matrix resorption are independent processes. As in pycnodysostosis cathepsin K deficient osteoclasts exhibit a modified morphology suggesting that cathepsin K knockout mice represent a valuable animal model for pycnodysostosis. So far no other extraskeletal organ abnormalities have been observed in cathepsin K deficient mice suggesting that inhibitors of cathepsin K may selectively affect bone remodelling and can be used for therapeutic interventions.

## ACKNOWLEDGMENTS

This work was supported by the Deutsche Forschungsgemeinschaft, Swiss National Science Foundation.

## REFERENCES

Bossard, M.J., *et al.*, 1996, Proteolytic activity of human osteoclast cathepsin K: Expression, purification, activation, and substrate identification. *J. Biol. Chem.* **271:** 12517-12524.

Brömme, D., and Okamoto K., 1995, The baculovirus cysteine proteinase has a cathepsin B-like S2-subsite specificity. *Biol. Chem. Hoppe Seylor* **376:** 379-384.

Delaisse, J.M., *et al.*, 1984, *In vitro* and *in vivo* evidence for the involvement of cysteine proteinases in bone resorption. *Biochem. Biophys. Res. Commun.* **125:** 441-447.

Dodds, R.A., *et al.*, 1998, Cathepsin K mRNA detection is restricted to osteoclasts during fetal mouse development. *J. Bone Miner. Res.* **13:** 673-682.

Drake, F.H., *et al.*, 1996, Cathepsin K, but not cathepsin B, L, or S, is abundantly expressed in human osteoclasts. *J. Biol. Chem.* **271:** 12511-12516.

Erlebacher, A., *et al.*, 1995, Towards a molecular understanding of skeletal development. *Cell* **80:** 371-378.

Edelson, J.G., *et al.*, 1992, Pycnodysostosis. Orthopedic aspects with a description of 14 new cases. *Clin. Orthoped. Rel. Res.* **280:** 264-276.

Everts, V., *et al.*, 1985, Phagocytosis of bone collagen by osteoclasts in two cases of pycnodysostosis. *Calcif. Tissue. Int.* **37:** 25-31.

Everts, V., *et al.*, 1992, Degradation of collagen in the bone-resorbing compartment underlying the osteoclast involves both cysteine proteinases and matrix metalloproteinases. *J. Cell. Physiol.* **150:** 221-231.

Everts, V., *et al.*, 1998, Cysteine proteinases and matrix metalloproteinases play distinct roles in the subosteoclastic resorption zone. *J. Bone Miner. Res.* **13:** 1420-1430.

Everts, V., *et al.*, 1999, *J. Bone Miner. Res.* **14:** S356.

Everts, V., *et al.*, 1999, Functional heterogeneity of osteoclasts: matrix metalloproteinases participate in osteoclastic resorption of calvarial bone but not in resorption of long bone. *FASEB J.* :1219-1230.

Gelb, B.D., *et al.*, 1996, Pycnodysostosis is a lysosomal disease caused by cathepsin K deficiency. *Science* **273:** 1236-1238.

Ho, N., *et al.*, 1999, Mutations of CTSK Result in Pycnodysostosis via a Reduction in Cathepsin K Protein. *J. Bone Miner. Res.* **14:** 1649-1653.

Inaoka ,T., *et al.*, 1995, Molecular cloning of human cDNA for cathepsin K: Novel cysteine proteinase predominantly expressed in bone. *Biochem. Biophys. Res. Commun.* **206:** 89-96.

Kafienah, W., *et al.*, 1998, Human cathepsin K cleaves native type I and II collagens at the N-terminal end of the triple helix. *Biochem.. J.* **331:** 727-732.

Li, Y.P., *et al.*, 1995, Cloning and complete coding sequence of a novel human cathepsin expressed in giant cells of osteoclastomas.*J. Bone Miner. Res.* **10:** 1197-1202.

Mano, H., *et al.*, 1996, Mammalian mature osteoclasts as estrogen target cells. *Biochem. Biophys. Res. Commun.* **223:** 637-642.

Meredith, S.C., *et al.*, 1978, Pycnodysostosis. A clinical, pathological, and ultramicroscopic study of a case. *J. Bone Joint Surg. Am.* **60:** 1122-1127.

Pennypacker, B., *et al.*, 1999, Bone density and strenght in cathepsin K null mice. *J. Bone Miner. Res.* **14:** S549.

Ratankoko, J., *et al.*, 1996, Mouse cathepsin K: cDNA cloning and predominant expression of the gene in osteoclasts and in some hypertrophic chondrocytes during mouse development. *FEBS Lett.* **393:** 301-313.

Saftig, P., *et al.*, 1998, Impaired osteoclastic bone resorption leads to osteopetrosis in Cathepsin K deficient mice. *Proc. Natl. Acad. Sci.* **95:** 13453-13458.

Sedano, H.D., *et al.*, 1968, Pycnodysostosis. Clinical and genetic considerations. *Am. J. Dis. Child.* **116:** 70-77.

Shapiro, F., *et al.*, 1980, Human osteopetrosis: A histological, ultrastructural, and biochemical study. *J. Bone Joint Surg.* **62A:** 384-399.

Shi, G.P., *et al.*, 1995, Molecular cloning of human cathepsin O, a novel endoproteinase and homologue of rabbit OC2. *FEBS Lett.* **357:** 129-134.

Tezuka, K., 1994, Molecular cloning of a possible cysteine proteinase predominantly expressed in osteoclasts. *J. Biol. Chem.* **269:** 1106-1109.

Thompson, S.K., *et al.*, 1997, Design of potent and selective human cathepsin K inhibitors that span the active site. *Proc. Natl. Acad. Sci. USA* **94:** 14249-14254.

Velasco, G., *et al.*, 1994, Human cathepsin O. Molecular cloning from a breast carcinoma, production of the active enzyme in *Escherichia coli*, and expression analysis in human tissues. *J.Biol .Chem .* **269:** 27136-27142.

Wang, Z.Q., *et al.*, 1992, Bone and haematopoetic defects in mice lacking c-fos. *Nature* **360:** 741-745.

Yoshida, H., *et al.*, 1990, The murine mutation osteopetrosis is in the coding region of the macrophage colony-stimulating factor gene. *Nature* **345:** 442-444.

# CERAMIDE AS AN ACTIVATOR LIPID OF CATHEPSIN D

Michael Heinrich, Marc Wickel, Supandi Winoto-Morbach, Wulf Schneider-Brachert[1], Thomas Weber[2,3], Josef Brunner[3], Paul Saftig[4], Christoph Peters[5], Martin Krönke[6] and Stefan Schütze[7]
*Institute of Immunology, University of Kiel, Germany*
*Current addresses:*
*[1]Institute of Medical Microbiology and Hygiene, University of Regensburg, Germany*
*[2]Memorial Sloan-Kettering Cancer Center, New York, NY-10021, U.S.A.*
*[3]Department of Biochemistry, Swiss Federal Institute of Technology, Zürich, Swizerland*
*[4]Department of Biochemistry II, University of Göttingen, Göttingen, Germany*
*[5]1rst Department of Internal Medicine, University of Freiburg, Freiburg, Germany*
*[6] Institute of Medical Microbiology and Hygiene, University of Cologne, Germany*
*7 Corresponding author*

Key words Acid Sphingomyelinase, Cathepsin D, Ceramide, TID-Ceramide

Abstract We have identified the aspartic protease cathepsin D as a novel intracellular target protein for the lipid second messenger ceramide. Ceramide specifically binds to and induces CTSD proteolytic activity. A-SMase deficient cells derived from Niemann-Pick patients show decreased CTSD activity that was reconstituted by transfection with A-SMase cDNA. Ceramide accumulation in cells derived from A-ceramidase defective Farber patients correlates with enhanced CTSD activity. These findings suggest that A-SMase-derived ceramide targets endolysosomal CTSD.

## 1. INTRODUCTION

Ceramide is a common intracellular second messenger for various cytokines, growth factors, Fas/APO-1 and other stimuli like chemotherapeutic drugs and stress factors. The production of ceramide is mediated either by hydrolysis of sphingomyelin engaging sphingomyelinase (SMase) or by de-novo synthesis involving ceramide-synthase (Spiegel and Merrill 1996). Ligand binding to the p55 TNF receptor (TR55), Interleukin-1 receptor 1 (IL-1 RI) and the Fas receptor results in activation of two SMases, a plasma membrane-bound neutral SMase (N-SMase) and an endo-lysosomal acid SMase (A-SMase) (Schütze *et al* 1992, Wiegmann *et al* 1994, Liu and Anderson, 1995, Cifone *et al* 1995). Each type of SMase generates the second messenger ceramide, however at different intracellular locations. Specific TNF receptor domains link to the respective SMases and to diverse signaling pathways (Wiegmann *et al* 1994, Adam-Klages *et al* 1996, Kreder *et al* 1999, Schwandner *et al* 1998, Wiegmann *et al* 1999). A role for A-SMase was suggested to transmit apoptotic signals in response to Fas/CD95, γ-irradiation, TNF and LPS (Cifone *et al* 1995, Santana *et al* 1996, Herr *et al* 1997, De Maria *et al* 1998, Monney *et al* 1998, Haimowitz-Friedmann *et al* 1998). To understand the role of ceramide during apoptosis and other cellular responses, it is critically important to characterize direct targets of ceramide action.

In the present study we sought for specific ceramide targets that colocalize with the subcellular site of A-SMase. Ceramide-affinity chromatography and D-*erythro*-ceramide-based photo-crosslinking revealed cathepsin D (CTSD) as a novel ceramide-binding protein (Schütze *et al* 1999, Wickel *et al* 1999, Heinrich *et al* 1999). Specific interaction of ceramide with CTSD lead to enhanced enzymatic activity. CTSD is endosomally active and involved in the proteolytic activation of proteins to be secreted. CTSD has been recently implicated in mediating apoptosis in response to TNF, IFNγ, CD95 (Deiss *et al* 1996), chemotherapeutic agents like etoposide and adriamycin (Wu *et al* 1998) and serum deprivation (Shibata *et al* 1998, Ohsawa *et al* 1999). Thus CTSD may link A-SMase to the secretory pathway and to apoptotic signaling events.

## 2. MATERIALS AND METHODS

Cells, mice and reagents U937, HeLa, and HEK 293 cells were obtained from ATTC and maintained in CLICK's RPMI culture medium (Biochrome). EBV-transfected human lymphoblasts JY and EBV transfected human lymphoblasts from Niemann-Pick patients (clones MS1418 and MS1271) were kindly provided by D. Green, La Jolla. Fibroblasts from Farber disease patients were kindly provides by K. Sandhoff and G. van Echten (Bonn). Cathepsin D deficient mice (C57BL/6 ctsd -/-) were generated as described (Saftig *et al* 1995).

## *Retroviral transfections with A-SMase and CTSD cDNA*

The cDNAs of human acid sphingomyelinase (A-SMase) and human prepro-CTSD were subcloned into the retroviral vector pLSXN or pBabe puro, respectively. Transient transfection of the retroviral producer line FLYA13 (Cosset *et al* 1995) was performed by calcium phosphate precipitation with pLXSN-ASMase or pLXSN-CTSD and pCMV/VSV-G expressing the G protein of vesicular stomatitis virus.

## *Ceramide-affinity chromatography and immunoblotting*

Generation of the Affinity Matrix: Activated CH-Sepharose 4B (Amersham Pharmacia Biotech) was resuspended in 1mM HCl and subsequently transferred to 30 %, 70 % and 100 % tetrahydrofuran (THF). 50 mg D-*erythro*-sphingosine (Sigma) in 6 ml anhydrous THF was coupled to activated CH-Sepharose 4B following addition of 100 µl N-ethylmorpholine. The reaction between the N-hydroxysuccinimide ester in the activated CH-Sepharose 4B and the amine in sphingosine result in the formation of an amide-bond and release of N-hydroxysuccinimide.

## *Affinity chromatography of cellular lysates*

U937 cells were homogenized in buffer H (150 mM KCl, 5 mM NaF, 1mM PMSF; 20 µM pepstatin, 20 µM leupeptin, 20 µM antipain [Boehringer Complete, 1 : 100] in 40 mM HEPES, pH 7.4). Lysates were cleared by centrifugation (5 min, 1000 xg), the supernatant adjusted to 0.075 % (v/v) Triton X-100 and after centrifugation for 1h at 100,000 xg, the supernatants, containing 25 - 30 mg protein were loaded onto the ceramide-affinity column. The column was washed and the bound proteins eluted with 100 µM D-*erythro*-ceramide in buffer H containing 0.075 % Triton X-100.

## *$^{125}$I-TID-ceramide UV-crosslinking and competition analysis*

N-[3-[[[2-(tributylstannyl)-4-[3-(triluoromethyl)-3H-diazerin-3-yl] benzyl] oxy] carbonyl] propanoyl] D-*erythro*-sphingosine (tin-precursor of TID-ceramide) was sythesized, purified by HPLC and radiolabeled as described (Weber and Brunner 1995).

## *Photoaffinity labeling*

2 µg of purified cathepsin D from human liver (Sigma) was incubated with 1 µCi (0.5 pmol) of $^{125}$I-TID ceramide in 20 µl buffer H containing

0.075 % Triton X-100 for 5 min at 37 °C in the absence or presence of competitor lipids. Photo-crosslinking was performed by UV-irradiation (2 min 100 W at 365 nm) using a UVP Model B100A Black Ray Lamp, Herolab, Wiesloch, Germany). Radiolabeled protein was analyzed after separation on 12.5 % SDS-PAGE by autoradiography using the FUJI BAS 1000 Bioimager (Raytest).

### *Functional cathepsin D assays*

Cellular cathepsin D activity was assayed by using the CellProbe GL*cathepsin D enzyme substrate (Coulter) and proteolysis of the peptide Glycine-Leucine-Rho 110 was estimated by flow cytometry. For estimating the activity of purified human liver CTSD (Sigma) or cellular CTSD, lysate-protein or human liver CTSD was incubated for indicated times with parathyroid hormone (PTH) at 37 °C in acidic buffer (100 mM sodium-acetate, 100 mM potassiumchloride, pH 4.2). Stimulation was performed with the various lipids indicated in the Figs either dissolved in DMSO or integrated into PC-liposomes to mimic in-vivo detergent-free conditions.

## 3. RESULTS

The interaction of ceramide with CTSD was revealed by the binding of CTSD to and specific elution from a ceramide affinity matrix. (Fig 1A) Cathepsin D eluted from the matrix when the biological active D-*erythro* ceramide, but did not elute when the biological inactive dihydroceramide, or galactocerebroside were employed as competitors. Of note, the free fatty acid palmitate was ineffective in replacing CTSD from the affinity matrix, while sphingosine is able to compete for ceramide-binding, suggesting that the sphingosine-(alkyl-)rather than the acyl chain in ceramide is mediating the CTSD-ceramide interaction.

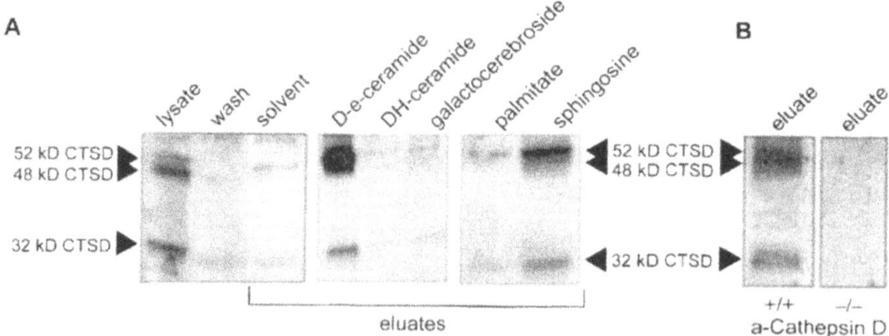


*Fig 1.* Cathepsin D binds to Ceramid-Sepharose

CTSD is the major aspartic protease of endo-lysosomes. The protease is synthesized and translocated into the ER as an inactive pre-proenzyme (52 kD) that is subsequently converted into an active, intermediate proenzyme (48 kD) in endosomal compartments (Gieselmann *et al* 1983, Rijnboutt *et al* 1992). Further cleavage in late endosomes and lysosomes generates the mature form of 32 kD (Fig 1C).

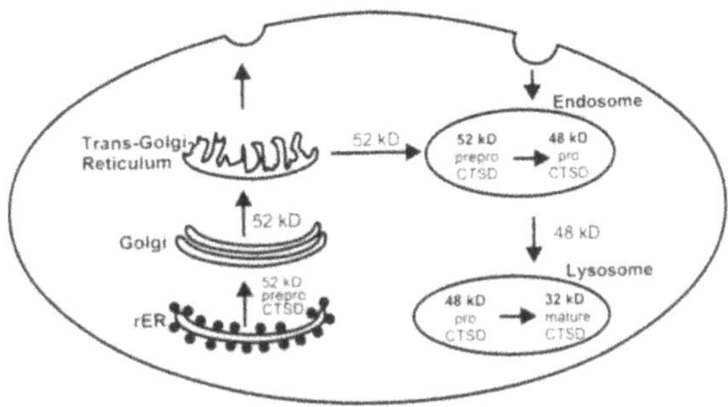


*Figure 1C.* Subcellular compartmentation of cathepsin D isoforms

Notably, all three isoforms of CTSD appeared in the ceramide-eluates, suggesting that the ceramide-binding site localizes within the 32 kD CTSD polypeptide. None of the CTSD isoforms could be eluted from the ceramide-sepharose matrix when loaded with lysates from kidney cells of CTSD-deficient mice (Fig 1B), providing further evidence for the identity of the

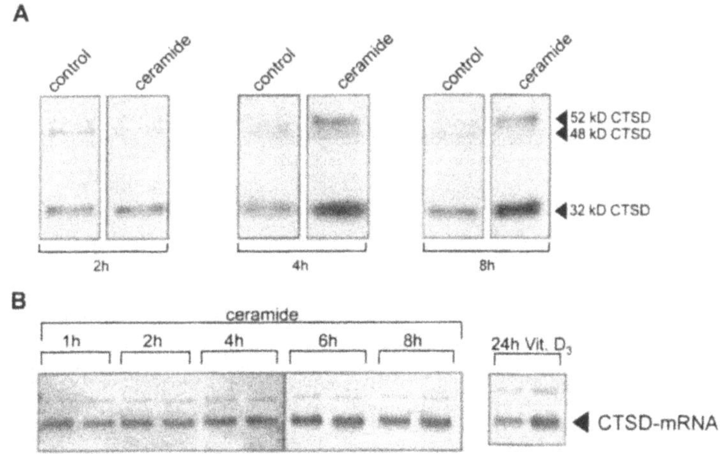


*Figure 2.* Ceramide induces posttranslational maturation of cathepsin D

eluted ceramide binding proteins with CTSD. The functional consequences of ceramide binding to CTSD was assayed by incubating U937 cells with synthetic D-*erythro*-ceramide, resulting in an increase of the mature 32 kD CTSD isoform after 4 h (Fig 2A). CTSD gene-expression was unaffected by ceramide treatment at this time point (Fig 2B), suggesting that ceramide induces posttranslational proteolytic processing of pre-pro-CTSD isoforms to yield mature, active CTSD.

Ceramide stimulates the enzymatic activity of CTSD in intact cells as assessed by employing a cell-permeable fluorescent dipeptide-substrate (Fig 3).The specificity of the fluorescent substrate for CTSD was revealed in fibroblasts from CTSD-deficient mice. In-vitro expression of recombinant 52 kD prepro CTSD additionally revealed that ceramide is capable to induce accelerated autoproteolysis of the prepro CTSD isoform (Heinrich *et al* 1999). As a possible explanation, ceramide binding to CTSD may lead to a conformational change in the inactive proenzyme, eventually resulting in refolding of the CTSD propeptide that masks the active site of the enzyme (Beyer and Dunn, 1996).

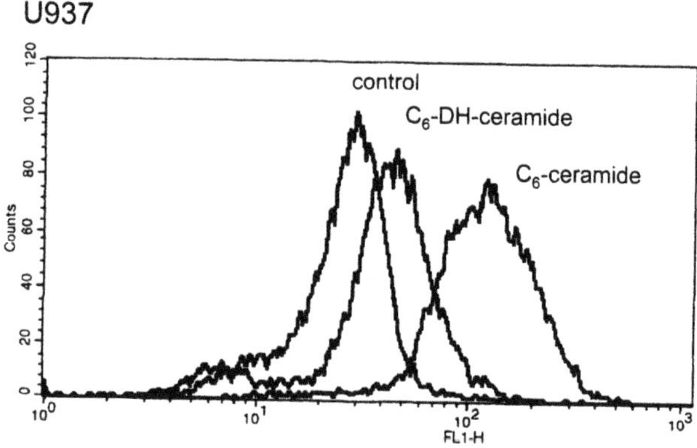


*Figure 3.* Ceramide stimulates CTSD enzymatic activity in U937 cells

The ceramide-specificity of binding to CTSD was demonstrated by competition analysis employing the UV-crosslinking ceramide-analog $^{125}$I-TID ceramide, showing that CTSD specifically binds the naturally occuring D-*erythro*-ceramide but not the functional inactive analoges D-*erythro*-dihydroceramide, D-*erythro*-dihydrosphingosine or other unrelated lipids, suggesting a specific role for ceramide in cathepsin D activation (Fig 4).

Notably, also D-*erythro*-sphingosine, but not the fatty acid palmitate competed for TID-ceramide binding to CTSD, again suggesting that the alkyl chain rather than the acyl chain in ceramide is important for the interaction with CTSD. In order to directly measure ceramide action on

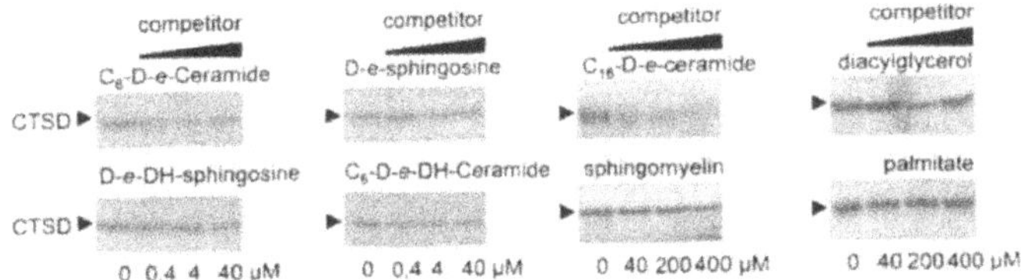


*Figure 4.* Specificity of TID-ceramide binding to cathepsin D

CTSD activity, a cell-free assay was established by use of parathyroid-hormone (PTH) as substrate and a monoclonal antibody specific for the 1-34 PTH peptide to detect proteolysis of the 1-84 PTH by immunoblotting. Fibroblasts from CTSD deficient mice again served as control for the specificity of the assay. Both, endogenous CTSD in cellular lysates as well as purified mature CTSD responded to D-*erythro*-ceramide but not to D-*erythro*-dihydroceramide with enhanced proteolytic activity (Fig 5A). Importantly, D-*erythro*-sphingosine is also able to activate CTSD, however, only when applied in DMSO or in Triton X-100 micelles. When sphingosine was presented to CTSD as a component of a liposomal lipid bilayer in the absence of detergent, sphingosine lost its ability to activate the enzyme (Fig 5B).

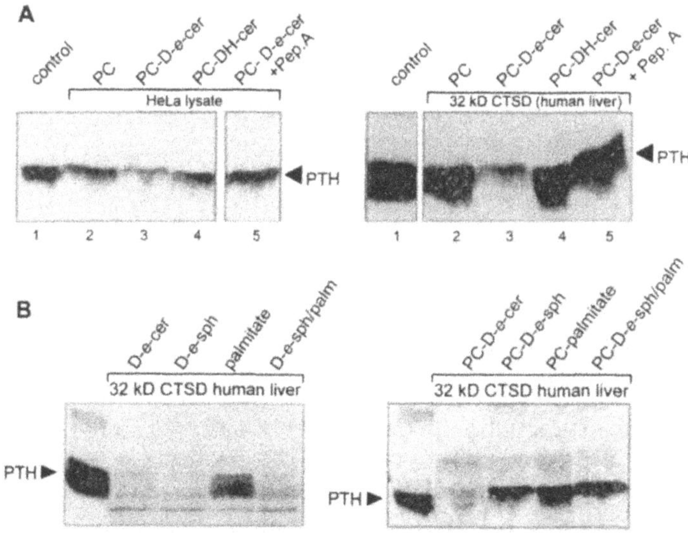


*Figure 5.* Specificity of ceramide activation of cathepsin D

In contrast, D-*erythro*-ceramide retained its stimulatory effect when incorporated in PC-liposomes. Since PC-liposomes mimic the conditions of lipid bilayers, these observations suggest that ceramide in its membrane-bound form is the physiological activator of CTSD. The acyl chain in ceramide is important for membrane association and the alkyl chain for interaction with the protein. This model implies that ceramide, through protrusion of the alkyl chain interacts with the hydrophobic cavity of cathepsin D, as recently suggested (Krönke 1997). Together, our results identify CTSD as the first protease that is directly and specifically activated by the lipid second messenger ceramide. A functional link between A-SMase, intracellular ceramide and CTSD activity becomes evident by employing cells that carry genetic defects in the A-SMase gene (Niemann-Pick type A), A-SMase transfection, and A-ceramidase-deficient fibroblasts from Farber disease patients: A-SMase deficient cells show decreased CTSD activity that was reconstituted by transfection with A-SMase cDNA, ceramide accumulation in cells derived fom A-ceramidase defective Farber patients correlates with enhanced CTSD activity. Thus endogenous ceramide levels in endo-lysosomal compartments regulate CTSD enzymatic activity (Fig 6). The aspartic protease cathepsin D represents the first ceramide binding protein that colocalizes with A-SMase within the endo-lysosomal compartment.

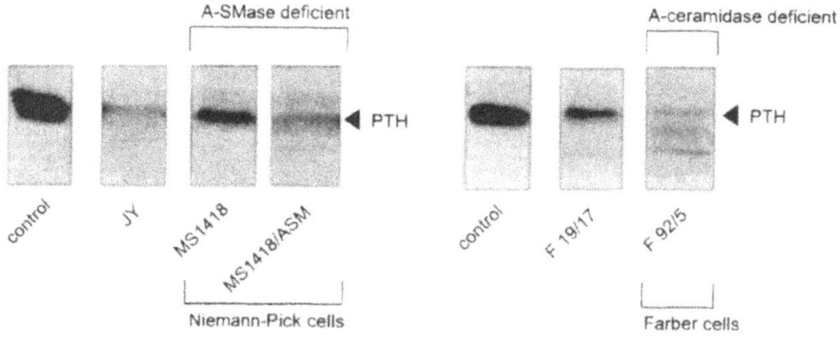


*Fig 6.* CTSD activity is related to endogenous ceramide levels

# 4. CONCLUSIONS

A regulatory function of cytokines like TNF and IFN-$\gamma$ on the modulation of CTSD isoform expression has previously been suggested (Deiss *et al* 1996). We have observed similar effects of TNF on CTSD maturation in U937, B-lymphoblasts and HeLa cells (data not shown). In light of the capability of TNF to stimulate ceramide production in endo-lysosomal compartments (Schütze *et al* 1992, Wiegmann *et al* 1994, 1999,

Schwandner *et al* 1998), it is tempting to speculate that TNF may induce CTSD maturation and activation via A-SMase generated ceramide. As to the question of how cytokines like TNF could signal A-SMase induced CTSD activation, recent evidence indicated that internalization of TR55 is required for A-SMase activation (Schütze *et al* 1999). TR55 bearing endosomes may fuse with trans-golgi vesicles containing A-SMase as well as the 52 kD prepro-CTSD. A-SMase-derived ceramide may then bind to and activate the inactive prepro-CTSD to generate the 48/32 kD active mature CTSD isoforms.

## ACKNOWLEDGMENTS

We thank Andrea Hethke and Gudrun Scherer for exellent technical assistance, R. Pohlmann for the generous gift of anti-murine cathepsin D antibodies, K. Sandhoff and G. van Echten for supply of A-SMase c-DNA and fibroblasts from Farber patients, and D. Green for supply of NPA-lymphoblasts. This work was supported by grants from the Deutsche Forschungsgemeinschaft (SFB 415), the Bundesministerium für Bildung und Forschung and the Swiss National Science Foundation.

## REFERENCES

Adam-Klages, S., Adam, D., Wiegmann, K., Struve, S., Kolanus, W., Schneider-Mergener, J., and Krönke, M., 1996, FAN, a novel WD-repeat protein, couples the p55 TNF-receptor to neutral sphingomyelinase. *Cell*, **86:** 937-947.

Beyer, B.M., and Dunn, B.M., 1996) Self-activation of recombinant human lysosomal procathepsin D at a newly engineered cleavage junction, „short" pseudocathepsin D. *J. Biol. Chem.* **271:** 15590-15596.

Cifone, M.G., De Maria, R., Roncaioli, P., Rippo, M.R., Azuma, M., Lanier, L.L., Santoni, A., and Testi, R., 1995, Apoptotic signaling through CD95 (Fas/APO-1) activates an acidic sphingomyelinase. *J. Exp. Med.* **177:** 1547-1552.

Cosset, F.-L., Takeuchi, Y., Battini, J.-L., Weiss, R.A., and Collins, M.K.L., 1995, High-titer packaging cells producing recombinant retroviruses resistant to human serum. *J. Virol.* **69:** 7430-7436.

Deiss,L.P., Galinka, H., Berissi, H., Cohen, O., and Kimchi, A., 1996, Cathepsin D protease mediates programmed cell death induced by interferon-γ, Fas/APO-1 and TNF-α. *EMBO J.* **15:** 3861-3870.

De Maria, R., Rippo, M.R., Schuchmann, E.H., and Testi, R., 1998, Acidic sphingomyelinase (ASM) is necessary for FAS-induced GD3 ganglioside accumulation and efficient apoptosis in lymphoid cells. *J. Exp. Med.***187:** 897-902.

Diment, S., Martin, K.L., and Stahl, P.D., 1989, Cleavage of parathyroid hormone in macrophage endosomes illustrates a novel pathway for intracellular processing of proteins. *J. Biol. Chem.* **264:** 13403-13406.

Faust, P.L., Kornfeld, S., and Chirgwin, J.M., 1985, Cloning and sequence analysis of cDNA for human cathepsin D. *Proc. Natl. Acad. Sci. USA* **82:** 4910-4914.

Gieselmann, V., Pohlmann, R., Hasilik, A., and von Figura, K, 1983, Biosynthesis and transport of cathepsin D in cultured human fibroblasts. *J. Cell Biol.*, **97:** 1-5.

Haimowitz-Friedmann, A., Cordon-Cardo, C., Bayoumy, S., Garzotto, M., McLouglin, M., Gallily, R., Edwards III, C.K., Schuchman, E.H., Fuks, Z., and Kolesnick, R., 1997, Lipopolysaccharide induces disseminated endothelial apoptosis requiring ceramide generation. *J. Exp. Med.* **186:** 1831-1841.

Heinrich, M., Wickel, M., Schneider-Brachert, W., Sandberg, C., Gahr, J., Schwandner, R., Weber, T., Brunner, J., Saftig, P., Peters, C., Krönke, M., and Schütze, S., 1999, Cathepsin D targeted by acid sphingomyelinase-derived ceramide. *EMBO J.* **18:** 5252-5263.

Herr, I., Wilhelm, D., Böhler, T., Angel, P., and Debatin, K.-M., 1997, Actvation of CD95 (APO-1/Fas) signaling by ceramide mediates cancer therapy-induced apoptosis. *EMBO J.* **16:** 6200-6208.

Kreder, D., Krut, O., Adam-Klages, S., Wiegmann, K., Scherer, G., Plitz, T., Jensen, J.-M., Proksch, E., Steinmann, J., Pfeffer, K., and Krönke, M ., 1999, Impaired neutral sphingomyelinase activation and cutaneous barrier repair in FAN-deficient mice. *EMBO J.* **18:** 2472-2479.

Krönke, M., 1997, The mode of ceramide action: the alkyl chain protrusion model. *Cytokine Growth Factor Rev.* **8:** 103-107.

Liu, P., and Anderson G.G.W., 1995, Compartmentalized production of ceramide at the cell surface. *J. Biol. Chem.* **270:** 27179-27185.

Monney, L., Olivier, R., Otter,I., Jansen, B., Poirier, G.G., and Borner, C., 1998, Role of an acidic compartment in tumor necrosis factor alpha-induced production of ceramide, activation of caspase-3 and apoptosis. *Eur. J. Biochem.* **251:** 295-303.

Ohsawa, Y., Isahara, K., Kanamori, S., Shibata, M., Kametaka, S., Gotow, T., Watanabe, T., Kominami, E., and Uchiyama, Y., 1999, An ultrastructural and immunohistochemical study of PC12 cells during apoptosis induced by serum depriviation with special reference to autophagy and lysosomal cathepsins. *Arch. Histol.. Cytol.* **61:** 395-403.

Perry, D. K., and Hannun, Y. A., 1998, The role of ceramide in cell signaling. *Biochim. Biophhys. Acta* **1436:** 233-243.

Rijnboutt, S., Stoorvogel, W., Geuze, H., Strous, G.J., 1992, Identification of subcellular compartments involved in biosynthetic processing of cathepsin D. *J. Biol. Chem.* **267:** 15665-15672.

Saftig, P., Hetman, M., Schmahl, W., Weber, K., Heine, L., Mossmann, H., Köster, A., Hess, B., Evers, M., von Figura, K., and Peters, C., 1995, Mice deficient for the lysosomal proteinase cathepsin D exhibit progressive atropy of the intestinal mucosa and profound destruction of lymphoid cells. *EMBO J.* **14:** 3599-3608.

Santana, P., Pena, L.A., Haimovitz-Friedmann, A., Martin, S., Green, D., Mc Loughlin, M., Cordon-Cardo, C., Schuchmann, E. H., Fuks, Z., and Kolesnick, R., 1996, Acid sphingomyelinase-deficient human lymphoblasts and mice are defective in radiation-induced apoptosis. *Cell* **86:** 189-199.

Schütze, S., Potthoff, K., Machleidt, T., Berkovic, D., Wiegmann, K., and Krönke, M., 1992, TNF activates NF-κB by phosphatidylcholine-specific phospholipase C-induced „acidic" sphingomyelin breakdown. *Cell* **71:** 765-776.

Schütze, S., Machleidt, T., Adam, D., Schwandner, R., Wiegmann, K., Kruse, M.-L., Heinrich, M., Wickel, M., and Krönke, M., 1999, Inhibition of receptor internalization by monodansylcadaverine selectively blocks p55 Tumor Necrosis Factor receptor death domain signaling. *J. Biol. Chem.* **274:** 10203-10212.

Schütze, S., Wickel, M., Winoto-Morbach, S., Heinrich, M., Weber, T., Brunner, J., and Krönke, M., 1999, Use of affinity chromatography and TID-ceramaide photoaffinity labeling for the detection of ceramide-binding proteins. *Methods in Enzymology* in press

Schwandner, R., Wiegmann, K., Bernado, K., Kreder, D, and Krönke, M.., 1998, TNF Receptor death domain-associated proteins TRADD and FADD signal acivation of acid sphingomyelinase. *J. Biol. Chem.* **273:** 5916-5922.

Spiegel, S., and Merrill, A. H. Jr., 1996, Sphingolipid metabolism and growth regulation. *FASEB J.* **10:** 1388-1397.

Shibata, M., Kanamori, S., Isahara, K., Ohsawa, Y., Konishi, A., Kametaka, S., Watanabe, T., Ebisu, S., Ishido, K., Kominami, E., and Uchiyama, Y., 1998, Participation of cathepsin D and B in apoptosis of PC12 cells following serum depriviation. *Biochem.. Biophys.. Res. Commun.* **251:** 199-203.

van Echten-Deckert, G., Klein, A., Linke, T., Heinemann, T., Weisgerber, J., and Sandhoff, K., 1997, Turnover of endogenous ceramide in cultured normal and Farber fibroblasts. *J. Lipid Res.* **38:** 2569-2579.

Weber, T., and Brunner, J., 1995, 2-(Tributylstannyl)-4-[3-(trifluoromethyl)-3H-diazirin-3-yl]benzyl alcohol: A building block for photolabeling and cross-linking reagents of very high specific radioactivity. *J. Am.. Chem. Soc.* **117:** 3084-3095

Wickel, M., Heinrich, M., Weber, T., Brunner, J., Krönke, M., and Schütze, S., 1999, Identification of intracellular ceramide target proteins by affinity chromaatography and TID-ceramide photoaffinity labeling. *Biochem.. Soc. Transactions* **27:** 393-400.

Wiegmann, K., Schütze, S., Machleidt, T., Witte, D., and Krönke, M., 1994, Functional dichotomy of neutral and acidic sphingomyelinase in tumor necrosis factor signaling. *Cell* **78:** 1005-1015.

Wiegmann, K, Schwandner, R., Krut, O., Yeh, W.-C., Mak, T.W., and Krönke, M., 1999, Requirement of FADD for Tumor Necrosis Factor-induced activation of acid sphingomyelinase. *J. Biol. Chem.* **274:** 5267-2570.

Wu, S., H., Saftig, P., Peters, C., and El-Deiry, W., 1998, Potential role for cathepsin D in p53-dependent tumor suppression and chemosensitivity. *Oncogene* **16:** 2177-2183.

# HUMAN CATHEPSIN X

*A novel cysteine protease with unique specificity*

Robert Ménard, Dorit K. Nägler[1], Rulin Zhang, Wendy Tam, Traian Sulea, and Enrico O. Purisima

*Biotechnology Research Institute, National Research Council of Canada, 6100 Avenue Royalmount, Montréal, Québec, Canada H4P 2R2,*

*[1]Present address: Institute of Pathology, Otto-von-Guericke University Magdeburg, Leipziger Str.44, 39120 Magdeburg, Germany.*

## 1. INTRODUCTION

Cysteine proteases constitute attractive targets for development of inhibitors as potential therapeutic agents due to their involvement in several pathological conditions. As is the case with many other gene families, the mining of databases from genome-sequencing projects has lead to the discovery of several new cysteine proteases of the papain family e.g. cathepsin V (Adachi *et al* 1998, Santamaría *et al* 1998a), cathepsin X (Nägler and Ménard 1998, Santamaría *et al* 1998b) and cathepsin F (Wang *et al* 1998, Nägler *et al* 1999) have all been cloned in the last two years. Although restricted expression patterns might exist, the increasing number of cysteine proteases is likely to render more challenging the task of designing inhibitors specific for a given enzyme. This is particularly important considering that the information available to date indicates relatively broad, overlapping specificities for these enzymes (Storer and Ménard 1996).

The primary structure of cathepsin X contains several unique features that clearly distinguish it from other human cysteine proteases. One of these distinctive features is a three amino acid residue insertion in a highly conserved region between the glutamine (Gln22) of the putative oxyanion hole and the active site cysteine (Cys31). This highly unusual insertion for a papain-like cysteine protease might confer special properties to the enzyme.

We have used molecular modeling to gain insights into the possible functional consequences of this insertion observed in the primary sequence, and to assist in the design of a good substrate for this protease. Kinetic characterization of substrate hydrolysis has been used to describe the specificity profile of this novel enzyme. The combination of functional characterization with structural model building has enabled us to propose the fine molecular details defining the specificity of cathepsin X.

## 2. EXPERIMENTAL

### 2.1 Materials

The vector (pPIC9) and *Pichia pastoris* strain GS115 were purchased from Invitrogen Corporation (San Diego, CA). The substrate Cbz-FR-MCA was purchased from IAF Biochem International Inc. (Laval, Qc). The substrates Abz-FRF(4$NO_2$) and Abz-FRF(4$NO_2$)A were from Enzyme Systems Products (Livermore, CA), and Abz-AFRSAAQ-EDDnp was obtained from Luiz Juliano (São Paolo, Brazil). Human cathepsins B and L were prepared as described previously (Nägler *et al* 1997, Carmona *et al* 1996). Human cystatin C was prepared as described earlier (Ekiel *et al* 1997).

### 2.2 Expression and Purification of Recombinant Cathepsin X

The human cathepsin X cDNA was cloned into the vector pPIC9, sequenced and expressed in the yeast *Pichia pastoris* as a prepro-α-factor fusion construct using the culture conditions recommended by Invitrogen. Procathepsin X was expressed and secreted at levels of approximately 5 mg/L of initial culture medium. After diafiltration against sodium acetate pH 5.0, and purification on a CM-sephadex column, mature cathepsin X was obtained by incubating the partially purified procathepsin X with a small amount of human cathepsin L. The sample was then dialyzed against phosphate buffer pH 7.0, and further purified on a DEAE column. Cathepsin X was stored at 4 °C in the elution buffer containing 100 μM of MMTS.

### 2.3 Kinetic Measurements

Kinetic experiments were performed as previously described (Nägler *et al* 1997, Ménard *et al* 1990). Inhibition studies with cystatin C have been carried out as described previously for inhibition of cathepsin L by its propeptide (Carmona *et al* 1996).

## 2.4 Molecular Modeling

The homology model of mature cathepsin X (residues Leu1-Pro240) was constructed using the COMPOSER and PROTEIN LOOPS modules in SYBYL 6.4 (Tripos, Inc., St. Louis, MO). Crystal structures of six cathepsins (PDB codes 1huc, 1cjl, 1mem, 1ppn, 1aec and 1ppo, respectively) were used as templates. Structural refinements were carried out in SYBYL 6.4 by energy minimization using the AMBER all-atom force-field (Weiner *et al* 1986). The structure was refined in three stages: i) the loop regions only, ii) including all side chains, and iii) allowing all atoms to relax. A conformational search was carried out for the loop Gln22-His23-Ile24-Pro25-Gln26-Tyr27-Cys28 using a Monte Carlo with energy minimization procedure (Li and Scheraga, 1987). The resulting structure was used to dock the substrate Cbz-FRF in the putative binding site of cathepsin X. The substrate binding mode was inferred from known complexes of cysteine proteases (PDB codes 1csb and 6pad).

# 3. RESULTS AND DISCUSSION

From the homology model for human cathepsin X, the three-residue insertion between the Gln22 of the putative oxyanion hole and the active site Cys31 is found on a loop corresponding to residues His23 to Tyr27, which we term the "mini-loop". The superposition of the cathepsin X model onto the cathepsin B crystal structure was striking in that the His23 of cathepsin X occupied a region in space which partially overlapped with His110 of cathepsin B, a residue which is considered to be responsible for the exopeptidase activity of the latter enzyme (Musil *et al* 1991). Complexation of cathepsin X with the substrate Cbz-FRF slightly refined the position of the histidine to optimally interact with the free carboxylate of the substrate $P_1$' residue. This finding suggests that cathepsin X may possess exopeptidase activity and motivated the choice of substrates for kinetic studies on substrate specificity. The results are presented in Table 1. It can be seen that

*Table 1.* Rate constant $k_{cat}/K_M$ for substrate hydrolysis by human cathepsin X at pH 5.0

| Substrate | $k_{cat}/K_M$ ( x $10^3$ $M^{-1}$ $s^{-1}$) |
|---|---|
| Abz-F-R-F(4NO2) | 123.2 ± 3.9 |
| Cbz-F-R-MCA | 0.064 ± 0.011 |
| Abz-A-F-R-S-A-A-Q-EDDnp | 0.030 ± 0.009 |

cathepsin X displays excellent carboxypeptidase activity against Abz-FRF(4NO$_2$), with a $k_{cat}/K_M$ value of 1.23 x $10^5$ $M^{-1}s^{-1}$ at pH 5.0. The activity of cathepsin X against the substrates Cbz-FR-MCA and Abz-AFRSAAQ-EDDnp were found to be extremely low, with $k_{cat}/K_M$ values of 64 $M^{-1}s^{-1}$ and 30 $M^{-1}s^{-1}$, respectively.

The positional specificity profiles of cathepsins X and B are compared in Fig 1. Both enzymes possess low endopeptidase activity against the extended IQF substrate, in comparison with their respective exopeptidase activities. The difference, however, is bigger for cathepsin X (4100-fold) than for cathepsin B (330-fold). The most striking difference is observed with the substrate Cbz-FR-MCA, which is as good as the exopeptidase substrate with cathepsin B, but is hydrolyzed 2000-fold slower than Abz-FRF(4NO$_2$) by cathepsin X. It is therefore clear that cathepsin X displays a much stricter exopeptidase activity than cathepsin B. It was also found that cystatin C is a very poor inhibitor of cathepsin X (no inhibition by cystatin C could be detected up to a concentration of 4 μM) as compared to cathepsin L or cathepsin B (Table 2). Cystatin C is a very potent inhibitor of cathepsin L ($K_i$ = 1.2 pM), and the dissociation constant of a cathepsin X – cystatin C complex must be at least 3x$10^6$-fold higher than for cathepsin L.

In the model of the complex between cathepsin X and the substrate Cbz-FRF, the C-terminal free carboxylate of the bound substrate is predicted to be engaged in hydrogen bond interactions with the imidazole δNH group of the protonated His23, the phenolic hydroxyl of Tyr27 and the indole NH group of Trp202 in the primed subsite region of cathepsin X. The interactions with His23 and Tyr27 from the mini-loop cannot be formed in any other member of the papain family of cysteine proteases. In addition, it

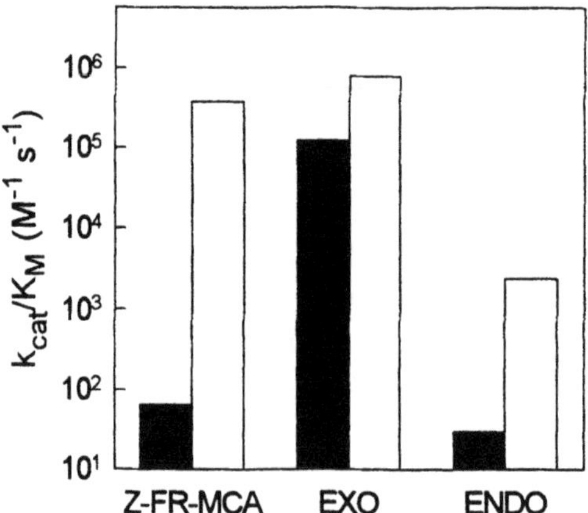


*Figure 1.* Positional specificity profile for cathepsin X (black bars) and cathepsin B (white bars). Z-FR-MCA: Cbz-FR-MCA substrate; EXO: carboxypeptidase substrate Abz-FRF(4NO$_2$) for cathepsin X, and dipeptidyl carboxypeptidase substrate Abz-FRF(4NO$_2$)A for cathepsin B; ENDO: endopeptidase substrate Abz-AFRSAAQ-EDDnp.

*Table 2.* Inhibition of human cathepsins X, B, and L by cystatin C at pH 5.0

| Enzyme | Ki (nM) |
|---|---|
| Cathepsin X | > 4000 |
| Cathepsin B | 1.1 ± 0.1 |
| Cathepsin L | 0.0012 ± 0.0002 |

appears that His23 and Tyr27 form a "wall" which occludes the $S_2$' subsite and prevents binding of an extra C-terminal residue of the substrate, or of the MCA moiety in $P_1$'. This can explain the poor activity of cathepsin X against the substrates Abz-AFRSAAQ-EDDnp and Cbz-FR-MCA, as well as the lack of significant inhibition by cystatin C.

The results of this study demonstrate that cathepsin X has a unique carboxypeptidase activity and can be considered to belong to a new subfamily within the papain family of cysteine proteases. The model of the cathepsin X structure consists of a two domain catalytic platform, a framework shared by papain-like enzymes, with functional diversity generated by the addition of a small structural element, the mini-loop. This relatively small modification to the endopeptidase platform results in a tremendous effect on the specificity of the enzyme. The marked differences between the primed subsites of cathepsin X relative to other members of the papain family of cysteine proteases will be of great value in designing specific inhibitors useful as research tools to investigate the physiological or pathological roles of the enzyme.

## REFERENCES

Adachi, W., *et al.*, 1998, Isolation and characterization of human cathepsin V: A major proteinase in corneal epithelium. *Invest. Ophthalmol. Visual Sci.* **39**: 1789-1796.

Carmona, E., *et al.*, 1996, Potency and selectivity of the cathepsin L propeptide as an inhibitor of cysteine proteases. *Biochemistry* **35**: 8149-8157.

Ekiel, I., *et al.*, 1997, NMR structural studies of human cystatin C dimers and monomers. *J. Mol. Biol.* **271**: 266-277.

Li, Z., and Scheraga, H.A., 1987, Monte Carlo-minimization approach to the multiple-minima problem in protein folding. *Proc. Natl. Acad. Sci. USA* **84**: 6611-6615.

Ménard, R., *et al.*, 1990, A protein engineering study of the role of aspartate 158 in the catalytic mechanism of papain. *Biochemistry* **29**: 6706-6713.

Musil, D., *et al.*, 1991, The refined 2.15 Å X-ray crystal structure of human liver cathepsin B: the structural basis for its specificity. *EMBO J.* **10**: 2321-2330.

Nägler, D.K., *et al.*, 1997, Major increase in endopeptidase activity of human cathepsin B upon removal of occluding loop contacts. *Biochemistry* **36**: 12608-12615.

Nägler, D.K., and Ménard, R., 1998, Human cathepsin X: A novel cysteine protease of the papain family with a very short proregion and unique insertions. *FEBS Lett.* **434**: 135-139.

Nägler, D.K., *et al.*, 1999, Full-length cDNA of human cathepsin F predicts the presence of a cystatin domain at the N-terminus of the cysteine protease zymogen. *Biochem. Biophys. Res. Commun.* **257**: 313-318.

Santamaría, I., *et al.*, 1998a, Cathepsin L2, a novel human cysteine proteinase produced by breast and colorectal carcinomas. *Cancer Res.* **58:** 1624-1630.
Santamaría, I., *et al.*, 1998b, Cathepsin Z, a novel human cysteine proteinase with a short propeptide domain and a unique chromosomal location. *J. Biol. Chem.* **273:** 16816-16823.
Storer, A.C., and Ménard, R., 1996, Recent insights into cysteine protease specificity: Lessons for drug design. *Perspect. Drug Discovery Des.* **6 :** 33-46.
Wang, B., *et al.*, 1998, Human cathepsin F. J. *Biol. Chem.* **273:** 32000-32008.
Weiner, S.J., *et al.*, 1986, An all atom force field for simulations of proteins and nucleic acids. *J. Comput. Chem.* **7:** 230-252.

# A NOVEL PROTEOLYTIC MECHANISM FOR TERMINATION OF THE $CA^{2+}$ SIGNALLING EVOKED BY PROTEINASE-ACTIVATED RECEPTOR-1 (PAR-1) IN RAT ASTROCYTES

Joachim J. Ubl and Georg Reiser
*Otto-von-Guericke Universität Magdeburg, Medizinische Fakultät, Institut für Neurobiochemie, Magdeburg, Germany*

## 1. INTRODUCTION

Among their myriad of roles, extracellular serine proteases, especially thrombin and trypsin can function as hormones to control cellular functions. Thrombin regulates the behavior of numerous cells, including astrocytes and neurones, by means of G protein-coupled proteinase-activated receptors (PAR). PAR-1, the prototype of this family, PAR-3, and PAR-4 can be activated by thrombin (Vu *et al* 1991, Ishihara *et al* 1997, Xu *et al* 1998). PAR-2 is activated by trypsin and mast cell tryptase, but not by thrombin (Molino *et al* 1997). Thrombin activates PAR-1 by binding to and cleaving its amino-terminal exodomain to unmask a new receptor amino terminus. This new amino terminus functions as a tethered ligand, which binds intramolecularly to a site within the same receptor molecule to effect transmembrane signalling. The synthetic peptide SFLLRN, which mimics the tethered ligand can function as receptor agonist, activating the receptor without proteolytic cleavage (Vu *et al* 1991).

A growing number of reports appearing during the last decade imply that thrombin and other serine proteases may play an important function in the central nervous sytem (Turgeon and Houenou 1997). In vitro studies revealed that thrombin stimulates DNA synthesis, induces proliferation, causes inhibition of stellation and rounding of astrocytes in culture (Cavanaugh *et al* 1990). Furthermore, we have previously shown that thrombin and thrombin receptor-activating peptides evoke $Ca^{2+}$ signals in C6 rat glioma cells (Czubayko and Reiser 1995) and primary cultures of rat astrocytes (Ubl and Reiser 1997).

The proteolytic and thus irreversible activation mechanism of PAR-1 stands in contrast to the reversible activation of classical G protein-coupled receptors (GPCR). Therefore the question arises, how PAR-1 is shut off. In endothelial cells, fibroblasts and blood-derived cell lines PAR-1 was rapidly phosphorylated and internalized, thus terminating thrombin signalling (Hoxie *et al* 1993, Ishii *et al* 1994, Woolkalis *et al* 1995). However, evidence exists that mechanisms other than receptor phosphorylation and trafficking exist.

We wanted to elucidate the mechanisms of PAR-1 signal termination and desensitization in rat astrocytes. Comparing the characteristics of the $Ca^{2+}$ response elicited by repeated stimulation of PAR-1 with thrombin and/or TRag, we detected that PAR-1-activation, most likely, induces activity of a trypsin-like protease which destroys the tethered ligand domain.

## 2. RESULTS

### 2.1 Characteristics of PAR-1-evoked $Ca^{2+}$ signals

Fura-2/AM-loaded rat astrocytes were briefly stimulated (60 s) with thrombin or the synthetic thrombin receptor agonist peptide (TRag: Ala-4-Fluoro-L-Phe-Arg-ß-Cyclohexyl-L-Ala-L-HomoArg-Tyr-$NH_2$). Experiments were carried out as described (Ubl and Reiser, 1997, Ubl *et al* 1998). Both agonists elicted a fast transient rise in $[Ca^{2+}]_i$ (Fig 1A insets). The transient increase was characterized by analysing the dose-dependence of the peak change in the fluorescence ratio (Δ F340 nm/F380 nm). Fig 1A gives the concentration-effect curves for thrombin (□) and TRag (O). Non-proteolytic activation with TRag is 1000-times less potent to induce $Ca^{2+}$ signals.

In most of the responding cells long-term superfusion with thrombin elicited a prolonged change of $[Ca^{2+}]_i$ characterized by a fast initial transient and an elevated plateau of $[Ca^{2+}]_i$ (Fig 1B) or in rare cases $[Ca^{2+}]_i$ oscillations (data not shown). Long-term $Ca^{2+}$ signalling depends on the presence of the protease since after omission of thrombin the $Ca^{2+}$ signal ceased. The $Ca^{2+}$ response could be restored by renewed addition of thrombin, albeit with an attenuated amplitude (Fig 1B). Similar patterns of long-term $[Ca^{2+}]_i$ changes, which also depend on the continuous presence of the agonist, were observed using TRag (data not shown). Taken together these results demonstrate that the presence and the concentration of thrombin and TRag are sensed by rat astrocytes via the duration and amplitude of the PAR-1-evoked $Ca^{2+}$ signal.

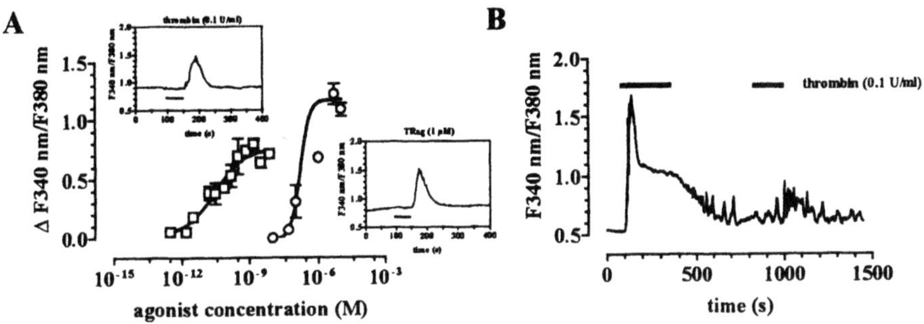


*Figure 1.* Characteristics of $Ca^{2+}$ signals in rat astrocytes due to PAR-1 activation.

## 2.2 Mechanism for terminating PAR-1 signalling

To elucidate the mechanism of PAR-1 signal termination and desensitization in rat astrocytes, we took advantage of the property of PAR-1 that the receptors can be activated proteolytically and non-proteolytically. We examined the $Ca^{2+}$ responses evoked by repeated homologous stimulation with either thrombin (0.1 U/ml) or TRag (1 μM) with an intervening wash of 5 min without agonists to reload intracellular $Ca^{2+}$ stores. Due to the different efficacy of thrombin and TRag on PAR-1, the response to a second PAR-1 activation was always compared to the mean amplitude of the initial $Ca^{2+}$ transient caused by the same agonist.

Irrespective of the activation mode, the second stimulation of PAR-1 always lead to a diminished amplitude of the $Ca^{2+}$ transient elicited (Fig 2A). However, the degree of attenuation depends on the agonists used and on the sequence of their addition. As summarized in Fig 2B, after a proteolytic activation of PAR-1 the subsequent $Ca^{2+}$ responses to thrombin or TRag were strongly similarly attenuated, by almost 70 %. However, with an initial non-proteolytic PAR-1 activation we obtained an unexpected result. Then we observed only a moderate desensitization (33 %) of a subsequent response to TRag, but a strong attenuation (64 %) of the $Ca^{2+}$ response elicited by thrombin as the second agonist. A plausible explanation for this difference is that after TRag activation the tethered ligand domain is altered in such a way that the receptor would be unresponsive to a subsequent challenge with thrombin, but still responsive to TRag.

One possibility to "silence" the tethered ligand domain is proteolytic cleavage within this domain. To verify our hypothesis, we used two approaches. Firstly, thermolysin, a metalloprotease known to cleave the

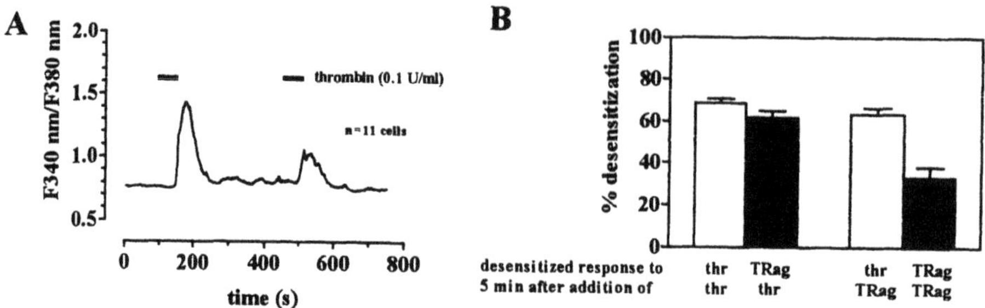


*Figure 2.* Desensitization of $[Ca^{2+}]_i$ in response to repeated PAR-1 stimulation. Roles of gingipains in the pathogenesis of periodontitis

PAR-1 within the tethered ligand domain was used. Rat astrocytes were superfused with thermolysin (5 U/ml) followed by thrombin (0.1 U/ml) or TRag (1 μM) for 60 s without an intervening wash. Thermolysin completely inhibited the $Ca^{2+}$ response to the addition of thrombin, but not to TRag (data not shown), clearly demonstrating that a cleavage of the tethered ligand domain leaves the cell unresponsive to a subsequent proteolytic activation, whereas the non-proteolytic receptor activation pathway was not altered. Additionally, we added thermolysin prior to thrombin and examined the response to a second PAR-1 stimulation with an intervening wash of 5 min. With this protocol thermolysin desensitized the subsequent TRag response by 28 %. This result implies that also an unspecific, non-activating cleavage of the N-terminus of PAR-1 can induce desensitization to TRag.

Secondly, a direct proof for our hypothesis came from the following experiments. Rat astrocytes were preincubated for 90 s with soybean trypsin inhibitor (SBTI; 100 ng/ml), stimulated for 60 s with TRag, followed by a 5 min superfusion with SBTI and a subsequent challenge with thrombin or TRag. Preincubation with SBTI did not alter the resting $[Ca^{2+}]_i$ and exerted no influence on the initial thrombin or TRag response (data not shown). However, the presence of SBTI increased the response to TRag by 31 % (n = 58) and to thrombin by 52 % (n = 28) obtained 5 min after non-proteolytic PAR-1 activation, compared to the controls (5 min superfusion without SBTI).

## 3. CONCLUSIONS

We used the change in $[Ca^{2+}]_i$ as a physiological parameter to investigate the regulation of PAR-1 signalling. PAR are activated by specific proteolysis of the receptor, generating an intramolecular tethered ligand, which binds to the receptor to effect transmembrane signalling. Since the ligand cannot diffuse away, the receptor becomes irreversibly activated. Therefore special mechanisms have to exist which shut off the protease-induced signals. Well characterized mechanisms for GPCR signal termination and desensitization include receptor phosphorylation and internalization. The same mechanisms seem to be involved in PAR-1 regulation since they were shown to inhibit PAR-1 signalling in a number of different cells (Brass *et al* 1994, Ishii *et al* 1994, Woolkalis *et al* 1995).

At first sight our results obtained after proteolytic cleavage of PAR-1 indicate that also in rat astrocytes receptor phosphorylation and/or internalization might be responsible for the signal attenuation observed. However, our observation that after non-proteolytic activation of PAR-1 the subsequent $Ca^{2+}$ response to thrombin was strongly attenuated, whereas the TRag-elicited $Ca^{2+}$ transient was only moderately decreased, implies an additional mechanism, which differentially affects a subsequent proteolytic or non-proteolytic activation.

A still enigmatic mechanism for shut-off and desensitization, specific for PAR-1, was already hypothezised by other groups (Norton *et al* 1993, Brass *et al* 1994, Hammes and Coughlin 1999), suggesting that additional mechanisms besides phosphorylation and/or endocytosis may terminate PAR-1 signalling. A recent work showed that activation of platelets with SFLLRN was associated with cleavage of PAR-1 and the release of a peptide, which could be detected by antibodies directed against the peptide released after thrombin activation (Ofosu *et al* 1998). This SFLLRN-induced PAR-1 cleavage could be prevented by serine protease inhibitors, but not by inhibitors of other proteases. The evidence from the literature and our own results reported here imply that modification of the tethered ligand may be involved in terminating thrombin signalling. Consequently, we hypothesized that an additional proteolytic cleavage modifies and thus inactivates the tethered ligand domain of PAR-1. In line with such a proteolytic mechanism are our experiments conducted with thermolysin, showing that cleavage of the tethered ligand inhibited the response to a subsequent proteolytic, but not to a non-proteolytic activation. The strongest evidence for an incapacitating proteolytic cleavage of PAR-1 after receptor activation came from our experiments showing that in the presence of SBTI the degree of desensitization after an initial TRag challenge was significantly reduced.

Taken together we have strong indications that further degradation (proteolysis) of the tethered ligand is involved in PAR-1 signal termination, avoiding self-activation of the receptor after removal of thrombin, as depicted in Fig 3. The cleavage site and identity of the hypothetical protease remain to be elucidated. Furthermore it is necessary to find out whether this mechanism is restricted to PAR-1. Limited or regulated proteolysis is a wide spread mechanism in regulating physiological (e. g. coagulation) or pathological (e. g. inflammation) responses. Therefore, it is exciting to

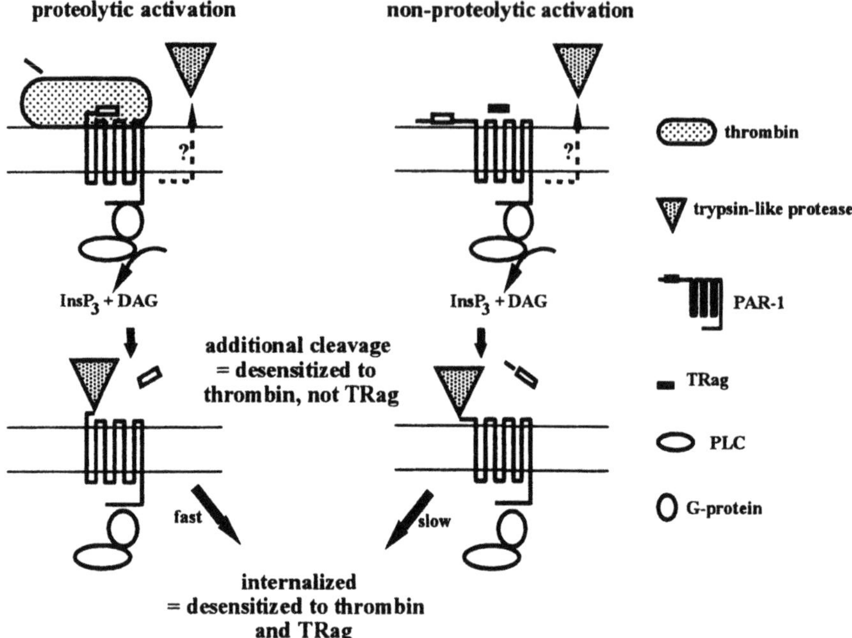


*Figure 3.* Novel model of PAR-1 signal termination involving a trypsin-like protease

speculate that proteolytic mechanisms similar to that suggested here for PAR-1 regulation, might apply even to cell surface receptors (GPCR, growth factor receptors, cytokine receptors) which are activated by agonists other than proteases.

## ACKNOWLEDGMENTS

This work was supported in part by Deutsche Forschungsgemeinschaft (436 RUS/17/24/98), LSA (grant 2923A), and Fonds der Chemischen Industrie.

## REFERENCES

Brass, L.F., *et al.*, 1994, Changes in the structure and function of the human thrombin receptor during receptor activation, internalization, and recycling. *J. Biol. Chem.* **269:** 2943-2952.

Cavanaugh, K.P., *et al.*, 1990, Reciprocal modulation of astrocyte stellation by thrombin and protease nexin-1. *J. Neurochem.* **54:** 1735-1743.

Czubayko, U., and Reiser, G., 1995, $[Ca^{2+}]_i$ oscillations in single rat glioma cells induced by thrombin through activation of cell surface receptors. *Neuroreport* **6:** 1249-1252.

Hammes, S.R., and Coughlin, S.R., 1999, Protease-activated receptor-1 can mediate responses to SFLLRN in thrombin-desensitized cells: Evidence for a novel mechanism for preventing or terminating signaling by PAR 1's tethered ligand. *Biochemistry* **38:** 2486-2493.

Hoxie, J.A., *et al.*, 1993, Internalization and recycling of activated thrombin receptors. *J. Biol. Chem.* **268:** 13756-13763.

Ishihara, H. *et al.*, 1997, Protease-activated receptor 3 is a second thrombin receptor in humans. *Nature* **386:** 502-506.

Ishii, K., *et al.*, 1994, Inhibition of thrombin receptor signaling by a G-protein coupled receptor kinase. Functional specificity among G-protein coupled receptor kinases. *J. Biol. Chem.* **269:** 1125-1130.

Molino, M., *et al.*, 1997, Interactions of mast cell tryptase with thrombin receptors and PAR-2. *J. Biol. Chem.* **272:** 4043-4049.

Norton, K.J., *et al.*, 1993, Immunologic analysis of the cloned platelet thrombin receptor activation mechanism: evidence supporting receptor cleavage, release of the N-terminal peptide, and insertion of the tethered ligand into a protected environment. *Blood* **82:** 2125-2136.

Ofosu, F.A., *et al.*, 1998, A trypsin-like platelet protease propagates protease-activated receptor-1 cleavage and platelet activation. *Biochem. J.* **336:** 283-285.

Turgeon, V.L., and Houenou, L.J., 1997, The role of thrombin-like (serine) proteases in the development, plasticity and pathology of the nervous system, *Brain Res. Rev.* **25:** 85-95.

Ubl, J.J., and Reiser, G., 1997, Characteristics of thrombin-induced calcium signals in rat astrocytes. *Glia* **21:** 361-369.

Ubl, J.J., *et al.*, 1998, Co-existence of two types of calcium-inducing protease-activated receptors (PAR-1 and PAR-2) in rat astrocytes and C6 glioma cells. *Neurosci* **86:** 597-609.

Vu, T.K., *et al.*, 1991, Molecular cloning of a functional thrombin receptor reveals a novel proteolytic mechanism of receptor activation, *Cell* **64:** 1057-1068.

Woolkalis, M.J., *et al.*, 1995, Regulation of thrombin receptors on human umbilical vein endothelial cells, *J. Biol. Chem.* **270:** 9868-9875.

Xu, W.-F., *et al.*, 1998, Cloning and characterization of human protease-activated receptor 4, *Proc. Natl. Acad. Sci. USA* **95:** 6642-6646.

# NATURAL AND SYNTHETIC INHIBITORS OF THE TUMOR-ASSOCIATED SERINE PROTEASE UROKINASE-TYPE PLASMINOGEN ACTIVATOR

Viktor Magdolen[1], Nuria Arroyo de Prada[1], Stefan Sperl[1,2], Bernd Muehlenweg[1], Thomas Luther[3], Olaf G. Wilhelm[1,4], Ulla Magdolen[1], Henner Graeff[1], Ute Reuning[1], and Manfred Schmitt[1]

[1] *Klinische Forschergruppe der Frauenklinik der Technischen Universität München, Klinikum rechts der Isar, Ismaninger Str. 22, D-81675 München, Germany;* [2] *Max-Planck-Institut für Biochemie, Am Klopferspitz 18A, D-82152 Martinsried, Germany;* [3] *Institut für Pathologie der Technischen Universität Dresden, Fetscherstr. 74, D-01307 Dresden, Germany;* [4] *Wilex Biotechnology GmbH, Grillparzer Str. 10B, D-81675 München, Germany*

## 1. INTRODUCTION

Cancer cell invasion requires the regulated expression of various proteolytic enzymes that degrade the surrounding extracellular matrix and dissociate cell-cell and/or cell-matrix attachments. Especially, the tumor cell surface-associated urokinase-type plasminogen activator system, consisting of the serine protease uPA, its receptor uPAR and its inhibitors PAI-1 and PAI-2, plays an important role in tumor cell invasion and metastasis (Reuning *et al* 1998, Schmitt *et al* 1997). Various normal and cancer cells synthesize uPA as a single-chain pro-enzyme (pro-uPA) with little or no intrinsic enzymatic activity. This zymogen is proteolytically converted to the active two-chain form *e.g.* by plasmin, another serine protease, or cysteine proteases such as the cathepsins B and L (Kobayashi *et al* 1991, Goretzki *et al* 1992). Binding of uPA to its receptor uPAR (CD87) focusses proteolytic activity to the tumor cell surface. In addition to uPAR, tumor cells also express (a) receptor(s) for plasmin(ogen) (Félez 1998). uPAR-bound uPA efficiently converts tumor cell-associated plasminogen into plasmin, an active serine protease with broad substrate specificity. Plasmin degrades a variety of components of the extracellular matrix (e.g. fibrin, fibronectin, or laminin) and activates some matrix metalloproteases (Murphy *et al* 1999, Hahn-Dantona *et al* 1999) which also break down macromolecules of the extracellular matrix such as the fiber-forming collagens I, II, III and/or the basement membrane protein collagen IV. Thus, the interplay between several tumor cell-associated proteolytic systems facilitates extracellular matrix degradation, tumor invasion, and metastasis.

## 2. NATURAL INHIBITORS OF THE UROKINASE-TYPE PLASMINOGEN ACTIVATOR (uPA)

The uPA/uPAR-system is under the control of the plasminogen activator inhibitors type 1 (PAI-1) and type 2 (PAI-2). These two inhibitors belong to the serine protease inhibitor superfamily (serpins) and share 55 % sequence homology (amino acid identity: 33 %). There is also structural similarity of the two proteins, proven by X-ray crystal structures of active mutants of PAI-1 (Sharp *et al* 1999) and PAI-2 (Harrop *et al* 1999). Both inhibitors

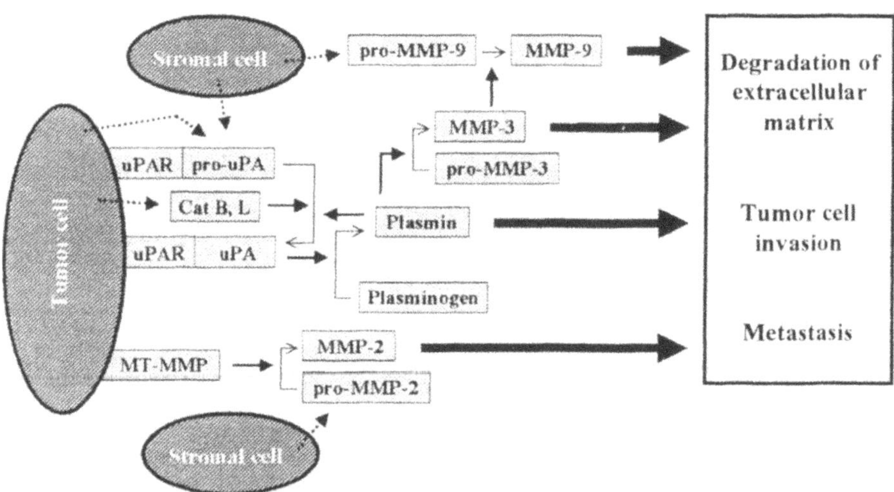


*Figure 1.* Tumor cell-associated proteolysis. uPA, which is secreted as an inactive pro-enzyme (pro-uPA) by stromal and tumor cells binds with high affinity to its tumor cell surface-associated receptor (uPAR, CD87). Receptor-bound pro-uPA is efficiently converted to uPA by cysteine proteases (*e.g.* cathepsin B or L [Cat B, L]) or by serine proteases (*e.g.* plasmin). uPA, in turn, activates plasminogen into plasmin which degrades a series of extracellular matrix components and may also activate certain matrix metalloproteases (*e.g.* pro-MMP-3). Proteolytic activity may also be focussed to the tumor cell surface by membrane-bound activators of proenzymes (*e.g.* activation of pro-MMP-2 by MT-MMPs). Degradation of the extracellular matrix in the vicinity of the tumor cells (tumor stroma) facilitates tumor cell invasion and metastasis.

interact with uPA and tPA (tissue-type plasminogen activator), forming 1:1 stoichiometric complexes with the respective target protease. Whereas PAI-1 recognizes and inhibits all active forms of the proteases (two-chain uPA as well as single- and two-chain tPA), PAI-2 only acts as an inhibitor for the two-chain forms of uPA and tPA (Egelund *et al* 1997, Ny and Mikus 1997).

PAI-1 is synthesized with a 21 amino acid signal peptide and secreted by the cell in a glycosylated form. Active PAI-1 is only metastable and spontaneously transforms into a latent, inactive conformation (half-life of approx. 2 h at 37 °C). The half-life of active PAI-1 can be increased up to more than 140 h by introducing four point mutations (Berkenpas *et al* 1995). *In vivo*, the biologically active conformation of PAI-1 is stabilized by vitronectin (VN), an extracellular matrix and plasma protein. Upon binding to VN the half-life of wild-type PAI-1 is doubled to 4 h (Kanse *et al* 1996). The interaction of PAI-1 with the adhesive glycoprotein VN apparently plays a major role in tumor cell invasion and metastasis. Adhesion of tumor cells to the extracellular matrix is promoted by the VN-binding protein uPAR as well as adhesion receptors of the integrin family such as $\alpha_v\beta_3$ and $\alpha_v\beta_5$. PAI-1 and uPAR compete for overlapping binding sites on VN; PAI-1 can also interfere with the interaction of VN with integrins (Kjøller *et al* 1997). Thus, the balance between the expression of uPAR/integrins on the tumor cell surface and PAI-1 appears to be a determinant for cell adhesion to or migration on VN (Loskutoff *et al* 1999). Furthermore, PAI-1 also interacts with extracellular matrix components heparin and fibrin, which emphasizes the importance of PAI-1 in such different processes as embryonic implantation and development, wound healing, and fibrinolysis (Egelund *et al* 1997, Loskutoff *et al* 1999, Eitzman and Ginsburg 1997).

PAI-2 lacks a cleavable signal peptide and is mainly present intracellularly in a non-glycosylated form. Only a small amount of PAI-2 (approx. 20 %) is glycosylated and secreted (Ny and Mikus 1997, Åstedt *et al* 1998). Under normal physiological conditions PAI-2 is not detectable in serum. However, during pregnancy PAI-2 is synthesized by trophoblast cells and released into the blood, achieving antigen levels of up to 260 ng/ml (Åstedt *et al* 1998). Besides the target proteases uPA and tPA, no further intra- or extracellular interaction partners of PAI-2 have been identified so far. Nevertheless, PAI-2 seems to be of relevance in Alzheimer's disease and inflammatory reactions and may also be involved in apoptosis (Ny and Mikus 1997, Åstedt *et al* 1998).

There are two other serpins, proteinase nexin-1 and protein C inhibitor, which - in addition to thrombin, trypsin, plasmin and/or activated protein C - are also capable to inhibit uPA (and tPA) under physiological conditions. However, these serpins react more slowly with the plasminogen activators as compared to PAI-1 and PAI-2 (Andreasen *et al* 1997, Sprengers and Kluft 1987).

## 3. CLINICAL RELEVANCE OF PAI-1 AND PAI-2

In recent years, several studies have been undertaken to identify and define new prognostic markers for the prediction of disease-free and overall survival of patients afflicted with solid malignant tumors. In a variety of malignancies, including cancer of the breast, ovary, cervix uteri, bladder, upper urogenital tract, kidney, head and neck, brain, lung, soft-tissue, stomach, colon, pancreas, esophagus and liver, a strong prognostic impact has been attributed to components of the uPA-system: uPA, uPAR, and/or PAI-1/-2 are statistically independent prognostic factors with the capacity to predict the probability of disease-free and/or overall survival (for summaries see Reuning *et al* 1998, Schmitt *et al* 1997).

In general, elevated tumor antigen levels of uPA and/or PAI-1 are associated with poor disease outcome and are conductive to tumor cell spread and metastasis (Reuning *et al* 1998, Schmitt *et al* 1997, Look and Foekens 1999). At first sight, it appears rather striking that elevated antigen levels of an inhibitor of uPA, PAI-1, in tumor tissue indicate poor prognosis for cancer patients. However, considering the additional functions of PAI-1, especially with respect to the modulation of the adhesive properties of tumor cells (Kjøller *et al* 1997, Loskutoff *et al* 1999, Eitzman and Ginsburg 1997), these observations may be explained by an involvement of PAI-1 in a balanced interference/induction of tumor cell attachment/migration. In contrast to PAI-1, elevated PAI-2 antigen levels predict a good prognosis of cancer patients (Look and Foekens 1999, Foekens *et al* 1995).

## 4. PAI-1 AND PAI-2 KNOCKOUT MICE

For analysis of the potential physiological and pathophysiological roles of PAI-1 and PAI-2, corresponding knockout mice have been generated. Homozygous PAI-$1^{-/-}$ mice are viable, fertile and do not show abnormalities in organogenesis and development (Carmeliet *et al* 1993a). PAI-1 knockout mice, however, display weak bleeding disorders, but appear more resistant against venous thrombosis (Carmeliet *et al* 1993b). Bajou and co-workers (Bajou *et al* 1998) observed that malignant murine keratinocytes, transplantated into PAI-1-deficient mice, were unable to invade surrounding tissue (local invasion). Furthermore, the PAI-1-deficient hosts failed to vascularize the implanted tumor cells. Upon intravenous injection of an adenoviral vector expressing human PAI-1 in these tumor-bearing mice, tumor cell invasion and associated angiogenesis were restored, demonstrating the relevance of PAI-1 expression in these processes.

*Table 1.* Properties of serpins PAI-1 and PAI-2

| | PAI-1 | PAI-2 |
|---|---|---|
| Function | Extracellular uPA/tPA inhibitor | Extracellular uPA/tPA inhibitor, intracellular regulator of apoptosis |
| Physiological relevance | Fibrinolysis, tumor invasion and metastasis, embryo implantation and development, wound healing | Placenta development, tumor invasion and metastasis, inflammatory processes, Alzheimer's disease, apoptosis |
| Structural determinants | 381 amino acids (mature protein)<br>Signal peptide (21 amino acids)<br>No cysteine<br>Extracellular 52 kDa glycosylated protein<br>P1-P1′: Arg348-Met349<br>X-ray structure of latent and center-cleaved PAI-1 as well as of an active PAI-1 mutant | 415 amino acids<br>No cleavable signal peptide<br>Six cysteines<br>47 kDa intracellular or 60 kDa extracellular glycosylated variant<br>P1-P1′: Arg380-Tyr381<br>X-ray struture of an active PAI-2 mutant<br>Spontaneous polymerization |
| Biosynthesis and activity | Synthesis as an active molecule in different cell types<br>Storage in platelets<br>Plasma level 5-85 ng/ml<br>Half life approx. 2 h<br>Reactivation possible *in vitro* | Below detection limit in serum under normal physiological conditions<br>During pregnancy approx. 260 ng/ml serum<br>No loss of activity in solution |
| Additional binding partners | Vitronectin (VN), heparin, fibrin<br>Binding to VN stabilizes active PAI-1 and increases its affinity towards thrombin approx. 1000-fold<br>Binding to heparin increases PAI-1 activity towards thrombin approx. 1000-fold | No further binding partners known |

Homozygous PAI-$2^{-/-}$ mice are also viable, fertile and do not display an obviously different phenotype when compared to wild-type mice (Dougherty *et al* 1999). However, up to now no tumorbiological relevant experiments such as injection of tumor cells in PAI-2-deficient mice and analysis of tumor growth, invasion and metastasis have been reported.

PAI-1- and PAI-2-deficient mice have been crossed and the resulting PAI-1/2 null mice analyzed. The mice are also viable and display a similar (mild) phenotype as observed in PAI-1-deficient mice. These observations indicate that PAI-1 and PAI-2 do not share essential functions. At least in the murine system, adaptive mechanisms may exist to substitute for the functions of PAI-1/2 (Dougherty *et al* 1999).

## 5. INTERACTION OF PAI-1 AND PAI-2 WITH uPAR-BOUND uPA

PAI-1 and PAI-2 are effective inhibitors of uPA in solution as well as bound to uPAR. Interaction of PAI-1 with the uPA/uPAR complex leads to internalization of the trimeric complex, which in turn stimulates cell proliferation (Blasi 1997). This internalization is mediated by other trans-membrane receptors and associated proteins (Nykjaer *et al* 1997, Casslén *et al* 1998, Conese and Blasi 1995). The uPA/PAI-1 complex is subsequently degraded in lysosomes, while uPAR, at least in part, is redistributed to the cell surface. In contrast to PAI-1, the trimeric PAI-2/uPA/uPAR complex is not internalized but processed on the cell surface resulting in a 70 kDa-fragment which contains PAI-2 and the active site of uPA and a 22 kDa-fragment which encompasses the amino-terminal fragment of uPA (ATF) harboring the uPAR binding site. The 70 kDa-fragment is released into the medium or internalized, whereas the enzymatically inactive 22 kDa-fragment remains bound to the receptor, thereby blocking the uPA binding site on uPAR (Ragno *et al* 1993, Ragno *et al* 1995). In cancerogenesis, the different fates of the trimeric uPA/uPAR/serpin complexes apparently correspond to the different biological functions of PAI-1 and PAI-2. PAI-1 bound to uPA/uPAR leads to stimulation of cell proliferation which promotes tumor growth and supports a rapid and efficient reorganization of the invasive front by triggering the redistribution of uPAR on the cell surface. PAI-2, on the other hand, acts as a true inhibitor of the plasminogen activation system, since it not only inhibits uPA activity but also prevents directed migration of tumor cells by blocking uPAR.

## 6. SYNTHETIC uPA-INHIBITORS

Compared to the multitude of reversible, synthetic inhibitors that have been developed towards several serine proteases (e. g. thrombin (Hauptmann and Sturzebecher 1999)), only a few low-molecular-weight uPA inhibitors are available. The common structural feature of these uPA inhibitors is a basic moiety containing an amidino- or guanidino-group, which binds to $Asp^{189}$ in the arginine-specific S1 pocket of uPA. The set of examined compounds contains substituted benzamidines and β-naphthamidines (Stürzebecher and Markwardt, 1978), amidinoindoles and 5-amidino-benzimidazoles (Tidwell *et al* 1983) as well as mono-substituted phenylguanidines (Yang *et al* 1990). Most of these compounds, however, display little or no selectivity for uPA, as they also inhibit related enzymes such as trypsin, thrombin, factor Xa, or plasmin. The inhibition constants of these inhibitors are usually in the low micromolar range (Table II). Billström *et al* (1995) observed a distinct decrease of tumor growth of DU 145 prostate adenocarcinoma cells in SCID mice after oral administration of the moderately selective uPA inhibitor p-aminobenzamidine at doses of 125-250 mg/kg/day. Side effects were negligible. The diuretic compound amiloride is a selective uPA inhibitor with an inhibition constant of 7 μM (Vassalli and Belin 1987), shown to reduce the size of prostate cancer xenografts in SCID mice. Similar effects were observed with p-aminobenzamidine (Jankun *et al* 1997).

At present, the most potent and selective active site-directed uPA inhibitors are substituted benzo[*b*]thiophen-2-carboxamidines, B-428 and B-623, with $IC_{50}$ values of 0.32 μM and 0.07 μM, respectively (Towle *et al* 1993). In the case of B-428 it has been demonstrated that the compound inhibits uPA-mediated processes such as proteolytic degradation of extracellular matrix as well as adhesion, migration, and invasion of tumor cells *in vitro* (Alonso *et al* 1996, Alonso *et al* 1998). Furthermore, a decrease in growth and metastasis of a tumor cell line overexpressing rat uPA was observed in a syngeneic model of rat prostate cancer (Rabbani *et al* 1995). The combination of this drug with the antiestrogen tamoxifen led to a significant reduction of primary tumor volume and metastasis in a syngeneic rat breast cancer model (Xing *et al* 1997). According to Swiercz and co-workers (1999), the uPA inhibitors amiloride, benzamidine, B-428, and B-623 caused a significant reduction of angiogenesis in a chicken embryo corioallantoic membrane model indicating that the uPA-system may also be involved in tumor angiogenesis.

## 7. CONCLUSIONS

In the present overview, the multifunctional capacity of the uPA-system in tumor biological processes was emphasized. In addition to its central role in pericellular proteolysis, uPA/uPAR-mediated activities contribute to many different processes like cell proliferation, adhesion, migration, and angiogenesis. Whereas extracellular PAI-2 solely acts as an inhibitor of uPA (and tPA), PAI-1 clearly exerts additional functions, e. g. involvement in modulation of uPAR- and integrin-mediated cell adhesion as well as in redistribution of uPAR on the tumor cell surface supporting the reorganization of the invasive front. Selective synthetic active site inhibitors of uPA may serve as novel therapeutic agents for anti-invasive and anti-proliferative cancer therapy.

*Table 2.* Synthetic up a active site inhibitors

| Inhibitor | Selectivity | Inhibitory effect (uPA) |
|---|---|---|
| 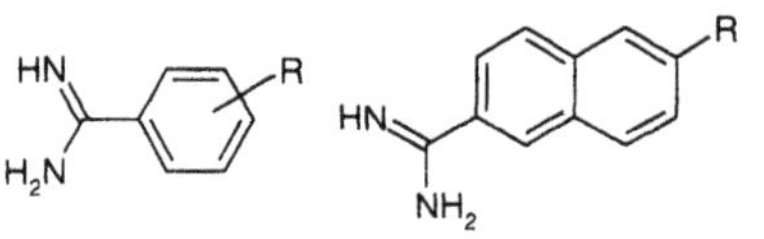 benzamidines and β-naphthamidines | $K_i$ for trypsin-, thrombin-, and plasmin-inhibition is in similar range | $K_i \approx 5\text{-}50\ \mu M$ |
| 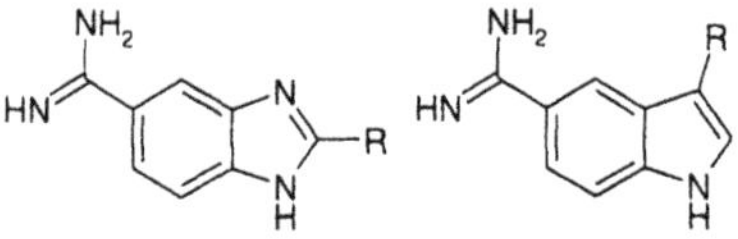 5-amidino-benzimidazoles and -indoles | $K_i$ for trypsin-, thrombin-, and plasmin-inhibition is in similar range | $K_i > 4\ \mu M$ |
| 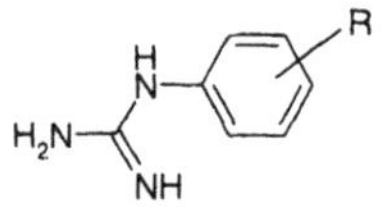 mono-subst. phenylguanidines | good selectivity for uPA | $K_i \geq 6\ \mu M$ |
| 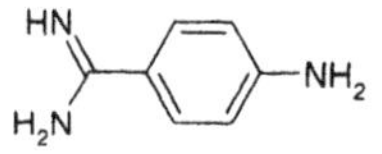 p-aminobenzamidine | moderately selective | $K_i = 82\ \mu M$ |
| 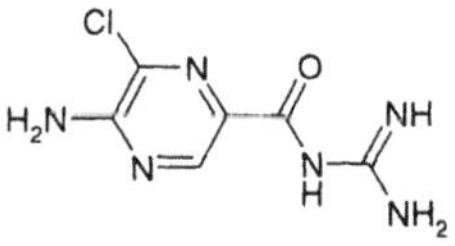 amiloride (diuretic) | selective uPA-inhibitor | $K_i = 7\ \mu M$ |
| 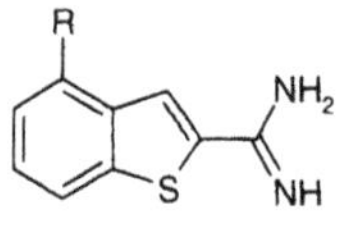 benzo[*b*]thiophen-2-carboxamidines | highly potent and selective | $IC_{50}$ [μM]:<br>B392: 3.7<br>B428: 0.32<br>B623: 0.07 |

# ACKNOWLEDGMENTS

This work was supported by grants of the Sonderforschungsbereich 469 and Graduiertenkolleg 333 of the Deutsche Forschungsgemeinschaft (DFG), of the Deutsche Krebshilfe e.V. (Dr. Mildred Scheel-Stiftung), and of the Hochschulsonderprogramm III of the Technische Universität München

## REFERENCES


Alonso, D.F., Farias, E.F., Ladeda, V., Davel, L., Puricelli, L., and Bal, d. K.J., 1996, Effects of synthetic urokinase inhibitors on local invasion and metastasis in a murine mammary tumor model. *Breast Cancer Res. Treat.* **40:** 209-223.

Alonso, D.F., Tejera, A.M., Farias, E.F., Bal, d. K.J., and Gómez, D E., 1998, Inhibition of mammary tumor cell adhesion, migration, and invasion by the selective synthetic urokinase inhibitor B428. *Anticancer Res.* **18:** 4499-4504.

Andreasen, P.A., Kjøller, L., Christensen, L., and Duffy, M.J., 1997, The urokinase-type plasminogen activator system in cancer metastasis: a review. *Int. J. Cancer* **72:** 1-22.

Åstedt, B., Lindoff, C., and Lecander, I., 1998, Significance of the plasminogen activator inhibitor of placental type (PAI-2) in pregnancy. *Semin. Thromb. Hemost.* **24:** 431-435.

Bajou, K., Noël, A., Gerard, R.D., Masson, V., Brünner, N., Holst-Hansen, C., Skobe, M., Fusenig, N.E., Carmeliet, P., Collen, D., and Foidart, J.M., 1998, Absence of host plasminogen activator inhibitor 1 prevents cancer invasion and vascularization. *Nat. Med.*. **4:** 923-928.

Berkenpas, M. B., Lawrence, D. A., and Ginsburg, D., 1995, Molecular evolution of plasminogen activator inhibitor-1 functional stability. *EMBO J.* **14:** 2969-2977.

Billström, A., Hartley-Asp, B., Lecander, I., Batra, S., and Åstedt, B., 1995, The urokinase inhibitor p-aminobenzamidine inhibits growth of a human prostate tumor in SCID mice. *Int. J. Cancer* **61:** 542-547.

Blasi, F., 1997, uPA, uPAR, PAI-1: key intersection of proteolytic, adhesive and chemotactic highways? *Immunol. Today* **18:** 415-417.

Carmeliet, P., Kieckens, L., Schoonjans, L., Ream, B., van Nuffelen, A., Prendergast, G., Cole, M., Bronson, R., Collen, D., and Mulligan, R.C., 1993a, Plasminogen activator inhibitor-1 gene-deficient mice. I. Generation by homologous recombination and characterization. *J. Clin. Invest.* **92:** 2746-2755.

Carmeliet, P., Stassen, J.M., Schoonjans, L., Ream, B., van den Oord, J J., De Mol, M., Mulligan, R.C., and Collen, D., 1993b, Plasminogen activator inhibitor-1 gene-deficient mice. II. Effects on hemostasis, thrombosis, and thrombolysis. *J. Clin. Invest.* **92:** 2756-2760.

Casslén, B., Gustavsson, B., Angelin, B., and Gåfvels, M., 1998, Degradation of urokinase plasminogen activator (uPA) in endometrial stromal cells requires both the uPA receptor and the low-density lipoprotein receptor-related protein/alpha2-macroglobulin receptor. *Mol. Hum. Reprod.* **4:** 585-593.

Conese, M., and Blasi, F., 1995, Urokinase/urokinase receptor system: internalization/ degradation of urokinase-serpin complexes: mechanism and regulation. *Biol. Chem. Hoppe Seyler* **376:** 143-155.

Dougherty, K.M., Pearson, J.M., Yang, A.Y., Westrick, R..J., Baker, M.S., and Ginsburg, D., 1999, The plasminogen activator inhibitor-2 gene is not required for normal murine development or survival. *Proc. Natl. Acad. Sci. USA* **96:** 686-691.

Egelund, R., Schousboe, S.L., Sottrup-Jensen, L., Rodenburg, K.W., and Andreasen, P.A., 1997, Type-1 plasminogen-activator inhibitor -conformational differences between latent, active, reactive-centre-cleaved and plasminogen- activator-complexed forms, as probed by proteolytic susceptibility. *Eur. J. Biochem.* **248:** 775-785.

Eitzman, D.T., and Ginsburg, D., 1997, Of mice and men. The function of plasminogen activator inhibitors (PAIs) in vivo. *Adv. Exp. Med. Biol.* **425**: 131-141.

Félez J., 1998, Plasminogen binding to cell surfaces. *Fibrinol. Proteol.* **12**: 183-189.

Foekens, J.A., Buessecker, F., Peters, H.A., Krainick, U., van Putten, W.L., Look, M.P., Klijn, J.G., and Kramer, M.D., 1995, Plasminogen activator inhibitor-2: prognostic relevance in 1012 patients with primary breast cancer. *Cancer Res.* **55**: 1423-1427.

Goretzki, L., Schmitt, M., Mann, K., Calvete, J., Chucholowski, N., Kramer, M., Günzler, W. A., Jänicke, F., and Graeff, H., 1992, Effective activation of the proenzyme form of the urokinase-type plasminogen activator (pro-uPA) by the cysteine protease cathepsin L. *FEBS Lett.* **297**: 112-118.

Hahn-Dantona, E., Ramos-DeSimone, N., Sipley, J., Nagase, H., French, D.L., and Quigley, J.P., 1999, Activation of proMMP-9 by a plasmin/MMP-3 cascade in a tumor cell model. Regulation by tissue inhibitors of metalloproteinases. *Ann. NY Acad. Sci.* **878**: 372-387.

Harrop, S.J., Jankova, L., Coles, M., Jardine, D., Whittaker, J.S., Gould, A.R., Meister, A., King, G.C., Mabbutt, B.C., and Curmi, P.M., 1999, The crystal structure of plasminogen activator inhibitor 2 at 2.0 Å resolution: implications for serpin function. *Structure Fold Des.* **7**: 43-54.

Hauptmann, J., and Stürzebecher, J., 1999, Synthetic inhibitors of thrombin and factor Xa: from bench to bedside. *Thromb. Res.* **93**: 203-241.

Jankun, J., Keck, R W., Skrzypczak-Jankun, E., and Swiercz, R., 1997, Inhibitors of urokinase reduce size of prostate cancer xenografts in severe combined immunodeficient mice. *Cancer Res.* **57**: 559-563.

Kanse, S.M., Kost, C., Wilhelm, O.G., Andreasen, P.A., and Preissner, K.T., 1996, The urokinase receptor is a major vitronectin-binding protein on endothelial cells. *Exp. Cell. Res.* **224**: 344-353.

Kjøller, L., Kanse, S.M., Kirkegaard, T., Rodenburg, K.W., Ronne, E., Goodman, S.L., Preissner, K.T., Ossowski, L., and Andreasen, P.A., 1997, Plasminogen activator inhibitor-1 represses integrin- and vitronectin-mediated cell migration independently of its function as an inhibitor of plasminogen activation. *Exp.Cell Res.* **232**: 420-429.

Kobayashi, H., Schmitt, M., Goretzki, L., Chucholowski, N., Calvete, J., Kramer, M., Günzler, W.A., Jänicke, F., and Graeff, H., 1991, Cathepsin B efficiently activates the soluble and the tumor cell receptor-bound form of the proenzyme urokinase-type plasminogen activator (pro-uPA). *J. Biol. Chem.* **266**: 5147-5152.

Look, M.P. and Foekens, J.A., 1999, Clinical relevance of the urokinase plasminogen activator system in breast cancer. *APMIS* **107**: 150-159.

Loskutoff, D.J., Curriden, S.A., Hu, G., and Deng, G., 1999, Regulation of cell adhesion by PAI-1. *APMIS* **107**: 54-61.

Murphy, G., Knäuper, V., Cowell, S., Hembry, R., Stanton, H., Butler, G., Freije, J., Pendás, A.M., and López-Otín, C., 1999, Evaluation of some newer matrix metalloproteinases. *Ann. NY Acad. Sci.* **878**: 25-39.

Ny, T., and Mikus, P., 1997 Plasminogen activator inhibitor type-2. A spontaneously polymerizing serpin that exists in two topological forms. *Adv. Exp. Med. Biol.* **425**: 123-130.

Nykjaer, A., Conese, M., Christensen, E.I., Olson, D., Cremona, O., Gliemann, J., and Blasi, F., 1997, Recycling of the urokinase receptor upon internalization of the uPA:serpin complexes. *EMBO J.* **16**: 2610-2620.

Rabbani, S. A., Harakidas, P., Davidson, D J., Henkin, J., and Mazar, A.P., 1995, Prevention of prostate-cancer metastasis in vivo by a novel synthetic inhibitor of urokinase-type plasminogen activator (uPA). *Int. J. Cancer* **63**: 840-845.

Ragno, P., Montuori, N., and Rossi, G., 1995, Urokinase-type plasminogen activator/type-2 plasminogen-activator inhibitor complexes are not internalized upon binding to the urokinase- type-plasminogen-activator receptor in THP-1 cells. Interaction of urokinase-type plasminogen activator/type-2 plasminogen-activator inhibitor complexes with the cell surface. *Eur. J. Biochem.* **233:** 514-519.

Ragno, P., Montuori, N., Vassalli, J.D., and Rossi, G., 1993, Processing of complex between urokinase and its type-2 inhibitor on the cell surface. A possible regulatory mechanism of urokinase activity. *FEBS Lett.* **323:** 279-284.

Reuning, U., Magdolen, V., Wilhelm, O.G., Fischer, K., Lutz, V., Graeff, H., and Schmitt, M., 1998, Multifunctional potential of the plasminogen activation system in tumor invasion and metastasis. Int. J. Oncol. **13:** 893-906.

Schmitt, M., Harbeck, N., Thomssen, C., Wilhelm, O.G., Magdolen, V., Reuning, U., Ulm, K., Höfler, H., Jänicke, F., and Graeff, H., 1997, Clinical impact of the plasminogen activation system in tumor invasion and metastasis: prognostic relevance and target for therapy. *Thromb. Haemost.* **78:** 285-296.

Sharp, A.M., Stein, P.E., Pannu, N.S., Carrell, R.W., Berkenpas, M.B., Ginsburg, D., Lawrence, D.A., and Read, R.J., 1999, The active conformation of plasminogen activator inhibitor 1, a target for drugs to control fibrinolysis and cell adhesion. *Structure* **7:** 111-118.

Sprengers, E.D., and Kluft, C., 1987, Plasminogen activator inhibitors. *Blood* **69:** 381-387.

Stürzebecher, J. and Markwardt, F., 1978, Synthetic inhibitors of serine proteinases. 17. The effect of benzamidine derivatives on the activity of urokinase and the reduction of fibrinolysis. *Pharmazie* **33:** 599-602.

Swiercz, R., Skrzypczak-Jankun, E., Merrell, M.M., Selman, S.H., and Jankun, J., 1999, Angiostatic activity of synthetic inhibitors of urokinase type plasminogen activator. *Oncol. Rep.* **6:** 523-526.

Tidwell, R.R., Geratz, J.D., and Dubovi, E.J., 1983, Aromatic amidines: comparison of their ability to block respiratory syncytial virus induced cell fusion and to inhibit plasmin, urokinase, thrombin, and trypsin. *J. Med. Chem.* **26:** 294-298.

Towle, M J., Lee, A., Maduakor, E.C., Schwartz, C.E., Bridges, A.J., and Littlefield, B.A., 1993, Inhibition of urokinase by 4-substituted benzo[b]thiophene-2- carboxamidines: an important new class of selective synthetic urokinase inhibitor. *Cancer Res.* **53:** 2553-2559.

Vassalli, J.D. and Belin, D., 1987, Amiloride selectively inhibits the urokinase-type plasminogen activator. *FEBS Lett.* **214:** 187-191.

Xing, R.H., Mazar, A., Henkin, J., and Rabbani, S.A., 1997, Prevention of breast cancer growth, invasion, and metastasis by antiestrogen tamoxifen alone or in combination with urokinase inhibitor B-428. *Cancer Res.* **57:** 3585-3593.

Yang, H., Henkin, J., Kim, K.H., and Greer, J., 1990, Selective inhibition of urokinase by substituted phenylguanidines: quantitative structure-activity relationship analyses. *J. Med. Chem.* **33:** 2956-2961.

# PROCESSING OF INTERLEUKIN-18 BY HUMAN VASCULAR SMOOTH MUSCLE CELLS

Elena Westphal, Ivar Friedrich*, Rolf-Edgar Silber*, Karl Werdan, and Harald Loppnow[1]

*Martin-Luther-Universität Halle-Wittenberg, Klinik und Poliklinik für Innere Medizin III, Forschungslabor, Magdeburger Str. 21, and *Klinik für Herz- und Thoraxchirurgie, Ernst-Grube-Straße 40, D-06097 Halle (Saale), GERMANY*

*[1]Corresponding author: PD Dr. H. Loppnow, Martin-Luther-Universität Halle-Wittenberg, Klinik und Poliklinik für Innere Medizin III, Forschungslabor, Magdeburger Str. 21,D-06097 Halle (Saale), GERMANY (Email: Harald.Loppnow@medizin.uni-halle.de)*

## 1. INTRODUCTION

Cells of the cardiovascular system are a potent source of various cytokines, including interleukin-1 (IL-1), IL-6, or chemokines (Windt and Rosenwasser 1984, Wagner *et al* 1985, Libby *et al* 1986, Gimbrone *et al* 1989, Loppnow and Libby 1989, Loppnow and Libby 1990, Wang *et al* 1991, Loppnow and Libby 1992, Schönbeck *et al* 1995, Loppnow *et al* 1998a). During normal development and physiological processes, as well as during pathological situations in the vessel wall or the heart, these mediators may contribute to regulation of inflammation, cell proliferation, hypertrophy, hyperplasia, contractility or cell death (Loppnow *et al* 1998b). The cytokines of the IL-1 family, in particular IL-1 and IL-18, are very potent activators of cells (Dinarello 1996). They stimulate production of other mediators or proliferation, and may interfere with regulation of cell death or contractility of cells. Both cytokines are produced as precursor proteins (Auron *et al* 1984, Lomedico *et al* 1984, March *et al* 1985, Gu *et al* 1995, Ghayur *et al* 1997). Only the mature proteins are active. The enzyme necessary for activation of both, IL-1 and IL-18, is the IL-1-converting enzyme (ICE) (Cerretti *et al* 1992, Thornberry *et al* 1992), also termed caspase-1. We have shown before that human vascular smooth muscle cells express caspase-1 but do not process IL-1ß precursor (Schönbeck *et al* 1997). Since IL-18 is another physiological substrate for caspase-1 we prepared recombinant IL-18 precursor (preIL-18) and analyzed the production and cleavage of this mediator in human vascular smooth muscle cells. Our results show that polymorphonuclear cells and monocytes, used for control, but not vascular smooth muscle cells, can process IL-18 precursor. These data indicate that inhibition of caspase-1 in smooth muscle cells is not restricted to cleavage of IL-1ß precursor.

# 2. MATERIAL AND METHODS

## 2.1 Isolation of human vessel wall cells and polymorphonuclear cells

We prepared human vascular smooth muscle cells from unused portions of the *vena saphena* by the methods described previously (Loppnow *et al* 1998a). Human polymorphonuclear cells were isolated from citrated blood by standard density centrifugation on ficoll.

## 2.2 Preparation of human recombinant IL-18 precursor

The IL-18 precursor cDNA was prepared by RT-PCR techniques using the following primers: sense, atatg gat cca tgg ctg ctg aac cag tag; antisense, ata tct gca gct agt ctt cgt ttt gaa cag. The PCR product was cloned into the vector pQE30 at the BamH1 and the Pst1 site. Finally, the plasmid was transformed into *E. coli* M15 and protein expression was initiated by IPTG stimulation. The recombinant protein was isolated by affinity chromatography on Ni-NTA according to the manufacturers instructions.

## 2.3 Processing assay

The cleavage of the IL-18 precursor was performed in the cleavage buffer described previously (Howard *et al* 1991). After the incubation samples were applied to SDS-PAGE, blotted onto nitrocellulose, and stained by Western blotting with the anti-IL-18 antibody.

# 3. RESULTS

Human vascular smooth muscle cells (SMC) express caspase-1. However, processing of preIL-18 by SMC has not been reported thus far. In order to investigate this question we prepared recombinant IL-18 precursor. The purified material was applied to SDS-PAGE and investigated in Western blot and protein staining. Fig 1 shows that the purified protein run at the expected size and that the protein was immunoreactive with the specific antibody directed against IL-18. This result and the correct base sequence of the used plasmid indicated that the produced IL-18 precursor was the expected protein and could be used in the processing assay.

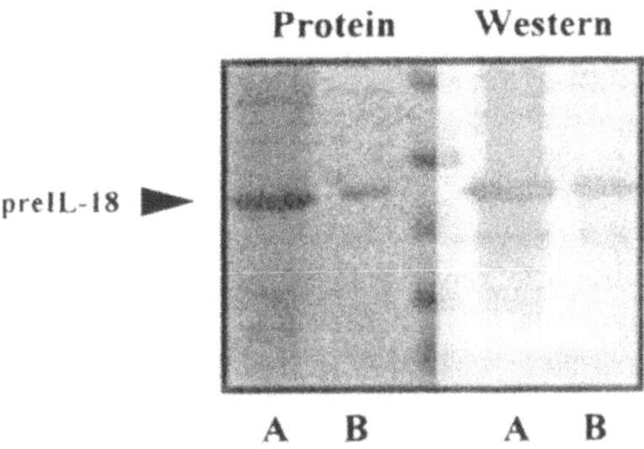


*Figure 1.* Characterization of the recombinant IL-18 precursor (preIL-18). PreIL-18 was prepared using the pQE30 system and purified by Ni-NTA affinity chromatography. Samples were run on SDS-PAGE and stained with coomassie (Protein) or analyzed in Western blot (Western). A, bacterial lysate; B, Ni-NTA-purified preIL-18.

The IL-18 precursor described above was used in processing assays with vascular smooth muscle cells, and for control purposes with human polymorphonuclear cells and monocytes. Incubation of the IL-18 precursor with recombinant caspase-1 resulted in appearance of the expected cleavage products (data not shown). The same bands were observed after cleavage of preIL-18 by human polymorphonuclear cells or monocytes (Fig 2). However, incubation of human vascular smooth muscle cells with the IL-18 precursor in the same system showed no processing of the precursor, in line with our previous results obtained with IL-1ß precursor.

These results show that although SMC express caspase-1, as described previously, they lack the capacity to process the IL-18 precursor, possibly due to the same inhibitory activity as described.

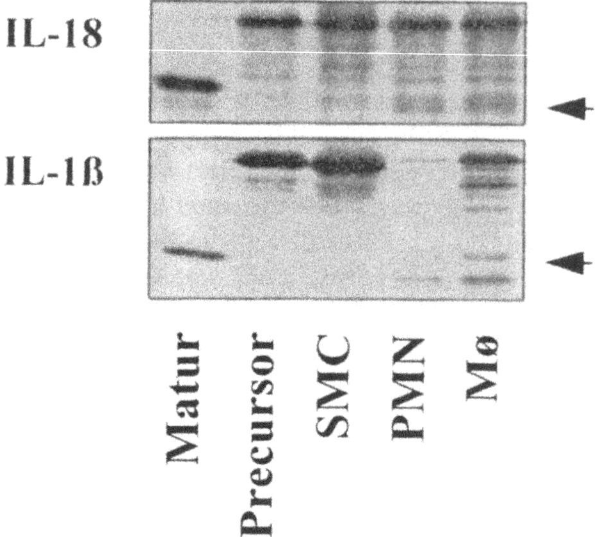


*Figure 2.* Cleavage of IL-18 precursor by human vascular smooth muscle cells. Recombinant IL-18 or IL-1β precursor were incubated with lysates of human vascular smooth muscle cells (SMC), and for control with polymorphonuclear cells (PMN) or monocytes (Mø), in processing buffer. The samples were run on SDS-PAGE and Western-Blot was performed. The relevant cleavage products are indicated by the arrows.

## 4. DISCUSSION

In this report we show that polymorphonuclear cells and monocytes, but not smooth muscle cells, can process IL-18 precursor. Interleukin-18 (Okamura *et al* 1995), also termed interferon-γ-(IFN-γ)-inducing factor (IGIF), was recently identified as a co-activator of IFN-γ production, for example in addition to IL-12. Besides IGIF-function a variety of other roles for IL-18 have been implicated, and IL-18 is thought to be a pleiotropic cytokine (Dinarello *et al* 1998). The IL-18 has amino acid sequence and protein folding similarity to the IL-1ß isoform (Bazan *et al* 1996). Like IL-1ß it does not contain a leader sequence and has to be cleaved behind aspartate by caspase-1 in order to obtain biological activity (Ghayur *et al* 1997, Gu *et al* 1997). It has been suggested that in cells also caspase-3 may process IL-18, however, the resulting cleavage products are inactive (Akita *et al* 1997). So far it has not been reported that cardiovascular or polymorphonuclear cells may process IL-18. Polymorphonuclear cells invading the vessel wall or other tissues may activate existing IL-18 precursor or even produce IL-18 themselves. Limited information exist regarding IL-1ß production or gene expression in PMN (Marucha *et al* 1990, Lord *et al* 1991). However, we did not detect biological IL-1 activity in supernatants of PMN in D10-assay or fibroblast assay (unpublished observation). The finding that PMN can process IL-18 precursor indicates that PMN may express caspase-1, which is in line with our previous observation, that these cells can process IL-1ß. Taken together, our data show that polymorphonuclear cells and monocytes, but not smooth muscle cells, can process IL-18 precursor, possibly by a caspase-1 dependent mechanism. These data indicate that vascular smooth muscle cells cannot process caspase-1 substrates by themselves. However, PMN have the capacity to process caspase-1 substrates, and thus, can interfere with regulation of inflammation, proliferation, contractility or even cell death in the vessel wall or other tissues.

## ACKNOWLEDGMENTS

The perfect technical assistance of Mrs. Claudia Pilowsky and Mrs. Susanne Koch is gratefully acknowledged. This work was supported by grants of the BMBF Forschungsverbund Halle (Project 06) and the Deutsche Forschungsgemeinschaft (Lo385/4-1) to H. Loppnow.

## REFERENCES

Akita, K., Ohtsuki, T., Nukada, Y., *et al.*, 1997, Involvement of caspase-1 and caspase-3 in the production and processing of mature human interleukin 18 in monocytic THP.1 cells. *J. Biol. Chem.***42:** 26595–26603.

Auron, P.E., Webb, A.C., Rosenwasser, L.J., Mucci, S.F., Rich, A., Wolff, S.M., and Dinarello, C.A., 1984, Nucleotide sequence of human monocyte IL-1 precursor cDNA. *Proc. Natl. Acad. Sci. USA* **81:** 7907-7911.

Bazan, J.F., Timans, J.C., Kastelein, R.A., 1996, A newly defined interleukin-1? *Nature* **379:** 591.

Cerretti, D.P., Kozlosky, C.J., Mosley, B., *et al.*, 1992, Molecular cloning of the interleukin-1ß-converting enzyme. *Science* **256:** 97–100.

Dinarello, C.A., 1996, Biologic basis for interleukin-1 in disease. *Blood* **87:** 2095–2147.

Dinarello, C.A., Novick, D., Puren, A.J., *et al.*, 1998, Overview of interleukin-18: more than an interferon-gamma inducing factor. *J. Leukoc. Biol.* **63:** 658–664.

Ghayur, T., Banerjee, S., Hugunin, M., *et al.*, 1997, Caspase-1 processes IFN-γ-inducing factor and regulates LPS-induced IFN-γ production. *Nature* **386:** 619–623.

Gimbrone, M.A., Jr., Obin, M.S., Brock, A.F., *et al.*, 1989, Endothelial interleukin-8: a novel inhibitor of leukocyte-endothelial interactions. *Science* **246:** 1601-1603.

Gu, Y., Kuida, K., Tsutsui, H., *et al.*, 1997, Activation of interferon-γ inducing factor mediated by interleukin-1ß-converting enzyme. *Science* **275:** 206–209.

Gu, Y., Wu, J., Faucheu, C., Lalanne, J.-L., Diu, A., Livingston, D.J., Su., and M.S.-S., 1995, IL-1ß-converting enzyme requires oligomerization for activity of processed forms in vivo. *EMBO J.* **14:** 1923–1931.

Howard, A.D., Kostura, M.J., Thornberry, N., *et al.*, 1991, IL-1-converting enzyme requires aspartic acid residues for processing of the IL-1β precursor at two distinct sites and does not cleave 31 KDa IL-1α. *J. Immunol.* **147:** 2964 – 2969.

Libby, P., Ordovas, J.M., Birinyi, L.K., Auger, K.S., and Dinarello, C.A., 1986, Inducible interleukin-1 gene expression in human vascular smooth muscle cells. *J. Clin. Invest.* **78:** 1432–1438.

Lomedico, P.T., Gubler, U., Hellmann, C.P., *et al.*, 1984, Cloning and expression of murine interleukin-1 cDNA in *Escherichia coli. Nature* **312:** 458–462.

Loppnow, H., and Libby, P., 1989, Adult human vascular endothelial cells express the IL-6 gene differentially in response to LPS or IL-1. *Cell. Immunol.* **122:** 493-503.

Loppnow, H., and Libby, P., 1990, Proliferating or interleukin-1-activated human vascular smooth muscle cells secrete copious interleukin-6. *J. Clin. Invest.* **85:** 731-738.

Loppnow, H., and Libby, P., 1992, Functional significance of human vascular smooth muscle cell-derived interleukin-1 in paracrine and autocrine regulation pathways. *Exp. Cell Res.* **198:** 283–290.

Loppnow, H., Bil, R., Hirt, S., Schönbeck, U., Herzberg, M., Werdan, K., Rietschel, E.T., Brandt, E., and Flad, H.-D., 1998a, Platelet-derived interleukin-1 induces cytokine production, but not proliferation of human vascular smooth muscle cells. *Blood* **91:** 134–141.

Loppnow, H., Werdan, K., Reuter, G., and Flad, H.-D., 1998b, The interleukin-1 and interleukin-1-converting enzyme families in the cardiovascular system. *Eur. Cyt. Net.* **9:** 670-675

Lord, P.C., Wilmoth, L.M., Mizel, S.B., and McCall, C.E., 1991, Expression of interleukin-1 alpha and beta genes by human blood polymorphonuclear leukocytes. *J. Clin. Invest.* **87:** 1312–1321.

March, C.J., Mosley, B., Larsen, A., *et al.*, 1985, Cloning, sequence and expression of two distinct human interleukin-1 complementary DNAs. *Nature* **315:** 641–647.

Marucha, P.T., Zeff, R.A., and Kreutzer, D.L., 1990, Cytokine regulation of IL-1ß gene expression in the human polymorphonuclear leukocyte. *J. Immunol.* **145:** 2932–2937.

Okamura-H, Tsutsi-H, Komatsu-T *et al.*, 1995, Cloning of a new cytokine that induces IFN-γ production by T cells. *Nature* **378:** 88–91.

Schönbeck, U., Brandt, E., Petersen, F., Flad, H.-D., and Loppnow, H., 1995, IL-8 specifically binds to endothelial but not to smooth muscle cells. *J. Immunol.* **154:** 2375–2383.

Schönbeck, U., Herzberg, M., Petersen, A., Wohlenberg, C., Gerdes, J., Flad, H.-D., and Loppnow, H., 1997, Human vascular smooth muscle cells express interleukin-1ß-converting enzyme (ICE), but inhibit processing of the interleukin-1ß precursor by ICE. *J. Exp. Med.* **185:** 1287-1294.

Thornberry, N.A., Bull, H.G., Calaycay, J.R., *et al.*, 1992, A novel heterodimeric cysteine protease is required for interleukin-1ß processing in monocytes. *Nature* **356:** 768-774.

Wagner, C.R., Vetto, R.M., and Burger, D.R., 1985, Expression of I-region-associated antigen (Ia) and interleukin-1 by subcultured human endothelial cells. *Cell. Immunol.* **93:** 91–104.

Wang, J.M., Sica, A., Peri, G., Walter, S., Padura, I.M., Libby, P., Ceska, M., Colotta, F., and Mantovani, A., 1991, Expression of monocyte chemotactic protein and IL-8 by cytokine-activated human vascular smooth muscle cells. *Arterioscler. Thromb.* **11:** 1166–1174.

Windt, M.R., and Rosenwasser, L.J., 1984, Human vascular endothelial cells produce interleukin-1. *Lymphokine Res.* **315:** 641–646.

# REVIEW:
# PEPTIDASES AND PEPTIDASE INHIBITORS IN THE PATHOGENESIS OF DISEASES

*Disturbances in the ubiquitin-mediated proteolytic system. Protease-Antiprotease imbalance in inflammatory reactions. Role of cathepsins in tumour progression*

Ute Bank[1], Sabine Krüger[2], Jürgen Langner[3] and Albert Roessner[2]
*[1]Institute of Immunology, [2]Institute of Pathology, Otto-von-Guericke-University, [3]Martin Luther University, Institute for Medical Immunology, Straße der OdF 6, D-06097 Halle*

## 1. INTRODUCTION

In this chapter we will review recent results in the field of involvement of proteolytic enzymes in aetiopathology of diseases. The multitude of physiological processes governed or influenced by proteolytic enzymes forms the basis for great variety of disorders and malfunctions, eventually developing into diseases which are more or less clearly the result of abnormal protease action. Table 1 contains a selection of diseases of this kind without the claim of being comprehensive, just to open the mind for the great number of disorders included.

In the table, the location of proteolysis defect in the upper part is confined primarily to the circulatory system, in the lower part it is the intercellular spaces or even the interior of cells. The proteinases involved (right column of Tab 1) only in some cases are specified as individuals. Again, the intention is rather to give an idea of the variety of proteinases or proteinase systems involved. The selected diseases in the table make clear that it is impossible to cover all of them in this chapter, rather we will concentrate on only some of them as indicated in the subheading.

*Table 1.* Diseases caused by disturbed or defective proteolysis

| Disease or Disease group | Proteinase or proteolytic system involved |
|---|---|
| Bleeding disorders | Blood coagulation proteases |
| Complement defects | Proteases of the complement activation cascade |
| Hypertension | Renin-ACE system |
| Pulmonary emphysema | Elastase/α-1 proteinase |
| Severe inflammation, septic shock | Proteinase/antiproteinase balance in blood |
| Wegeners disease | Proteinase III |
| Rheumatoid diseases | Matrix-Metallo Proteinases |
| Periodontal diseases | PMN leukocyte proteinases; bacterial proteinases |
| Muscular dystrophies | Ubiquitin proteasome system, lysosomal proteinases |
| Chronic degenerative CNS diseases | Ubiquitin proteasome system, lysosomal proteinases |
| Tumor development | Cathepsins; MMPs |

# 2. DISEASES INVOLVING THE UBIQUITIN-MEDIATED PROTEOLYTIC SYSTEM

In complexity of structure and sophistication of function the ubiquitin-mediated proteolytic system (UMPS) is more and more considered as a counterpart to the ribosome, the protein synthesis organelle, and similarly subject to finely tuned regulations just being recognized. Perhaps it is not to great a surprise, with better understanding of structure and function of this proteolytic system, also to learn of failures of some of its components that result in pathological states presenting with clinical pictures. This part of the chapter is dealing with some of them. For the reader not as familiar in this field the UMPS shall be introduced in the following paragraph. There are in literature many excellent reviews on the structure as well as on the different aspects of function of the UMPS (e.g. Ciechanover and Schwartz 1994, Coux *et al* 1996, Baumeister *et al* 1997, Peters *et al* 1998), therefore here we will only very briefly outline the groups of enzymes involved and the order of their action in UMPS.

## 2.1 Enzymology of the UMPS

Generally, cells have two main proteolytic pathways, the lysosomal and the proteasomal one. Degradation of exterior material, phagocytosed or endocytosed (also receptor-mediated) takes place in lysosomes, and under stress situations or starvation intracellular proteins also are degraded there. The UMPS, however, seems to be the system for the highly selective turnover of intracellular proteins under basal metabolic conditions. The degradative reactions of UMPS are performed by the 26S proteasome, whereas the substrate specificity (resulting in the selectivity of turnover) most probably is brought in by the ubiquitination reactions that are highly ATP-dependent. The small protein ubiquitin serves as a label for proteins to be degraded, and three groups of enzymes (plus ATP) work consecutively to form polyubiquitinated proteins that are then degraded in the 26S proteasome under release of reutilizable ubiquitin. The first step is the ATP-dependent activation of a ubiquitin molecule becoming attached to the so-called E1 enzyme under formation of a high-energy thioester bond to the C terminal Gly76 of ubiquitin.

The second step is the transfer of ubiquitin to a ubiquitin conjugating enzyme, E2. Finally, ubiquitin is transferred to a target protein destined to be degraded, mostly catalyzed by a third protein, E3. After several repetitions of the latter reaction, a polyubiquitinated target protein results that carries up to ten or more ubiquitin molecules bound via isopeptide bonds between Lys48 and Gly76 (or also Lys23, Lys63) of ubiquitin. Now the substrate is ready for degradation in the 26S proteasome. The 19S cap of 26S proteasomes seems to perform the de-ubiquitination and the unwinding of the protein structure. The polypeptide chain then is threaded into the central channel of the proteasome, where the proteolytic attack of the target protein follows resulting in peptides (for antigen presentation, cf. Goldberg and Rock 1992) and/or amino acids (Ehring *et al* 1996), under involvement of cytosolic exopeptidases. Protein turnover is a highly selective process, the half-lives of individual cellular proteins vary between a few minutes and several days. This selectivity results on the one hand from signals in the substrate proteins in form of special amino acid sequences (see below), on the other hand it is influenced by the enzymes of the ubiquitination pathway. As to the latter is to mention that except E1, that seems present in only one active form in eucaryotes, the E2s and E3s exist in related families (in yeast 13 E2s, in mammals 20 – 30 are estimated) or in unrelated groups (three to four E3s are known at present). Additionally, there are indications of oligomerization between different members of E2s resulting in a "combinatorial expansion of specificities" as well as involvement of additional "specificity factors", all resulting in the great diversification of specificity of the ubiquitination reactions. Signals in the proteins that control their degradation include the N-terminal amino acid (N-end rule), the presence of a PEST-motif or of a "destruction box" present in B-type cyclins (review in Ciechanover and Schwartz 1994). Also the phosphorylation of a specific Ser or Thr residue within such destruction domains seems part of the signal, that after all resides within an extended region of 50 - 60 amino acids.

The central importance of the UMPS in cell biology becomes obvious when reviewing the special substrates degraded by this pathway:

- The cyclins (Jentsch 1992) and cyclin-dependent kinase inhibitors (Sheaff and Roberts 1996), thus posing proteolysis at a very central place of cell cycle regulation (King *et al* 1996).

- Many oncoproteins have very short half-lives, and their degradation was found ATP-dependent. P53, c-myc, c-myb, c-fos, E1A are among the ones studied in more detail. These two groups of proteins degraded by the UMPS bring this proteolytic pathway in close connection to tumorigenesis, a topic that here will be treated only very briefly.

- Several transcription factors have been found to be degraded via the UMPS, including Mat $\alpha$2 and Gon4 in yeast and AP-1, c-jun, c-fos, MyoD (Hochstrasser and Kornitzer 1998). The NF-$\kappa$B activation pathway is regulated via degradation of the inhibitor I$\kappa$B in ubiquitin-dependent proteolysis (reviewed in Chen and Maniatis 1998).

- Membrane proteins (transporter enzymes in yeast, growth factor receptors and others in mammalian cells) are ubiquitinated after ligand binding and then directed into the endocytic pathway or the UMPS (Hicke 1997)

- The degradative pathway of cytosolic proteins resulting in T cell epitopes to be presented via MHC class I molecules to $CD8^+$ T lymphocytes uses the 26S proteasome with or without prior ubiquitination of the respective substrate proteins. This pathway is of utmost importance for the defense of viral infections and for the immune surveillance of neoplastic cells (Goldberg and Rock 1992).

- Experiments with yeast mutants in E2 family members revealed the involvement of ubiquitinylating enzymes in DNA repair (review in l.c. Schaffner *et al* 1998).

- Degradation of misfolded or otherwise abnormal proteins as produced under stress conditions is thought to be largely mediated by the UMPS (Jentsch 1992).

Considering these many groups of highly important substrates of the UMPS makes clear that any dysfunction of one of its components may result in far-reaching effects in cell physiology, even bringing about states with clinical significance. It seems, however, that on the one hand the UMPS is equipped with a high functional redundancy that renders defects remaining hidden, or, on the other hand, that defects when inborn are lethal. This may found the reason for our still very fragmentary and incomplete knowledge of diseases resulting from disturbances in the UMPS.

## 2.2 Diseases caused by dysfunction of the UMPS

Among the few human disorders known to result from a defect in the UMPS, Angelman Syndrome (AS) is the one most clearly characterized. This rare inborn neurological disease is characterized by mental retardation, seizures, ataxia, abnormal gait and inappropriate laughter. Genetic studies revealed abnormalities of chromosome 15 (region 15 q 11-13) in AS patients (Williams *et al* 1995), and in 1997 two groups found frame shift mutations in a gene (UME3A) located in this region coding for an E3 enzyme of the UMPS in a special subgroup of AS patients (Kishino *et al* 1997, Matsuura *et al* 1997). The E3 enzyme truncated due to these mutations had been named earlier E6-associated protein (E6-AP), E6 being a protein of certain human papilloma viruses forming a complex with E6-AP and this complex then has an E3 ubiquitin-protein ligase activity that inactivates p53, the notable tumor suppressor gene product. How the loss of this special ubiquitin ligase leads to AS symptoms is still not at all understood, but there is a Drosophila gene called *bendless* that may be helpful in unraveling this problem. The gene product of bendless is an E2-type ubiquitin-conjugating enzyme with a role in synaptic connectivity among a subset of neurons and seems involved in the development of the visual system, mutant flies present with abnormalities in axon guidance and neuronal connectivity (Oh *et al* 1994).

In mice a strain designated $a^{18H}$ with a mutation in another ubiquitin-protein ligase gene *(Itch)* has been described, coding for a novel member of E3 enzymes (Perry *et al* 1998). The mutated mice develop a spectrum of immunological diseases varying (in dependence on the genetic background) from inflammatory disease of large intestine to pulmonary chronic interstitial inflammation with alveolar proteinosis, inflammation of the glandular stomach and skin scars due to constant itching, combined with hyperplasia of lymphocytes, hematopoietic cells and the forestomach epithelium. As to the biochemical pathways involved in development of this multitude of symptoms the discussion invokes cytokine pathways involving GM-CSF, Kit and interferons $\alpha/\beta$. It is easy to forecast finding patients with similar symptoms due to mutations in the human gene homologue, the more so since ligand dependent polyubiquitination of receptor tyrosine kinases of several growth factors e.g. PDGF $\alpha$ and $\beta$, EGF, CSF-1, FGF with subsequent internalization forms a pathway of growth factor activity regulation (Mori *et al* 1995, Miyazawa *et al* 1994) and degradation.

Other plasma membrane molecules degraded in ubiquitinated form include the T cell antigen receptor $\zeta$ subunit, the c-kit receptor and also membrane channel molecules (Hicke 1997). The epithelial $Na^+$ channel (ENaC) plays a major role in sodium transport in kidney and other epithelia (colon) as well as in regulating blood pressure. It is composed of three homologous subunits ($\alpha\beta\gamma$) that all consist of two transmembrane domains, a large extracellular domain and two short cytoplasmic tails. Each of the three subunits possesses two conserved proline-rich sequences (P1 and P2) in their C-terminal intracellular tails forming an interaction region for a ubiquitin-protein ligase (E3 enzyme) related to the E6-AP mentioned above and regulating the normal turnover of ENaC via proteasomal degradation. This E3 enzyme has been named Nedd 4 and contains (in human) four so called WW domains, regions which, like SH3 domains, bind to proline-rich sequences and thus bring the active site (the so-called hect domain = homologous to the E6-associated protein carboxy terminus) in close proximity to ENaC, facilitating its ubiquitination. Genetic linkage analysis in patients with Liddle`s syndrome, a hereditary form of systemic hypertension characterized by hyperactivity of ENaC has revealed the locus encoding β or γ subunits of the sodium channel linked to the disease. Mutations introducing premature stop codons or frameshift mutations in these patients lead to deletion of the P2 domains in β or γ subunits of ENAC resulting in inhibited degradation due to abrogation of binding of Nedd and hindered ubiquitination. Accordingly, in this disease the defect is not in one of the components of the UMPS but in a substrate of special relevance to be degraded by this system.

A similar situation is found in cystic fibrosis, a disease caused by a mutation in the gene encoding an epithelial chloride-channel, called cystic fibrosis transmembrane conductance regulator (CFTR). The mutation in most of the cases is deletion of a phenylalanine (ΔF508) from a cytoplasmically localized position of CFTR. ΔF508 CFTR molecules fail to fold correctly, resulting in retention and degradation in a pre-Golgi compartment that has been defined later as the UMPS (Ward et al 1995). The mutated CFTR molecules, however, when forced under special cell culture conditions to be expressed at the cell membrane will form functional cAMP-activated chloride channels, the mutation therefore does not necessarily interfere with the function of CFTR but rather serves to target it exclusively into the UMPS degradative pathway.

The detailed study of this and other proteins originating in the ER and being degraded by proteasomes in cytosol as evidenced by using inhibitors of the proteasome resulting in reduced degradation of e.g. MHC class I heavy chains or T cell receptor α chains, resulted in the conception that the translocation machinery of the ER can be used for export and for import of proteins bidirectionally (Sommer and Wolf 1997, Brodsky and McCracken 1997).

One other aspect should still be mentioned in connection with ubiquitin-dependent internalization of plasma membrane proteins: in yeast and mammalian cells, a number of such proteins will only be mono- or di-ubiquitinated and their degradation then takes place not via proteasome but in the endocytic compartment. An additional signal is phosphorylation of special serine residues (Hicke 1997). All the about 10 mammalian signaling receptors that have been shown to be ubiquitinated at the cell surface span the membrane a single time and are themselves tyrosine kinases or are linked to tyrosine kinases. It remains to be shown if this is a discriminative characteristic leading to their degradation in the endocytic compartment.

Muscle protein turnover is a physiological process where protein substrate amounts of many grams every day are degraded and resynthesized. Investigations in several rat models but also of human muscle biopsies led to the conclusion of enhanced UMPS activity as reason of increased protein degradation in a number of diseases like renal tubular defects, acute and chronic uremia, neuromuscular diseases and muscle immobilization, burns, diabetes, sepsis, AIDS and cancerous cachexia, starvation and eating disorders (Mitch and Goldberg 1996, Hasselgren and Fischer 1997). The biochemical adaptations in the UMPS observed under these conditions include increase of levels of mRNA encoding proteasome subunits and ubiquitin as well as of the rate of ubiquitin conjugation due to increased activity of E2 and E3 enzymes. Additionally, from experimental and clinical studies in this field much could be learned about regulatory factors of the UMPS, including glucocorticoids, insulin, thyroid hormone and cytokines (TNF, IL-1, IL-6, IFN). There exist rather specific inhibitors for the proteasome (Mitch and Goldberg 1996, Rock *et al* 1994) whose clinical application in consideration of other essential cellular processes (e.g. cell division, see below) inhibited in parallel still does not seem to have been tested clinically.

By far not as clearly understood is the role of ubiquitinated proteins found in intraneuronal or glia cell inclusions in neurodegenerative diseases of different origin, including Pick's, Alzheimer's, Creutzfeldt-Jabob and Parkinson disease, Amyotrophic lateralsclerosis, progressive supranuclear palsy and others (Alves-Rodrigues *et al* 1998). Among the proteins colocalizing with ubiquitin immunoreactivity are the intermediate filament network, vimentin, heat shock proteins, neurofilaments, the tau protein, ataxin 3, huntingtin, and prion protein. The factors supposed leading to occurrence of ubiquitinated inclusions in no case are really clear. Reactive oxygen species and other stress factors but also neurotoxic compounds may inhibit the normal elimination of ubiquitin-protein deposits or may lead to structural changes in the substrate proteins rendering them inaccessible to UMPS degradation. The latter may be true in Huntington's disease, where the proteins huntingtin and ataxin-3 have long polyglutamine extensions promoting the formation of nuclear aggregations causing formation of insoluble fibrils. Similarly in some cases of Parkinson's disease a mutation in the protein α-synuclein (Ala53Thr) might be the cause of aggregate formation instead of degradation via UMPS. Such observations show aggregations of aberrant ubiquitinated proteins as a characteristic of many neuropathologies, and as a common age-related phenomenon of the brain, but it remains yet unresolved if they are cause or consequence of the diseases in which they are observed.

The presence of 26S proteasomes in nuclei has been demonstrated already many years ago, and some nuclear proteins as possible substrates have mentioned just in the last paragraph. The UMPS, however, is involved in cell cycle control much more directly than these observations suggest (King *et al* 1996). Namely, there are two steps in the cell cycle directly influenced by ubiquitin-dependent proteolysis. The first one is the progression of a cell from G1 to S that utilizes an E2 called cyclin-dependent kinase(CDC)-34. This ubiquitin conjugating enzyme participates in the destruction of several proteins, including the G1 cyclins CLN2 and CLN3, the crucial substrate, however, is an inhibitor of a cyclin-dependent kinase called SIC1. This protein has to be degraded for cell cycle to progress form G1 to S as studies in yeast have revealed, substrate-specific phosphorylation of SIC1 by CDC28 being the triggering reaction for ubiquitination of SIC1 by CDC34, followed by destruction of the inhibitor. The active S-phase CDK then initiates DNA replication. The second reaction in cell cycle involving UMPS is mitosis, where the metaphase-anaphase transition is regulated by a large E3 complex, known as the cyclosome or the anaphase-promoting complex (APC). APC becomes activated during G2 by maturation promoting factor (MPF), a cyclin-dependent kinase composed of cyclin B and CDC2. APC then degrades two non-cyclin proteins (CUT2 and PDS1) as well as cyclin B and the cell cycle can promote through anaphase to telophase.

It is easy to imagine that deregulations of the cell cycle may result in deleterious proliferative states like neoplasia. Regulation of cell cycle progression according to these results very intimately involves the UMPS. Thus it is not surprising to find the UMPS also involved in deviations of this regulation, namely in tumorigenesis. There are two observations bringing special cases of this process in connection with this degradative system. The first one is the tumor suppressor p53, being regulated by the UMPS and its involvement in viral tumorigenesis. Some cancers of the cervix uteri are caused by human papilloma viruses (HPV16 and 18), obviously due to p53 ubiquitination by means of the above-mentioned E6/E6-AP protein complex and followed by p53 degradation (Huibregtse *et al* 1998). As a consequence, the "safeguard of the genome", p53, is lacking and this paves the way to tumor development. The other example is provided by the observation in colorectal carcinomas of an inverse correlation of the levels of the cell cycle inhibitor p27 and the median survival time of patients (Loda *et al* 1997). It could be shown that low p27 expression in tumor material was due to high degradative activity of UMPS for p27. This supports the suggestion that aggressive colorectal carcinomas may originate from tumor cell clones lacking p27 due to increased proteasome-mediated degradation and very directly points to the pathogenetic potential of the UMPS. In the years coming, very probably we will see still more examples of this kind of disturbances under involvement of this special degradative system.

# 3. PROTEASE-ANTIPROTEASE-IMBALANCE IN INFLAMMATORY DISEASES

The causal involvement of proteases in the pathopysiology of diseases is often related to a disturbed balance between the proteolytic activities and the specific inhibitory potential. However, the latter is of relevance in most of the cases given in table 1. Impressive examples for protease-antiprotease-imbalances with wide spreading consequences are found in inflammatory processes. At foci of inflammation, endopeptidases, under physiological conditions mostly active lysosomales, are released in large amounts from infiltrating leukocytes as well as several specific tissue cells belonging to the reticoendothelial system. Especially two classes of those proteases - namely PMN-derived proteases and matrix metalloproteinases - are thought to play a crucial role at sites of inflammation by degrading matrix proteins and processing regulatory relevant bioactive polypeptides.

## 3.1 Imbalance between PMN-derived neutral serine proteases and their inhibitors in locally restricted inflammatory processes

It is a well-known fact that acute inflammatory processes are accompanied by a massive infiltration of activated polymorphonuclear granulocytes (PMN) into the damaged or infected tissues. These cells are characterized by a broad proteolytical equipment - cell surface bound proteases as well as lysosomal serine, cysteine and metalloproteinases. Among these enzymes the neutral serine proteases elastase, cathepsin G and proteinase 3 seem of particular relevance, because they are active at physiological pH values and are, in contrast to the other above mentioned protease groups, stored in granules and released therefrom as mature active enzymes in very high amounts during activation or disruption of neutrophils (for review see: Bieth 1999, Salvesen 1999, Hoidal 1999). However, the major physiological function of these enzymes was believed to be the intralysosomal degradation of phagocytozed microorganisms and tissue debries. Beyond this, an involvement in the migration of neutrophils through the endothelial cell layer, the basement membrane or through the tissue is controversially discussed (Steadman 1997).

Several highly potent endogenous inhibitors protect cells and tissues from the deleterious effects of excess extracellular PMN-protease activity:

The acute phase protein $\alpha_1$-protease inhibitor ($\alpha_1$-PI; synonymous $\alpha_1$-antitrypsin) is a highly potent irreversible plasma inhibitor of these enzymes. Another prominent member of the familiy of serine proteases inhibitors - shortly designated as serpins - $\alpha_1$-antichymotrypsin ($\alpha_1$-ACT) is one of the physiological inhibitors of cathepsin G. A further excellent inhibitor of elastase and cathepsin G, but not of proteinase 3, the secretory leukocyte proteinase inhibitor (SLPI) was originally described as inhibitor in the respiratory epithelium, but it is also produced by other cells types and is present in the plasma and other body fluids (Jin 1997, Ohlsson 1997, Denison 1999, Wiedow 1998 ). Elafin, primarily detected in the skin and the lung, is a low molecular weight elastase inhibitor with a unique structure, not related to the serpin family (Tsunemi 1992; for review: Molhuizen 1995). A smaller fraction of elastase was shown to be bound to the 750kDa $\alpha_2$-macroglobulin ($\alpha_2$-M). The inhibitory potency of $\alpha_2$-M remains controversial, not least since recent data indicated that $\alpha_2$-M complexed elastase retained its cartilage destroying activity in rheumatoid arthritis (Moore 1999). Beyond these inhibitors, several cytoplasmatic protease inhibitors, e. g. from disrupted or dying neutrophils, are believed to gain also importance in later phases of the inflammatory process (Scott 1999).

Although this strong inhibitory potential exists, there is no doubt that PMN-proteases from cells released can escape from inhibition at sites of inflammation. However, numerous animal and in vitro studies provide several lines of evidence for a crucial role of these proteases in the proteolysis of host proteins at sites of inflammation. Despite of the wealth of available information about the biochemistry of the PMN-serine proteases and their natural inhibitors, the mechanisms how these proteases can act in a surrounding full of highly potent inhibitors, have been elucidated just in the last few years. Additionally, the detection of free active elastase and cathepsin G at sites of inflammation as well as the detection of specific matrix protein digestion products indicated clearly that the effective protease inhibitor concentration may be insufficient under certain circumstances (Hawkins 1968, Jochum 1999). The following paragraphs summarize shortly the recent knowlegde concerning these mechanisms.

### 3.1.1 Proteinase release during "frustrated phagocytosis"

Excessive amounts of granule proteins, inclusive proteolytic enzymes, are released during the so called „frustrated phagocytosis" at surfaces inside the body, such as the glomerular basement membrane or the joint cartilage. It has been suggested that between the attached PMNs and the surface a protease inhibitor-free lumen is formed, allowing an unrestricted action of released PMN-proteases by spatial segregation from the surrounding inhibitory potential. Moreover, histological investigations in rheumatoid arthritis revealed that with progressive erosion clefts are formed in the cartilage, where extracellular proteases are not accessible for high molecular weight inhibitors (Janoff 1976).

### 3.1.2 Membrane rebinding of released proteinases

Elastase, proteinase 3 and cathepsin G released into the surrounding medium have been shown to re-bind to variable extent to the cell surface of neutrophils possibly because of their cationic property (Owen 1995, Bangalore 1994). Recent reports suggest that the membrane binding of elastase is differentially regulated by cytokines (Cepinskas 1999). Once attached to the plasma membrane or membrane pieces, these enzymes have been shown to be remarkably resistant to inhibition by the naturally occurring protease inhibitors (Owen 1995). Thus, the membrane binding of the enzymes focuses and preserves their catalytical activity in the immediate pericellular microenvironment. The same holds true for the inhibition of elastase bound to joint cartilage, which was also found to be poorly inhibited by high molecular weight serine protease inhibitors (Kawabata 1996).

### 3.1.3 Local effects of proteinase inhibitors

Whereas most of the serpins are abundantly present in the circulation, in inflamed extravascular sites the concentrations of functional PMN protease inhibitors ($\alpha_1$-PI, $\alpha_2$-M and $\alpha_1$-ACT) have been described to be much less abundant. Despite of the upregulated production of these inhibitor proteins by hepatocytes during an acute phase reaction, it is likely that the local concentrations at sites of inflammation do not reach values high enough for the complete inhibition of serine protease activities. However, whether a local $\alpha_1$-PI or SLPI production, recently shown to be inducible in lung epithelial cells or neuronal cells, is really sufficient to augment the effective antiprotease defence, is unclear as yet (Boutten 1998, Carlson 1988). Moreover, in *Heliobacter pylori* infection, the local SLPI induction was found to be impaired (see Nilius *et al* this book ch. 45).

### 3.1.4 Inactivation of proteinase inhibitors

In addition, the local antiprotease deficiency is strenghtened, because naturally occuring inhibitors were found to be inactivated by oxidative modification or proteolytic degradation at sites of inflammation. Travis and Salvesen discovered in the early 80ies that the oxidation of the methionine$^{358}$ residue of the $\alpha_1$-PI molecule led to a dramatic slowing of the association with elastase (Travis and Salvesen 1983). Moreover, the affinity of the protease-inhibitor interaction was found to be considerably reduced. Similar to $\alpha_1$-PI, the other above mentioned serine protease inhibitors were found to be susceptible to oxidative inactivation (Wu 1999, Vogelmeier 1997, Abbink 1991). Thus, oxygen free radicals amplify the PMN-protease catalyzed proteolysis in a synergistic manner. Moreover, the oxidative modification of serpins has been described to lead to an increased susceptibility to proteolytic cleavage. However, many of the endogenous serine protease inhibitors have been shown to be proteolytically inactivated: SLPI is cleaved by proteinase 3 (Rao 1993) but was originally described to be more resistant against inactivation than $\alpha_1$-PI (Axelsson 1988). $\alpha_1$-ACT is inactivated by elastase, $\alpha_1$-PI is inactivated by stromelysin or matrilysin, $\alpha_2$-M by elastase (Winyard 1991, Sires 1994, Baumstark 1970) to mention only some examples. Particularly in bacterial infections, also bacterial proteases were suggested to be involved in the rapid inactivation of endogenous protease inhibitors (Draper 1998, Nelson 1999). Taken together, these mechanisms may lead to a dramatic reduction of the local inhibitory potential just at the time,when the PMN protease release reaches maximal levels.

### 3.1.5 Genetic variants of proteinase inhibitors

PMN-protease mediated tissue injury was found to be more pronounced in patients with dysfunctional genetic variants of the acute phase protein $\alpha_1$-PI. Whereas the vast majority of humans have the normal M form, some carry the so-called Z gene variant (heterozygote PiSZ with moderate deficiency, homozygote PiZZ with severe deficiency). The mutantion affects the conformation of the reactive center loop. The substitution of the positively charged lysine$^{342}$ residue for a negatively charged glutamic acid in this loop led to an aggregation of $\alpha_1$-PI in the endoplasmatic reticulum of the hepatocytes and subsequently to a reduced secretion into the plasma (for review see: Lomas 1993, Norman 1997). Comparable less active genetic variants were described for other serpins such as C1-inhibitor and antithrombin. The genetically determined $\alpha_1$-PI-defiency was found to result in a more rapid progression and aggravation of inflammatory diseases associated with PMN-mediated tissue destruction, such as vasculitis or inflammtory lung disorders (Esnault 1997, Norman 1997).

## 3.2 Actions of inefficiently inactivated PMN proteases

As mentioned at the beginning of this part of the review, a major physiological function of PMN-serine proteases is commonly thought to be the intralysosomal degradation of engulfed cell debries or microoroganisms.

Considering the acidic pH value within the phagolysosomal compartment, the question was raised, to what extent such neutral proteases indeed may act under these conditions. Consequently, it was hypothesized that the extracellularly released neutral serine proteases may be of particular regulatory relevance at local sites of inflammation (Henson 1992). Until now, special emphasis was given to the deleterious potential of PMN serine proteases at foci of inflammation attributed to the matrix degrading activity of these proteases. A broad spectrum of inflammatory disorders was identified to be accompanied by tissue destruction and remodelling caused by excess of extracellular PMN-protease activity. Most prominent examples are the adult respiratory distress syndrome (ARDS), pulmonary emphysema, cystic fibrosis, rheumatoid arthritis, autoimmune vasculitis and periodontal disease (for review see Heinzelmann, 1999; Pillinger 1995, Edwards 1997, Birrer 1993).

Besides the degradation of several extracellular matrix proteins, these enzymes have been reported to cause the detachment and lysis of intact cells (Ballieux 1994, Venaille 1998). Moreover, the induction of apoptosis by these proteases has been described (Yang 1996, Taekema-Roelvink 1998, Bird 1999). On the other hand, however, there is increasing evidence that the extracellular release of PMN-proteases has not only detrimental, destructive aspects. With respect to the matrix protein-degrading activity of PMN-serine proteases it has to be considered that the elimination of damaged tissue structures is one of the intrinsic preconditions for wound healing processes. Moreover, besides the degradation of matrix proteins numerous bioactive polypeptides have been identified as substrates for PMN-proteinases, supporting the hypothesis of an active regulatory role in inflammatory processes (Jochum 1999, Machovich 1990). However, in most cases the putative pathophysiological consequences of the processing of bioactive proteins are difficult to assess, or are double-egded with respect to certain situations. Whereas the cleavage of several proteins of the coagulation and fibrinolysis cascades by PMN-proteases was suggested to cause the induction of life-threatening conditions such as the disseminated intravascular coagulation (Horbach 1999; Samis 1998), the activation of the complement factors has been considered to be benefical in light of the body's fight against pathogens (Doering 1994).

Most interesting are results accumulated just in the last couple of years concerning the processing of several components of the cytokine network. Here, PMN-serine proteases were found to interfere in manifold manner. Various active proinflammatory cytokines such as tumor necrosis factor-α (TNF-α), the neutrophil-activation peptide-2 (NAP-2), interleukin (IL)-8, (GRO-α) or IL-6 (IL-6) have been identified as substrates of PMN-serine proteases (Scuderi 1991, Nortier 1991, Padrines 1994, Leavell 1997, Bank

1999). The degradation of these cytokines by these enzymes was found to be accompanied by a loss of bioactivity and is thereby limiting the activating potency of these factors in inflammation. Since most of these proinflammatory cytokines are potent stimulators of neutrophil protease release, their inactivation could be considered in sense of a direct feedback mechanism. Interestingly, the combined inactivation of IL-6 and TNF-α was recently shown to result in the complete prevention of the acute phase response induction (Bopst 1998). On the other hand, recently published reports indicate that active PMN-derived serine proteases play a crucial role in the proteolytic control of cytokine or growth factor availability by catalyzing the release of mature active cytokine molecules from the membrane or of matrix bound inactive precursors. This was reported for TNF-α, IL-1ß, IL-18, basic FGF (fibroblast growth factor), NAP-2, transforming growth factor-α and several forms of latent transforming growth factor-ß (Mueller 1990, Brandt 1991, Car 1991, Robache-Gallea 1995, Csernok 1996, Coeshot 1999, Rifkin 1999). In this activation process proteinase 3 seems to play an outstanding role, even though it shares many characteristics with neutrophil elastase. However, in spite if the narrow substrate specificities of proteinase 3 and elastase, only proteinase 3 is highly potent in processing the inactive proforms of TNF-α and IL-1ß to the bioactive proteins. It has been assumed that this pathway, alternatively to the release of the active cytokines by the specific converting enzymes TACE and ICE, is particularly operative at sites of inflammation (Coeshott 1999).

Beyond that, PMN-proteases have also been suggested to be involved in the solubilization of ligand-binding ectodomains from membrane-bound cytokine receptors, such as the 75 kDa TNF-receptor, the IL-6 receptor gp80 chain, or CD25, the IL-2 receptor α chain (Porteu 1991, Bank 1999). This seems to represent an alternative shedding mechanism, complementary to the activation-dependent shedding by specific membrane-bound secretases under certain conditions. Importantly, in contrast to the shedding by these specific cellular metalloproteinases (predominantly not identified until now) the release of ligand-binding cytokine receptor ectodomains by PMN serine proteases is not dependent on the activation status of the receptor-bearing cell. A number of other membrane-bound proteins such as the T cell antigens CD2, CD4 and CD8, the adhesion molecules ICAM-1, the monocytic LPS-receptor CD14, and the anti-adhesive protein CD43, were reported to be cleaved specifically by PMN-serine proteases (Doering 1995, Bank 1999, Champagne 1998, Le-Barillec 1999, Remold-O'Donnell 1995) Taken together, these fascinating results indeed suggest that PMN-proteases possess a regulatory role by participating in the multiple control mechanims for the action of polypeptide effector molecules, particularly at local sites of inflammation.

## 3.3 Imbalances between matrix-metalloproteinases and their natural inhibitors

Matrix metalloproteinases (MMP) are key enzymes in normal and pathological tissue. Like the PMN serine proteases they are thought to play a prominent role in inflammation-associated tissue destruction, since macromolecules of the extracellular matrix are the main MMP substrates. As already described in detail for the lysosomal serine proteases, it is again the balance between the MMPs and their natural inhibitors, which is important for a tightly regulated physiological action of these proteases (cf. Kekow *et al* this book ch. 47).

Extracellular MMP activities are controlled by a family of reversible-binding inhibitors, endogenous tissue inhibitors of metalloproteinases - called TIMPs. Until now, four members of this inhibitor family have been characterized, which possess close structural similarities (for review see Gomez 1997). Imbalances between MMPs and TIMPs were found to cause an accelated degradation of matrix proteins in inflammation, chronic degenerative diseases and in tumor invasion. The following shortly summarized mechanisms have been suggested to result in a disturbed protease-inhibitor-balance:

### 3.3.1 Increased expression and/or release of MMPs

Under certain pathophysiolocal conditions an increased expression and release of MMPs was observed. In inflammation, this is closely associated with an elevated invasion of activated leukocytes into the tissue. In cancer an overexpression of MMPs by tumor cells in the invasion front has been reported (recent review Kleiner 1999). As suggested in one of the following chapters (Gottschalk *et al* this book ch. 49), the pathogenesis of endometriosis, a benign gynecological disorder, accompanied with the emigration of endometrium cells, is caused by an overexpression of MMP-1 (and a simultaneous TIMP deficiency).

### 3.3.2 MMP activation cascades

Since the MMPs are secreted as inactive proenzymes, they have to become activated by limited proteolysis. Therefore, the MMP activation cascades are critical elements in the control of MMP activities. In inflammatory processes, for instance, elevated activities of MMP activating proteases, among them the neutrophil serine proteases, were detected, which consequently causes an local excess of active MMPs (Ferry 1997, Capodici 1989).

### 3.3.3 Influence of cytokines on MMP secretion

While the secretion of MMPs and TIMPs is often coordinately regulated, some cytokines and other bioactive factors have been shown to differentially regulate the expression and release of MMPs and TIMPs. To mention only one very recently published example of a cytokine-induced imbalance, IL-8 was recently reported to inhibit the local TIMP-1 expression in arteriosclerotic plaques, thereby leading to a local imbalance between MMPs and TIMPs (Moreau 1999) .

In cancer, the putative relationship of a reduced TIMP expression with an increased invasive and metastatic potential is a matter of controversial discussion (Gomez 1997). On the other hand, the age-related impairment of a TIMP-1 and -2 upregulation in skin wounds, is believed to be causal for a retarded wound healing and might explain the predisposition of the elderly for chronic wound heling disorders (Ashcroft 1997)

### 3.3.4 Inactivation of TIMPs

Like the serpins, TIMPs were found to be inactivated by oxidative modification or proteolytical degradation. Because of their special structural features (functionally important disulphide bonds), TIMPs were found to be highly susceptible to inactivation by several reactive oxygen metabolites produced in large amounts at sites of inflammation. (Striclin 1992, Fears 1996, Shabani 1998). Moreover, various non-target proteases have been described to catalyze the degradation of TIMPs, which is associated with a subsequent loss of their MMP inhibitory and anti angiogenic potency. Very recently cathepsin B was found to cause the fragmentation of the two major MMP inhibitors TIMP-1 and TIMP-2 (Kostoulas 1999). Interestingly, neutrophil elastase was demonstrated to be the only protease which is capable of inactivating TIMP-1 complexed with pro-MMP-9. The complexation of TIMPs was described to stabilize their structur and, thereby, potentiating their inhibitory activity toward active MMPs (Itho 1995).

## 3.4 Protease inhibitors in therapy

Since there was already obtained early strong evidence that under pathophysiological conditions a complex network of active serine proteases and MMPs causes unspecific tissue breakdown or the induction of biochemical disregulations, much effort has been made to develop strategies for limiting excessive proteases activities (Travis 1991, Wieczorek 1999). Interestingly, first clinical trials have been done already in the seventies (Varava 1976). In view of the strong tissue destructive potency of PMN-derived elastase (plus cathepsin G and proteinase 3) the clinical studies focused on the augmention of the anti-elastase defense shield. Two main strategies were followed: the development of synthetic low molecular weight inhibitors or the supplementation of natural inhibitors produced as recombinant proteins, probably in a modified form. Indeed, in numerous experimental and animal studies the protective potency of a supplementation with naturally occuring serine protease inhibitors such as $\alpha_1$-PI and SLPI, was clearly demonstrated. Consequently, the $\alpha_1$-PI replacement has beenin use in the therapy of PMN-dominated acute inflammatory processes or chronic degenerative diseases already for a couple of years. However, while some preliminary results are promising (Doering 1999, Stiskal 1999), the antiprotease therapy has been realized to be associated with some unresolved problems. This includes their real availability at sites of excessive protease activity, their pharmacokinetics, their specificity, the potential antigenicity, and other aspects. To take up one major point, it became evident that the way of application is of particular importance for a successful restoration of the protease-antiprotease balance at local sites of PMN protease induced tissue destruction. Especially good results are obtained with the local application of elastase inhibitors, for instance as aerosols in degenerative and inflammatory lung diseases such as empysema or cystic fibrosis (Doering 1999, Dirksen 1999). On the other hand, it was also reported that the aerosol therapy is most beneficial for well ventilated lung tissue, but may be insufficient to neutralize elastase in poorly ventilated, highly inflamed areas as are seen in cystic fibrosis (Stolk 1995) . Another example of a locally restricted application is the usage of serine protease inhibitors in peritoneal lavage (Berling 1998). Additionally, as a new approach of synthetic elastase inhibitors the binding of inhibitor molecules on wound dressings was recently suggested (Edwards 1999) .

Until now, no substantial side effects of the $\alpha_1$-PI replacement therapy have been described (Vogelmeier 1997a). Nevertheless, using recombinant natural proteinase inhibitors in long-time therapy of $\alpha_1$-PI-deficiency, the *in vivo* formation of anti-inhibitor antibodies, as known for the therapeutic application of recombinant cytokines, could raise problems.

To overcome the above described problem of a rapid oxidative inactivation of natural inhibitors in the replacement therapy, it has soon been suggested to use genetically modified, nonoxidizible inhibitor mutants which retain their inhibitory properties to a large extent (Courtney 1985, Travis 1991). With respect to the oxidative inactivation of protease inhibitors, it is most interesting that recent studies indicate that rSLPI not only induced an increase of the elastase inhibitor potential, but also improved the antioxidant protection by raising glutathione levels *in vivo* (Gillisen 1993, Vogelmeier, 1996). However, not only for this reason, rSLPI seems to provide unique therapeutic opportunities: this protein was recently reported to have - independent of its inhibitory function - a broad spectrum antibiotic activity that includes antiretroviral, bactericidal, and antifungal activity (McNeely 1997, Tomee 1998, Song 1999). However, the same has recently been found to hold true for the serine protease inhibitor Elafin (Simpson 1999).

With respect to the limitation of excessive MMP activities in degenerative and malignant diseases experimental studies as well as clinical trials have produced encouraging results, which are, however, prone to some pitfalls (Blavier 1999, Drummond 1999, Greenwald 1999, Kumagai 1999). Surprisingly, there is only one MMP inhibitor clinically available for therapeutic use to date. However, experimental studies and first clinical trials using synthetic as well as natural MMP inhibitors have confirmed a reduced tendency of tumor spread and metastasis (for review see Wojtowicz-Praga 1997, Jones 1999, Blavier 1999) or revealed a protective effect in degenerative diseases such as asthma or rheumatoid arthritis (Kumagai 1999, Allaire 1998, DeBri 1998).

Since most of the endogenous inhibitors, like $\alpha_1$-PI or TIMP-1 are inhibitors of a set of proteases, it is a point at issue whether the therapy with such „broad-spectrum“ inhibitors might have unexpected side effects by blocking not only detrimental but also physiologically essential proteolytic pathways. This is one of the reasons for the effort made in the development of highly potent, but more specific synthetic inhibitors for an effective therapeutical control of protease activities (Shinguh 1998, Wieczorek 1999, Jones 1999). Additionally, specific inhibitory substances from microorganisms, plants or animals (such as annelidae) are currently under investigation for their utility in reducing excessive proteolytic activities. And not least, the gene transfer via viral vectors is currently dicussed as one of the future concepts of inhibitor therapy in patients with $\alpha_1$-PI deficiency (Blank 1994, Allaire 1998, Eckman 1999). Despite of the recent successes in using protease inhibitors for limiting excessive protease activities, future work focuses on the circumvention of the present limitations of this therapeutic strategy (Greenwald 1999)

# 4. CYSTEINE PROTEINASES IN TUMOR PROGRESSION

The transition from a benign to a malignant tumor is defined by the ability of tumor cells to invade local tissues at the primary site and to cross natural barriers, e. g. the extracellular matrix, basement membranes, intercellular junctions, and interstitial stroma (Mignatti and Rifkin, 1993). During metastasis, single tumor cells must detach from the tumor mass to become mobile. At first, they penetrate the basement membrane and invade the stroma of the surrounding tissue (invasion). Capillary endothelial cells invade the tumor (angiogenesis), and tumor cells enter the blood vessels (intravasation). Some of these malignant cells arrest at a distant vascular bed, extravasate, and form metastatic colonies at sites distant from the primary tumor (Stracke and Liotta, 1995).

The degradation of extracellular matrix (ECM) components is a crucial step at all these stages. A variety of proteolytic enzymes can degrade such components, e. g. collagen, fibronectin and proteoglycans. Besides the matrix metalloproteases (MMPs) (Polette *et al* 1998 and Kahari *et a,* 1990), the cysteine proteases cathepsin L and cathepsin B (Berquin and Sloane 1994, Yan *et al* 1998) are reported to be associated with metastasis. These proteases appear to act in a cascade manner together with serine proteases and the aspartic protease cathepsin D.

## 4.1 Cathepsins B and L

Cathepsins B and L are lysosomal endopeptidases which belong to the cysteine protease class. They are widely distributed and participate in the intralysosomal degradation of molecules taken up from the extracellular environment.

In tumor cells, alterations at one or more of the levels that regulate cysteine protease biosynthesis lead to an upregulation, membrane association, and secretion of cathepsin B. Extensive studies on the cysteine protease expression in different cancer types, including melanoma and carcinoma of the lung, colon, prostate and breast revealed an increased expression of cathepsins B and L (Kos *et al* 1997, Krepela *et al* 1998, Murnane *et al* 1991, Foekens *et al* 1998, Friedrich *et al* 1999). In some of these tumor entities, both cathepsin B and cathepsin L as well as their inhibitors reached prognostic significance. Our own results showed that in chondrosarcoma of bone, overexpression of cathepsin B was associated with a high rate of local recurrence and decreased recurrence-free survival (Haeckel, personal communication). Cathepsin L did not reach statistical significance in chondrosarcomas, whereas in breast cancer an elevated cathepsin L level was a strong and independent prognostic factor, with an impact comparable to that of axillary lymph node status and grading (Thomssen *et al* 1997). The literature on the expression of cathepsins B and L in tumors is contradictory. Sampling and methodological differences, the cellular composition, the extent of macrophage infiltration, and the state of preservation of the tissue seem to be responsible for the conflicting results. Further studies on larger patient populations using specific methods are needed to confirm the results obtained so far (Elliott and Sloane 1996).

Furthermore, the localization of cathepsins B and L alters as the grade of malignancy of human tumors increases (Keppler *et al* 1994, Sloane *et al* 1994). Proenzyme and/or mature cathepsins B and L can be secreted, associated with the plasma membrane, or redistributed to vesicular compartments. Studies using immunohistochemistry and *in situ* hybridization found that cathepsins B and L are located at the invasive edges; furthermore, these studies demonstrated the existence of secreted and membrane-bound cathepsins B and L in different tumor entities (Frosch *et al* 1999) and tumor cell lines (Kobayashi *et al* 1993). It has been shown that under slightly acidic conditions, for example in peritumoral environment, secretion of active cathepsin B is enhanced (Rozhin *et al* 1994). In the case of cathepsin L, it was suggested that an overproduction of the enzyme saturates the lysosomal pathway, resulting in secretion of procathepsin L (Prence *et al* 1990).

Many studies investigating the role of proteases in tumor progression have focused on the invasive steps of the metastatic cascade, including the degradation of basement membrane elements. Both cathepsin B and cathepsin L are able to degrade extracellular matrix elements either directly or indirectly by activating other proteases such as the receptor bound pro-uPA (Kobayashi *et al* 1991). In *in vitro* invasion assays, Kobayashi *et al* (1992) and Kolkhorst *et al* (1998) tested several cell lines treated with protease inhibitors (E-64, CA074, Batimastat) or anticatalytical antibodies. Decreased protease activity resulted in a reduced invasion of treated cells in most tumor cell types. Inhibition of migration through matrigel by E-64 shows that cysteine proteases are mainly responsible for these processes. Treatment with CA074 emphasized the special role of cathepsin B in the invasion steps of breast carcinoma cells, melanoma cells and ovarian cancer cells. Overexpression of the specific cysteine proteinase inhibitor cystatin C in a stable transformed melanoma cell line resulted in decreased motility and invasion (Sexton and Cox 1997). Similar results were obtained by our group for human osteosarcoma cells transfected with antisense cathepsin B. The reduced level of active cathepsin B led to a markedly lower invasion and motility compared with the parental cells (Krueger *et al* 1999 and this book ch. 44). The application of antisense oligonucleotides against cathepsin L underlined the results obtained for cathepsin B in human osteosarcomas (Krueger, unpublished) and demonstrated that several proteases act sequentially or simultaneously in the dissolution of basement membranes.

The spectrum of natural substrates of cysteine proteases, together with the morphological findings and the results of the *in vitro* findings in functional assays, suggest an important role of cathepsin B and cathepsin L in malignant tumor diseases. The importance of proteases and their receptors has triggered an extensive search for anti-protease drugs for cancer therapy. The first results obtained in clinical trials (Rasmussen and McCann 1997, Parsons *et al* 1997) are promising.

# 5. CONCLUSION

With this review we wanted to present three aspects of proteinases in diseases, namely (i) the newly energing involvement of the ubiquitin mediated proteolysis system in a growing number of clinical manifestations; (ii) the importance of the fine tuning of the delicate balance of PMN proteases and their inhibitors in inflammatory diseases and (iii) the very special effects exerted by lysosomal cysteine proteinases on tumour metastasis and progression. In all three topics during the last few years have been published surprising new results leading to better understanding of the normal and pathologic functions influenced by proteolytic enzymes. It is easy to forecast the continuation of this process, and this makes one curious what might be the next steps of discovery in this field.

# REFERENCES


Abbink, J.J., Kamp, A.M., Nieuwenhuys, E.J., Nuijens, J.H., Swaak ,A.J., and Hack, C.E., 1991, Predominant role of neutrophils in the inactivation of alpha 2- macroglobulin in arthritic joints. *Arthritis Rheum.* **34:** 1139-1150.

Allaire, E., Forough, R., Clowes, M., Starcher, B., and Clowes, A.W., 1998, Local overexpression of TIMP-1 prevents aortic aneurysm degeneration and rupture in a rat model. *J. Clin. Invest* **102:** 1413-1420.

Alves-Rodrigues, A., Gregori, L., and Figueiredo-Pereira, M.E., 1998, Ubiquitin, cellular inclusions and their role in neurodegeneration. *Trends Neurosci.* **21:** 516-520.

Ashcroft, G.S., Herrick, S.E., Tarnuzzer, R.W., Horan, M.A., Schultz, G.S., and Ferguson, M.W., 1997, Human ageing impairs injury-induced in vivo expression of tissue inhibitor of matrix metalloproteinases (TIMP)-1 and -2 proteins and mRNA. *J. Pathol.* **183:** 169-176.

Axelsson, L., Linder, C., Ohlsson, K., and Rosengren, M., 1988, The effect of the secretory leukocyte protease inhibitor on leukocyte proteases released during phagocytosis. *Biol. Chem. Hoppe Seyler* Suppl. **369:** 89-93.

Ballieux, B.E., Hiemstra, P.S., Klar-Mohamad, N., Hagen, E.C., van Es, L.A., van der Woude, F.J., and Daha, M.R., 1994, Detachment and cytolysis of human endothelial cells by proteinase 3. *Eur. J. Immunol.* **24:** 3211-3215.

Bangalore, N., Travis, J., 1994, Comparison of properties of membrane bound versus soluble forms of human leukocytic elastase and cathepsin G. *Biol. Chem. Hoppe Seyler* **375:** 659-666.

Bank, U., Kupper, B., Reinhold, D., Hoffmann, T., and Ansorge, S., 1999, Evidence for a crucial role of neutrophil-derived serine proteases in the inactivation of interleukin-6 at sites of inflammation. *FEBS Lett.* **461:** 235-240.

Barrett, A.J., Rawlings, N.D., and Woessner, J.F., 1998, *Handbook of proteolytic enzymes.* Academic Press, San Diego

Baumeister, W., Cejka, Z., Kania, M., and Seemuller, E., 1997, The proteasome: a macromolecular assembly designed to confine proteolysis to a nanocompartment. *Biol. Chem.* **378:** 121-130.

Baumstark, J.S., 1970, Studies on the elastase-serum protein interaction. II. On the digestion of human a2-macroglobulin, an elastase inhibitor, by elastase. *Biochim. Biophys. Acta* **207:** 318-330.

Berling, R., Borgstrom, A., and Ohlsson, K., 1998, Peritoneal lavage with aprotinin in patients with severe acute pancreatitis. Effects on plasma and peritoneal levels of trypsin and leukocyte proteases and their major inhibitors. *Int. J. Pancreatol.* **24:** 9-17.

Berquin, I.M., Sloane, B.F., 1994, Cysteineprotease and tumor progression. *Perspect.Drug Discov.* **2:** 371-388.

Bieth, J.G., 1998, Leukocyte elastase. In *Handbook of proteolytic enzymes*. (A.J. Barrett, N.D. Rawlings and J.F. Woessner, eds). Academic Press, San Diego, pp. 54-60.

Bird, P.I., 1999, Regulation of pro-apoptotic leucocyte granule serine proteinases by intracellular serpins. *Immunol. Cell Biol.* **77:** 47-57.

Birrer, P., 1993, Consequences of unbalanced protease in the lung: protease involvement in destruction and local defense mechanisms of the lung. *Agents Actions Suppl.* **40:** 3-12.

Blank, C.A., Brantly, M., 1994, Clinical features and molecular characteristics of alpha 1-antitrypsin deficiency. *Ann. Allergy* **72:** 105-120.

Blavier, L., Henriet, P., Imren, S., and Declerck, Y.A., 1999, Tissue inhibitors of matrix metalloproteinases in cancer. *Ann. N. Y .Acad. Sci.* **878:** 108-119.

Bopst, M., Haas, C., Car, B., and Eugster, H.P., 1998, The combined inactivation of tumor necrosis factor and interleukin-6 prevents induction of the major acute phase proteins by endotoxin. *Eur. J. Immunol.* **28:** 4130-4137.

Boutten, A., Venembre, P., Seta, N., Hamelin, J., Aubier, M., Durand, G., and Dehoux, M.S., 1998, Oncostatin M is a potent stimulator of alpha1-antitrypsin secretion in lung epithelial cells: modulation by transforming growth factor-beta and interferon-gamma. *Am. J. Respir. Cell Mol. Biol.* **18:** 511-520.

Brandt, E., Van Damme, J., and Flad, H.D., 1991, Neutrophils can generate their activator neutrophil-activating peptide 2 by proteolytic cleavage of platelet-derived connective tissue- activating peptide III. *Cytokine* **3:** 311-321.

Brodsky, J.L., McCracken, A.A., 1997, ER-associated and proteasome-mediated protein degradation: How two topologically restricted events came together. *Trends cell Biol.* **7:** 151-156.

Capodici, C., Muthukumaran, G., Amoruso, M.A., and Berg, R.A., 1989, Activation of neutrophil collagenase by cathepsin G. *Inflammation* **13:** 245-258.

Car, B.D., Baggiolini, M., and Walz, A., 1991, Formation of neutrophil-activating peptide 2 from platelet-derived connective-tissue-activating peptide III by different tissue proteinases. *Biochem. J.* **275:** 581-584.

Carlson, J.A., Rogers, B.B., Sifers, R.N., Hawkins, H.K., Finegold, M.J., and Woo, S.L., 1988, Multiple tissues express alpha 1-antitrypsin in transgenic mice and man. *J. Clin. Invest* **82:** 26-36.

Catanese, J., Kress, L.F., 1984, Enzymatic inactivation of human plasma C1-inhibitor and alpha 1- antichymotrypsin by Pseudomonas aeruginosa proteinase and elastase. *Biochim. Biophys. Acta* **789:** 37-43

Cepinskas, G., Sandig, M., and Kvietys, P.R., 1999, PAF-induced elastase-dependent neutrophil transendothelial migration is associated with the mobilization of elastase to the neutrophil surface and localization to the migrating front. *J. Cell Sci.* **112:** 1937-1945.

Champagne, B., Tremblay, P., Cantin, A., and St.Pierre, Y., 1998, Proteolytic cleavage of ICAM-1 by human neutrophil elastase. *J. Immunol.* **161:** 6398-6405.

Chen, Z.J., Maniatis, T., 1998, Role of the ubiquitin-proteasome pathway in NFkB activation. In *Ubiquitin and the biology of the cell*. (J.M. Peters, J.R. Harris and D. Finley, eds). Plenum Press, London/New York, pp. 303-322.

Cichy, J., Potempa, J., Chawla, R.K., and Travis, J., 1995, Stimulatory effect of inflammatory cytokines on alpha 1- antichymotrypsin expression in human lung-derived epithelial cells. *J. Clin. Invest* **95:** 2729-2733.

Ciechanover, A., Schwartz, A.L., 1994, The ubiquitin-mediated proteolytic pathway: mechanisms of recognition of the proteolytic substrate and involvement in the degradation of native cellular proteins. *FASEB J.* **8:** 182-191.

Coeshott, C., Ohnemus, C., Pilyavskaya, A., Ross, S., Wieczorek, M., Kroona, H., Leimer, A.H., and Cheronis, J., 1999, Converting enzyme-independent release of tumor necrosis factor alpha and IL-1beta from a stimulated human monocytic cell line in the presence of activated neutrophils or purified proteinase 3. *Proc. Natl. Acad. Sci. USA* **96:** 6261-6266.

Courtney, M., Jallat, S., Tessier, L.H., Benavente, A., Crystal, R.G., and Lecocq, J.P., 1985, Synthesis in E. coli of alpha 1-antitrypsin variants of therapeutic potential for emphysema and thrombosis. *Nature* **313:** 149-151.

Coux, O., Tanaka, K., and Goldberg, A.L., 1996, Structure and functions of the 20S and 26S proteasomes. *Annu. Rev. Biochem.* **65:** 801-847.

Csernok, E., Szymkowiak, C.H., Mistry, N., Daha, M.R., Gross, W.L., and Kekow, J., 1996, Transforming growth factor-beta (TGF-beta) expression and interaction with proteinase 3 (PR3) in anti-neutrophil cytoplasmic antibody (ANCA)- associated vasculitis. *Clin. Exp. Immunol.* **105:** 104-111.

de Bri, E., Lei, W., Svensson, O., Chowdhury, M., Moak, S.A., and Greenwald, R.A., 1998, Effect of an inhibitor of matrix metalloproteinases on spontaneous osteoarthritis in guinea pigs. *Adv. Dent. Res.* **12:** 82-85.

Denison, F.C., Kelly, R.W., Calder, A.A., and Riley, S.C., 1999, Secretory leukocyte protease inhibitor concentration increases in amniotic fluid with the onset of labour in women: characterization of sites of release within the uterus. *J. Endocrinol.* **161:** 299-306.

Desrochers, P.E., Mookhtiar, K., Van Wart, H.E., Hasty, K.A., and Weiss, S.J., 1992, Proteolytic inactivation of alpha 1-proteinase inhibitor and alpha 1- antichymotrypsin by oxidatively activated human neutrophil metalloproteinases. *J. Biol. Chem.* **267:** 5005-5012.

Dirksen, A., Dijkman, J.H., Madsen, F., Stoel, B., Hutchison, D.C., Ulrik, C.S., Skovgaard, L.T., Kok-Jensen, A., Rudolphus, A., Seersholm, N., Vrooman, H.A., Reiber, J.H., Hansen, N.C., Heckscher, T., Viskum, K., and Stolk, J., 1999, A Randomized Clinical Trial of alpha(1)-Antitrypsin Augmentation Therapy. *Am. J. Respir. Crit Care Med.* **160:** 1468-1472.

Doring, G., 1994, The role of neutrophil elastase in chronic inflammation. *Am.J.Respir.Crit Care Med.* **150:** S114-S117.

Doring, G., Frank, F., Boudier, C., Herbert,S., Fleischer,B., and Bellon,G., 1995, Cleavage of lymphocyte surface antigens CD2, CD4, and CD8 by polymorphonuclear leukocyte elastase and cathepsin G in patients with cystic fibrosis. *J. Immunol.* **154:** 4842-4850.

Draper, D., Donohoe, W., Mortimer, L., and Heine, R.P., 1998, Cysteine proteases of Trichomonas vaginalis degrade secretory leukocyte protease inhibitor. *J. Infect. Dis.* **178:** 815-819.

Drummond, A.H., Beckett, P., Brown, P.D., Bone, E.A., Davidson, A.H., Galloway, W.A., Gearing, A.J., Huxley, P., Laber, D., McCourt, M., Whittaker, M., Wood, L.M., and Wright, A., 1999, Preclinical and clinical studies of MMP inhibitors in cancer. *Ann. N. Y. Acad. Sci.* **878:** 228-235.

Eckman, E.A., Mallender, W.D., Szegletes, T., Silski, C.L., Schreiber, J.R., Davis, P.B., and Ferkol, T.W., 1999, In vitro transport of active alpha(1)-antitrypsin to the apical surface of epithelia by targeting the polymeric immunoglobulin receptor. *Am. J. Respir. Cell Mol. Biol.* **21:** 246-252.

Edwards, J.V., Bopp, A.F., Batiste, S., Ullah, A.J., Cohen, I.K., Diegelmann, R.F., and Montante, S.J., 1999, Inhibition of elastase by a synthetic cotton-bound serine protease inhibitor: in vitro kinetics and inhibitor release. *Wound. Repair Regen.* **7:** 106-118.

Edwards, S.W., Hallett, M.B., 1997, Seeing the wood for the trees: the forgotten role of neutrophils in rheumatoid arthritis. *Immunol. Today* **18:** 320-324.

Ehring, B., Meyer, T.H., Eckerskorn, C., Lottspeich, F., and Tampe, R., 1996, Effects of major-histocompatibility-complex-encoded subunits on the peptidase and proteolytic activities of human 20S proteasomes. Cleavage of proteins and antigenic peptides. *Eur. J. Biochem.* **235:** 404-415.

Elliott, E., Sloane, B.F., 1996, The cysteine protease cathepsin B in cancer. *Perspect. Drug Discov.* **6:** 12-32.

Esnault, V.L., Audrain, M.A., and Sesboue, R., 1997, Alpha-1-antitrypsin phenotyping in ANCA-associated diseases: one of several arguments for protease/antiprotease imbalance in systemic vasculitis. *Exp. Clin. Immunogenet.* **14:** 206-213.

Fantuzzi, G., 1998, Lecture at the: Second Joint Meeting of the International Society foe Interferon and Cytokine Research and the International Cytokine Society, Jerusalem.

Ferry, G., Lonchampt, M., Pennel, L., de Nanteuil, G., Canet, E., and Tucker, G.C., 1997, Activation of MMP-9 by neutrophil elastase in an in vivo model of acute lung injury. *FEBS Lett.* **402:** 111-115.

Foekens, J.A., Kos, J., Peters, H.A., Krasovec, M., Look, M.P., Cimerman, N., Meijer-van Gelder, M.E., Henzen-Logmans, S.C., van Putten, W.L., and Klijn, J.G., 1998, Prognostic significance of cathepsins B and L in primary human breast cancer. *J. Clin. Oncol.* **16:** 1013-1021.

Frears, E.R., Zhang, Z., Blake, D.R., O'Connell, J.P., and Winyard, P.G., 1996, Inactivation of tissue inhibitor of metalloproteinase-1 by peroxynitrite. *FEBS Lett.* **381:** 21-24.

Friedrich, B., Jung, K., Lein, M., Turk, I., Rudolph, B., Hampel, G., Schnorr, D., and Loening, S.A., 1999, Cathepsins B, H, L and cysteine protease inhibitors in malignant prostate cell lines, primary cultured prostatic cells and prostatic tissue. *Eur. J. Cancer* **35:** 138-144.

Frosch, B.A., Berquin, I., Emmert-Buck, M.R., Moin, K., and Sloane, B.F., 1999, Molecular regulation, membrane association and secretion of tumor cathepsin B. *APMIS* **107:** 28-37.

Gillis, S., Furie, B.C., and Furie, B., 1997, Interactions of neutrophils and coagulation proteins. *Semin. Hematol.* **34:** 336-342.

Gillissen, A., Birrer, P., McElvaney, N.G., Buhl, R., Vogelmeier, C., Hoyt, R.F., Jr., Hubbard, R.C., and Crystal, R.G., 1993, Recombinant secretory leukoprotease inhibitor augments glutathione levels in lung epithelial lining fluid. *J. Appl. Physiol* **75:** 825-832.

Goldberg, A.L., Rock, K.L., 1992, Proteolysis, proteasomes and antigen presentation. *Nature* **357:** 375-379.

Gomez, D.E., Alonso, D.F., Yoshiji, H., and Thorgeirsson, U.P., 1997, Tissue inhibitors of metalloproteinases: structure, regulation and biological functions. *Eur. J. Cell Biol.* **74:** 111-122.

Greenwald, R.A., 1999, Thirty-six years in the clinic without an MMP inhibitor. What hath collagenase wrought? *Ann. N. Y. Acad. Sci.* **878:** 413-419.

Hasselgren, P.O., Fischer, J.E., 1997, The ubiquitin-proteasome pathway: review of a novel intracellular mechanism of muscle protein breakdown during sepsis and other catabolic conditions. *Ann. Surg.* **225:** 307-316.

Hawkins, D., Cochrane, C.G., 1968, Glomerular basement membrane damage in immunological glomerulonephritis. *Immunology* **14:** 665-681.

Heinzelmann, M., Mercer-Jones, M.A., and Passmore, J.C., 1999, Neutrophils and renal failure. *Am. J. Kidney Dis.* **34:** 384-399.

Henson, P.M., Henson, J.E., Fittschen, C., Bratton, D., and Riches, D.W.H., 1992, Degranulation and secretion by phagocytic cells. In *Inflammation: Basic principles and clinical correlates.* (J.I. Gallin, I.M. Goldstein and R. Snyderman, eds). Raven Press, New York, pp. 511-540.

Hicke, L., 1997, Ubiquitin-dependent internalization and down-regulation of plasma membrane proteins. *FASEB J.* **11:** 1215-1226.

Hochstrasser, M., Kornitzer, D., 1998, Ubiquitin degradation of transcription regulators. In *Ubiquitin and the biology of the cell.* (J.M. Peters, J.R. Harris and D. Finley, eds). Plenum Press, London/New York, pp. 279-302.

Hoidal, J.R., 1998, Myeloblastin. In *Handbook of proteolytic enzymes.* (A.J. Barrett, N.D. Rawlings and J.F. Woessner, eds). Academic Press, San Diego, pp. 62-65.

Horbach, D.A., van Oort, E., Lisman, T., Meijers, J.C., Derksen, R.H., and de Groot, P.G., 1999, Beta2-glycoprotein I is proteolytically cleaved in vivo upon activation of fibrinolysis. *Thromb. Haemost.* **81:** 87-95.

Huibregtse, J.M., et al., 1998, Ubiquitination of the p53 tumor suppressor. In *Ubiquitin and the biology of the cell.* . (J.M. Peters, J.R. Harris and D. Finley, eds). Plenum Press, London/New York, pp. 323-543.

Itoh, Y., Nagase, H., 1995, Preferential inactivation of tissue inhibitor of metalloproteinases-1 that is bound to the precursor of matrix metalloproteinase 9 (progelatinase B) by human neutrophil elastase. *J. Biol .Chem.* **270:** 16518-16521.

Janoff, A., Feinstein, G., Malemud, C.J., and Elias, J.M., 1976, Degradation of cartilage proteoglycan by human leukocyte granule neutral proteases--a model of joint injury. I. Penetration of enzyme into rabbit articular cartilage and release of 35SO4-labeled material from the tissue. *J. Clin. Invest* **57:** 615-624.

Jentsch, S., 1992, Ubiquitin-dependent protein degradation: a cellular perspective. *Trends cell Biol.* **2:** 98-103.

Jin, F.Y., Nathan, C., Radzioch, D., and Ding, A., 1997, Secretory leukocyte protease inhibitor: a macrophage product induced by and antagonistic to bacterial lipopolysaccharide. *Cell* **88:** 417-426.

Jochum, M., Billing, A.G., Frohlich, D., Schildberg, F.W., MacHleidt, W., Cheronis, J.C., Gippner-Steppert, C., and Fritz, H., 1999, Proteolytic destruction of functional proteins by phagocytes in human peritonitis. *Eur. J. Clin. Invest* **29:** 246-255.

Johnson, D., Travis, J., 1979, The oxidative inactivation of human alpha-1-proteinase inhibitor. Further evidence for methionine at the reactive center. *J. Biol .Chem.* **254:** 4022-4026.

Jones, L., Ghaneh, P., Humphreys, M., and Neoptolemos, J.P., 1999, The matrix metalloproteinases and their inhibitors in the treatment of pancreatic cancer. *Ann. N. Y. Acad. Sci.* **880:** 288-307.

Kahari, V.M., Saarialho-Kere, U., 1999, Matrix metalloproteinases and their inhibitors in tumour growth and invasion. *Ann. Med.* **31:** 34-45.

Kawabata, K., Moore, A.R., and Willoughby, D.A., 1996, Impaired activity of protease inhibitors towards neutrophil elastase bound to human articular cartilage. *Ann. Rheum. Dis.* **55:** 248-252.

Keppler, D., Waridel, P., Abrahamson, M., Bachmann, D., Berdoz, J., and Sordat, B., 1994, Latency of cathepsin B secreted by human colon carcinoma cells is not linked to secretion of cystatin C and is relieved by neutrophil elastase. *Biochim. Biophys. Acta* **1226:** 117-125.

King, R.W., Deshaies, R.J., Peters, J.M., and Kirschner, M.W., 1996, How proteolysis drives the cell cycle. *Science* **274:** 1652-1659.

Kishino, T., Lalande, M., and Wagstaff, J., 1997, UBE3A/E6-AP mutations cause Angelman syndrome. *Nat. Genet.* **15:** 70-73.

Kleiner, D.E., Stetler-Stevenson, W.G., 1999, Matrix metalloproteinases and metastasis. *Cancer Chemother.Pharmacol.* Suppl. **43:** S42-S51.

Kobayashi, H., Schmitt, M., Goretzki, L., Chucholowski, N., Calvete, J., Kramer, M., Gunzler, W.A., Janicke, F., and Graeff, H., 1991, Cathepsin B efficiently activates the soluble and the tumor cell receptor-bound form of the proenzyme urokinase-type plasminogen activator (Pro-uPA). *J. Biol. Chem.* **266:** 5147-5152.

Kobayashi, H., Ohi, H., Sugimura, M., Shinohara, H., Fujii, T., and Terao, T., 1992, Inhibition of in vitro ovarian cancer cell invasion by modulation of urokinase-type plasminogen activator and cathepsin B. *Cancer Res.* **52:** 3610-3614.

Kobayashi, H., Moniwa, N., Sugimura, M., Shinohara, H., Ohi, H., and Terao, T., 1993, Effects of membrane-associated cathepsin B on the activation of receptor-bound prourokinase and subsequent invasion of reconstituted basement membranes. *Biochim. Biophys. Acta* **1178:** 55-62.

Kolkhorst, V., Sturzebecher, J., and Wiederanders, B., 1998, Inhibition of tumour cell invasion by protease inhibitors: correlation with the protease profile. *J.Cancer Res. Clin. Oncol.* **124:** 598-606.

Kos, J., Stabuc, B., Schweiger, A., Krasovec, M., Cimerman, N., Kopitar-Jerala, N., and Vrhovec, I., 1997, Cathepsins B, H, and L and their inhibitors stefin A and cystatin C in sera of melanoma patients. *Clin. Cancer Res.* **3:** 1815-1822.

Kostoulas, G., Lang, A., Nagase, H., and Baici, A., 1999, Stimulation of angiogenesis through cathepsin B inactivation of the tissue inhibitors of matrix metalloproteinases. *FEBS Lett.* **455:** 286-290.

Krepela, E., Prochazka, J., Karova, B., Cermak, J., and Roubkova, H., 1998, Cysteine proteases and cysteine protease inhibitors in non-small cell lung cancer. *Neoplasma* **45:** 318-331.

Krüger, S., et al., 1999, Inhibitory effects of antisense cathepsin B cDNA transfection on invasion and motility in a human osteosarcoma cell line. *Cancer Res.* **59:** 6010-6014.

Kumagai, K., Ohno, I., Okada, S., Ohkawara, Y., Suzuki, K., Shinya, T., Nagase, H., Iwata, K., and Shirato, K., 1999, Inhibition of matrix metalloproteinases prevents allergen-induced airway inflammation in a murine model of asthma. *J. Immunol.* **162:** 4212-4219.

Le Barillec, K., Si-Tahar, M., Balloy, V., and Chignard, M., 1999, Proteolysis of monocyte CD14 by human leukocyte elastase inhibits lipopolysaccharide-mediated cell activation. *J. Clin. Invest* **103:** 1039-1046.

Leavell, K.J., Peterson, M.W., and Gross, T.J., 1997, Human neutrophil elastase abolishes interleukin-8 chemotactic activity. *J. Leukoc. Biol.* **61:** 361-366.

Loda, M., Cukor, B., Tam, S.W., Lavin, P., Fiorentino, M., Draetta, G.F., Jessup, J.M., and Pagano, M., 1997, Increased proteasome-dependent degradation of the cyclin-dependent kinase inhibitor p27 in aggressive colorectal carcinomas. *Nat. Med.* **3:** 231-234.

Lomas, D.A., Carrell, R.W., 1993, A protein structural approach to the solution of biological problems: alpha 1-antitrypsin as a recent example. *Am.J.Physiol* **265:** L211-L219.

Machovich, R., Owen, W.G., 1990, The elastase-mediated pathway of fibrinolysis. *Blood Coagul. Fibrinolysis* **1:** 79-90.

Matsuura, T., Sutcliffe, J.S., Fang, P., Galjaard, R.J., Jiang, Y.H., Benton, C.S., Rommens, J.M., and Beaudet, A.L., 1997, De novo truncating mutations in E6-AP ubiquitin-protein ligase gene (UBE3A) in Angelman syndrome. *Nat. Genet.* **15:** 74-77.

McNeely, T.B., Shugars, D.C., Rosendahl, M., Tucker, C., Eisenberg, S.P., and Wahl, S.M., 1997, Inhibition of human immunodeficiency virus type 1 infectivity by secretory leukocyte protease inhibitor occurs prior to viral reverse transcription. *Blood* **90:** 1141-1149.

Mignatti, P., Rifkin, D.B., 1993, Biology and biochemistry of proteinases in tumor invasion. *Physiol Rev.* **73:** 161-195.

Mitch, W.E., Goldberg, A.L., 1996, Mechanisms of muscle wasting. The role of the ubiquitin-proteasome pathway. *N. Engl. J. Med.* **335:** 1897-1905.

Mitsuhashi, H., Asano, S., Nonaka, T., Masuda, K., and Kiyoki, M., 1997, Potency of truncated secretory leukoprotease inhibitor assessed in acute lung injury models in hamsters. *J. Pharmacol. Exp. Ther.* **282:** 1005-1010.

Miyazawa, K., Toyama, K., Gotoh, A., Hendrie, P.C., Mantel, C., and Broxmeyer, H.E., 1994, Ligand-dependent polyubiquitination of c-kit gene product: a possible mechanism of receptor down modulation in M07e cells. *Blood* **83:** 137-145.

Molhuizen, H.O., Schalkwijk, J., 1995, Structural, biochemical, and cell biological aspects of the serine proteinase inhibitor SKALP/elafin/ESI. *Biol. Chem. Hoppe Seyler* **376:** 1-7.

Moore, A.R., Appelboam, A., Kawabata, K., Da Silva, J.A., D'Cruz, D., Gowland, G., and Willoughby, D.A., 1999, Destruction of articular cartilage by alpha 2 macroglobulin elastase complexes: role in rheumatoid arthritis. *Ann. Rheum. Dis.* **58:** 109-113.

Moreau, M., Brocheriou, I., Petit, L., Ninio, E., Chapman, M.J., and Rouis, M., 1999, Interleukin-8 mediates downregulation of tissue inhibitor of metalloproteinase-1 expression in cholesterol-loaded human macrophages: relevance to stability of atherosclerotic plaque. *Circulation* **99:** 420-426.

Mori, S., Claesson-Welsh, L., Okuyama, Y., and Saito, Y., 1995, Ligand-induced polyubiquitination of receptor tyrosine kinases. *Biochem. Biophys. Res. Commun.* **213:** 32-39.

Mueller, S.G., Paterson, A.J., and Kudlow, J.E., 1990, Transforming growth factor alpha in arterioles: cell surface processing of its precursor by elastases. *Mol. Cell Biol.* **10:** 4596-4602.

Murnane, M.J., Sheahan, K., Ozdemirli, M., and Shuja, S., 1991, Stage-specific increases in cathepsin B messenger RNA content in human colorectal carcinoma. *Cancer Res.* **51:** 1137-1142.

Nelson, D., Potempa, J., Kordula, T., and Travis, J., 1999, Purification and characterization of a novel cysteine proteinase (periodontain) from Porphyromonas gingivalis. Evidence for a role in the inactivation of human alpha1-proteinase inhibitor. *J. Biol. Chem.* **274:** 12245-12251.

Norman, M.R., Mowat, A.P., and Hutchison, D.C., 1997, Molecular basis, clinical consequences and diagnosis of alpha-1 antitrypsin deficiency. *Ann. Clin. Biochem.* **34:** 230-246.

Nortier, J., Vandenabeele, P., Noel, E., Bosseloir, Y., Goldman, M., and Deschodt-Lanckman, M., 1991, Enzymatic degradation of tumor necrosis factor by activated human neutrophils: role of elastase. *Life Sci.* **49:** 1879-1886.

Oh, C.E., McMahon, R., Benzer, S., and Tanouye, M.A., 1994, bendless, a Drosophila gene affecting neuronal connectivity, encodes a ubiquitin-conjugating enzyme homolog. *J. Neurosci.* **14:** 3166-3179.

Ohlsson, S., Tufvesson, B., Polling, A., and Ohlsson, K., 1997, Distribution of the secretory leucocyte proteinase inhibitor in human articular cartilage *Biol. Chem.* **378:** 1055-1058.

Owen, C.A., Campbell, M.A., Sannes, P.L., Boukedes, S.S., and Campbell, E.J., 1995, Cell surface-bound elastase and cathepsin G on human neutrophils: a novel, non-oxidative mechanism by which neutrophils focus and preserve catalytic activity of serine proteinases. *J.Cell Biol.* **131:** 775-789.

Padrines, M., Wolf, M., Walz, A., and Baggiolini, M., 1994, Interleukin-8 processing by neutrophil elastase, cathepsin G and proteinase-3. *FEBS Lett.* **352:** 231-235.

Parsons, S.L., Watson, S.A., Brown, P.D., Collins, H.M., and Steele, R.J., 1997, Matrix metalloproteinases. *Br. J. Surg.* **84:** 160-166.

Perry, W.L., Hustad, C.M., Swing, D.A., O'Sullivan, T.N., Jenkins, N.A., and Copeland, N.G., 1998, The itchy locus encodes a novel ubiquitin protein ligase that is disrupted in a18H mice. *Nat. Genet.* **18:** 143-146.

Peters, J.M., Harris, J.R., Finley, D., (eds.), 1998, *Ubiquitin and the biology of the cell.* Plenum Press. London/New York.

Pillinger, M.H., Abramson, S.B., 1995, The neutrophil in rheumatoid arthritis. *Rheum. Dis. Clin. North Am.* **21:** 691-714.

Polette, M., Birembaut, P., 1998, Membrane-type metalloproteinases in tumor invasion. *Int. J. Biochem. Cell Biol.* **30:** 1195-1202.

Porteu, F., Brockhaus, M., Wallach, D., Engelmann, H., and Nathan, C.F., 1991, Human neutrophil elastase releases a ligand-binding fragment from the 75-kDa tumor necrosis factor (TNF) receptor. Comparison with the proteolytic activity responsible for shedding of TNF receptors from stimulated neutrophils. *J. Biol. Chem.* **266:** 18846-18853.

Prence, E.M., Dong, J.M., and Sahagian, G.G., 1990, Modulation of the transport of a lysosomal enzyme by PDGF. *J. Cell Biol.* **110:** 319-326.

Rao, N.V., Marshall, B.C., Gray, B.H., and Hoidal, J.R., 1993, Interaction of secretory leukocyte protease inhibitor with proteinase-3. *Am. J. Respir. Cell Mol. Biol.* **8:** 612-616.

Rasmussen, H.S., McCann, P.P., 1997, Matrix metalloproteinase inhibition as a novel anticancer strategy: a review with special focus on batimastat and marimastat. *Pharmacol. Ther.* **75:** 69-75.

Remold-O'Donnell, E., Parent, D., 1995, Specific sensitivity of CD43 to neutrophil elastase. *Blood* **86:** 2395-2402.

Rifkin, D.B., Mazzieri, R., Munger, J.S., Noguera, I., and Sung, J., 1999, Proteolytic control of growth factor availability. *APMIS* **107:** 80-85.

Robache-Gallea, S., Morand, V., Bruneau, J.M., Schoot, B., Tagat, E., Realo, E., Chouaib, S., and Roman-Roman, S., 1995, In vitro processing of human tumor necrosis factor-alpha. *J. Biol. Chem.* **270:** 23688-23692.

Rock, K.L., Gramm, C., Rothstein, L., Clark, K., Stein, R., Dick, L., Hwang, D., and Goldberg, A.L., 1994, Inhibitors of the proteasome block the degradation of most cell proteins and the generation of peptides presented on MHC class I molecules. *Cell* **78:** 761-771.

Rozhin, J., Sameni, M., Ziegler, G., and Sloane, B.F., 1994, Pericellular pH affects distribution and secretion of cathepsin B in malignant cells. *Cancer Res.* **54:** 6517-6525.

Sallenave, J.M., Si-Tah,.M., Cox, G., Chignard, M., and Gauldie, J., 1997, Secretory leukocyte proteinase inhibitor is a major leukocyte elastase inhibitor in human neutrophils. *J. Leukoc. Biol.* **61:** 695-702.

Salvesen, G.S., 1998, Cathepsin G. In *Handbook of proteolytic enzymes.* (A.J. Barrett, N.D. Rawlings and J.F. Woessner, eds). Academic Press, San Diego, pp. 60-62..

Samis, J.A., Kam, E., Nesheim, M.E., and Giles, A.R., 1998, Neutrophil elastase cleavage of human factor IX generates an activated factor IX-like product devoid of coagulant function. *Blood* **92:** 1287-1296.

Schaffner, M., Smith, S., and Jentsch, S., 1999, The ubiquitin-conjugatin system. In *Ubiquitin and the biology of the cell.* London, New York: Plenum Press, pp. 65-98.

Scott, F.L., Hirst, C.E., Sun, J., Bird, C.H., Bottomley, S.P., and Bird, P.I., 1999, The intracellular serpin proteinase inhibitor 6 is expressed in monocytes and granulocytes and is a potent inhibitor of the azurophilic granule protease, cathepsin G. *Blood* **93:** 2089-2097.

Scuderi, P., Nez, P.A., Duerr, M.L., Wong, B.J., and Valdez, C.M., 1991, Cathepsin-G and leukocyte elastase inactivate human tumor necrosis factor and lymphotoxin. *Cell Immunol.* **135:** 299-313.

Sexton, P.S., Cox, J.L., 1997, Inhibition of motility and invasion of B16 melanoma by the overexpression of cystatin C. *Melanoma Res.* **7:** 97-101.

Shabani, F., McNeil, J., and Tippett, L., 1998, The oxidative inactivation of tissue inhibitor of metalloproteinase-1 (TIMP-1) by hypochlorous acid (HOCI) is suppressed by anti-rheumatic drugs. *Free Radic. Res.* **28:** 115-123.

Sheaff, R.J., Roberts, J.M., 1996, End of the line: proteolytic degradation of cyclin-dependent kinase inhibitors *Chem. Biol.* **3:** 869-873.

Shinguh, Y., Yamazaki, A., Inamura, N., Fujie, K., Okamoto, M., Nakahara, K., Notsu, Y., Okuhara, M., and Ono, T., 1998, Biochemical and pharmacological characterization of FR134043, a novel elastase inhibitor. *Eur. J. Pharmacol.* **345:** 299-308.

Simpson, A.J., Maxwell, A.I., Govan, J.R., Haslett, C., and Sallenave, J.M., 1999, Elafin (elastase-specific inhibitor) has anti-microbial activity against gram-positive and gram-negative respiratory pathogens. *FEBS Lett.* **452:** 309-313.

Sires, U.I., Murphy, G., Baragi, V.M., Fliszar, C.J., Welgus, H.G., and Senior, R.M., 1994, Matrilysin is much more efficient than other matrix metalloproteinases in the proteolytic inactivation of alpha 1-antitrypsin. *Biochem. Biophys. Res. Commun.* **204:** 613-620.

Sloane, B.F., Moin, K., Sameni, M., Tait, L.R., Rozhin, J., and Ziegler, G., 1994, Membrane association of cathepsin B can be induced by transfection of human breast epithelial cells with c-Ha-ras oncogene. *J. Cell Sci.* **107:** 373-384.

Sommer, T., Wolf, D.H., 1997, Endoplasmic reticulum degradation: reverse protein flow of no return. *FASEB J.* **11:** 1227-1233.

Song, X., Zeng, L., Jin, W., Thompson, J., Mizel, D.E., Lei, K., Billinghurst, R.C., Poole, A.R., and Wahl, S.M., 1999, Secretory leukocyte protease inhibitor suppresses the inflammation and joint damage of bacterial cell wall-induced arthritis. *J. Exp. Med.* **190:** 535-542.

Steadman, R., St John, P.L., Evans, R.A., Thomas, G.J., Davies, M., Heck, L.W., and Abrahamson, D.R., 1997, Human neutrophils do not degrade major basement membrane components during chemotactic migration. *Int. J. Biochem. Cell Biol.* **29:** 993-1004.

Stiskal, J.A., O'Brien, K.K., Kelly, E.N., and Dunn, M.S., 1999, Reducing neutrophil elastase-induced lung injury. *Can. Respir.J.* **6:** 138-140.

Stolk, J., Camps, J., Feitsma, H.I., Hermans, J., Dijkman, J.H., and Pauwels, E.K., 1995, Pulmonary deposition and disappearance of aerosolised secretory leucocyte protease inhibitor. *Thorax* **50:** 645-650.

Stracke, M., Liotta, L., 1995, Molecular mechanisms of tumor cell metastasis. In *The molecular basis of cancer.* Sanders, Philadelphia, pp. 233-247.

Stricklin, G.P., Hoidal, J.R., 1992, Oxidant-mediated inactivation of TIMP. *Matrix* Suppl. **1:** 325.

Taekema-Roelvink, M.E., van Kooten, C., Janssens, M.C., Heemskerk, E., and Daha, M.R., 1998, Effect of anti-neutrophil cytoplasmic antibodies on proteinase 3- induced apoptosis of human endothelial cells. *Scand. J. Immunol.* **48:** 37-43.

Thomssen, C., Oppelt, P., Janicke, F., Ulm, K., Harbeck, N., Hofler, H., Kuhn, W., Graeff, H., and Schmitt, M., 1998, Identification of low-risk node-negative breast cancer patients by tumor biological factors PAI-1 and cathepsin L. *Anticancer Res.* **18:** 2173-2180.

Tomee, J.F., Koeter, G.H., Hiemstra, P.S., and Kauffman, H.F., 1998, Secretory leukoprotease inhibitor: a native antimicrobial protein presenting a new therapeutic option? *Thorax* **53:** 114-116.

Travis, J., Salvesen, G.S., 1983, Human plasma proteinase inhibitors. *Annu. Rev. Biochem.* **52:** 655-709.

Travis, J., Fritz, H., 1991, Potential problems in designing elastase inhibitors for therapy. *Am. Rev. Respir. Dis.* **143:** 1412-1415.

Tsunemi, M., Kato, H., Nishiuchi, Y., Kumagaye, S., and Sakakibara, S., 1992, Synthesis and structure-activity relationships of elafin, an elastase- specific inhibitor. *Biochem. Biophys. Res. Commun.* **185:** 967-973.

Varava, G.N., Barabash, R.D., Levitskii, A.P., Sukmanskii, V.B., and Konovets, V.M., 1976, Effect of electrogingivectomy and elastase inhibitor on the enzymatic activity of the gingiva during the treatment of parodontosis. *Stomatologiia* **55:** 20-23.

Venaille, T.J., Ryan, G., and Robinson, B.W., 1998, Epithelial cell damage is induced by neutrophil-derived, not pseudomonas-derived, proteases in cystic fibrosis sputum. *Respir. Med.* **92:** 233-240.

Vogelmeier. C., Gillissen, A., and Buhl, R., 1996, Use of secretory leukoprotease inhibitor to augment lung antineutrophil elastase activity. *Chest* **110:** 261S-266S.

Vogelmeier, C., Biedermann, T., Maier, K., Mazur, G., Behr, J., Krombach, F., and Buhl, R., 1997, Comparative loss of activity of recombinant secretory leukoprotease inhibitor and alpha 1-protease inhibitor caused by different forms of oxidative stress. *Eur. Respir. J.* **10:** 2114-2119.

Vogelmeier, C., Kirlath, I., Warrington, S., Banik, N., Ulbrich, E., and Du Bois, R.M.. 1997a, The intrapulmonary half-life and safety of aerosolized alpha1-protease inhibitor in normal volunteers. *Am. J. Respir. Crit Care Med.* **155:** 536-541.

Ward, C.L., Omura, S., and Kopito, R.R., 1995, Degradation of CFTR by the ubiquitin-proteasome pathway. *Cell* **83:** 121-127.

Wieczorek, M., Gyorkos, A., Spruce, L.W., Ettinger, A., Ross, S.E., Kroona, H.S., Burgos-Lepley, C.E., Bratton, L.D., Drennan, T.S., Garnert, D.L., Von Burg, G., Pilkington, C.G., and Cheronis, J.C., 1999, Biochemical characterization of alpha-ketooxadiazole inhibitors of elastases. *Arch. Biochem. Biophys.* **367:** 193-201.

Wiedow, O., Schroder, J.M., Gregory, H., Young, J.A., and Christophers, E., 1990, Elafin: an elastase-specific inhibitor of human skin. Purification, characterization, and complete amino acid sequence. *J. Biol. Chem.* **265:** 14791-14795.

Wiedow, O., Harder, J., Bartels, J., Streit, V., and Christophers, E., 1998, Antileukoprotease in human skin: an antibiotic peptide constitutively produced by keratinocytes. *Biochem. Biophys. Res. Commun.* **248:** 904-909.

Williams, C.A., Zori, R.T., Hendrickson, J., Stalker, H., Marum, T., Whidden, E., and Driscoll, D.J., 1995, Angelman syndrome. *Curr. Probl. Pediatr.* **25:** 216-231.

Winyard, P.G., Zhang, Z., Chidwick, K., Blake, D.R., Carrell, R.W., and Murphy, G., 1991, Proteolytic inactivation of human alpha 1 antitrypsin by human stromelysin. *FEBS Lett.* **279:** 91-94.

Wojtowicz-Praga, S.M., Dickson, R.B., and Hawkins, M.J., 1997, Matrix metalloproteinase inhibitors. *Invest New Drugs* **15:** 61-75.

Wu, S.M., Pizzo, S.V., 1999, Mechanism of hypochlorite-mediated inactivation of proteinase inhibition by alpha 2-macroglobulin. *Biochemistry* **38:** 13983-13990.

Yan, S., Sameni, M., and Sloane, B.F., 1998, Cathepsin B and human tumor progression. *Biol. Chem.* **379:** 113-123.

# THE ROLE OF PROTEOLYSIS IN ALZHEIMER'S DISEASE

Nigel M. Hooper, Alison J. Trew, Edward T. Parkin and Anthony J. Turner
*School of Biochemistry and Molecular Biology, University of Leeds, Leeds LS2 9JT, UK*

Key words proteinase, presenilin, lipid rafts, secretase, cholesterol.

Abstract Alzheimer's disease is characterised by the progressive deposition of the 4 kDa β-amyloid peptide (Aβ) in extracellular senile plaques in the brain. Aβ is derived by proteolytic cleavage of the amyloid precursor protein (APP) by various proteinases termed secretases. α-Secretase is inhibited by hydroxamate-based zinc metalloproteinase inhibitors such as batimastat with $I_{50}$ values in the low micromolar range, and displays many properties in common with the secretase that releases angiotensin converting enzyme. A cell impermeant biotinylated derivative of one such inhibitor completely blocked the release of APP from the surface of neuronal cells, indicating that α-secretase cleaves APP at the cell-surface. A range of hydroxamate-based compounds have been used to distinguish between α-secretase and tumour necrosis factor-α convertase, a member of the ADAMs (a disintegrin and metalloproteinase-like) family of zinc metalloproteinases. Recent data suggests that the presenilins may be aspartyl proteinases with the specificity of γ-secretase. Although APP and the presenilins are present in detergent-insoluble, cholesterol- and glycosphingolipid-rich lipid rafts, they do not behave as typical lipid raft proteins, and thus it is unclear whether these membrane domains are the sites for proteolytic processing of APP.

## 1. INTRODUCTION

Alzheimer's disease is the most common human neurodegenerative disorder affecting 10 million individuals worldwide. Approximately one person in 20 over the age of 65 and one in every five over 80 is afflicted with Alzheimer's disease. By the year 2025 it is predicted that there will be over 22 million Alzheimer's sufferers. The disease is characterised by the progressive deposition of the 4 kDa β-amyloid peptide (Aβ) in extracellular

*Cellular Peptidases in Immune Functions and Diseases 2*
Edited by Langner and Ansorge, Kluwer Academic/Plenum Publishers, 2000 

senile plaques in selected regions of the brain. Aβ is a 40-43 amino acid peptide derived by proteolytic cleavage of the β-amyloid precursor protein (APP), a type I integral membrane glycoprotein (Checler 1995, Price and Sisodia 1998, Selkoe 1998). APP is subject to proteolytic processing by proteinases termed secretases. Cleavage of APP at the N-terminus of the Aβ peptide by β-secretase and at the C-terminus by one or more γ-secretases (Fig 1) constitutes the amyloidogenic pathway for processing of APP. In addition, APP can be processed in the non-amyloidogenic pathway by α-secretase which cleaves within the Aβ domain (Fig 1) preventing deposition of this intact amyloidogenic peptide. Regulation of the balance of APP processing by the amyloidogenic and non-amyloidogenic pathways through either selective inhibition of β- and γ-secretases or activation of α-secretase can all be considered as potential therapeutic approaches to the treatment of Alzheimer's disease.

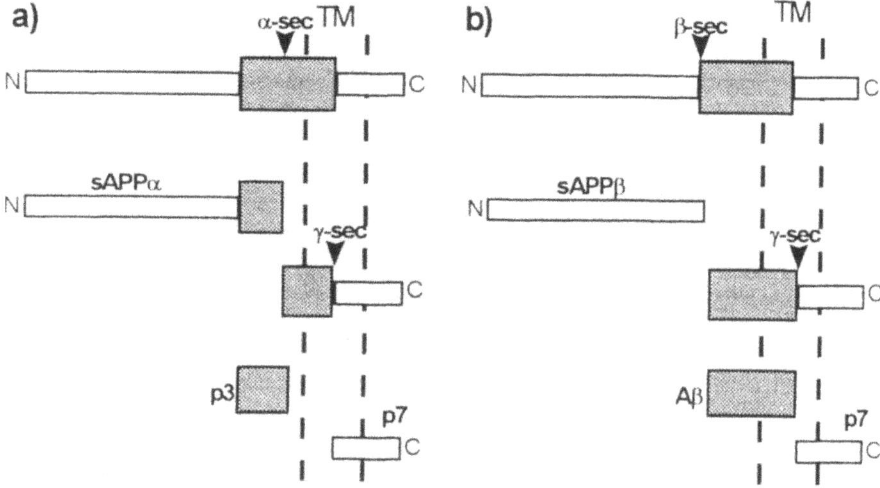


Figure 1. Schematic representation of APP and its processing by α-, β- and γ-secretases APP (open bar), which is attached to the cell surface by a transmembrane region (TM), is processed via either (a) the non-amyloidogenic or (b) amyloidogenic pathways. In the non-amyloidogenic pathway α-secretase (α-sec) cleaves APP within the Aβ peptide releasing a large soluble fragment of APP (sAPPα). In the amyloidogenic pathway β-secretase (β-sec) and γ-secretase (γ-sec) generate the Aβ peptide (grey box), and release the soluble sAPPβ.

## 2. α-SECRETASE

α-Secretase cleaves APP within the Aβ domain between Lys16 and Leu17 (Esch *et al* 1990), some 12 residues on the extracellular side of the membrane, thereby precluding formation of the Aβ peptide. This non-amyloidogenic cleavage of APP releases the large ectodomain of APP (sAPPα) which has neuroprotective properties (Barger and Harmon 1997, Furukawa *et al* 1996, Mattson *et al* 1999) and memory-enhancing effects (Meziane *et al* 1998). α-Secretase activity is known to be stimulated by phorbol esters and other agents that activate protein kinase C, including the muscarinic agonist carbachol (Buxbaum *et al* 1992, Caporaso *et al* 1992), resulting in the regulated secretion of sAPPα. Although α-secretase has yet to be isolated, this proteinase appears to be an integral membrane protein (Roberts *et al* 1994, Sisodia 1992). Of a range of class-specific proteinase inhibitors examined, only the zinc-chelating agent, 1,10-phenanthroline caused significant inhibition of α-secretase activity (Roberts *et al* 1994). More recently, we have shown that α-secretase is inhibited by hydroxamate-based zinc metalloproteinase inhibitors such as batimastat, BB2116 and marimastat (Parvathy *et al* 1998a) (Table 1). The release of APP by α-secretase from both SH-SY5Y and IMR-32 neuronal cells was blocked by batimastat with $I_{50}$ values of approx. 3 μM. In contrast, neither the level of full-length APP nor its cleavage by β-secretase were affected. Batimastat, marimastat and BB2116 are active site directed inhibitors that co-ordinate to the essential zinc ion and were originally designed as inhibitors of matrix metalloproteinases. However, α-secretase does not appear to be a matrix metalloproteinase as batimastat, marimastat and BB2116 inhibit collagenase with $I_{50}$ values in the low nM range, and several potent matrix metalloproteinase inhibitors have no effect on the activity of α-secretase.

*Table 1.* Inhibition profiles of α-secretase, ACE secretase and TNF-α convertase with hydroxamate-based compounds. Data from Parvathy *et al*., 1997; 1998a; 1998b. n.d., not determined.

| | $I_{50}$ (μM) | | |
|---|---|---|---|
| Compound | α-secretase | ACE secretase | TNF-α convertase |
| Batimastat | 3.3 | 1.6 | 0.019 |
| Compound 1 | 17.9 | 35.0 | 0.037 |
| Compound 4 | > 20 | > 100 | 0.18 |
| Marimastat | 1.2 | 8.3 | n.d. |
| BB2116 | 7.7 | 3.5 | n.d. |

Previously, we had noted similarities between α-secretase and the secretase that releases angiotensin converting enzyme (ACE) from the cell-surface (Oppong and Hooper 1993, Parvathy *et al* 1997). ACE plays a key role in the control of blood pressure homeostasis and inhibitors of the enzyme are used clinically in the treatment of hypertension and congestive heart failure. Like APP, ACE is a Type I integral membrane glycoprotein which is subject to a post-translational proteolytic processing event that releases a soluble form of the protein from the cell-surface. α-Secretase and

ACE secretase display a similar inhibition profile with a range of hydroxamate-based zinc metalloproteinase inhibitors (Parvathy *et al* 1998a) (Table 1). The two secretases share a number of other properties in common. Both cleave their respective substrates between a basic and a hydrophobic residue (Lys-Leu and Arg-Leu) and are stimulated by phorbol esters. Furthermore, both secretases appear to be integral membrane proteins, resistant to removal from the membrane by high salt, and solubilized by the detergents Triton X-100 and CHAPS but not by octyl glucoside. Thus α-secretase and ACE secretase appear to be closely related integral membrane zinc metalloproteinases. The cellular location of the α-secretase cleavage of APP has been the subject of much debate. There is conflicting data as to whether α-secretase cleaves APP at the cell-surface, within an intracellular compartment, or at both locations (discussed in Checler 1995, Evin *et al* 1994, Gasparini *et al* 1998). In order to address this we have exploited our recent observation that α-secretase is inhibited by hydroxamate-based zinc metalloproteinase inhibitors (Parvathy *et al* 1998a) to determine the relative contribution of cell-surface to intracellular α-secretase activity by using a cell-impermeant biotinylated derivative of one such inhibitor (Parvathy *et al* 1999). These results clearly show that the constitutive α-secretase cleavage of APP occurs exclusively at the cell-surface. This observation has significant implications for the regulation of the cellular processing of APP in the light of recent reports that β- and γ-secretase can potentially act on APP in the secretory pathway (Chyung *et al* 1997, Cook *et al* 1997, Hartmann *et al* 1997, Wild-Bode *et al* 1997) before it reaches the cell-surface where α-secretase acts. Thus there must be additional, as yet unidentified, control mechanisms in place under normal physiological conditions to ensure that APP is processed primarily by the non-amyloidogenic pathway to produce the neuroprotective sAPPα.

## 3. ADAMS PROTEASES

It is now apparent that the release of a range of structurally and functionally diverse membrane proteins, including APP, ACE, tumour necrosis factor (TNF)-α, L-selectin, transforming growth factor-α, interleukin-6 receptor and TNF receptors I and II, is blocked by inhibitors such as batimastat and BB2116 (Arribas *et al* 1996, Hooper *et al* 1997). The enzyme that cleaves pro-TNF-α, named TNF-α convertase, has been isolated, cloned and sequenced (Black *et al* 1997, Moss *et al* 1997), and the structure of its catalytic domain determined (Maskos *et al* 1998). TNF-α convertase, a membrane-bound enzyme, belongs to the reprolysin, adamalysin or ADAMs (A Disintegrin And Metalloproteinase) family of zinc metalloproteinases that are similar to, but distinct from, the matrix metalloproteinases (Black and White 1998, Wolfsberg and White, 1996). Recent data suggests that this enzyme is involved in the release of a number of cell-surface proteins (Peschon *et al* 1998). The phorbol ester-induced release of sAPPα is deficient in TNF-α convertase knockout non-neuronal cells, although the uninduced, constitutive release of sAPPα was unaffected (Buxbaum *et al* 1998, Merlos-Suarez *et al* 1998). Although TNF-α convertase was shown to cleave a synthetic peptide spanning the α-secretase cleavage site in APP, the kinetics of this cleavage were not reported, and it was not shown that the enzyme could cleave full-length, membrane-bound APP (Buxbaum *et al* 1998). Previously we have shown that the constitutive release of sAPPα from the human neuroblastoma SH-SY5Y cell line is inhibited by batimastat, but not by the structurally related compound 4 (Parvathy *et al* 1998a), whereas TNF-α convertase is inhibited by both compounds (Parvathy *et al* 1998b) (Table 1). Reverse-transcription PCR and Northern blot analysis has revealed that TNF-α convertase is expressed in SH-SY5Y cells. However, although the carbachol-stimulated release of sAPPα from the SH-SY5Y cells was completely blocked by batimastat, compound 4 had no inhibitory effect (Trew, Turner and Hooper, unpublished). Batimastat inhibits TNF-α convertase with an $IC_{50}$ of 19 nM (Parvathy *et al* 1998b) and a $K_i$ of 11 nM (Roghani *et al* 1999), with maximal inhibition being observed at 100 nM in a cell-based assay (Lum *et al* 1999). In contrast both the constitutive and carbachol-stimulated release of sAPPα from the SH-SY5Y cells were inhibited by batimastat with $IC_{50}$ values of 0.3-1.2 μM (Trew, Turner and Hooper, unpublished). These observations indicate that TNF-α convertase is not involved in the regulated secretion of sAPPα from the SH-SY5Y cells. The observation that batimastat acts with similar potency on both the constitutive and carbachol-stimulated release of sAPPα suggests that a single α-secretase, possibly the same as ACE secretase, is involved in the constitutive and regulated cleavage of APP. Recently it has been reported that another ADAM proteinase, ADAM 10, is involved in the constitutive and regulated secretion of sAPPα from human embryonic kidney cells (Lammich *et al* 1999). It remains to be determined whether ADAM 10 or any other ADAM proteinases are responsible for the α-secretase cleavage of APP in neuronal cells.

## 4. PRESENILINS AND γ-SECRETASE

The generation of the Aβ peptide from the membrane-associated fragment resulting from β-secretase cleavage of APP requires the action of γ-secretase which cleaves at the C-terminus of the Aβ peptide sequence (Fig 1). Originally, it was proposed that γ-secretase acted in an acidic, endosomal compartment following endocytosis of cell-surface APP and its cleavage by β-secretase, with the secretion of the released Aβ40 peptide (Evin *et al* 1994). However, more recent results indicate that Aβ, in particular the longer, more amyloidogenic form Aβ42, is generated in the ER within the secretory pathway (Cook *et al* 1997, Hartmann *et al* 1997, Wild-Bode *et al* 1997). Whether this alternative processing of APP in the secretory pathway occurs in all cells or just neurons is unclear. It is also currently unclear whether the γ-secretase that produces Aβ40 is distinct from the activity that generates Aβ42 (Citron *et al* 1996, Klafki *et al* 1996, Yamazaki *et al* 1997). Recent data using difluoro ketone-based inhibitors suggests that γ-secretase may be an aspartyl proteinase (Wolfe *et al* 1998, Wolfe *et al* 1999a). Because the proposed site of cleavage by γ-secretase lies within the membrane-spanning domain of APP (see Fig 1), it is unclear how this enzyme acts.

The presenilin genes which have been shown to be major loci for early onset familial Alzheimer's disease, encode the proteins presenilin-1 (PS-1) and presenilin-2 (PS-2) (Levy-Lahad *et al* 1995, Rogaev *et al* 1995, Sherrington *et al* 1995). The presenilin proteins are highly homologous (67 % sequence identity) multiple membrane spanning proteins that are also highly similar to SEL-12, a *Caenorhabditis elegans* protein involved in the Notch signalling pathway (Levitan *et al* 1996, Li and Greenwald 1996). Full-length PS-1 and PS-2 exhibit a molecular mass of approx. 50 kDa upon electrophoresis under denaturing conditions. In most systems studied to date though, the full-length molecules are proteolytically cleaved within the large hydrophilic loop domain encoded by exon 10 to yield a C-terminal fragment (CTF) of ~17-20 kDa and an N-terminal fragment (NTF) of ~25-35 kDa. In human brain tissue we found that PS-1 existed almost exclusively as its cleaved fragments, while PS-2 was present predominantly as the full-length protein (Parkin *et al* 1999a). This cleavage of the full-length presenilins is highly regulated and the resulting heterodimers form a stable complex which is the biological and pathological functional unit (Capell *et al* 1998, Steiner *et al* 1998, Thinakaran *et al* 1998, Tomita *et al* 1998). However, the proteinase (termed presenilinase) involved in cleaving the full-length presenilins into the functional NTF and CTF complex has yet to be identified. As the action of presenilinase is one of activation, inhibition of this enzyme may provide a novel therapeutic approach to the treatment of Alzheimer's disease.

Mutations in the presenilins associated with familial Alzheimer's disease lead to altered processing of APP and the accumulation of the more amyloidogenic Aβ42 peptide (Citron *et al* 1998). Neuronal cells derived from PS-1 knock-out mouse embryos, which are phenotypically similar to Notch knock-out mice, have been shown to secrete 3-4-fold less Aβ than wild-type neuronal cells (De Strooper *et al* 1998), implicating a role for PS-1 in the γ-secretase cleavage of APP. Recently it has been reported that mutation of either of two conserved transmembrane aspartate residues in PS-1, Asp257 and Asp385, substantially reduces the production of Aβ and increases the amounts of the C-terminal fragments of APP that are the substrates of γ-secretase (Wolfe *et al* 1999b). In addition, mutating either of the Asp residues to Ala prevented the normal endoproteolysis of PS-1. These results indicate that the two transmembrane aspartate residues are critical for both PS-1 endoproteolysis and γ-secretase activity, and suggest that PS-1 is either a unique diaspartyl cofactor for γ-secretase or is itself γ-secretase, an autoactivated intramembranous aspartyl proteinase.

## 5. LIPID RAFTS

A recent study has shown that cholesterol depletion inhibits the generation of Aβ in hippocampal neurons due to inhibition of β-secretase (Simons *et al* 1998). The cholesterol transporting protein ApoE4 is a known risk factor for Alzheimer's disease, further implicating alterations in cholesterol metabolism in regulating the proteolytic processing of APP. There are differing reports as to whether APP is present in the detergent-insoluble, cholesterol- and glycosphingolipid-rich membrane microdomains (DIGs) or lipid rafts (Hooper 1999, Simons and Ikonen 1997). A minor proportion of APP was detected in DIGs isolated from hippocampal neurons (Simons *et al* 1998) and cortical neurons (Bouillot *et al* 1996), depletion of cellular cholesterol was shown to reduce the association of APP with DIGs (Simons *et al* 1998), and α-secretase, at least in non-neuronal cells, has been reported to be located in DIGs/caveolae (Ikezu *et al* 1998). A minor proportion of APP, as well as a small proportion of the NTF and CTF of PS-1, were found in DIGs isolated from rat brain grey matter (Lee *et al* 1998). Using a method in which the DIGs fraction is not contaminated with known non-DIG marker proteins, we were unable to detect APP in DIGs isolated from mouse cerebellum or SH-SY5Y cells (Parkin *et al* 1997). In agreement with this result is a previous observation in which APP was shown to be totally excluded from the detergent-insoluble fraction in hippocampal neurons (Tienari *et al* 1996) and a recent report which failed to detect APP in DIGs isolated from SH-SY5Y cells (Morishima-Kawashima and Ihara 1998). Although more recently we have shown that a minor proportion of APP is present in DIGs from mouse cerebral cortex, where it behaves as an atypical DIGs protein (Parkin *et al* 1999b). Using an identical isolation procedure we have demonstrated also that full-length PS-2 and the NTF and CTF of PS-1 are present in DIGs isolated from human and mouse cerebral cortex (Parkin *et al* 1999a). Thus cholesterol- and glycosphingolipid-rich lipid rafts may play a role in regulating the lateral separation of APP and the presenilins within the lipid bilayer, and alterations in the cellular and/or extracellular levels of cholesterol may affect the distribution of the proteins in these structures, allowing them to associate/disassociate from their various proteinases.

## 6. CONCLUSIONS

Although the deposition of the Aβ peptide appears to be a key step in the pathogenesis of Alzheimer's disease, still little is known about the proteases involved in cleaving APP. α-Secretase is almost certainly a zinc metalloproteinase and may be a member of the ADAMs family, although it appears to be distinct from TNF-α convertase. Recent data suggests that PS-1 may be γ-secretase, although further work is required to confirm this, while the identity of β-secretase remains a mystery. The unequivocal identification and subsequent characterisation of the APP secretases is important for the development of therapeutic strategies to control the accumulation of Aβ in the brain and the subsequent pathological effects of Alzheimer's disease.

## ACKNOWLEDGMENTS

We thank the Medical Research Council of Great Britain and the Wellcome Trust for financial support of this work.

## REFERENCES

Arribas, J., Coodly, L., Vollmer, P., Kishimoto, T. K., Rose-John, S. and Massague, J., 1996, Diverse cell surface protein ectodomains are shed by a system sensitive to metalloprotease inhibitors. *J. Biol. Chem.* **271:** 11376-11382.

Barger, S. W. and Harmon, A. D., 1997, Microglial activation by Alzheimer amyloid precursor protein and modulation by apolipoprotein E. *Nature* **388:** 878-881.

Black, R. A., Rauch, C. T., Kozlosky, C. J., Peschon, J. J., Slack, J. L., Wolfson, M. F., Castner, B. J., Stocking, K. L., Reddy, P., S., S. *et al.*, 1997, A metalloproteinase disintegrin that releases tumour-necrosis factor-α from cells. *Nature* **385:** 729-733.

Black, R. A. and White, J. M., 1998, ADAMs: focus on the protease domain. *Curr. Opin. Cell Biol.* **10:** 654-659.

Bouillot, C., Prochiantz, A., Rougon, G. and Allinquant, B., 1996, Axonal amyloid precursor protein expressed by neurons *in vitro* is present in a membrane fraction with caveolae-like properties. *J. Biol. Chem.* **271:** 7640-7644.

Buxbaum, J. D., Oishi, M., Chen, H. I., Pinkas-Kramarski, R., Jaffe, E. A., Gandy, S. E. and Greengard, P., 1992, Cholinergic agonists and interleukin 1 regulate processing and secretion of the Alzheimer β/A4 amyloid protein precursor. *Proc. Natl. Acad. Sci. USA* **89:** 10075-10078.

Buxbaum, J. D., Liu, K.-N., Luo, Y., Slack, J. L., Stocking, K. L., Peschon, J. J., Johnson, R. S., Castner, B. J., Cerretti, D. P. and Black, R. A., 1998, Evidence that tumor necrosis factor α converting enzyme is involved in regulated α-secretase cleavage of the Alzheimer amyloid protein precursor. *J. Biol. Chem.* **273:** 27765-27767.

Capell, A., Grunberg, J., Pesold, B., Diehlmann, A., Citron, M., Nixon, R., Beyreuther, K., Selkoe, D. J. and Haass, C., 1998, The proteolytic fragments of the Alzheimer's disease-associated presenilin-1 form heterodimers and occur as a 100-150kDa molecular mass complex. *J. Biol. Chem.* **273:** 3205-3211.

Caporaso, G. L., Gandy, S. E., Buxbaum, J. D., Ramabhadran, T. V. and Greengard, P., 1992, Protein phosphorylation regulates secretion of Alzheimer β/A4 amyloid precursor protein. *Proc. Natl. Acad. Sci. USA* **89:** 3055-3059.

Checler, F., 1995, Processing of the β-amyloid precursor protein and its regulation in Alzheimer's disease. *J. Neurochem.* **65:** 1431-1444.

Chyung, A. S. C., Greenberg, B. D., Cook, D. G., Doms, R. W. and Lee, V. M.-Y., 1997, Novel β-secretase cleavage of β-amyloid precursor protein in the endoplasmic reticulum/intermediate compartment of NT2N cells. *J. Cell Biol.* **138:** 671-680.

Citron, M., Diehl, T. S., Gordon, G., Biere, A. L., Seubert, P. and Selkoe, D. J., 1996, Evidence that the 42- and 40-amino acid forms of amyloid β protein are generated from the β-amyloid precursor protein by different protease activities. *Proc. Natl. Acad. Sci. USA* **93:** 13170-13175.

Citron, M., Eckman, C.B., Diehl, T.S., Corcoran, C., Ostaszewski, B.L., Xia, W., Levesque, G., Hyslop, P.S.G., Younkin, S.G. and Selkow, D.J., 1998, Additive effectis of PS1 and APP mutations on secretion of the 42-residue amyloid β-protein. *Neurobiol. Dis.* **5:** 107-116.

Cook, D.G., Forman, M. S., Sung, J.C., Leight, S., Kolson, D.L., Iwatsubo, T., Lee, V.M.-Y. and Doms, R. W.,1997, Alzheimer's Aβ(1-42) is generated in the endoplasmic reticulum/intermediate compartment of NT2N cells. *Nature Med.* **3:** 1021-1023.

De Strooper, B., Saftig, P., Craessaerts, K., Vanderstichele, H., Guhde, G., Annaert, W., Von Figura, K. and Van Leuven, F., 1998, Deficiency of presenilin-1 inhibits the normal cleavage of amyloid precursor protein. *Nature* **329:** 387-390.

Esch, F. S., Keim, P. S., Beattie, E. C., Blacher, R. W., Culwell, A. R., Oltersdorf, T., McClure, D. and Ward, P. J., 1990, Cleavage of amyloid β peptide during constitutive processing of its precursor. *Science* **248:** 1122-1124.

Evin, G., Beyreuther, K. and Masters, C. L., 1994, Alzheimer's disease amyloid precursor protein (AβPP): proteolytic processing, secretases and βA4 amyloid production. *Amyloid: Int. J. Exp. Clin. Invest.* **1:** 263-280.

Furukawa, K., Sopher, B. L., Rydel, R. E., Begley, J. G., Pham, D. G., Martin, G. M., Fox, M. and Mattson, M. P., 1996, Increased activity-regulating and neuroprotective efficacy of α-secretase-derived secreted amyloid precursor protein conferred by a C-terminal heparin-binding domain. *J. Neurochem.* **67:** 1882-1896.

Gasparini, L., Racchi, M., Binetti, G., Trabucchi, M., Solerte, S. B., Alkon, D., Etcheberrigaray, R., Gibson, G., Blass, J., Paoletti, R. *et al.*, 1998, Peripheral markers in testing pathophysiological hypotheses and diagnosing Alzheimer's disease. *FASEB J.* **12:** 17-34.

Hartmann, T., Bieger, S. C., Bruhl, B., Tienari, P. J., Ida, N., Allsop, D., Roberts, G. W.,

Masters, C. L., Dotti, C. G., Unsicker, K. *et al.* (1997). Distinct sites of intracellular production for Alzheimer's disease Aβ40/42 amyloid peptides. *Nature Med.* **3**, 1016-1020.

Hooper, N. M., Karran, E. H. and Turner, A. J., 1997, Membrane protein secretases. *Biochem. J.* **321:** 265-279.

Hooper, N. M., 1999, Detergent-insoluble glycosphingolipid/cholesterol-rich membrane domains, lipid rafts and caveolae. *Molecular Membrane Biology* **16:** 145-156.

Ikezu, T., Trapp, B. D., Song, K. S., Schlegel, A., Lisanti, M. P. and Okamoto, T., 1998, Caveolae, plasma membrane microdomains for α-secretase-mediated processing of the amyloid precursor protein. *J. Biol. Chem.* **273:** 10485-10495.

Klafki, H.-W., Abramowski, D., Swoboda, R., Paganetti, P. A. and Staufenbiel, M., 1996, The carboxyl termini of β-amyloid peptides 1-40 and 1-42 are generated by distinct γ-secretase activities. *J. Biol. Chem.* **271:** 28655-28659.

Lammich, S., Kojro, E., Postina, R., Gilbert, S., Pfeiffer, R., Jasionowski, M., Haass, C. and Fahrenholz, F., 1999, Constitutive and regulated α-secretase cleavage of Alzheimer's amyloid precursor protein by a disintegrin metalloprotease. *Proc. Natl. Acad. Sci. USA* **96:** 3922-3927.

Lee, S.-J., Liyanage, U., Bickel, P. E., Xia, W., Lansbury, P. T. J. and Kosik, K. S., 1998, A detergent-insoluble membrane compartment contains Aβ *in vivo*. *Nature Med.* **4:** 730-734.

Levitan, D., Doyle, T. G., Brousseau, D., Lee, M. K., Thinakaran, G., Slunt, H. H., Sisodia, S. S. and Greenwald, I., 1996, Assessment of normal and mutant human presenilin function in *Caenorhabditis elegans*. *Proc. Natl. Acad. Sci. USA* **93:** 14940-14944.

Levy-Lahad, E., Wasco, W., Poorkaj, P., Romano, D. M., Oshima, J., Pettingell, W. H., Yu, C., Jondro, P. D., Schmidt, S. D., Wang, K. *et al.*, 1995, Candidate gene for the chromosome 1 familial Alzheimer's disease locus. *Science* **269:** 973-977.

Li, X. and Greenwald, I., 1996, Membrane topology of the C. elegans SEL-12 presenilin. *Neuron* **17:** 1015-1021.

Lum, L., Wong, B. R., Josien, R., Becherer, J. D., Erdjument-Bromage, H., Schlondorff, J., Tempst, P., Choi, Y. and Blobel, C. P., 1999, Evidence for a role of a tumor necrosis factor-alpha (TNF-alpha)- converting enzyme-like protease in shedding of TRANCE, a TNF family member involved in osteoclastogenesis and dendritic cell survival. *J Biol Chem* **274:** 13613-13618.

Maskos, K., Fernandez-Catalan, C., Huber, R., Bourenkov, G. P., Bartunik, H., Ellestad, G. A., Reddy, P., Wolfson, M. F., Rauch, C. T., Castner, B. J. et al., 1998, Crystal structure of the catalytic domain of human tumor necrosis factor-α-converting enzyme. *Proc. Natl. Acad. Sci. USA* **95**: 3408-3412.

Mattson, M. P., Guo, Z. H. and Geiger, J. D., 1999, Secreted form of amyloid precursor protein enhances basal glucose and glutamate transport and protects against oxidative impairment of glucose and glutamate transport in synaptosomes by a cyclic GMP-mediated mechanism. *J. Neurochem.* **73:** 532-537.

Merlos-Suarez, A., Fernandez-Larrea, J., Reddy, P., Baselga, J. and Arribas, J., 1998, Pro-tumor necrosis factor-α processing activity is tightly controlled by a component that does not affect notch processing. *J. Biol. Chem.* **273:** 24955-24962.

Meziane, H., Dodart, J.-C., Mathis, C., Little, S., Clemens, J., Paul, S. M. and Ungerer, A., 1998, Memory-enhancing effects of secreted forms of the β-amyloid precursor protein in normal and amnestic mice. *Proc. Natl. Acad. Sci. USA* **95:** 12683-12688.

Morishima-Kawashima, M. and Ihara, Y., 1998, The presence of amyloid β-protein in the detergent-insoluble membrane compartment of human neuroblastoma cells. *Biochemistry* **37:** 15247-15253.

Moss, M. L., Jin, S.-L. C., Milla, M. E., Burkhart, W., Carter, H. L., Chen, W.-J., Clay, W. C., Didsbury, J. R., Hassler, D., Hoffman, C. R. et al., 1997, Cloning of a disintegrin metalloprotease that processes precursor tumour-necrosis factor-α. *Nature* **385:** 733-736.

Oppong, S. Y. and Hooper, N. M., 1993, Characterization of a secretase activity which releases angiotensin-converting enzyme from the membrane. *Biochem. J.* **292:** 597-603.

Parkin, E. T., Hussain, I., Turner, A. J. and Hooper, N. M., 1997, The amyloid precursor protein is not enriched in caveolae-like, detergent-insoluble membrane microdomains. *J. Neurochem.* **69:** 2179-2188.

Parkin, E. T., Hussain, I., Karran, E. H., Turner, A. J. and Hooper, N. M., 1999a, Characterisation of detergent-insoluble complexes containing the familial Alzheimer's disease-associated presenilins. *J. Neurochem.* **72:** 1534-1543.

Parkin, E. T., Turner, A. J. and Hooper, N. M. (1999b). Amyloid precursor protein, although partially detergent-insoluble in mouse cerebral cortex, behaves as an atypical lipid raft protein. *Biochem. J.* in press.

Parvathy, S., Oppong, S. Y., Karran, E. H., Buckle, D. R., Turner, A. J. and Hooper, N. M., 1997, Angiotensin converting enzyme secretase is inhibited by metalloprotease inhibitors and requires its substrate to be inserted in a lipid bilayer. *Biochem. J.* **327:** 37-43.

Parvathy, S., Hussain, I., Karran, E. H., Turner, A. J. and Hooper, N. M., 1998a, Alzheimer's amyloid precursor protein α-secretase is inhibited by hydroxamic acid-based zinc metalloprotease inhibitors: similarities to the angiotensin converting enzyme secretase. *Biochemistry* **37:** 1680-1685.

Parvathy, S., Karran, E. H., Turner, A. J. and Hooper, N. M., 1998, The secretases that cleave angiotensin converting enzyme and the amyloid precursor protein are distinct from tumour necrosis factor-α convertase. *FEBS Lett.* **431:** 63-65.

Parvathy, S., Hussain, I., Karran, E. H., Turner, A. J. and Hooper, N. M., 1999, Cleavage of the Alzheimer's amyloid precursor protein by α-secretase occurs at the surface of neuronal cells. *Biochemistry* **38:** 9728-9734.

Peschon, J. J., Slack, J. L., Reddy, P., Stocking, K. L., Sunnarborg, S. W., Lee, D. C., Russell, W. E., Castner, B. J., Johnson, R. S., Fitzner, J. N. *et al.*, 1998, An essential role for ectodomain shedding in mammalian development. *Science* **282:** 1281-1284.

Price, D. L. and Sisodia, S. S., 1998, Mutant genes in familial Alzheimer's disease and transgenic models. *Annu. Rev. Neurosci.* **21:** 479-505.

Roberts, S. B., Ripellino, J. A., Ingalls, K. M., Robakis, N. K. and Felsenstein, K. M., 1994, Non-amyloidogenic cleavage of the β-amyloid precursor protein by an integral membrane metalloendopeptidase. *J. Biol. Chem.* **269:** 3111-3116.

Rogaev, E. I., Sherrington, R., Rogaeva, E. A., Levesque, G., Ikeda, M., Liang, Y., Chi, H., Lin, C., Holman, K., Tsuda, T. *et al.*, 1995, Familial Alzheimer's disease in kindreds with missense mutations in a gene on chromosome 1 related to the Alzheimer's disease type 3 gene. *Nature* **376:** 775-778.

Roghani, M., Becherer, J. D., Moss, M. L., Atherton, R. E., Erdjument-Bromage, H., Arribas, J., Blackburn, R. K., Weskamp, G., Tempst, P. and Blobel, C. P., 1999, Metalloprotease-disintegrin MDC9: intracellular maturation and catalytic activity. *J Biol Chem* **274:** 3531-3540.

Selkoe, D. J., 1998, The cell biology of β-amyloid precursor protein and presenilin in Alzheimer's disease. *Trends Cell Biol.* **8:** 447-453.

Sherrington, R., Rogaev, E. I., Liang, Y., Rogaeva, E. A., Levesque, G., Ikeda, M., Chi, H., Lin, C., Li, G., Holman, K. *et al.*, 1995, Cloning of a gene bearing missense mutations in early-onset familial Alzheimer's disease. *Nature* **375:** 754-760.

Simons, K. and Ikonen, E., 1997, Functional rafts in cell membranes. *Nature* **387:** 569-572.

Simons, M., Keller, P., De Strooper, B., Beyreuther, K., Dotti, C. G. and Simons, K., 1998), Cholesterol depletion inhibits the generation of β-amyloid in hippocampal neurons. *Proc. Natl. Acad. Sci. USA* **95:** 6460-6464.

Sisodia, S., 1992, β-amyloid precursor protein cleavage by a membrane-bound protease. *Proc. Natl. Acad. Sci. USA* **89:** 6075-6079.

Steiner, H., Capell, A., Pesold, B., Citron, M., Kloetzel, P. M., Selkoe, D. J., Romig, H., Mendla, K. and Haass, C., 1998, Expression of Alzheimer's disease--associated presenilin-1 is controlled by proteolytic degradation and complex formation. *J. Biol. Chem.* **273:** 32322-32331.

Thinakaran, G., Regard, J. B., Bouton, C. M. L., Harris, C. L., Price, D. L., Borchelt, D. R. and Sisodia, S. S., 1998, Stable association of presenilin derivatives and absence of presenilin interactions with APP. *Neurobiol. Dis.* **4:** 438-453.

Tienari, P. J., de Strooper, B., Ikonen, E., Simons, M., Weidemann, A., Czech, C., Hartmann, T., Ida, N., Multhaup, G., Masters, C. L. et al., 1996, The β-amyloid domain is essential for axonal sorting of amyloid precursor protein. *EMBO J.* **15:** 5218-5229.

Tomita, T., Tokuhiro, S., Hashimoto, T., Aiba, K., Saido, T. C., Maruyama, K. and Iwatsubo, T., 1998, Molecular dissection of domains in mutant presenilin 2 that mediate overproduction of amyloidogenic forms of amyloid β peptides. *J. Biol. Chem.* **273:** 21153-21160.

Wild-Bode, C., Yamazaki, T., Capell, A., Leimer, U., Steiner, H., Ihara, Y. and Haass, C., 1997, Intracellular generation and accumulation of amyloid β-peptide terminating at amino acid 42. *J. Biol. Chem.* **272:** 16085-16088.

Wolfe, M. S., Citron, M., Diehl, T. S., Xia, W., Donkor, I. O. and Selkoe, D. J., 1998, A substrate-based difluoro ketone selectively inhibits Alzheimer's γ-secretase activity. *J. Med. Chem.* **41:** 6-9.

Wolfe, M. S., Xia, W., Moore, C. L., Leatherwood, D. D., Ostaszewski, B., Rahmati, T., Donkor, I. O. and Selkoe, D. J., 1999a, Peptidomimetic probes and molecular modeling suggest that Alzheimer's γ-secretase is an intramembrane-cleaving aspartyl protease. *Biochemistry* **38:** 4720-4727.

Wolfe, M. S., Xia, W., Ostaszewski, B. L., Diehl, T. S., W.T., K. and Selkoe, D. J., 1999b, Two transmembrane aspartates in presenilin-1 required for presenilin endoproteolysis and γ-secretase activity. *Nature* **398:** 513-517.

Wolfsberg, T. G. and White, J. M. 1996, ADAMs in fertilization and development. *Dev. Biol.* **180:** 389-401.

Yamazaki, T., Haass, C., Saido, C. T., Omura, S. and Ihara, Y., 1997, Specific increase in amyloid β-protein 42 secretion ratio by Calpain inhibition. *Biochemistry* **36:** 8377-8383.

# OBSERVING PROTEASES IN LIVING CELLS

Kamiar Moin*[†], Lisa Demchik[†], Jianxin Mai*, Jan Duessing[¶], Christoph Peters[¶] and Bonnie F. Sloane*[†]

**Department of Pharmacology, Wayne State University School of Medicine †Barbara Ann Karmanos Cancer Institute, Detroit, Michigan 48201, USA, and*

*¶Albert Ludwigs University, Freiburg, Germany*

*To whom all correspondence should be addressed: Dr. Bonnie F. Sloane, Department of Pharmacolog, Wayne State University School of Medicine, 540 East Canfield, Detroit, Michigan 48201, USA*

Abstract The lysosomal cysteine protease cathepsin B has been implicated in tumor progression and metastasis in part due to its altered trafficking. In order to analyze the trafficking of cathepsin B in living cells, we utilized enhanced green fluorescent protein (EGFP) fused to various cathepsin B constructs for transfecting two cell lines: an invasive human breast adenocarcinoma cell line (BT20) and a cathepsin B deficient mouse embryonic fibroblast cell line (MEF T -/-). The cells were transiently transfected with four cathepsin B-EGFP fusion constructs: full-length preprocathepsin B-EGFP, cathepsin B pre region-EGFP, cathepsin B prepro region-EGFP, and cathepsin B prepro region-EGFP with a mutation of the glycosylation site in the pro region. The full length construct showed vesicular distribution throughout the cells in both cell lines. In both BT20 and MEF T -/- cells, preregion-EGFP was localized in a ring tightly associated with the cell nucleus, suggesting distribution to the endoplasmic reticulum. The distribution of the prepro region-EGFP construct was similar except that it also included some patchy areas adjacent to the nucleus. This suggested that the cathepsin B prepro region-EGFP might have entered the Golgi. Distribution of the mutated cathepsin B prepro region-EGFP was similar to that of wild-type prepro region-EGFP in the MEF T -/-. In the invasive BT20 cells, however, the mutated prepro region-EGFP showed a vesicular distribution throughout the cytoplasm and in cell processes. This distribution is similar to that of endogenous cathepsin B in these cells. Our results suggest that: 1) tumor cells have an alternative mechanism for trafficking of cathepsin B which is independent of the mannose-6-phosphate receptor pathway, and 2) the pro region of cathepsin B may contain the sorting sequence necessary for its trafficking via this pathway.

# 1. INTRODUCTION

Proteolytic enzymes have been implicated in the invasive processes that accompany malignant progression. One such enzyme is the cysteine protease cathepsin B. In normal cells, cathepsin B is involved in intracellular protein degradation within lysosomes. Trafficking of lysosomal enzymes to this compartment has been shown to occur mainly via the mannose-6-phosphate receptor (MPR) pathway (for review see Kornfeld, 1987). Alternative pathways for lysosomal enzyme trafficking, however, have been shown to occur in the slime mold (Cardelli *et al* 1986), yeast (Conibear *et al* 1998), and mammalian cells (McIntyre and Erickson, 1991). In tumors, the localization of cathepsin B often changes in parallel with malignant transformation. There can be constitutive secretion of procathepsin B, inducible secretion of active cathepsin B, association of cathepsin B (activity/protein) with plasma membrane/endosomal fractions, and association of cathepsin B (protein) with vesicles at the cell periphery and with the plasma membrane (for review, see Sloane *et al* 1994). This altered trafficking would be consistent with cathepsin B playing a role in tumor invasion. Previous studies by us indicated that cathepsin B from normal and tumor cells was not different structurally (Moin *et al* 1992). In a recent study, however, we did show that cathepsin B in tumors existed as different isoforms as compared to that from normal tissues, and that conversion of cathepsin B from single chain to double chain may not occur or may occur more slowly in tumors (Moin *et al* 1999).

Whether cathepsin B trafficking in tumor cells occurs via an alternative pathway is not known. Previous studies on localization and trafficking of cathepsin B used immunocytochemical methods and fixed cells (Calkins *et al* 1998, Sameni *et al* 1995, Campo *et al* 1994, Rempel *et al* 1994, Erdel *et al* 1990, Krepela *et al* 1987). These types of studies are not capable of analyzing the dynamics of cathepsin B trafficking because only the protein that has already been synthesized can be observed. The ideal method to study the dynamics of protein trafficking is in living cells. Until recently such studies were beyond our reach. The discovery of the green fluorescent protein (GFP) from the jellyfish, *Aequoria victoria,* has now made intracellular localization and trafficking studies in living systems possible (for review see Tsien *et al* 1998). GFP is a naturally fluorescent protein that can be fused with any protein to form a chimera that can then be expressed. The expressed chimera can be excited to fluoresce in living systems without any treatment thus allowing *in vivo* tracking of events (Tsien *et al* 1998). Herein, we generated several constructs representing different cathepsin B sequences fused with enhanced GFP (EGFP) to study the dynamics of cathepsin B trafficking in a human tumor cell line and a mouse embryo fibroblast cell line lacking the cathepsin B gene. We show that there may be an alternative pathway for cathepsin B trafficking in tumor cells.

# 2. MATERIALS AND METHODS

## *Cell culture*

BT20 human breast carcinoma cells and MEF -/- mouse embryo fibroblast cathepsin B double knockout cells were grown in DMEM and MEM, respectively, following standard cell culture protocols. Ten-percent fetal calf serum was added to the media in all cases.

## *Construction of plasmids for expression of CB variants-GFP fusion proteins*

Plasmid pEGFP-N1, containing enhanced green fluorescent protein gene, was purchased from Clontech (Palo Alto, CA). Three cathepsin B cDNA fragments: 51 bp corresponding to the prepeptide, 237 bp corresponding to the prepropeptide and 987 bp corresponding to the full-length single-chain protein, excluding the last 10 amino acids at the carboxyl terminus, were amplified by polymerase chain reaction using a human cathepsin B cDNA plasmid as a template. The last 10 amino acid residues at the C-terminus are normally removed during enzyme maturation (Sloane *et al* 1994). Therefore, to insure that cathepsin B-GFP fusion remained intact, we designed a construct that did not include these residues. The prepeptide, prepropeptide and full length cathepsin B cDNA fragments were cloned into the pEGFP-N1 vector in the correct orientation and with the correct reading frame to express CB variant-EGFP fusion proteins (Fig 1). To obtain the cathepsin B cDNA fragments, three primer pairs were synthesized according to the cathepsin B cDNA sequence. The primer pair for the 51-bp prepeptide cDNA fragment was composed of the following sequences: 5' primer-5'ATTTGAATTCATGTGGCAGCTCTGGGCCTCC3' (containing the underlined EcoRI site); 3' primer-5'ACCCGGATCCCGGGCATTGGCCAA CACCAG3' (containing the underlined BamHI site). The primer pair for the 237-bp prepropeptide cDNA fragment was composed of the following sequences: 5' primer-5'ATTTGAATTCATGTGGCAGCTCTGGGCCTCC3 (containing the underlined EcoRI site); 3' primer-5'ACCCGGATCC TTCAGGTCCTCGGTAAACAT3' (containing the underlined BamHI site). The primer pair for the 987-bp full-length fragment was composed of the following sequences: 5' primer-5'GGGGTTCTCGAGAATGTGGCAGCT CTGGGCCTCC3' (containing the underlined XhoI site); 3' primer-

5'CGGTGCGTGGAATTCCAGCCACCAC3' (containing the underlined EcoRI site). The plasmid containing the 51-bp cDNA fragment was designated as plasmid pCBpre-GFP; the plasmid containing the 237-bp cDNA fragment as plasmid pCBprepro-GFP; and the plasmid containing the 987-bp cDNA fragment as pPreproCB-GFP. The plasmid pCBpreproN38Q-GFP, containing a mutation from N38 to Q38 in the CB propeptide, was constructed through polymerase chain reaction and the overlap extension technique using cathepsin B cDNA as a template. Two primer pairs were used to amplify two separate fragments corresponding to the 5' or 3' end of the prepropeptide cDNA sequence with overlapping at the mutation site. The primer pair for the 5' end fragment was composed of the following sequences: 5'primer-5'ATTTGA ATTCATGTGGCGCTCTGGGCCT CC3' (containing the underlined EcoRI site); 3' primer-5'CCACGTGGT CTGCCGTTTGTT-GACATAGTT3' (containing the un- derlined mutated codon). The primer pair for the 3' end fragment was com- posed of the following sequences: 5'primer-5'AACAAACGGCAGACCACGTGGCAG GCCGGG3' (containing the underlined mutated codon); 3' primer-5'ACCC-GGATCCTTCAGGTCCTCGGTAAACAT3' (containing the underlined BamHI site). After overlap extension, the N38Q mutated prepropeptide cDNA fragments were generated and subsequently cloned into the pEGFP-N1 plasmid. The identity and orientation of the constructs were confirmed by DNA sequencing.

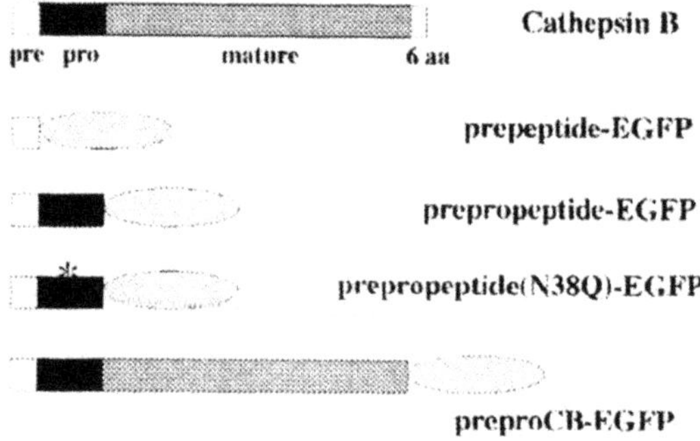


*Figure 1.* Cathepsin B-EGFP fusion constructs. The four cathepsin B-EGFP constructs generated were: prepeptide-EGFP, cathepsin B pre region fused to EGFP; prepropeptide-EGFP, cathepsin B prepro region fused to EGFP; prepropeptide(N38Q)-EGFP, cathepsin B prepro region with N38Q mutation fused to EGFP; preproCB-EGFP, full length cathepsin B sequence fused to EGFP.

## *Expression of CB variant-EGFP fusion proteins*

Approximately 2 x $10^5$ MEF -/- cells were plated in 6-cm dishes. The cells were then transfected with 24-36 μg of either the control plasmid or the

EGFP fusion constructs utilizing calcium phosphate precipitation method according to the procedure described by Graham and van der Eb (1973). The BT20 cells were transfected by electroporation utilizing a Gene Pulser II apparatus equipped with Capacitance Extender Plus (BioRad Inc., Hercules, CA) according to the procedure described by Baum *et al.* (1994). Approximately 6 x $10^6$ cells were transfected with 50 μg DNA at 0.276 volts.

# 3. RESULTS

## *GFP expression in living cells*

We chose to study the BT20 human breast carcinoma cells because we had previously shown that cathepsin B in these cells exhibited peripheral and cell surface distribution (Sameni *et al* 1995). The MEF -/- fibroblasts were chosen because they lack cathepsin B and therefore trafficking pathways would not be overwhelmed by endogenous cathepsin B. As a control, EGFP alone was expressed in both the invasive BT20 human breast carcinoma cells and the cathepsin B knockout MEF T -/- mouse fibroblasts (Fig 2). In both cell lines, EGFP was distributed diffusely throughout the cells. This was expected since EGFP does not have a signal peptide and is synthesized on free ribosomes in the cytoplasm (for review, see Tsien *et al* 1998).

## *Expression of cathepsin B-EGFP chimeras in living cells.*

In order to study the trafficking of the cysteine protease cathepsin B in living cells and to explore the role that specific regions of cathepsin B may play in this process, we fused EGFP with various cathepsin B sequences (see Materials and Methods). Expression of the full-length preprocathepsin B-EGFP construct resulted in a vesicular distribution of the gene product in both MEF T-/- and BT20 cells (Fig 3). This is similar to the distribution of mature cathepsin B in normal and tumor cells (Calkins *et al* 1998, Sameni *et al* 1995), suggesting that fusion of cathepsin B to EGFP did not hinder the normal trafficking of cathepsin B in either fibroblasts or tumor cells. Expression of only the CB prepeptide (CB prepeptide-EGFP) resulted in appearance of fluorescence in a ring or patchy areas around the nuclei of both BT20 and MEF -/- cells (Fig 4). This pattern of staining is consistent with that of the rough endoplasmic reticulum (RER). Fusion of the prepro sequence of cathepsin B to EGFP resulted in a pattern of EGFP expression that was the same in both MEF -/- and BT20 cells; in this case there was polarized fluorescence around the nucleus, resembling staining for Golgi

(Fig 5). Expression of unglycosylated cathepsin B prepropeptide-EGFP, containing a mutation in the glycosylation site in the pro sequence of cathepsin B (N38Q), resulted in an interesting localization pattern. In the MEF T-/- fibroblasts, the EGFP fusion product appeared as a polarized patch adjacent to the nucleus (Fig 6b). This was expected since proteins with altered glycosylation have been shown to not exit from the RER (Tao *et al* 1994). In contrast, in the BT20 cells, the EGFP fusion product appeared in perinuclear vesicles as well as in vesicles in cell processes (Fig 6a). This resembled the distribution of mature cathepsin B (Fig 3a), suggesting that perhaps there is an alternative mechanism for trafficking of cathepsin B in BT20 cells (see below).

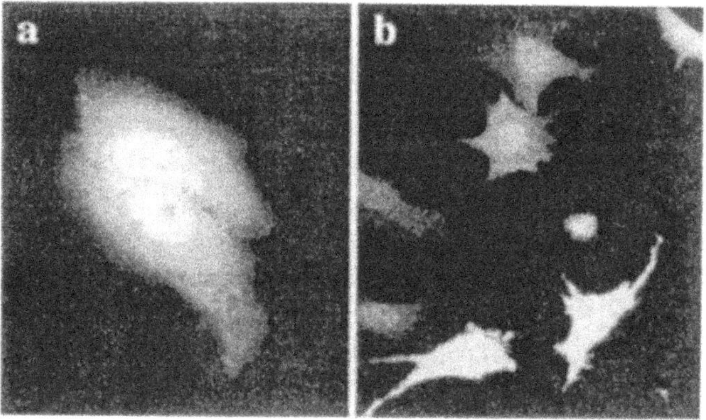


*Figure 2. Expression of EGFP in BT20 and MEF -/- cells.* BT20 (panel a) and MEF -/- cells (panel b) were transfected with EGFP alone. EGFP was distributed in a diffuse cytoplasmic pattern.

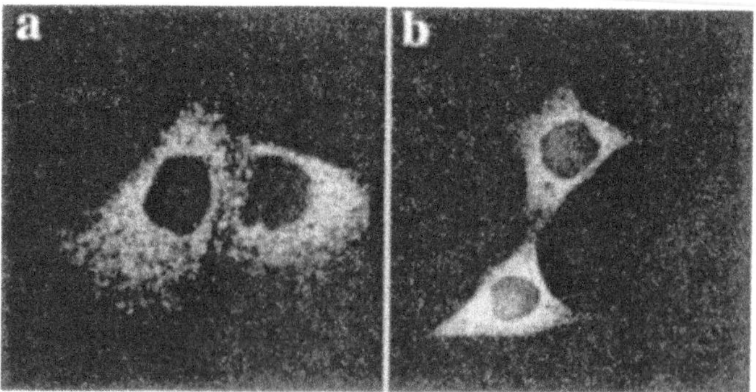


*Figure 3. Expression of full-length cathepsin B-EGFP.* Mature cathepsin B-EGFP is expressed in a vesicular pattern throughout both BT20 (panel a) and MEF -/- (panel b) cells. Note the high level of expression in the perinuclear region in the MEF -/- cells.

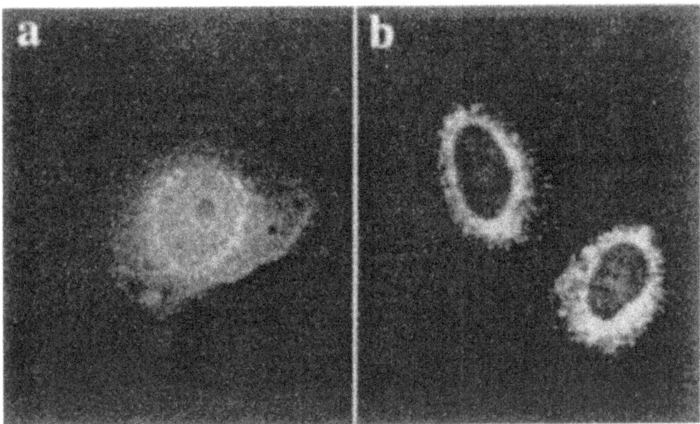


*Figure 4. Expression of cathepsin B prepeptide-EGFP*. EGFP fused to the prepeptide region of cathepsin B was present closely associated with the cell nucleus suggesting localization in the rough endoplasmic reticulum. Panel a, BT20 cells; panel b, MEF -/- cells.

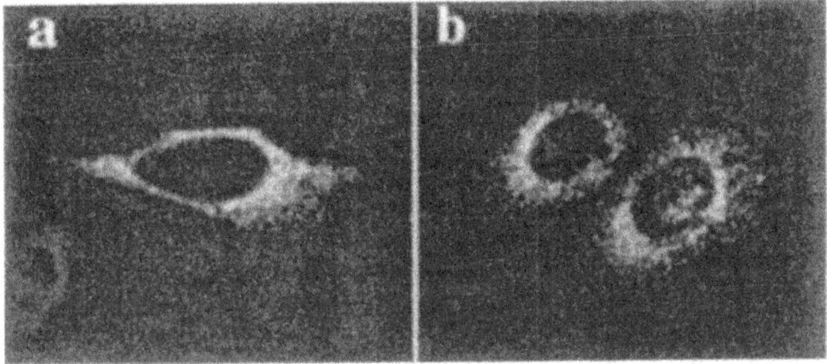


*Figure 5. Expression of cathepsin B prepropeptide-EGFP*. Fusion of prepro region of cathepsin B to EGFP resulted in the distribution of EGFP to regions resembling Golgi. Panel a, BT20; panel b, MEF -/-.

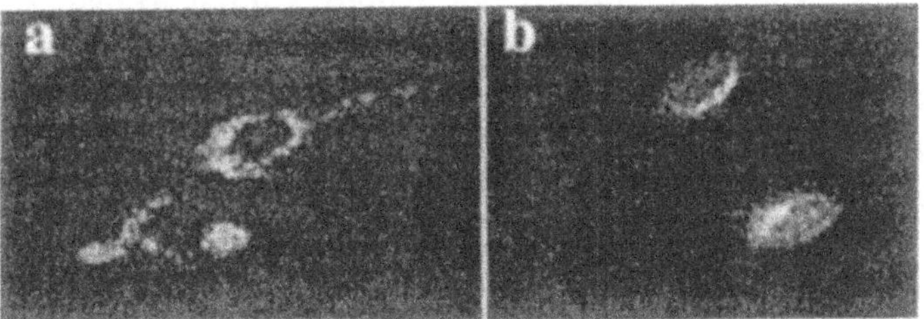


*Figure 6. Expression of EGFP containing the mutated prepro region of cathepsin B*. Expression of mutated cathepsin B prepro-EGFP resulted in different EGFP distribution in the BT20 cells as compared to the MEF -/-. Panel a, BT20; panel b, MEF -/-.

# 4. CONCLUSIONS

The lysosomal enzyme cathepsin B has been implicated in tumor progression and metastasis in part due to increases in its secretion, activity, mRNA and protein levels in tumors and tumor cells (for review, see Sloane *et al* 1994). In normal cells, cathepsin B, a lysosomal enzyme, is assumed to be transported to this compartment via the MPR pathway (for review, see Kornfeld 1989). In fact, the distribution of cathepsin B in normal cells is vesicular and perinuclear, consistent with a lysosomal distribution (Calkins *et al* 1998, Sameni et al 1995). On the other hand, in previous studies involving tumors and tumor cells, we and others have shown that the distribution of cathepsin B is shifted towards the cell periphery (Calkins *et al* 1998, Sameni *et al* 1995, Rempel *et al* 1994, Sloane *et al* 1994, Erdel *et al* 1990, Krepela *et al* 1987). This is consistent with the enzyme being secreted from these cells.

All of the previous studies were of the immunocytochemical type involving fixed cells. Such studies do not provide information with regard to the dynamics of intracellular protein trafficking. This is because the cargoes under investigation are already at their final destination in the cells. In order to gain insight into the trafficking events as they unfold it is necessary to conduct studies in living cells. To do this, we generated four cathepsin B-EGFP chimeras and transfected a cathepsin B knockout cell line, MEF -/- mouse fibroblast, as well as a highly invasive breast carcinoma cell line, BT20. EGFP is an ideal tracker for our studies because it is an unglycosylated cytosolic protein without a signal peptide (for review see Tsien *et al* 1998). Therefore, it is not expected to enter the vesicular protein processing and trafficking pathways (for review see Kornfeld 1989). This is emphasized by the control experiment depicted in Figure 2, in which EGFP is shown to be expressed as a diffuse green fluorescence throughout the cytoplasm. On the other hand, expression of the full-length cathepsin B-EGFP chimera resulted in a distribution pattern similar to that observed for cathepsin B in the many immunocytochemical studies performed earlier (Fig 3 or Calkins *et al* 1998, Sameni *et al* 1995). This observation suggests that EGFP did not interfere with cathepsin B trafficking. Although other investigators have also reported that GFP does not interfere with normal trafficking of proteins in cell (Tarasova *et al* 1997), we will need to confirm our present observations by performing colocalization studies with markers for various vesicular compartments.

When we fused the pre sequence of cathepsin B to EGFP, we noticed a shift in localization of EGFP (Fig 4) from that of EGFP alone (Fig 2). In both MEF -/- and BT20 cells, preCB-EGFP was confined to patchy areas around the nucleus. This was expected as the cathepsin B presequence has been presumed to be a signal peptide which directs the enzyme to the lumen of the RER (Qian *et al* 1991, Cao *et al* 1994, Berquin *et al* 1995). Since there is no glycosylation site on EGFP, preCB-EGFP would not be glycosylated and therefore would remain in this compartment without further transport to the Golgi (Kornfeld 1989, Tao *et al* 1994). On the other hand, expression of EGFP fused to the prepropeptide of cathepsin B resulted in its distribution to polarized patches around the nucleus in both cell types (Fig 5), suggesting that the EGFP chimera was transported to the Golgi. As stated above, transport to the Golgi signifies glycosylation. Although EGFP has no glycosylation sites (Tsien *et al* 1998), the propeptide of cathepsin B does contain a glycosylation site (Chan *et al* 1986, Cao *et al* 1994). Currently, it is not known if this site is glycosylated under normal circumstances. Cathepsin B glycosylation does occur on the mature enzyme as shown by carbohydrate analyses of the mature protein (Takahashi *et al* 1984). To confirm the involvement of the glycosylation site of the prosequence of cathepsin B in sorting of EGFP to the Golgi, we eliminated the glycosylation site by point mutation (Fig 1). Expression of EGFP with the mutated (N38Q) prosequence of cathepsin B presented the most intriguing results. In the MEF -/- cells, EGFP expression was almost identical to that of the EGFP fused to the cathepsin B prepeptide (Figs. 4 and 5). This was expected since there was no glycosylation of the protein. Interestingly, in BT20 cells, the pattern of expression of EGFP was similar to that of mature cathepsin B-EGFP, i.e., vesicular expression throughout the cell including in the cell periphery and cell processes (Fig 6). This unexpected observation suggested that in the BT20 cells there may be an alternative mechanism for cathepsin B trafficking that does not involve the classical carbohydrate recognition markers. If true, the implication is that in tumor cells there is a system involved in trafficking and secretion of cathepsin B that recognizes the enzyme's primary prosequence. The experiments performed in the present study utilized a cytomegalovirus (CMV) promoter and transient transfection. To better understand cathepsin B trafficking, we need to perform stable transfections and make observations over time. To do so, we are currently in the process of developing inducible constructs to perform four-dimensional studies, i.e., three-dimensional studies in real time.

## ACKNOWLEDGMENTS

This study was supported by US Public Health Service Grant CA 56586. The Zeiss LSM-310 confocal microscope is part of a core facility of the Barbara Ann Karmanos Cancer Institute, and the Center in Molecular and Cellular Toxicology with Human Applications, Wayne State University, Detroit, Michigan.

# REFERENCES


Baum, C., Forster, P., Hegewisch-Becker. S., and Harbers, K., 1994, An optimized electroporation protocol applicable to a wide range of cell lines. Bio Techniques **17:** 1057-62.

Berquin, I., Cao, L., Fong, D., and Sloane, B.F., 1995, Identification of two new exons and multiple transcription start points in the 5'-untranslated region of the human cathepsin-B-encoding gene. *Gene* **159:** 143-149.

Calkins, C.C., Sameni, M., Koblinski, J., Sloane, B.F., and Moin, K., 1998, Differential localization of cysteine protease inhibitors and a target cysteine protease, cathepsin B, by immuno-confocal microscopy. *J. Histochem. Cytochem.* **46:** 745-752.

Campo, E., Munoz, J., Miquel, R., Palacin, A., Cardesa, A., Sloane, B.F., and Emmert-Buck, M.R., 1994, Cathepsin B expression in colorectal carcinomas correlates with tumor progression and shortened patient survival. *Am. J. Pathol.* **145:** 301-309.

Cardelli, J.A., Golumbeski, G.S., and Dimond, R., 1986, Lysosomal enzymes in Dictyostelium discoideum are transported to lysosomes at distinctly different rates. *J. Cell Biol.* **102:** 1264-70.

Chen, S.J., Segundo, B.S., McCormick, M.B., and Steiner, D.F., 1986, Nucleotide and predicted amino acid sequence of cloned human and mouse preprocathepsin B cDNAs. *Pro. Natl. Acad. Sci. USA* **83:** 7721-25.

Conibear, E., Stevens, T.H., 1988, Multiple sorting pathways between the late Golgi and the vacuole in yeast. *Biochim.. Biophys. Acta,* **1404:** 211-30.

Erdel, M., Trefz, G., Spiess, E., Habermaas, S., Spring, H., Lah, T., and Ebert, W., 1990, Localization of cathepsin B in two human lung cancer cell lines. *J. Histochem. Cytochem.* **38:** 1313-1321.

Graham, F.I., van der Eb, A.J., 1973, A new technique for the assay of infectivity of human adenovirus 5 DNA. *Virology* **53:** 456-67.

Kornfeld, S., 1987, Trafficking of lysosomal enzymes. *FASEB J.* **1:** 462-468.

Krepela, E., Bartek, J., Skalkova, D., Vicar, J., Rasnick, D., Taylor-Papadimitriou, J., and Hallowes, R.C., 1987, Cytochemical and biochemical evidence of cathepsin B in malignant, transformed and normal breast epithelial cells. *J. Cell Sci.* **87:** 145-154.

McIntyre GF, Erickson AH., 1991. Procathepsins L and D are membrane-bound in acidic microsomal vesicles. *J. Biol. Chem.* **266:** 15438-45.

Moin, K., Cao, L., Day, N.A., Koblinski, J.E., and Sloane, B.F., 1999, Tumor cell membrane cathepsin B. *Biol. Chem.* **379:** 1093-1099.

Moin, K., Day, N.A., Sameni, M., Hasnain, S., Hirama, T., and Sloane, B.F., 1992, Human tumor cathepsin B: comparison with normal liver cathepsin B. *Biochem. J.* **285:** 427-434.

Qian, F., Chan, S., Bajkowski, A., Steiner, D., and Frankfater, A., 1991, Characterization of the cathepsin B gene and multiple mRNAs in human tissues: evidence for alternative splicing of cathepsin B pre-mRNA. *DNA and Cell Biol.* **12:** 299-309.

Sameni, M., Elliott, E., Ziegler, G., Fortgens, P.H., Dennison, C., and Sloane.BF., 1995, Cathepsin B and D are localized at the surface of human breast cancer cells. *Pathol.. Oncol. Res.* **1:** 43-53.

Sloane, B., Moin, K., and Lah, T.T., 1994, Regulation of lysosomal endopeptidases in malignant neoplasia. In *Aspects of the Biochemistry and Molecular Biology of Tumors.* (T.G. Pretlow and T.P. Pretlow, eds.) Academic Press, New York, pp. 411-466.

Takahashi, T., Schmidt, P.G., and Tang, J., 1984, Novel carbohydrate structures of cathepsin B from porcine spleen. *J. Biol. Chem.* **259:** 6059-62.

Tao, K., Stearns, N.A., Dong, J., Wu, Q., Sahagian, G., 1994, The proregion of cathepsin L is required for proper folding, stability, and ER exit. *Arch. Biochem. Biophys.* **311:** 19-27.

Tarasova, N.I., Stauber, R.H., Choi, J.K, Hudson, E.A., Czerwinski, G., Miller, G.L., Pavlakis, G.N., Michejda, C.J., and Wank, S.A., 1997, Visualization of G protein-coupled receptor trafficking with the aid of Green Fluorescent Protein. *JBC* **23**: 14817-24.

Tsien, R.Y., 1998, The green fluorescent protein. *Ann. Rev. Biochem.* **67**: 509-44.

# THE ROLE OF CYSTEINE PROTEASES IN INTRACELLULAR PANCREATIC SERINE PROTEASE ACTIVATION

Markus M. Lerch[1], Walter Halangk[2], Burkhard Krüger[3]
*Department of Medicine B, Westfälische Wilhelms-Universität Münster[1], Division of Experimental Surgery, Otto-von-Guericke-Universität Magdeburg[2], and Division of Medical Biology, Department of Pathology, Rostock University[3], Germany*

Key words acute pancreatitis, autodigestion, cathepsin B, trypsinogen, trypsin

Abstract Autodigestion by proteolytic enzymes is thought to represent the critical mechanism by which acute pancreatitis is initiated. Where and why pancreatic proteases, which are physiologically stored and secreted as inactive precursor zymogens, are activated within the pancreas has remained controversial. Here we present data which indicate that: the lysosomal protease cathepsin B can activate trypsinogen in vitro in a manner that is similar to trypsinogen activation by enterokinase; that cathepsin B colocalizes with trypsinogen in the secretory compartment of the rat pancreas and of the human pancreas; that trypsinogen activation begins in a secretory compartment that is distinct from mature zymogen granules; and that the inhibition of cathepsin B can either increase or decrease premature trypsinogen activation depending on the concentration of the inhibitor, its specificity and its site of action in the pancreatic acinar cell. These observations elucidate some of the complex relations between cysteine and serine proteases in the pancreas with respect to their mechanisms of activation, their subcellular sites of action, and their possible role in the onset of pancreatitis.

## 1. INTRODUCTION

Much of the protein secretion of the exocrine pancreas consists of digestive pro-enzymes, called zymogens, that require cleavage of an activation peptide by a protease. After entering the small intestine, the pancreatic zymogen trypsinogen is first processed to trypsin by the intestinal protease, enterokinase. Trypsin then activates the other pancreatic zymogens. Under physiological conditions pancreatic proteases remain inactive during synthesis, intracellular transport, and secretion from acinar cells. They become active only when they reach the small intestine and get in contact with brushborder enterokinase. One century ago it was suggested that the pancreas of patients who had died during episodes of acute pancreatitis had undergone an autodigestion by its own digestive enzymes (Chiari 1896). Since then many attempts have been made to prove or disprove the role of a premature and intracellular zymogen activation as the initial or initiating event in the course of acute pancreatitis.

## 2. THE ROLE OF TRYPSINOGEN ACTIVATION IN THE ONSET OF PANCREATITIS

Although small amounts of trypsinogen are probably activated within the pancreatic acinar cell under physiological condition, a number of protective mechanisms normally prevent cell damage from tryptic activity. The protective mechanisms within the zymogen granules include: a) the presence of large amounts of pancreatic trypsin inhibitor, b) an acidic pH within the distal secretory pathway that is below the optimum for most proteolytic enzymes, and c) proteases that can degrade other already active proteases. A premature activation of large amounts of trypsinogen, however, could potentially overwhelm these protective mechanisms. This may lead to damage of zymogen-confining membrane and release of activated zymogens into the cytosol. The suggestion that zymogen activation plays a central role in the pathogenesis of pancreatitis is based on the following observations: a) pancreatic trypsin and elastase activities both increase early in the course of experimental pancreatitis (Bialek *et al* 1991), b) the activation peptides of trypsinogen and pro-carboxypeptidase A1 (PCA1), which are cleaved from the respective proenzyme during the process of activation, are released into the serum early in the course of acute pancreatitis (Schmidt *et al* 1992), c) a pretreatment with gabexate mesilate, a serine protease inhibitor, reduces the incidence of ERCP-induced pancreatitis (Cavallini *et al* 1996), d) serine protease inhibitors reduce injury in experimental pancreatitis (Lasson *et al* 1984) and e) forms of the hereditary variety of pancreatitis associated with mutations in the trypsinogen gene that could render trypsinogen more prone to premature activation or may render active trypsin more resistant to degradation by other proteases (Whitcomb *et al* 1996). These observations provide compelling evidence that the premature and intracellular zymogen activation plays a critical role in initiating acute pancreatitis.

## 3. PREMATURE TRYPSINOGEN ACTIVATION BEGINS IN CYTOPLASMIC VACUOLES

Virtually all direct evidence for zymogen activation within pancreatic acinar cells is derived from experimental models of the disease and not from human studies. Most of these cell and animal models make use of the fact that stimulation of the pancreas with concentrations of secretagogue that are at least two orders of magnitude greater than those required to generate a maximal secretory response results in a blockage of enzyme secretion and in the rat, mouse, and dog, in a mild and edematous variety of disease (Lerch *et al* 1994). In addition to in vivo models, isolated functional subunits of the pancreas, known as acini, have been used to examine early events in pancreatitis. When exposed to supramaximal concentrations of the gut hormone cholecystokinin (or its peptide analogue caerulein), the intact pancreas (in vivo) or isolated acini have a similar response. First, secretion decreases from the maximum level elicited by physiologic stimulus (i.e. a meal). This effect is known as high dose inhibition of secretion. Although the mechanism of the decreased secretion is unknown, it may be the result of disrupting the apical actin cytoskeleton (Jungermann *et al* 1995). Second, activation of proteases is detected after supramaximal secretagogue stimulation in vivo or in acini (Grady *et al* 1998, Saluja *et al* 1997). Initial in vivo studies did not define the site of zymogen activation, and were consistent with protease activation within the acinar cell, the pancreatic duct, or the interstitial space. Subsequent studies in isolated acini, however, demonstrated that activation does take place within the acinar cell (Grady *et al* 1998, Krüger *et al* 1998, Saluja *et al* 1999). The identity of the subcellular compartment in which protease activation begins has been recently described. When acini are loaded with a membrane permeable protease substrate that releases a fluorescent probe only when proteolytically cleaved and after supramaximal stimulation of the cells, the fluorescence generated by trypsin activation first appeared in a membrane confined compartment near the apical pole of acinar cells. The most immediate explanation for this finding would have been that intracellular protease activation begins in pancreatic zymogen granules, the physiological storage compartment for digestive enzymes. However, in subcellular fractionation experiments tryptic activity was concentrated to the fraction that contained cytoplasmic vacuoles and not to the fraction that contained mature zymogen granules (Krüger *et al* 1998). In vivo studies (Hofbauer *et al* 1998, Otani *et al* 1998) confirmed that activation takes place in a vesicular compartment that does not correspond to mature zymogen granules (Fig 1a).

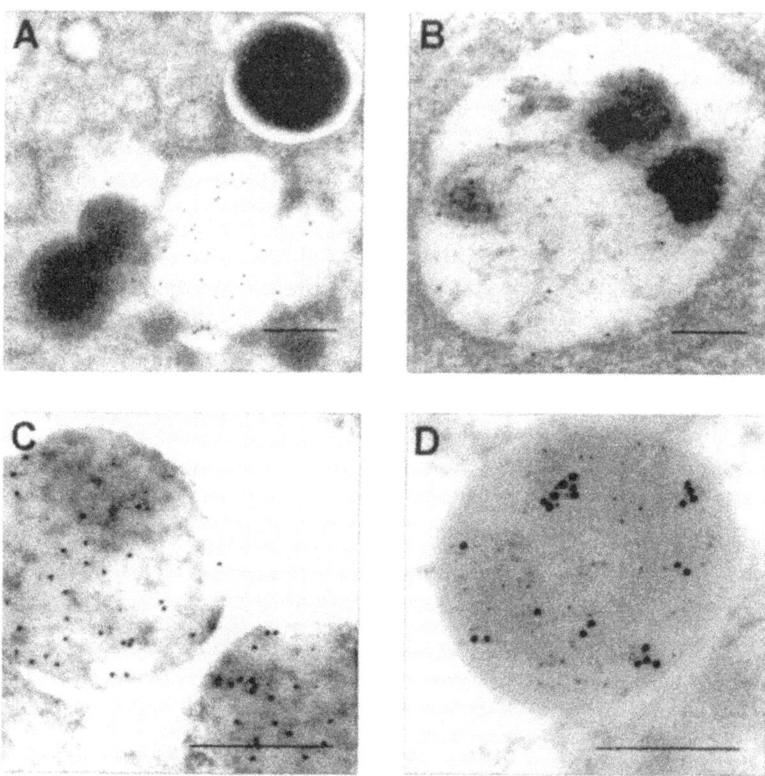


*Figure 1A.* In this experiment acute pancreatitis was induced in the rat by infusion of the secretagogue caerulein at supramaximal concentrations (10µg/kg/h) for only 30 minutes. The pancreas was rapidly removed and ultrathin cryosections were immuno-gold labeled with an antibody directed against TAP, the activation peptide of trypsinogen, which is detectable only after activation of trypsinogen. The antibody detects neither trypsinogen nor trypsin but is specific for TAP. Note that the gold label (15nm) is found exclusively over cytoplasmic vacuoles and not over zymogen granules or the cytosol. 1B) in this experiment rat pancreas from animals with secretagogue-induced pancreatitis was labeled with polyclonal antibodies against cathepsin B (15nm gold) and monoclonal antibodies against trypsinogen (5nm gold). The colocalization of both labels in cytoplasmic vacuoles supports the hypothesis that cathepsin B represents the trigger for intracellular trypsinogen activation. 1C) The detection of cathepsin B (10nm gold) together with trypsinogen (5nm gold) in mature zymogen granules of the pancreas from untreated control rats indicates that this colocalization does occur under physiological conditions and may thus be unrelated to the onset of pancreatitis. 1D) A similar colocalization of cathepsin B (15nm gold) with trypsinogen (5nm gold) can also be found in the zymogen granules of human pancreas (unperfused organ donor pancreas). Bars indicate 100nm.

# 4. THE ROLE OF LYSOSOMAL ENZYMES IN PREMATURE TRYPSINOGEN ACTIVATION

A number of experimental observations have suggested that cathepsin B has an important role in zymogen activation (Steer *et al* 1987). These include the following: a) Cathepsin B has been reported to activate trypsinogen in vitro (Figarella *et al* 1988), b) cathepsin B was found to be redistributed to a zymogen-granule enriched subcellular compartment (Saluja *et al* 1987), and c) lysosomal enzymes colocalize with digestive zymogens during the early course of experimental pancreatitis (Watanabe *et al* 1984). Although this cathepsin-hypothesis appears attractive from a cell biological point of view and valid alternative hypotheses are lacking, it has received much criticism and a number of experimental observations appear to be incompatible with its assumptions: a) a colocalization of cathepsins with digestive zymogens has not only been observed in the initial phase of acute pancreatitis but also under physiological control conditions and in the absence of pancreatitis (Tooze *et al* 1991, Willemer *et al* 1990), b) a redistribution of cathepsin B into a zymogen-enriched subcellular compartment can be induced in vivo by experimental conditions that interfere with lysosomal sorting and are neither associated with nor followed by the development of acute pancreatitis (Lerch *et al* 1993), c) the administration of lysosomal protease inhibitors in vivo does not prevent the onset of acute experimental pancreatitis (Steer *et al* 1993) and in vitro results with these inhibitors were inconclusive and contradictory (Leach *et al* 1991, Saluja *et al* 1997). As a consequence some autors have even suggested a protective role for cathepsin B against a premature zymogen activation in the pancreas (Gorelick *et al* 1992).

In the current project we have tried to resolve some of the conflicting observations and controversies regarding the role of cathepsin B in premature trypsinogen activation. In the initial experiment we have incubated purified bovine trypsinogen in a buffered solution (20μg/ml acetate buffer, 20 mM $CaCl_2$, pH5.5) for three hours and no autoactivation was observed. When, on the other hand, a small amount of purified cathepsin B (40 μg/ml) was added to the solution an activation of trypsinogen occurred over the same period (Fig 2A). On Western blots of the same protein preparations the cleavage products obtained by either enterokinase or cathepsin B were similar which indicates that the processing of trypsinogen by cathepsin B also involves the release of the activation peptide (Fig 2B).

In a next experiment we have investigated whether and where cathepsin B colocalizes with trypsinogen in the exocrine pancreas. When we used the animal model of acute secretagogue-induced pancreatitis in the rat cathepsin B was detected in the same cytoplasmic vacuoles in which we have previously found the activation peptide of trypsinogen (Fig 1A and B). This indicates that premature intracellular trypsinogen activation could be induced by colocalized cathepsin B. Surprisingly, a colocalization of trypsinogen and cathepsin B was also detected in mature zymogen ganules, a compartment that is not involved in premature trypsinogen activation (Fig 1C), and this observation could be confirmed by EM-immunolocalization studies in the pancreas of human organ donors (Fig 1D). This indicates that in the rat, as well as in humans, a significant proportion of lysosomal enzymes enters the regulated secretory pathway without leading to pathophysiological alterations and without inducing premature activation of serine proteases. Our next experiments indicated that the relation between the two classes of proteases is even more complex than thought previously.

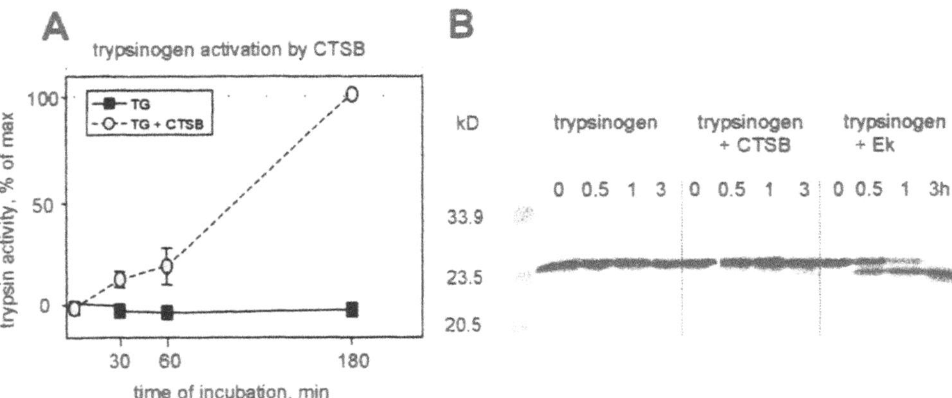


*Figure 2A.* Purified trypsinogen was kept for 180 min at 37 °C (TG) and no autoactivation occured. In the presence of cathepsin B (TG + CTSB) a rapid activation of trypsinogen to active trypsin was found. 2B) In vitro activation of purified bovine trypsinogen by either cathepsin B (CTSB) or enterokinase (Ek) resulted in comparable cleavage products which indicates that a similar processing has occurred (16 % SDS-Gel, non-reducing conditions). Note that optimal conditions were chosen for enterokinase incubation (40 mU, Tris-buffer, pH 7.8, 1 mM $CaCl_2$) whereas conditions for cathepsin B activation were chosen to process only about one percent of the trypsinogen in solution, an amount that corresponds to the in vivo situation inside acinar cells.

When we studied the effect of various inhibitors of cathepsin B in a model of premature trypsinogen activation in living pancreatic acini (Krüger *et al* 1998) we found that E64 at concentrations of 0.1mM significantly increased premature protease activation whereas at 1mM it decreased trypsinogen activation (Fig 3A). While some of these differential effects can be explained by the limited specificity of E64 for cathepsin B (at higher concentrations it also inhibits trypsin activity, Fig 3B) and by the limited membrane permeablity for the inhibitor they cannot explain the increase in trypsin activation seen at lower E64 concentrations. When we used a membrane permeant analogue of E64 named E64d we found an almost complete reduction of the caerulein-induced accumulation of TAP in living acini (Fig 3C). The presence of TAP, as opposed to evidence of trypsin activity, indicates the trypsinogen activation that has occurred over time rather then actual activity at a given point in time. These observations indicate that cathepsin B may have different functions in respect to serine

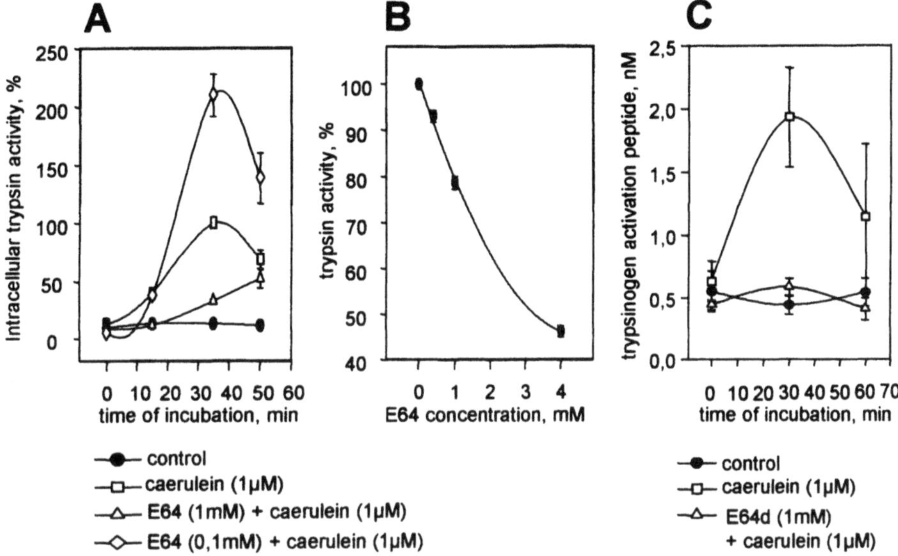


*Figure 3A.* When isolated acini were incubated in the presence of a fluorogenic substrate for trypsin (Ile-Pro-Arg-rhodamine 110) the exposure to supramaximal concentrations of caerulein induced a rapid cleavage of the substrate with consecutive release of measurable fluorescence which indicates intracellular trypsin activity. Low concentrations of the cathepsin B inhibitor E64 (0.1mM) increased this intracellular trypsin activity, whereas higher concentrations (1mM) reduced it. 3B) One of the confounding factors of this experiment is the specificity of E64 which also inhibits trypsin at higher concentrations in vitro. 3C) When the inhibitor E64d, which is more membrane permeant, was used the release of TAP in living acini was reduced to control levels.

protease activation and that these functions can depend on the compartment in which trypsinogen and cathepsin B have colocalized. This involves the secretory compartment in which we have shown that premature trypsinogen activation begins, but also the cytosol where completely different conditions in terms of electrolyte concentration and pH prevail and, according to the data presented here, could also involve the extracellular space.

## ACKNOWLEDGMENTS

Supported by grants from the DFG and BMBF (IZKF Münster). We thank H. Kirschke and E. Weber (Halle/Saale) for providing cathepsin B antibodies.

## REFERENCES

Bialek, R., Willemer, S,. Arnold, R., *et al.*, 1991, Evidence of intracellular activation of serine proteases in acute cerulein-induced pancreatitis in rats. *Scand. J. Gastroenterol.* **26:** 190-196.

Cavallini, G., Tittobello, A., Frulloni, L., *et al.*, 1996, Gabexate for the prevention of pancreatic damage related to endoscopic retrograde cholaniopancreatography. *N. Engl. J. Med.* **335:** 919-923.

Chiari, H., 1896, Über die Selbstverdauung des menschlichen Pankreas. *Z. Heilk.* **17:** 69-96.

Figarella, C., Miszczuk-Jamska, B., Barrett, A., 1988, Possible lysosomal activation of pancreatic zymogens: activation of both human trypsinogens by cathepsin B and spontaneous acid activiation of human trypsinogen-1. *Biol. Chem. Hoppe-Seyler* **369:** 293-298.

Gorelick, F.S., Modlin, I.M., Leach, S.D., *et al.*, 1992, Intracellular proteolysis of pancreatic zymogens. *Yale J. Biol. Med.* **65:** 407-20.

Grady, T., Mah'Moud, M., Otani, T., Rhee, S., *et al.*, 1998, Zymogen proteolysis within the pancreatic acinar cell is associated with cellular injury. *Am. J. Physiol.* **275:** G1010-G1017.

Hofbauer, B., Saluja, A.K., Lerch, M.M., *et al.*, 1998, Intra-acinar cell activation of trypsinogen during caerulein-induced pancreatitis in rats. *Am. J. Phys.* **275:** G352-G362.

Jungermann, J., Lerch, M.M., Weidenbach, H., *et al.*, 1995, Disassembly of the rat pancreatic acinar cell cytoskeleton during supramaximal secretagogue stimulation. *Am. J. Physiol.* **268:** G328-G338.

Krüger, B., Lerch, M.M., Tessenow, W., 1998, Direct detection of premature protease activiation in living pancreatic acinar cells. *Lab. Invest.* **78:** 763-764.

Lasson, A., Ohlsson, K., 1984, Protease inhibitors in acute pancreatitis: correlation between biochemical changes and clinical course. *Scand. J. Gastroenterol.* **19:** 779-786.

Leach, S.D., Modlin, I.M., Scheele, G.A., at al., 1991, Intracellular activation of digestive zymogens in rat pancreatic acini. Stimulation by high does of cholecystokinin. *J. Clin. Invest.* **87:** 362-366.

Lerch, M.M., Saluja, A.K., Dawra, R., *et al.*, 1993, The effect of chloroquine administration on two experimental models of acute pancreatitis. *Gastroenterology* **104:** 1768-1779.

Lerch, M.M., Adler, G., 1994, Experimental animal models of acute pancreatitis. *Int. J. Pancreatol.* **15:** 159-170.

Otani, T., Chepilko, S.M., Grendell, J.H., *et al.*, 1998, Codistribution of TAP and the granule membrane protein GRAMP-92 in rat caerulein-induced pancreatitis. *Am. J. Physiol.* **275:** G999-G1009.

Saluja, A.K., Sadamitsu, H., Saluja, M., *et al.*, 1987, Subcellular redistribution of lysosomal enzymes during caerulein-induced pancreatitis. *Am. J Physiol.* **253:** G508-G516

Saluja, A.K., Donovan, E.A., Yamanaka, K., *et al.*, 1997, Cerulein-induced in vitro activation of trypsinogen inrat pancreatic acini is mediated by cathepsin B. *Gastroenterology* **113:** 304-310.

Saluja, A.K., Bhagat, L., Lee, H.S., *et al.*, 1999, Secretagogue-induced digestive enzyme activation and cell injury in rat pancreatic acini. *Am. J. Physiol.* **276:** G835-G842.

Schmidt, J., Fernandez-del Castillo, C., Rattner, D.W., et al, 1992, Trypsinogen-activation peptides in experimental rat pancreatitis: prognostic implications and histopathologic correlates. *Gastroenterology* **103:** 1009-1016.

Steer, M.L., Meldolesi, J., 1987, The cell biology of experimental pancreatitis. *N. Engl. J. Med.* **316:** 144-150.

Steer, M.L., Saluja, A.K., 1993, Experimental acute pancreatitis: studies of the early events that lead to cell injury. In *The pancreas: Biology, Pathobiology, and Disease.* (V.L.W. Go., E.P. DiMagno, J.D. Gardner, E. Lebenthal, W.A. Reber, and G.A. Scheele, eds.), Raven Press, New York, pp. 489-500.

Tooze, J., Hollinshead, M., Hensel, G., *et al.*, 1991, Regulated secretion of mature cathepsin B from rat exocrine pancreatic cells. *Eur. J. Cell. Biol.* **56:** 187-200.

Watanabe, O., Baccino, F.M., Steer, M.L., *et al.*, 1984, Supramaximal caerulein stimulation and ultrastructure of rat pancreatic acinar cell: early morphological changes during development of experimental pancreatitis. *Am. J. Physiol.* **246:** G457-G467.

Whitcomb, D.C., Gorry, M.C., Preston, R.A., *et al.*, 1996, Hereditary pancreatitis is caused by a mutation on the cationic trypsinogen gene. *Nat. Genet.* **14:** 141-145.

Willemer, S., Bialek, R., Adler, G., 1990, Localization of lysosomal and digestive enzymes in cytoplasmic vacuoles in caerulein-pancreatitis. *Histochemistry* **94:** 161-70.

# PEPTIDASES IN THE ASTHMATIC AIRWAYS

Vincent H.J. van der Velden[1,2], Brigitta A.E. Naber[1], Peter Th. W. van Hal[2], Shelley E. Overbeek[2], Henk C. Hoogsteden[2], Marjan A. Versnel[1]
*Departments of [1]Immunology and [2]Pulmonary Medicine, Erasmus University and University Hospital Rotterdam-Dijkzigt, Rotterdam, The Netherlands*

## 1. INTRODUCTION

Asthma is clinically characterized by reversible airway obstruction and airway hyperresponsiveness. Nowadays, it is thought that these symptoms are the result of a chronic inflammation of the airways, characterized by an influx of leukocytes and increased levels of inflammatory mediators. This inflammation is caused, at least partially, by peptides such as cytokines, chemokines, and neuropeptides. Degradation of peptides by peptidases is an important mechanism to modulate peptide-mediated inflammation. Peptidases with the catalytic site exposed at the external surface (reviewed in van der Velden and Hulsmann 1999b) are widely distributed on the surface of many different cell types.

Initial studies on the role of peptidases in the airways have focused on the modulation of neurogenic inflammation. Neurogenic inflammation is the result of stimulation of sensory nerves, resulting in the release of neuropeptides (such as substance P (SP) and neurokinin A (NKA)). These neuropeptides may subsequently cause bronchoconstriction, mucus secretion, vasodilation, microvascular leakage, and the recruitment of leukocytes (van der Velden and Hulsmann 1999a), and thus mimic many of the pathophysiological features of asthma. In addition, neuropeptides also have many immunomodulatory functions (reviewed in van der Velden and Hulsmann 1999a). Thus, neuropeptides not only evoke constriction of the bronchus by direct effects on smooth muscle cells, submucosal glands, and

*Cellular Peptidases in Immune Functions and Diseases 2*
Edited by Langner and Ansorge, Kluwer Academic/Plenum Publishers, 2000

blood vessels, but may also contribute to the initial and chronic phase of the airway inflammation observed in asthmatics.

Although the apparent increased effects of sensory neuropeptides in asthma may be due to several mechanisms, studies using laboratory animals have indicated that peptidases play a major role in limiting neurogenic inflammatory responses (Kohrogi *et al* 1989, Martins *et al* 1991, Nadel, 1990). Thus far, most attention has been given to the role of neutral endopeptidase 24.11 (NEP). NEP (identical to common acute lymphoblastic leukemia antigen (CALLA) or CD10) is a membrane-bound metalloenzyme which cleaves peptide-bonds at the $NH_2$-terminal side of hydrophobic amino acids, thereby being able to inactivate a variety of small peptides, including SP, NKA, bradykinin, endothelin, and bombesin-like peptides (BLP). It has been demonstrated that inhibition of NEP, either by drugs or by environmental factors such as ozone, results in potentiation of neuropeptide-induced effects in the airways (reviewed in van der Velden and Hulsmann, 1999b). In contrast, administration of aerosolized recombinant NEP may prevent neuropeptide-mediated cough (Kohrogi *et al* 1989). Besides modulating neurogenic inflammation, NEP also regulates fMLP-mediated chemotaxis of neutrophils (Shipp *et al* 1991a), proliferation and maturation of B cells (Salles *et al* 1992, Salles *et al* 1993), and BLP-mediated lung development (Sunday *et al* 1992).

In addition to NEP, several other peptidases are able to hydrolyze (neuro)peptides and therefore may be involved in the modulation of peptide-mediated effects. Aminopeptidase N (APN) and dipeptidyl peptidase IV (DPP IV) are of particular interest, since both enzymes are membrane-bound molecules (and thus able to cleave extracellular peptides) that have been characterized on non-hematopoietic cells as well as hematopoietic cells. APN, which is identical to CD13, preferentially cleaves neutral amino acids from the $NH_2$-terminal side of peptides, including enkephalins, fMLP, tachykinins, and cytokines like interleukin (IL)-1β, IL-2, IL-6 and IL-8 (Hoffmann *et al* 1993, Kanayama *et al* 1995). Its general function is to reduce cellular responses to peptides, but APN may also be involved in processing major histocompatibility (MHC) class II-bound peptides (Larsen *et al* 1996) and in the degradation of type IV collagen (Fujii *et al* 1995). On many cells, APN is co-expressed with NEP and it is thought that initial cleavage by NEP may precede APN activity.

DPP IV, which is identical to CD26, is a serine protease which preferentially cleaves Xaa-Pro and less frequently Xaa-Ala dipeptides from the $NH_2$-terminus of polypeptides. Among the possible substrates for DPP IV are SP, bradykinin, and certain chemokines. DPP IV may also be able to degrade cytokines, like IL-1β, IL-2 and IL-6, although preceding cleavage by an endopeptidase may be required (Hoffmann *et al* 1993). In addition,

DPP IV may also function as an adhesion molecule to fibronectin (Cheng *et al* 1998) and is involved in T-cell activation (Fleischer, 1994).

As many peptide-mediated processes take place in the human airways, a thorough understanding of the expression and activity of the peptidases in the human airways may add to our understanding of normal lung development and function. In addition, it may probably help to explain pathological changes, including the development of lung tumors and the pathophysiological features of asthma. In this report, we will review the present knowledge about peptidases in the healthy and asthmatic airways. We will focus on our own studies and will discuss these in the context of other recent data.

## 2. EXPRESSION AND ACTIVITY OF MEMBRANE-BOUND PEPTIDASES IN THE HEALTHY AND ASTHMATIC AIRWAYS

### 2.1 Aminopeptidase N

Using immunohistochemistry and enzyme histochemistry we have analyzed the distribution of APN in the human bronchus (van der Velden *et al* 1998b). APN was widely distributed, being present on blood vessels, glandular ducts, connective tissue, perichondrium, and nerves (Fig 1A and B). Many of these sites also possess NEP activity (Baraniuk *et al* 1995), which is in accordance with the proposed sequential inactivation of peptides by NEP and APN. Thus, NEP and APN may collaborate on the surface of these cells to modulate the cell's response towards peptide-mediated signals. In contrast to NEP, no expression of APN was found on smooth muscle cells and bronchial epithelial cells. This is in accordance with the lack of effect of APN-inhibitors on tachykinin-induced smooth muscle contraction (Djokic *et al* 1989, Dusser *et al* 1989, Sekizawa *et al* 1987a, Sekizawa *et al* 1987b).

In addition to structural airway cells, mononuclear phagocytes, eosinophilic granulocytes and certain dendritic cells expressed APN (van der Velden *et al* 1998b). APN expression on granulocytes has been shown to be involved in the modulation of chemotactic responses (Kanayama *et al* 1995, Shipp *et al* 1991a), whereas APN expression on mononuclear phagocytes and dendritic cells may be involved in processing of MHC class II-bound peptides (Larsen *et al* 1996). In addition, preliminary data indicate that APN may affect the maturation and/or differentiation of dendritic cells (V.H.J. van der Velden, unpublished data).

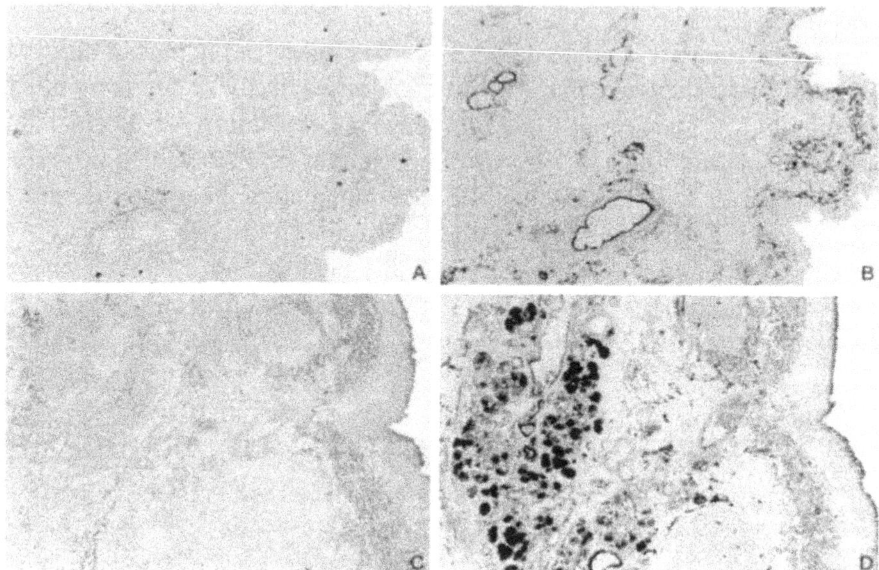


*Figure 1.* Distribution of APN and DPP IV in human bronchial tissue. APN expression was determined using immunohistochemistry. A: negative control; B: CLB-CD13, an antibody specific for APN. DPP IV activity was determined using enzyme histochemistry. C: negative control; D: gly-pro-MNA (specific substrate for DPP IV). See reference van der Velden *et al* 1998b for technical details.

Comparison of APN expression in bronchial biopsies of allergic asthmatics and healthy subjects revealed an increased number of APN-expressing cells in the bronchial epithelium of asthmatics (Fig 2). The number of APN-expressing cells correlated with the number of L25-positive dendritic cells and with the number of eosinophils found in the bronchial epithelium. Double-stainings showed that both cell types indeed expressed APN, which is in accordance to the known distribution of CD13 on these cells. Therefore, it seems likely that the increase in APN-positive cells in the bronchial epithelium reflects the increase of dendritic cells and eosinophils observed in the bronchial epithelium of asthmatics.

Alternatively, the increased number of APN-expressing cells in the bronchial epithelium of asthmatic patients could be due to an upregulation or induction of APN on certain leukocytes. Previous studies have shown that APN can be upregulated on the surface of mononuclear phagocytes by interleukin (IL)-4 (van Hal *et al* 1994). Asthma is considered a Th2-like disease and increased levels of IL-4 have been detected in asthmatic airways (Humbert *et al* 1996, Walker *et al* 1994). To determine whether increased expression of APN on mononuclear phagocytes is a feature of asthma, it would be of interest to compare the expression of APN on alveolar macrophages of healthy individuals and allergic asthmatics. A recent study has indicated that APN expression can also be induced on T lymphocytes after adhesion to epithelial cells (Riemann *et al* 1997). However, it is not likely that this explains the increased number of APN-expressing cells in the asthmatic patients, since there was no increase in the number of T lymphocytes in the bronchial epithelium of asthmatics compared to healthy controls.

The lack of difference in the number of APN-positive cells in the lamina propria is probably due to the observation that the majority of APN is expressed on non-hematopoietic cells. It was therefore not possible to quantify APN expression on infiltrating leukocytes accurately.

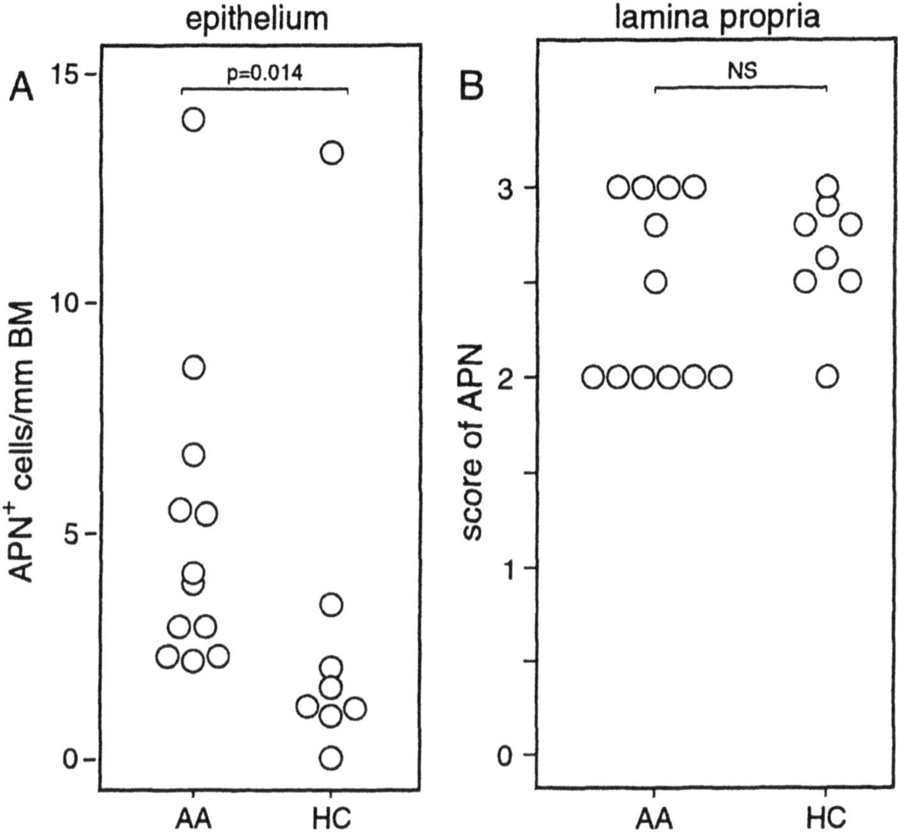


*Figure 2.* APN expression in bronchial biopsies obtained from healthy controls (HC) and allergic asthmatics (AA). See reference van der Velden *et al* 1998b for technical details.

## 2.2 Dipeptidyl peptidase IV

Immunohistochemical and enzyme histochemical analysis of the distribution of DPP IV in the human bronchus revealed that DPP IV was strongly present in serosal submucosal glands and moderately expressed on blood vessels, predominantly post-capillary venules (Fig 1C and D) (van der Velden *et al* 1998b). DPP IV expression in submucosal glands did not seem to be restricted to the cell membrane, but appeared to be located intracellularly as well. It is not clear whether the DPP IV of submucosal glands is secreted in the bronchial lumen. However, DPP IV activity can be detected in bronchoalveolar lavage (BAL) fluid (van der Velden *et al* 1999). DPP IV has also been found in submandibular and parotid glands and a role for DPP IV in the secretion or reabsorption process of secretory proteins and peptides has been suggested (Fukasawa *et al* 1981, Sahara and Suzuki, 1984). In glandular endometrial epithelial cells from cows, a DPP IV molecule missing the signal sequence has been detected. Further studies are required to determine the characteristics of the DPP IV molecule in serosal submucosal glands.

Endothelial cells were shown to express all peptidases examined (NEP, APN, DPP IV), but the distribution amongst arteries, capillaries and venules showed some marked differences (van der Velden *et al* 1998b). The site-restricted presence of different peptidases may represent a mechanism to control blood flow and plasma leakage at specific locations. In addition, recent studies indicate that lung endothelial DPP IV promotes adhesion and metastasis of breast cancer cells via tumor cell surface-associated fibronectin (Cheng *et al* 1998).

DPP IV expression could also be found on T cells (double labeling with CD3). DPP IV has been shown to be a marker for activated T cells (Fleischer, 1994) and plays an important role in T cell responses. Comparison of DPP IV expression between bronchial biopsies of healthy controls and allergic asthmatics did not reveal significant differences (van der Velden *et al* 1998b), suggesting that the number of activated T cells did not differ. Several other studies have shown that the airways of allergic asthmatics contain increased numbers of activated, but not total, T cells (Corrigan *et al* 1988, Wilson *et al* 1992). This apparent discrepancy may be explained by recent studies indicating that DPP IV is predominantly expressed on Th0 and Th1 cells (Willheim *et al* 1997), whereas many T cells in the airways of allergic asthmatics show a Th2-like phenotype (Humbert *et al* 1996, Walker *et al* 1994).

## 2.3 Neutral endopeptidase

The expression of NEP in the human bronchus has been described by Baraniuk et al. (Baraniuk *et al* 1995). NEP was found on epithelial cells, smooth muscle cells, submucosal glands, and endothelial cells. In our study (van der Velden *et al* 1998b), we used enzyme histochemistry to determine the distribution of NEP in the human bronchus. A very weak NEP activity was observed, but attribution of this activity to a certain cell type was difficult. Low levels of activity were observed in the bronchial epithelium and submucosal glands. In contrast, we observed strong NEP activity in the guinea pig trachea, especially in the tracheal epithelium, as has also been described previously (Kummer and Fischer, 1991). This may indicate that (epithelial) NEP is much more important in the modulation of neurogenic inflammatory reactions in the guinea pig than in humans. Alternatively, this may suggest that peptidergic mechanisms are less prominent in humans compared to guinea pigs. Indeed, peptidergic innervation of human airways seems sparse, whereas a dense network of tachykinin-containing peptidergic nerves can be found in the airways of rodents (Howarth *et al* 1991, Lundberg *et al* 1984).

Since NEP activity in the human bronchus was hard to detect, we were not able to determine whether there was a difference in NEP activity between bronchial biopsies of healthy controls and allergic asthmatics. In another study it was found that asthmatics treated with steroids expressed significantly more NEP on their bronchial epithelium than did nonsteroid-treated asthmatics (Sont *et al* 1997). However, in this study no comparison was made between non-asthmatic subjects and asthmatic patients. Therefore, it remains to be established whether this difference was due to a reduced NEP expression in the nonsteroid-treated patients that could be reversed by the use of steroids, or that inhaled steroids increased the normal expression of NEP. The latter possibility is supported by the increased upregulation of NEP on human bronchial epithelial cells by steroids *in vitro* (see below) and the lack of an obvious difference in NEP activity between control subjects and mildly asthmatic subjects found in an *in vivo* study (Cheung *et al* 1993).

## 3. ACTIVITY OF SOLUBLE PEPTIDASES IN THE HEALTHY AND ASTHMATIC AIRWAYS

Although peptidases are normally membrane-bound enzymes, soluble forms can be detected in body fluids. These soluble counterparts may either be derived from shedding of membrane-bound peptidases or may be formed by post-translational cleavage of the membrane-bound form. NEP activity in serum probably arises from shedding of the entire membrane-bound peptidase (Soleilhac *et al* 1996). Increased serum activity of NEP has been observed in underground miners exposed to coal dust particles (Soleilhac *et al* 1996) and in patients with adult respiratory distress syndrome (ARDS) (Johnson *et al* 1985) or sarcoidosis (Almenoff *et al* 1986). Although the source of the increased NEP levels remains to be determined, it has been suggested that increased NEP levels may reflect local tissue damage with subsequent shedding of membrane-bound NEP (Johnson *et al* 1985, Soleilhac *et al* 1996). Alternatively, NEP might be released from activated granulocytes sequestered in the lung and leak into the bloodstream (Johnson *et al* 1985). Serum APN activity predominantly comprises a circulating isoform of the CD13 antigen (Favaloro *et al* 1993). DPP IV activity in serum (which may enhance antigen-induced T cell proliferation) has recently been shown to originate, at least in part, from the DPPL-T antigen expressed on the surface of activated T cells (Duke-Cohan *et al* 1996). There is evidence that serum DPP IV activity is decreased in patients with malignancies and in auto-immune and inflammatory disorders (reviewed in van der Velden and Hulsmann, 1999b). In contrast to serum, until recently little was known about the presence of peptidases in BAL fluid. We therefore aimed to analyze the presence of NEP, APN, and DPP IV in BAL fluid and to determine the activities of these peptidases in serum and BAL fluid of subjects with asthma.

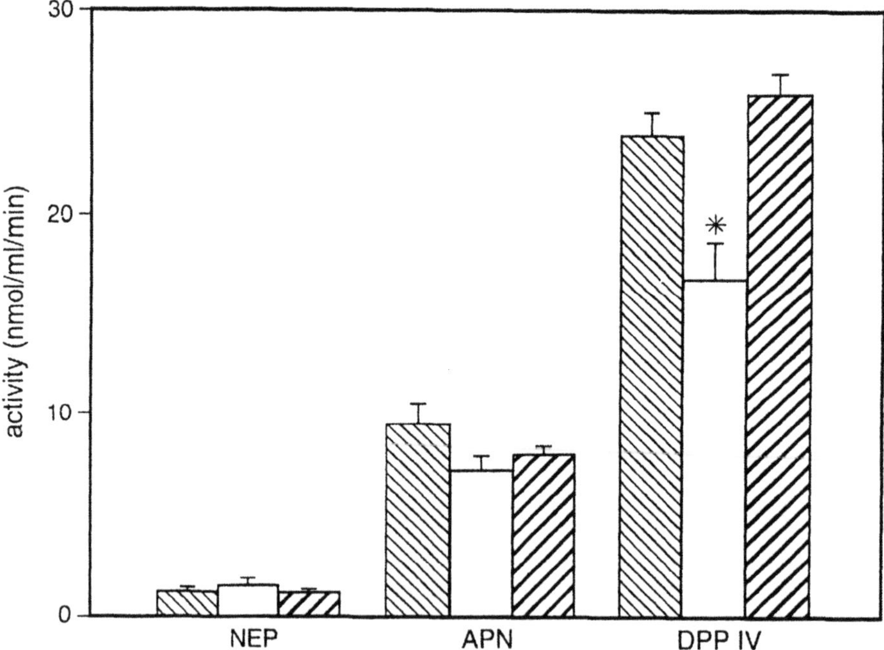


*Figure 3.* NEP, APN, and DPP IV activity in serum from healthy non-smokers (shaded bars), smokers (white bars), and allergic asthmatics (heavily shaded bars). *: $p < 0.05$ compared to healthy non-smokers.

NEP, APN, and DPP IV activity were determined in serum and BAL fluid of healthy controls using chromogenic assays (van der Velden *et al* 1999). All three peptidases could be detected both in serum and in BAL fluid, but activities (expressed per ml) were much higher in serum (Fig 3 and

4). However, if peptidase activities in BAL fluid are expressed as undiluted epithelial lining fluid (assuming that the epithelial lining fluid is diluted 100-fold in BAL fluid), NEP activity in BAL fluid is approximately 10-fold higher than in serum, whereas APN activity is of similar magnitude and DPP IV activity is two- to threefold higher in serum. Together with the lack of correlation between peptidase activities in BAL fluid and serum, these findings suggest that the presence of peptidases in these two compartments is regulated independently of each other and that NEP is locally released in the airways.

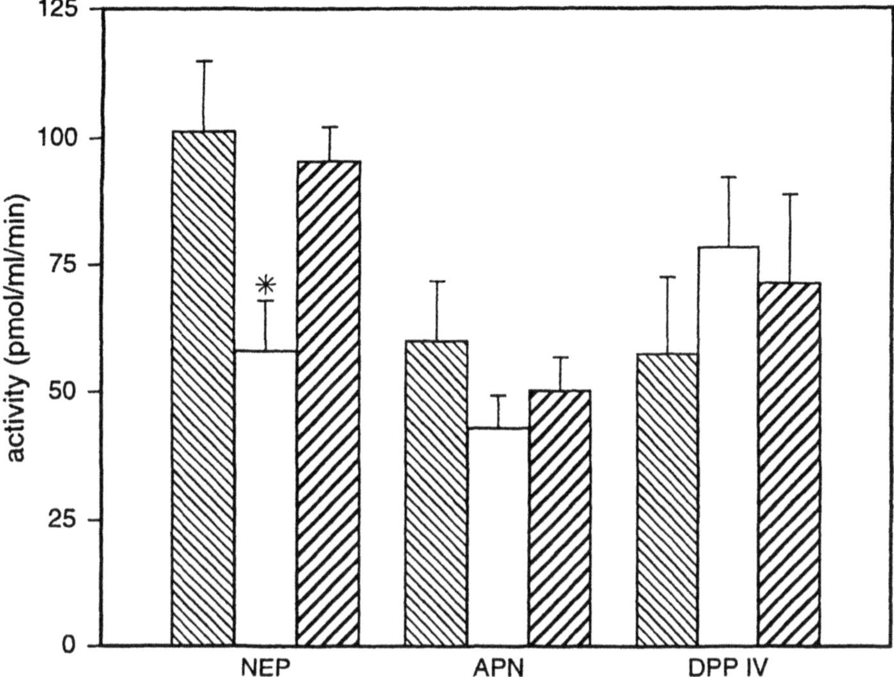


*Figure 4.* NEP, APN, and DPP IV activity in BAL fluid from healthy non-smokers (shaded bars), smokers (white bars), and allergic asthmatics (heavily shaded bars). *: $p < 0.05$ compared to healthy non-smokers.

Comparison of peptidase activities in serum and BAL fluid from healthy controls and stable asthmatics did not reveal significant differences in NEP, APN and DPP IV activity. The lack of difference in NEP activity in serum and BAL fluid between healthy controls and allergic asthmatics may indicate that NEP activity is not altered in the airways of asthmatics. It may also indicate that there is little tissue damage in the airways of stable asthmatic patients. One could speculate that tissue damage occurs during asthmatic exacerbations and that this may cause increased peptidase activities in serum (Johnson *et al* 1985, Soleilhac *et al* 1996). Remarkably, our preliminary data suggest that also during and up to 5 days after an asthmatic exacerbation no increase in peptidase activities can be observed in serum.

In addition to increased peptidase levels in serum due to tissue damage, increased serum peptidase activities may reflect activation of granulocytes sequestered in the airways (Johnson *et al* 1985). In contrast to ARDS, which is characterized by strongly increased numbers of neutrophils in the airways, the numbers of (eosinophilic) granulocytes in the asthmatic patients are relatively low. Thus, if NEP and/or APN were released from granulocytes sequestered in the airways of asthmatic airways, the absolute amounts will probably be low. Furthermore, we and others did not observe a correlation between peptidase activities in serum or BAL fluid and cell numbers of leukocytes (Juillerat-Jeanneret *et al* 1997, van der Velden *et al* 1999), suggesting that there is no predominant hematopoietic source of the soluble peptidases in healthy controls or asthmatic patients. During other pathological conditions (such as neoplasms, infections or sarcoidosis) increased numbers of granulocytes or lymphocytes in the airways may, however, significantly contribute to the activities of APN and DPP IV in BAL fluid (Juillerat-Jeanneret *et al* 1997).

We cannot rule out the possibility that similar NEP activity in BAL fluid (and serum) from healthy controls and asthmatics is the result of a reduced membrane-bound NEP activity (either due to reduced expression or inactivation of the enzyme) together with increased shedding of the enzyme. To determine whether inactivation of NEP occurs in asthmatics, data on NEP activity in BAL fluid should be compared with ELISA data. Analysis of soluble ICAM-1 or cytokeratin-19 levels may indicate whether increased shedding or epithelial injury occurs in the airways of asthmatics compared to healthy controls.

Cigarette smoke has been shown to inhibit NEP activity in laboratory animals. Our study shows that in humans, cigarette smoke reduces NEP activity in BAL fluid (Fig 4). Since NEP modulates the growth and differentiation of bronchial epithelial cells by hydrolyzing BLP, reduced NEP activity may promote BLP-mediated proliferation and facilitate the development of lung carcinomas (Cohen *et al* 1996, Ganju *et al* 1994, Shipp *et al* 1991b). In accordance to our observation, increased levels of BLP have been found in the lower respiratory tract of asymptomatic smokers (Aguayo *et al* 1989). Furthermore, recent studies indicate that human lung cancers show low or absent NEP activity (Cohen *et al* 1996) and that NEP levels are severely reduced in BAL fluid of lung cancer patients (Cohen *et al* 1999). We hypothesize that cigarette smoke facilitates the development of lung carcinomas at least in part by inhibiting NEP activity. Further studies need to be performed to demonstrate that cell surface NEP activity in humans is also inhibited by cigarette smoke and to prove that the reduced NEP activity is due to inactivation of the enzyme rather than decreased presence of the peptidase itself.

# 4. MODULATION OF PEPTIDASES BY GLUCOCORTICOIDS

Glucocorticoids are widely used in the treatment of asthma and are able to reduce inflammatory reactions in the airways (reviewed in van der Velden, 1998). Glucocorticoids are potent inhibitors of cytokine production by a variety of cells, thereby suppressing inflammatory responses. In addition, the anti-inflammatory actions of glucocorticoids may be mediated by modulation of peptidase activities.

## 4.1 *In vitro* studies

Studies using laboratory animals have shown that NEP present on the bronchial epithelium plays a major role in the hydrolysis of (neuro)peptides, and thereby in modulating neurogenic inflammation (Devillier *et al* 1988, Farmer and Togo, 1990, Fine *et al* 1989, Frossard *et al* 1989, Grandordy *et al* 1988). We and others (Borson and Gruenert, 1991, Lang and Murlas, 1993) therefore analyzed the effect of glucocorticoids on the activity and expression of peptidases on human bronchial epithelial cells. Our studies show that stimulation of the human bronchial epithelial cell line BEAS 2B cells with the synthetic glucocorticoid dexamethasone (DEX) strongly increased NEP expression and activity (Fig 5). This DEX-induced increase was observed both in the absence and in the presence of cytokines (van der Velden *et al* 1998a), indicating that glucocorticoids may also increase NEP during inflammatory processes. The effect of dexamethasone was abolished by the glucocorticoid receptor antagonist RU38486, indicating that the effect was mediated via the glucocorticoid receptor. The synthetic testosterone analogue R1881 had no effect on NEP activity, suggesting that the effect is specific for glucocorticoids and not for steroid hormones in general. In accordance to our data, an *in vivo* study has shown increased expression of NEP on the bronchial epithelium of steroid-treated asthmatics compared to nonsteroid-treated asthmatics (Sont *et al* 1997). Part of the beneficial effects of glucocorticoid treatment in asthma may thus be mediated via upregulation of NEP by human bronchial epithelial cells, thereby limiting neurogenic inflammation.

## 4.2 *In vivo* studies

To investigate whether glucocorticoids also modulated peptidases *in vivo*, stable allergic asthmatics were treated with placebo or the inhaled synthetic glucocorticoid fluticasone propionate (500 μg twice daily) for 12 weeks (van

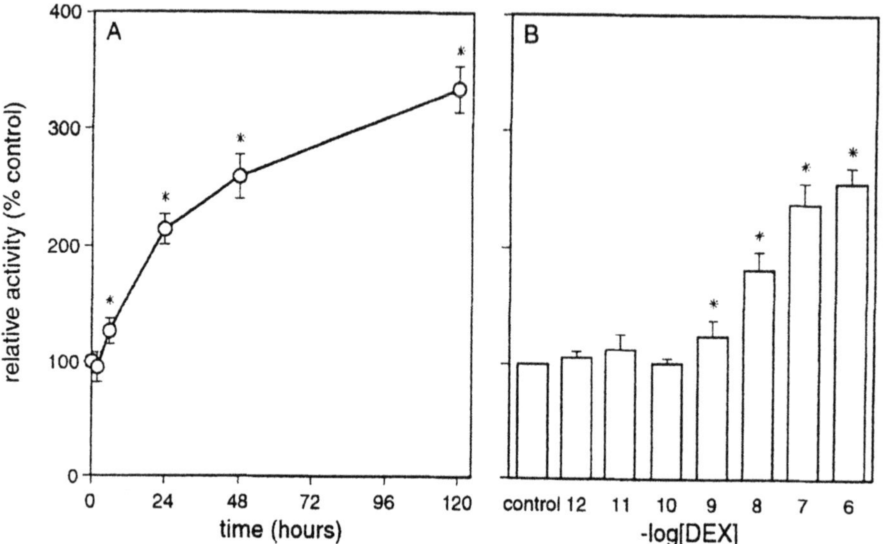


*Figure 5.* Effect of DEX on NEP activity by the human bronchial epithelial cell line BEAS 2B. A: DEX ($10^{-6}$ M) stimulated NEP in a time-dependent manner (mean ± SEM; n = 2 - 10). B: DEX stimulated NEP activity in a dose-dependent manner (mean ± SEM; n = 4; 24 h incubation) Activity of unstimulated cells = 100 %. $^*$: $p < 0.05$ compared to unstimulated cells.

der Velden *et al* 1999). Venous blood samples were taken and BAL was performed before and after this treatment period and at both visits lung function was measured and a methacholine dose-response curve was constructed. Treatment of allergic asthmatics with fluticasone propionate for 12 weeks significantly improved their lung function, whereas no improvement was observed in the patients with placebo. In contrast, no significant effects were observed on peptidase activities in BAL fluid (Fig 6) or serum. To our knowledge, no other studies thus far have determined the effects of glucocorticoids on peptidase activities in serum or BAL fluid. As indicated above, *in vitro* studies have shown that glucocorticoids upregulate the expression of NEP on human bronchial epithelial cells. In addition, treatment of asthmatic patients with inhaled glucocorticoids increases NEP expression by the bronchial epithelium (Sont *et al* 1997). Thus, (inhaled) glucocorticoids increase the surface membrane expression of NEP on bronchial epithelial cells, but do not affect soluble NEP activity in BAL fluid. This may indicate that NEP activity in BAL fluid is not derived from bronchial epithelial cells. Alternatively, shedding of NEP from the surface of bronchial epithelial cells may not be a random process but may be affected by glucocorticoids (which upregulate surface membrane expression but possibly reduce the relative amount of NEP shedded from the membrane).

Analysis of NEP activity in culture supernatants of human bronchial epithelial cells stimulated with or without steroids will give more insight in this phenomenon.

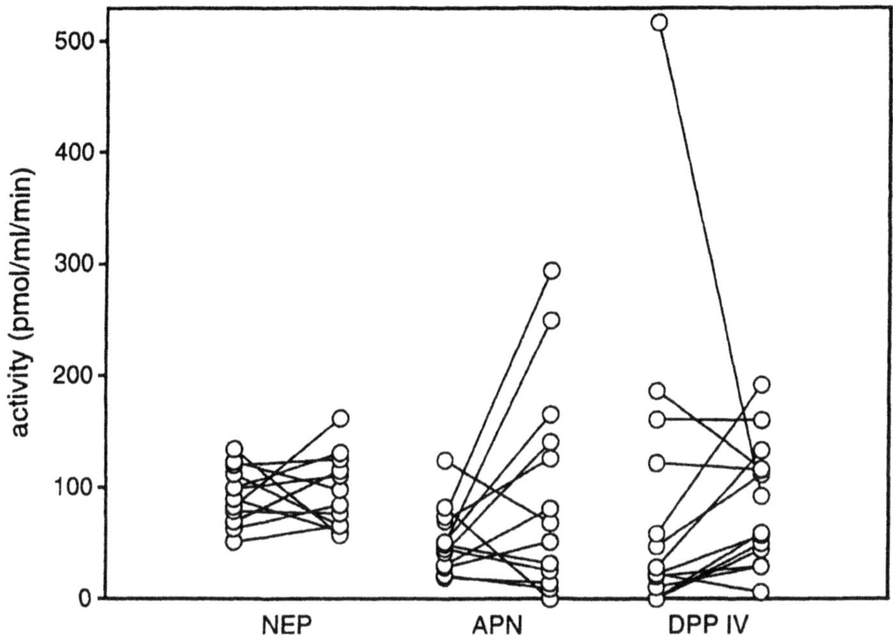


*Figure 6.* Effect of fluticasone propionate on peptidase activities in BAL fluid. Allergic asthmatics were treated for 12 weeks with the inhaled glucocorticoid fluticasone propionate. Before and after this period BAL was performed and peptidase activites were analyzed.

## 5. CONCLUSIONS

Studies on the role of peptidases in the pathogenesis of asthma have not been able to convincingly demonstrate a dysfunction of these enzymes in the airways of stable asthmatics. Although asthmatic airways are more responsive to tachykinin-induced bronchoconstriction and nasal congestion (Joos *et al* 1987, Joos *et al* 1994), no apparent reduction in NEP activity could be found in stable mild asthmatic patients (Cheung *et al* 1993). Our studies indicate that peptidase activities in BAL fluid and serum do not differ remarkably between healthy controls and allergic asthmatics. In addition, we did not observe major differences in the expression of APN and DPP IV between bronchial biopsies of asthmatics and healthy controls. No data are currently available on the expression of NEP in the airways of asthmatics compared to healthy subjects, although some data may suggest a reduced NEP expression in the bronchial epithelium, but not the lamina propria, from nonsteroid-treated asthmatics (Sont *et al* 1997). It seems therefore unlikely that there is a generally reduced activity of peptidases in the airways of stable asthmatic patients. To further determine whether peptidases and neuropeptides contribute to asthma, *in vivo* studies using selective neurokinin receptor antagonists should be performed both in the presence and absence of peptidase inhibitors.

## ACKNOWLEDGMENTS

We gratefully acknowledge Mr. T.M. van Os for preparing the figures. This investigation was supported by a grant from The Netherlands Asthma Foundation (grant 32.92.73).

## REFERENCES

Aguayo, S.M., Kane, M. A., King, T.E., Jr., Schwarz, M.I., Grauer, L., and Miller, Y.E., 1989, Increased levels of bombesin-like peptides in the lower respiratory tract of asymptomatic cigarette smokers. *J. Clin. Invest.* **84:** 1105-1113.

Almenoff, J., Skovron, M.L., and Teirstein, A.S., 1986, Thermolysin-like serum metalloendopeptidase. A new marker for active sarcoidosis that complements serum angiotensin-converting enzyme. *Ann. NY Acad. Sci.* **465:** 738-743.

Baraniuk, J.N., Ohkubo, K., Kwon, O.J., Mak, J., Ali, M., Davies, R., Twort, C., Kaliner, M., Letarte, M., and Barnes, P. J.,1995, Localization of neutral endopeptidase (NEP) mRNA in human bronchi. *Eur. Respir. J.* **8:** 1458-1464.

Borson, D.B., and Gruenert, D.C., 1991, Glucocorticoids induce neutral endopeptidase in transformed human tracheal epithelial cells. *Am. J. Physiol.* **260:** L83-L89.

Cheng, H.C., Abdel-Ghany, M., Elble, R.C., and Pauli, B.U., 1998, Lung endothelial dipeptidyl peptidase IV promotes adhesion and metastasis of rat breast cancer cells via tumor cell surface-associated fibronectin. *J. Biol. Chem.* **273:** 24207-24215.

Cheung, D., Timmers, M.C., Zwinderman, A.H., den Hartigh, J., Dijkman, J.H., and Sterk, P.J., 1993, Neutral endopeptidase activity and airway hyperresponsiveness to neurokinin A in asthmatic subjects in vivo. *Am. Rev. Respir. Dis.* **148:** 1467-1473.

Cohen, A.J., Bunn, P.A., Franklin, W., Magill-Solc, C., Hartmann, C., Helfrich, B., Gilman, L., Folkvord, J., Helm, K., and Miller, Y.E., 1996, Neutral endopeptidase: variable expression in human lung, inactivation in lung cancer, and modulation of peptide-induced calcium flux. *Cancer Res.* **56:** 831-839.

Cohen, A.J., Franklin, W.A., Magill, C., Sorenson, J., and Miller, Y.E., 1999, Low neutral endopeptidase levels in bronchoalveolar lavage fluid of lung cancer patients. *Am. J. Respir. Crit. Care Med.* **159:** 907-910.

Corrigan, C.J., Hartnell, A., and Kay, A.B., 1988, T lymphocyte activation in acute severe asthma. *Lancet* **1:** 1129-1132.

Devillier, P., Advenier, C., Drapeau, G., Marsac, J., and Regoli, D., 1988, Comparison of the effects of epithelium removal and of an enkephalinase inhibitor on the neurokinin-induced contractions of guinea-pig isolated trachea. *Br. J. Pharmacol.* **94:** 675-684.

Djokic, T.D., Dusser, D.J., Borson, D.B., and Nadel, J.A. (1989). Neutral endopeptidase modulates neurotensin-induced airway contraction. *J. Appl. Physiol.* **66:** 2338-2343.

Duke-Cohan, J.S., Morimoto, C., Rocker, J.A., and Schlossman, S.F., 1996, Serum high molecular weight dipeptidyl peptidase IV (CD26) is similar to a novel antigen DPPT-L released from activated T cells. *J. Immunol.* **156:** 1714-1721.

Dusser, D.J., Jacoby, D.B., Djokic, T.D., Rubinstein, I., Borson, D.B., and Nadel, J.A., 1989, Virus induces airway hyperresponsiveness to tachykinins: role of neutral endopeptidase. *J. Appl. Physiol.* **67:** 1504-1511.

Farmer, S.G., and Togo, J., 1990, Effects of epithelium removal on relaxation of airway smooth muscle induced by vasoactive intestinal peptide and electrical field stimulation. *Br. J. Pharmacol.* **100:** 73-78.

Favaloro, E.J., Browning, T., and Nandurkar, H., 1993, The hepatobiliary disease marker serum alanine aminopeptidase predominantly comprises an isoform of the haematological myeloid differentiation antigen and leukaemia marker CD-13/gp150. *Clin.. Chim.. Acta.* **220:** 81-90.

Fine, J.M., Gordon, T., and Sheppard, D., 1989, Epithelium removal alters responsiveness of guinea pig trachea to substance P. *J. Appl. Physiol.* **66:** 232-237.

Fleischer, B., 1994, CD26: a surface protease involved in T-cell activation. [Review]. *Immunol. Today* **15:** 180-184.

Frossard, N., Rhoden, K.J., and Barnes, P.J., 1989, Influence of epithelium on guinea pig airway responses to tachykinins: role of endopeptidase and cyclooxygenase. *J. Pharmacol. Exp. Ther.* **248:** 292-298.

Fujii, H., Nakajima, M., Saiki, I., Yoneda, J., Azuma, I., and Tsuruo, T., 1995, Human melanoma invasion and metastasis enhancement by high expression of aminopeptidase N/CD13. *Clin. Exp. Metastasis* **13:** 337-344.

Fukasawa, K.M., Fukasawa, K., Sahara, N., Harada, M., Kondo, Y., and Nagatsu, I., 1981, Immunohistochemical localization of dipeptidyl aminopeptidase IV in rat kidney, liver, and salivary glands. *J. Histochem. Cytochem.* **29:** 337-343.

Ganju, R.K., Sunday, M., Tsarwhas, D.G., Card, A., and Shipp, M.A., 1994, CD10/NEP in non-small cell lung carcinomas. Relationship to cellular proliferation. *J. Clin. Invest.* **94:** 1784-1791.

Grandordy, B.M., Frossard, N., Rhoden, K.J., and Barnes, P.J., 1988, Tachykinin-induced phosphoinositide breakdown in airway smooth muscle and epithelium: relationship to contraction. *Mol. Pharmacol.* **33:** 515-519.

Hoffmann, T., Faust, J., Neubert, K., and Ansorge, S., 1993, Dipeptidyl peptidase IV (CD26) and aminopeptidase N (CD13) catalyzed hydrolysis of cytokines and peptides with N-terminal cytokine sequences. *FEBS Lett.* **336:** 61-64.

Howarth, P.H., Djukanovic, R., Wilson, J.W., Holgate, S.T., Springall, D R., and Polak, J.M., 1991, Mucosal nerves in endobronchial biopsies in asthma and non-asthma. *Int. Arch. Allergy Appl. Immunol.* **94:** 330-333.

Humbert, M., Durham, S.R., Ying, S., Kimmitt, P., Barkans, J., Assoufi, B., Pfister, R., Menz, G., Robinson, D.S., Kay, A.B., and Corrigan, C.J., 1996, IL-4 and IL-5 mRNA and protein in bronchial biopsies from patients with atopic and nonatopic asthma: evidence against "intrinsic" asthma being a distinct immunopathologic entity. *Am. J. Respir. Crit. Care Med.* **154:** 1497-1504.

Johnson, A.R., Coalson, J.J., Ashton, J., Larumbide, M., and Erdos, E.G., 1985, Neutral endopeptidase in serum samples from patients with adult respiratory distress syndrome. Comparison with angiotensin-converting enzyme. *Am. Rev. Respir. Dis.* **132:** 1262-1267.

Joos, G., Pauwels, R., and van der Straeten, M., 1987, Effect of inhaled substance P and neurokinin A on the airways of normal and asthmatic subjects. *Thorax* **42:** 779-783.

Joos, G F., Germonpre, P.R., Kips, J.C., Peleman, R A., and Pauwels, R.A., 1994, Sensory neuropeptides and the human lower airways: present state and future directions. *Eur. Respir. J.* **7:** 1161-1171.

Juillerat-Jeanneret, L., Aubert, J.-D., and Leuenberger, P., 1997, Peptidases in human bronchoalveolar lining fluid, macrophages, and epithelial cells: Dipeptidyl (amino)peptidase IV, aminopeptidase N, and dipeptidyl (carboxy)peptidase (angiotensin-converting enzyme). *J. Lab. Clin. Med.* **130:** 603-614.

Kanayama, N., Kajiwara, Y., Goto, J., el Maradny, E., Maehara, K., Andou, K., and Terao, T., 1995, Inactivation of interleukin-8 by aminopeptidase N (CD13). *J. Leukoc. Biol.* **57:** 129-134.

Kohrogi, H., Nadel, J.A., Malfroy, B., Gorman, C., Bridenbaugh, R., Patton, J.S., and Borson, D.B., 1989, Recombinant human enkephalinase (neutral endopeptidase) prevents cough induced by tachykinins in awake guinea pigs. *J. Clin. Invest.* **84:** 781-786.

Kummer, W., and Fischer, A., 1991, Tissue distribution of neutral endopeptidase 24.11 ('enkephalinase') activity in guinea pig trachea. *Neuropeptides* **18:** 181-186.

Lang, Z., and Murlas, C.G., 1993, Dexamethasone increases airway epithelial cell neutral endopeptidase by enhancing transcription and new protein synthesis. *Lung* **171:** 161-172.

Larsen, S.L., Pedersen, L.O., Buus, S., and Stryhn, A., 1996, T cell responses affected by aminopeptidase N (CD13)-mediated major histocompatibility complex class II-bound peptides. *J. Exp. Med.* **184:** 183-189.

Lundberg, J.M., Hokfelt, T., Martling, C.R., Saria, A., and Cuello, C., 1984, Substance P-immunoreactive sensory nerves in the lower respiratory tract of various mammals including man. *Cell Tissue Res.* **235:** 251-261.

Martins, M.A., Shore, S.A., and Drazen, J M., 1991, Peptidase modulation of the pulmonary effects of tachykinins. *Int. Arch. Allergy Appl. Immunol.* **94:** 325-329.

Nadel, J.A., 1990, Neutral endopeptidase modulation of neurogenic inflammation in airways. *Eur. Respir. J. Suppl.* **12:** 645s-651s.

Riemann, D., Kehlen, A., Thiele, K., Löhn, M., and Langner, J., 1997, Induction of aminopeptidase N/CD13 on human lymphocytes after adhesion to fibroblast-like synoviocytes, endothelial cells, epithelial cells, and monocytes/macrophages. *J. Immunol.* **158:** 3425-3432.

Sahara, N., and Suzuki, K., 1984, Ultrastructural localization of dipeptidyl peptidase IV in rat salivary glands by immunocytochemistry. *Cell Tissue Res.* **235:** 427-432.

Salles, G., Chen, C.Y., Reinherz, E.L., and Shipp, M A., 1992, CD10/NEP is expressed on Thy-1low B220+ murine B-cell progenitors and functions to regulate stromal cell-dependent lymphopoiesis. *Blood* **80:** 2021-2029.

Salles, G., Rodewald, H.R., Chin, B.S., Reinherz, E.L., and Shipp, M.A., 1993, Inhibition of CD10/neutral endopeptidase 24.11 promotes B-cell reconstitution and maturation in vivo. *Proc. Natl. Acad. Sci. U S A* **90:** 7618-7622.

Sekizawa, K., Tamaoki, J., Graf, P.D., Basbaum, C.B., Borson, D.B., and Nadei, J.A., 1987a, Enkephalinase inhibitor potentiates mammalian tachykinin-induced contraction in ferret trachea. *J. Pharmacol. Exp. Ther.* **243:** 1211-1217.

Sekizawa, K., Tamaoki, J., Nadel, J. A., and Borson, D.B., 1987b, Enkephalinase inhibitor potentiates substance P- and electrically induced contraction in ferret trachea. *J. Appl. Physiol.* **63:** 1401-1405.

Shipp, M.A., Stefano, G.B., Switzer, S.N., Griffin, J.D., and Reinherz, E.L., 1991a, CD10 (CALLA)/neutral endopeptidase 24.11 modulates inflammatory peptide-induced changes in neutrophil morphology, migration, and adhesion proteins and is itself regulated by neutrophil activation. *Blood* **78:** 1834-1841.

Shipp, M.A., Tarr, G.E., Chen, C.Y., Switzer, S.N., Hersh, L.B., Stein, H., Sunday, M E., and Reinherz, E.L., 1991b, CD10/neutral endopeptidase 24.11 hydrolyzes bombesin-like peptides and regulates the growth of small cell carcinomas of the lung. *Proc. Natl. Acad. Sci. U S A* **88:** 10662-10666.

Soleilhac, J.M., Lafuma, C., Porcher, J.M., Auburtin, G., and Roques, B.P., 1996, Characterization of a soluble form of neutral endopeptidase-24.11 (EC 3.4.24.11) in human serum: enhancement of its activity in serum of underground miners exposed to coal dust particles. *Eur. J. Clin. Invest.* **26:** 1011-1017.

Sont, J.K., van Krieken, J.H., van Klink, H.C., Roldaan, A.C., Apap, C.R., Willems, L.N., and Sterk, P.J., 1997, Enhanced expression of neutral endopeptidase (NEP) in airway epithelium in biopsies from steroid- versus nonsteroid-treated patients with atopic asthma. *Am. J. Respir. Cell Mol. Biol.* **16:** 549-556.

Sunday, M.E., Hua, J., Torday, J.S., Reyes, B., and Shipp, M.A., 1992, CD10/neutral endopeptidase 24.11 in developing human fetal lung. Patterns of expression and modulation of peptide-mediated proliferation. *J. Clin. Invest.* **90:** 2517-2525.

van der Velden, V.H.J., 1998, Glucocorticoids: mechanisms of action and anti-inflammatory potential in asthma. *Mediators Inflamm.* **7:** 229-237.

van der Velden, V.H.J., and Hulsmann, A.R., 1999a, Autonomic innervation of human airways: structure, function, and pathophysiology in asthma [In Process Citation]. *Neuroimmunomodulation* **6:** 145-159.

van der Velden, V.H J., and Hulsmann, A.R., 1999b, Peptidases: structure, function and modulation of peptide-mediated effects in the human lung. *Clin. Exp. Allergy* **29:** 445-456.

van der Velden, V.H.J., Naber, B.A., van Hal, P.Th.W., Overbeek, S.E., Hoogsteden, H.C., and Versnel, M.A., 1999, Peptidase activities in serum and bronchoalveolar lavage fluid from allergic asthmatics - comparison with healthy non-smokers and smokers and effects of inhaled glucocorticoids. *Clin. Exp. Allergy* **29:** 813-823.

van der Velden, V.H J., Naber, B.A.E., van der Spoel, P., Hoogsteden, H.C., and Versnel, M.A., 1998a, Cytokines and glucocorticoids modulate human bronchial epithelial cell peptidases. *Cytokine* **10:** 55-65.

van der Velden, V.H.J., Wierenga-Wolf, A.F., Adriaansen-Soeting, P.W.C., Overbeek, S.E., Moller, G.M., Hoogsteden, H.C., and Versnel, M.A., 1998b, Expression of aminopeptidase N and dipeptidyl peptidase IV in the healthy and asthmatic bronchus. *Clin. Exp. Allergy* **28:** 110-120.

van Hal, P.Th.W., Hopstaken-Broos, J.P., Prins, A., Favaloro, E.J., Huijbens, R.J., Hilvering, C., Figdor, C.G., and Hoogsteden, H.C., 1994, Potential indirect anti-inflammatory effects of IL-4. Stimulation of human monocytes, macrophages, and endothelial cells by IL-4 increases aminopeptidase-N activity (CD13; EC 3.4.11.2). *J. Immunol.* **153:** 2718-2728.

Walker, C., Bauer, W., Braun, R. K., Menz, G., Braun, P., Schwarz, F., Hansel, T. T., and Villiger, B. (1994). Activated T cells and cytokines in bronchoalveolar lavages from patients with various lung diseases associated with eosinophilia. *Am. J. Respir. Crit. Care Med.* **150:** 1038-1048.

Willheim, M., Ebner, C., Baier, K., Kern, W., Schrattbauer, K., Thien, R., Kraft, D., Breiteneder, H., Reinisch, W., and Scheiner, O. (1997). Cell surface characterization of T lymphocytes and allergen-specific T cell clones: correlation of CD26 expression with T(H1) subsets. *J. Allergy Clin. Immunol.* **100:** 348-355.

Wilson, J. W., Djukanovic, R., Howarth, P. H., and Holgate, S. T. (1992). Lymphocyte activation in bronchoalveolar lavage and peripheral blood in atopic asthma. *Am. Rev. Respir. Dis.* **145:** 958-960.

# INACTIVATION OF INTERLEUKIN-6 BY NEUTROPHIL PROTEASES AT SITES OF INFLAMMATION

*Protective effects of soluble IL-6 receptor chains*

Ute Bank, Bernadette Küpper, Siegfried Ansorge
*Institute of Immunology, Otto-von-Guericke-Universität Magdeburg, Leipziger Straße 44, D-39120 Magdeburg*

Key words interleukin-6, PMN, elastase, cathepsin G, proteinase 3, soluble receptors

Abstract In contrast to the excessively elevated immunochemically detectable concentrations of interleukin-6 (IL-6) in inflammatory exudates, the IL-6 bioactivities are significantly reduced, suggesting an inactivation of IL-6 at sites of inflammation. Since high amounts of proteases are released by invading neutrophils (PMN) in close temporal correlation to elevated IL-6 concentrations at sites of inflammation, this study focused on effects of the PMN-derived proteases elastase (NE), proteinase 3 (PR 3) and cathepsin G (Cat G) on the bioactivity and molecular integrity of IL-6. Here, we demonstrate that these enzymes play a crucial role in the initiation of the degradation and subsequent inactivation of IL-6 at sites of inflammation. Soluble IL-6 receptor subunits elicit a protective effect against the IL-6 inactivation by Cat G, only. Possible consequences of the proteolytical IL-6 inactivation for local inflammatory processes will be discussed.

## 1. INTRODUCTION

The neutrophil serine proteases NE, PR3 and Cat G are mainly known to be involved in the degradation of matrix proteins and the killing of microorganisms at local sites of inflammation. Beyond that, numerous bioactive, non-matrix proteins were identified as substrates of these enzymes under pathophysiological conditions. (Bieth 1990, Salvesen 1999,

Hoidal 1999). This suggests a regulatory relevance of these enzymes, especially in localized inflammatory processes. Interestingly, recent reports provided evidences that the PMN-proteases are capable of modulating the bioactivities of cytokines at local sites of inflammation, too (Padrines *et al.* 1994, Coeshott *et al.* 1999).

Previously, we reported a close temporal correlation of elevated IL-6 levels and the release of high amounts of proteases by infiltrating PMN in the inflammatory processes *in vivo*, suggesting protease-cytokine interactions (Bank *et al.* 1995). This led to the hypothesis, that these enzymes may have the capacity to modulate the bioactivity of IL-6 at sites of inflammation. In the present paper, we provide evidences that serine proteases released by degranulating PMN play a crucial role in the initiation of the proteolytical IL-6 inactivation. It will be shown that the binding of IL-6 to the soluble receptor subunits sIL-6R and sgp130 prevents the degradation of IL-6 by Cat G (and partially by PR 3), but not the elastase catalyzed cleavage. Furthermore, biological consequences of the proteolytical inactivation of the pleiotropic inflammatory cytokine IL-6 will be discussed.

## 2. MATERIALS AND METHODS

### *Comparative determination of IL-6 concentrations and bioactivities in inflammatory exudates*

Synovial fluids, ascites, pleural effusions, and CSF were obtained from patients with acute inflammatory diseases as approved by the local ethics commitee. Cell-free supernatants of these exudates were prepared by a two-step centrifugation procedure (15 min, 700 x g 10 min, 900 x g). The IL-6 concentrations in these samples were detected by Elisa (R & D Systems), the IL-6 bioactivities assayed by determinating the proliferation rate of IL-6 sensitive B9 hybridoma cells (ATCC). In both assays, purified human lymphocyte-derived IL-6 (Boehringer Mannheim), adjusted to the international IL-6 standard (NIBSC), was used as standard.

### *Exposure of IL-6 to PMN-proteases and analysis of IL-6 proteolyis products*

PMN of healthy donors (purity > 98 %, viability 95 – 99 %), lysates and subcellular fractions of PMN were prepared as previously described (Bank 1997, Sengelov 1993). Glycosylated (CHO derived) as well as unglycosylated rhIL-6 (from E. coli) was incubated in the presence of $10^7$

vital activated PMN, lysates and subcellular fractions, or with purified NE (ICN), Cat G (ICN) and PR 3 (IBL), respectively. IL-6 proteolysis products were analyzed by Western blotting using an affinity purified, polyclonal anti-human IL-6 rabbit IgG (Genzyme) and chromogenic or chemoluminescence detection systems. Cleavage site analysis was performed by N-terminal sequencing (ABI Procise system, Edman-degradation) and by MALDI-TOF mass spectrometry (Hewlett Packard G2025A LD-TOF). The sequences of IL-6 fragments were predicted using the General Protein Mass Analysis for Windows 2.13 software.

# 3. RESULTS

## 3.1 Reduced IL-6 bioactivities in inflammatory exudates

Because of the high proteolytic potential at sites of inflammation, the IL-6 bioactivities in various biological fluids derived from sites of acute inflammation, were analyzed in direct comparison to the immunochemically detectable IL-6 concentrations (Fig1). Whereas elevated IL-6 concentrations up to 459 ng/ml (60129 IU/ml) were detectable by ELISA, in 76 % of all samples (N = 115) the bioactivity of IL-6 was considerably below the predicted values. In Figure 1 the individual relative bioactivity data values (related to the ELISA data) as well as the average concentrations of IL-6 in various kinds of inflammatory exudates are presented.

## 3.2 IL-6 inactivation by PMN-derived serine proteases

The reduced IL-6 bioactivities in the majority of inflammatory exudates supported the hypothesis that proteolytic enzymes at local sites of inflammation may affect the IL-6 bioactivity. Since especially high concentrations of active PMN-proteases (up to 40µg/ml) were detectable in the inflammatory exudates (Bank 1995), effects of these proteases on the molecular integrity and biological activity of IL-6 were analyzed *in vitro*.

Initial experiments revealed IL-6 is rapidly cleaved and inactivated in the presence of vital activated PMN or PMN lysates. The majority of the IL-6 inactivating activity was found to be localized in the granule and membrane fraction of neutrophils. Among several group specific protease inhibitors only the serine protease inhibitor diisopropyl fluorophosphate (DFP) abolished the IL-6 degradation. Analysing the individual capacity of each of the three serine PMN-proteases NE, PR3 and Cat G to inactivate IL-6, it became evident that each of the proteases is capable of cleaving IL-6, independent of its glycosylation status. The hydrolysis of IL-6 by NE led to the formation of two immunoreactive fragments, initially (Fig 2, lane 4). PR 3, which was described to have a NE-like substrate specificity, was found to release an immunoreactive 16.2 kDa (Fig 2, lane 7). Cat G initially catalyzes the hydrolysis of unglycosylated IL-6 into two intermediate fragments with apparent molecular weights of 9 and 12 kDa. (Fig 2, lane 10). By mass

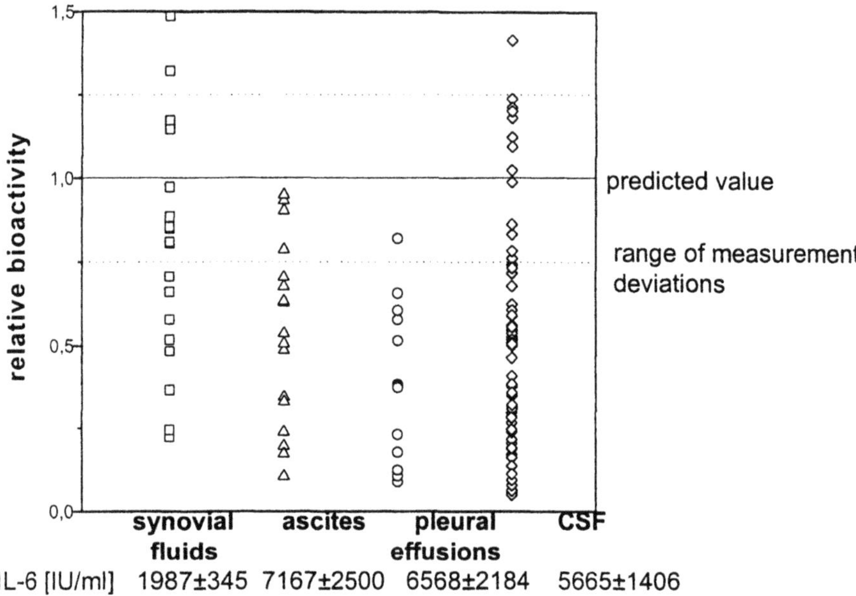


*Figure 1.* The relative IL-6 bioactivity in inflammatory exudates calculated on the basis of the immunochemically detectable IL-6 concentration and given for each kind of inflammatory exudate separately.

spectrometry and N-terminal sequencing, the initially hydrolyzed peptide bonds were detected to be $Val^{11}$-$Ala^{12}$ and $Leu^{19}$-$Thr^{20}$ (NE), $Phe^{78}$-$Asn^{79}$ (Cat G) and $Ala^{145}$-$Ser^{146}$ (PR 3). Determinating the growth promoting activity of protease-treated (in comparison to untreated) IL-6, the cleavage of IL-6 was found to be accompanied by a time-dependent loss of its bioactivity. No significant difference between the inactivation kinetics or of glycosylated rh IL-6 forms was observable (not shown). Importantly, further experiments showed that IL-6 is also inactivated in the presence of inflammatory exudates with high proteolytic potential (not shown).

### 3.3 Influence of soluble IL-6 receptor subunits on the hydrolysis of IL-6 by PMN serine proteases

Since the soluble forms of the IL-6 receptor subunits gp80 (sIL-6R) and gp130 (sgp 130) have been discussed to function as carrier proteins of IL-6 and to protect its molecular integrity *in vivo*, possible effects of the binding of IL-6 to the soluble receptors were investigated. IL-6 was preincubated in the presence of a molar surplus of the ligand-binding rh sIL-6R (Fig 2, lanes 5, 8 and 11) or in the presence of rh sIL-6R plus the soluble form of the signal transducing receptor subunit gp130 (ternary IL-6/IL-6 receptor complex; Fig 2, lanes 6, 9 and 12). The IL-6/IL-6 receptor complexes were exposed to purified NE, PR 3 or Cat G. As shown in Fig 2, only the Cat G catalyzed cleavage of IL-6 was prevented completely (Fig 2, lane 10 - 12). The cleavage catalyzed by proteinase 3 seems to be temporally delayed (Fig 2, lane 7 - 9), the IL-6 degradation by NE was not affected (Fig 2, lane 4-6).

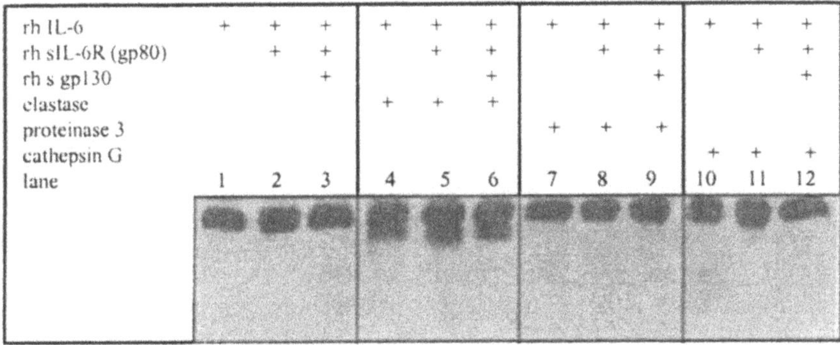


*Figure 2.* Hydrolysis of free or receptor bound IL-6 by purified PMN proteases. rh IL-6 (0,25μg) were preincubated with soluble receptors as indicated, thereafter incubated in the presence of purified NE, PR3 or Cat G for 30 min and subjected to Western blotting analysis using the affinity-purified, polyclonal anti-IL-6 IgG as primary antibody.

## 4. CONCLUSIONS

Here we provide evidence, that NE, PR3 and Cat G play a crucial role in the proteolytical degradation of IL-6, which may be the cause for the reduced IL-6 bioactivities detectable in inflammatory exudates from local sites of inflammation. Each of these proteases is capable of degrading IL-6 very rapidly, independently of its glycosylation status. Structure-function analysis revealed that especially the peptide bonds initially attacked by PR 3 and Cat G are located in sequence regions, which are essential for the structural stability and biological activitity of the IL-6 molecule (Somers 1997) Moreover, the $Phe^{78}$-residue in P1-position of the Cat G cleaved peptide bond is mainly involved in the binding of IL-6 to the gp80 receptor subunit. This explains the protective effect of the sIL-6R binding on the Cat G catalyzed IL-6 cleavage. Beyond that, the formation of ternary complexes with the signal transducing receptor subunit sgp130 was found to result in a slightly delayed IL-6 degradation by PR 3. In contrast, the formation of the two N-terminal truncated IL-6 fragments by NE is not influenced in the presence of the soluble receptor subunits, probably the structural instabile N-terminal loop sequence of IL-6 is not involved in the receptor binding .

Since IL-6 has been reported to elicit both, proinflammatory and antiinflammatory effects, it is difficult to assess, whether the proteolytical inactivation is rather limiting or promoting the inflammatory process. Due to the previously shown neutrophil-stimulating activity of IL-6 (Bank 1997), the inactivation of this cytokine by synergistic action of PMN-serine proteases might act as a direct feedback mechanism terminating IL-6-induced activation signals. With respect to the function of IL-6 as key cytokine for the induction of antiinflammatory mediators, the rapid inactivation of IL-6 might result in a perpetuation and further strengthening of the local inflammatory process. This assumption is supported by preliminary data implicating an impaired induction of antiinflammatory cytokines at local sites of neurotrauma. Taken together, these data support that PMN-derived serine proteases are important tools for the regulation of the biological activity of cytokines at local sites of inflammation.

## REFERENCES

Bank, U., *et al.*, 1997, Effekt of interleukin-6 (IL-6) and transforming growth factor- (TGF-b) on neutrophil elastase release. *Inflammation* **19:** 83-99.

Bieth, J.G., 1999, In *Handbook of Proteolytic Enzymes.* (A.J. Barrett, N.D. Rawlings and J.F. Woessner, eds.), Academic Press, San Diego, pp. 42-50, 54-60.

Coeshott C., 1999, Converting enzyme-independent release of tumor necrosis factor alpha and IL-1beta from a stimulated human monocytic cell line in the presence of activated neutrophils or purified proteinase 3. *Proc. Natl. Acad. Sci. USA* **96:** 6261-6266.

Hoidal, J.R., 1999, In *Handbook of Proteolytic Enzymes.* (A.J. Barrett, N.D. Rawlings and J.F. Woessner, eds.) Academic Press, San Diego, pp. 62-65.

Padrines, M., *et al.*, 1994, Interleukin-8 processing by neutrophil elastase, cathepsin G and proteinase 3. *FEBS Lett.* **352:** 231-235.

Salvesen, G.S., 1999, In *Handbook of Proteolytic Enzymes.* (A.J. Barrett, N.D. Rawlings and J.F. Woessner, eds.), Academic Press, San Diego, pp. 60-62.

Sengelov, H., 1993, Subcellular localization and dynamics of Mac-1 (alpha m beta 2) in human neutrophils. *J. Clin. Invest.* **92:** 1467-1476.

Somers, W., *et al.*, 1997, 1.9 A crystal structure of interleukin 6: implications for a novel mode of receptor dimerization and signaling. *EMBO J.* **16:** 989-997.

Tilg, H., *et al.*, 1997, IL-6 and APPs: anti-inflammatory and immunosuppression mediators. *Immunol. Today* **18:** 428-432.

# ANTISENSE INHIBITION OF CATHEPSIN B IN A HUMAN OSTEOSARCOMA CELL LINE

Sabine Krueger[1], Carsten Haeckel[1], Frank Buehling[2]
*[1]Institute of Pathology, [2] Institute of Experimental Internal Medicine, Otto-von-Guericke Universität Magdeburg, Leipziger Str. 44, 39120 Magdeburg*

## 1. INTRODUCTION

Tumor invasion and metastasis are complex, multistep processes involving cellular detachment from primary sites, adhesion to extracellular/cellular matrices, local proteolytic degradation of matrix proteins and migration. Several proteases are involved in these complex processes. Besides matrix metalloproteases (MMP's) (Liotta and Stetler-Stevenson, 1990) and the urokinase-type plasminogen activator (u-PA) (Schmidt *et al* 1992), cathepsin B (cat B) (Sloane *et al.* 1990) is considered to play important roles in the different destructive occurrences. Besides direct proteolysis of components of the basement membrane and extracellular matrix, cat B participates in enzyme activation cascades, including other cathepsins and uPA (Kobayashi *et al* 1991).

To block not only the catalytic activity of the cysteine protease cat B (e.g. by anticatalytical antibodies), the osteosarcoma cell line MNNG/HOS was stably transfected with antisense cat B constructs, resulting in direct suppression of cat B expression in the tumor cells. We selected cell clones with the lowest enzyme expression to functionally characterize them in in-vitro assays of invasion and motility, compared with the parental cell line MNNG/HOS.

## 2. MATERIALS AND METHODS

### *Cell culture*

The human osteosarcoma cell line MNNG/HOS (ATCC, Rockville, USA) was grown under standard conditions in RPMI 1640, 10 % FBS, and antibiotics/antimycotics (PAA, Germany). Prior to functional assays, 0.02 % EDTA was used to detach cells instead of trypsinisation.

### *Preparation of the expression vector*

The 1053bp cathepsin B cDNA fragment was RT-PCR amplified using the synthetic primers: CBvw 5'-GATCTAGGATCCGGCTTC-3' and CBrw 5'-CCCACG-GCAGATTAGATC-3'. The amplified cDNA fragment was directly cloned into the eucaryotic pcDNA3.1/V5/His-TOPO vector (Invitrogen). Sense and antisense clones were distinguished by sequence analysis.

### *Transfection and selection*

The parental MNNG/HOS cells (~ 5 x $10^5$) were transfected with 2 µg of expression vector using Lipofectin® (Gibco BRL) according to the manufacturer's instructions. Twenty-four hours after transfection, 500 µg/ml Geneticin was added to the cultures. Five weeks later 20 stable antisense cathepsin B clones were isolated and characterized by enzyme quantification.

### *Quantification of cathepsin B by ELISA*

Cathepsin B in cell extracts was detected using the Cathepsin B ELISA from Jozef Stefan Institute, Ljubljana, Slovenia according to the manufacturer's instructions.

### *Northern blot analysis*

Total RNA was isolated from cell culture using Trizol reagent (Gibco BRL) according to the manufacturer's instructions. Ten µg of RNA was electrophoresed on 1 % agarose gels, transferred to a nylon membrane, and probed with a 740bp cathepsin B fragment labeled with [$^{32}$P]dCTP. After stripping, Northern blots were reprobed with ß-actin.

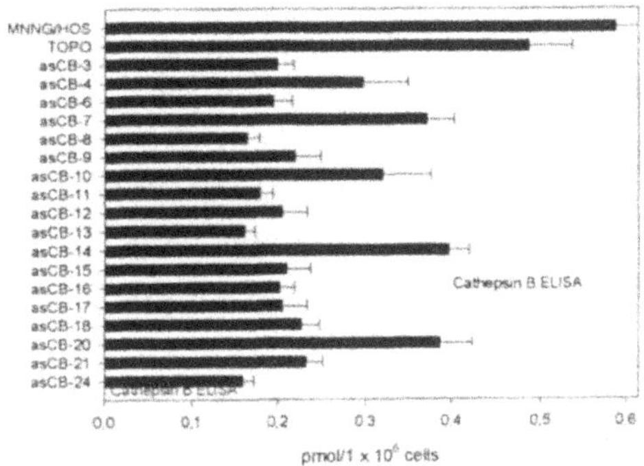


*Figure 1.* Quantification of cathepsin B protein by ELISA in cell lysates of 1 x $10^6$ parental or antisense-transfected cells per ml (mean ± SD). Compared with MNNG/HOS and MNNG/TOPO, cathepsin B concentrations were significantly reduced in all transfected cell clones (two-tailed, unpaired $t$-test; $p < 0.05$).

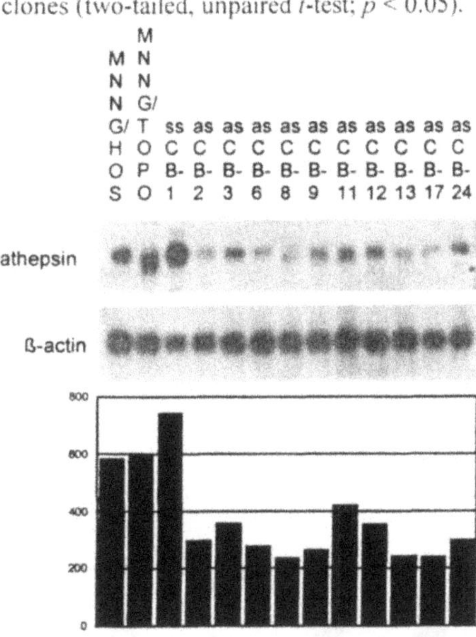


*Figure 2.* Northern blot analysis of the 2.2 kb cathepsin B transcript in MNNG/HOS and transfectants. Both cathepsin B and actin specific signals on X-ray films were quantified using the ImageMaster® (Pharmacia, Germany). Cathepsin B band intensities were normalized to the corresponding actin bands (bar graph).

## *In vitro invasion and motility*

Tumor cells were labeled with (methyl $^3$H)-thymidin for 24 h. Cellular motility was evaluated in uncoated transwell chambers® (8 µm pore size). Chemotaxis with collagen I (5 µg/ml in the lower compartment) was additionally determined. Invasion assays were done on matrigel-precoated filters (100 µg/cm$^2$). In case of invasion, 8 x 10$^5$ cells/well were incubated on the reconstituted basement membrane for 72 h. For motility, 1 x 10$^5$ cells/well were seeded onto the filters for 24 h. Cells harvested from the lower sites of the filters were measured in a liquid scintilation counter.

# 3. RESULTS

MNNG/HOS cells were stably transfected with expression vectors containing the cathepsin B cDNA fragment in antisense orientation downstream of a CMV promotor. To determine whether antisense cathepsin B reduced the corresponding enzyme expression, we characterized MNNG/HOS and the transfectants by Northern blotting and ELISA. In screening, as cat B clones 3, 8, 13, 17 and 24 showed the lowest expression (30 – 50 %) of cathepsin B protein (Fig 1).

Comparable results were found at the mRNA level. Northern blot analysis revealed a considerable decrease of approximately 50 % in the cathepsin B mRNA level of the antisense-transfected cell lines 3, 8, 13, 17 and 24 (Fig 2). In functional assays, cathepsin B transfectants showed a significantly reduced invasion and motility compared with the parental cell line MNNG/HOS. Whereas 3.5 % of the MNNG/HOS cells passed the

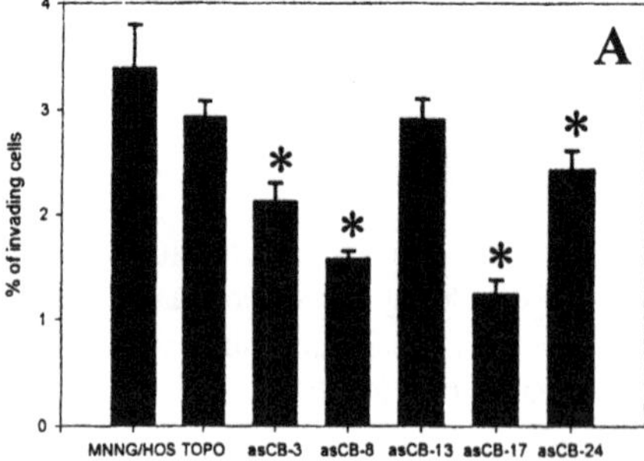


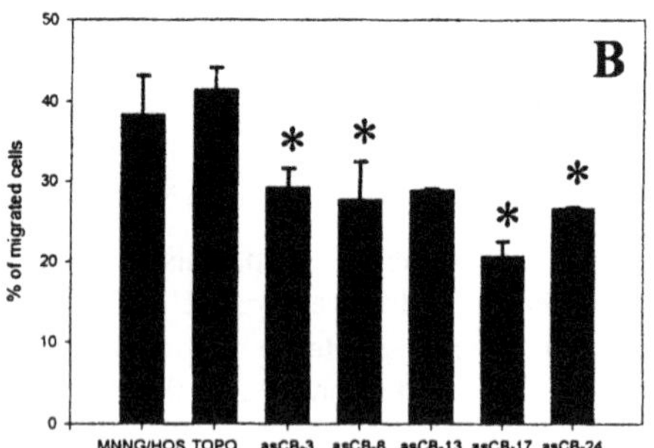


*Figure 3.* Invasion (A) and motility (B) of MNNG/HOS, MNNG/TOPO and antisense clones measured in Matrigel-coated or uncoated Transwell chambers. (* statistically significant; two-tailed, unpaired $t$-test: $p < 0.05$; (mean ± SD))

reconstituted Matrigel matrix, only 2.1 % of asCB-3, 1.58 % of asCB-8 and 1.25 % of asCB-17 were detectable on the lower side of the filters (Fig 3A). As shown in Fig 3B, the antisense-transfected cell clones passed the filter barrier more slowly (only 20 - 29 %) than did the untransfected MNNG/HOS (38 %) and MNNG/ TOPO cells (41 %).

Since the antisense-transfected cell clones exhibited reduced migration and invasion, we attempted to clarify if there were differences in the adhesive potential to protein matrices between transfected clones and MNNG/HOS. Overall, the adhesion of the cell clones to Matrigel was lower than to collagen I, but the antisense clones showed no tendency to adhere more slowly than did MNNG/HOS (results not shown).

## 4. CONCLUSION

The correlation between cathepsin B expression levels and cancer aggressiveness (Campo *et al* 1994, Sinha *et al* 1995) as well as the results of *in vitro* (Kobayashi *et al* 1992, Redwood *et al* 1992) and *in vivo* (Van Noorden *et al* 1998) models suggest that cathepsin B is an important component of the invasive phenotype of many tumor types. In the study presented here, the osteosarcoma cell line MNNG/HOS was stably transfected with antisense constructs of cathepsin B. Using antisense technology (for review see Hélène and Toulmé 1990), both catalytic and possible non-catalytic functions of cathepsin B were affected.

The findings obtained in the *in vitro* invasion or motility assays (Fig 3) are consistent with the results obtained by the cysteine protease inhibitor E-64 and the specific cathepsin B inhibitor CA074 in ovarian cancer cells and bladder tumor cells (Kobayashi *et al* 1992, Redwood *et al* 1992). Kolkhorst *et al* (1998) found that in cell lines with higher expression of cathepsin B, CA074 efficiently inhibited cellular invasion, whereas in other cell lines, cysteine protease inhibitors were less effective, indicating that proteases such as uPA and MMPs play a more important role in these tumor cells. Several laboratories dealing with different tumor entities (Mohanam *et al* 1997, Watanabe *et al* 1996, Mikkelsen *et al* 1995) have demonstrated that a network of proteolytic enzymes, together with the corresponding receptors and endogenous inhibitors, play a role in tumor malignancy.

Our results suggest that cathepsin B is involved in invasion and metastazation of osteosarcoma cells. Besides the demonstrable direct proteolytic activities of cathepsin B against extracellular matrix components, it is the interaction with other proteases in a complex proteolytic cascade which appears to be important for tumor progression in osteosarcomas.

# REFERENCES

Campo, E., *et al.*, 1994, Cathepsin B expression in colorectal carcinomas correlates with tumor progression and shortened patient survival. *Am. J. Pathol.* **145:** 301-309.

Hélène, C., and Toulmé, J-J., 1990, Specific regulation of gene expression by antisense, sense and antigen nucleic acids. *Biochim. Biophys. Acta.* **1049:** 99-125.

Kobayashi, H., *et al.*, 1991, Cathepsin B efficently activates the soluble and the tumor cell receptor-bound form of the proenzyme urokinase-type plasminogen activator (pro-uPA). *J. Biol. Chem.* **266:** 5147-5152.

Kobayashi, H., *et al.*, 1992, Inhibition of *in vitro* ovarian cancer cell invasion by modulation of urokinase plasminogen activator and cathepsin B. *Cancer Res.* **52:** 3610-3614.

Kolkhost, V., *et al.*, 1998, Inhibition of tumor cell invasion by protease inhibitors: correlation with the protease profile. *J. Cancer Res. Clin. Oncol.* **124:** 598-606.

Liotta, L.A., and Stetler-Stevenson, W.G., 1990, Metalloproteinases and cancer invasion. Semin. *Cancer Biol.* **2:** 99-106.

Mikkelsen, T., *et al.*, 1995, Immunolocalization of cathepsin B in human glioma: implications for tumor invasion and angiogenesis. *J. Neurosurg.* **83:** 285-290.

Mohanam, S., *et al.*, 1997, In vitro inhibition of human glioblastoma cell line invasiveness by antisense uPA receptor. *Oncogene* **14:** 1351-1359.

Redwood, S.M., *et al.*, 1992, Abrogation of the invasion of human bladder tumor cells by using protease inhibitors. *Cancer* **69:** 1212-1219.

Schmidt, M., *et al.*, 1992, Tumor-associated urokinase-type plasminogen activator: biological and clinical significance. *Biol. Chem. Hoppe-Seyler* **373:** 611-622.

Sinha, A.A., *et al.*, 1995, Cathepsin B in angiogenesis of human prostate: an immunohistochemical and immunoelectron microscopic analysis. *Anat. Rec.* **241:** 353-362.

Sloane, B.F., *et al.*, 1990, Cathepsin B and its endogenous inhibitors: the role in tumor malignancy. *Cancer and Metastasis Reviews* **9:** 333-352.

Van Noorden, C.J.F., *et al.*, 1998, Heterogeneous suppression of experimentally induced colon cancer metastasis in rat liver lobes by inhibition of extracellular cathepsin B. *Clin. Exp. Metastasis* **16:** 159-167.

Watanabe, M., *et al.*, 1996, Inhibition of metastasis in human gastric cancer cells transfected with tissue inhibitor of metalloproteinase 1 gene in nude mice. *Cancer.* Suppl. **77:** 1676-1680.

# PROTEASE-PROTEASE INHIBITOR BALANCE IN THE GASTRIC MUCOSA

*Influence of* Helicobacter pylori *infection*

Manfred Nilius[1], Thomas Vahldieck[2], Ina Repper[1], Armin Sokolowski [3], Paul Janowitz[2], Peter Malfertheiner[1]
*[1]Center of Internal Medicine, Dept. Gastroenterology, Hepatology and Infectious Diseases*
*[2]Dept. Internal Medicine, Teaching Hospital Burg; [3]Dept. Clinical Chemistry*
*Otto-von-Guericke-University Magdeburg, Leipziger Str. 44, D-39120 Magdeburg, Germany*

## 1. INTRODUCTION

In *Helicobacter pylori* infection the bacteria colonizing the gastric epithelium are able to induce gastritis and peptic ulceration. Although the infection, if untreated, is chronic, it is characterized by induction of an acute inflammatory reaction in the mucosa indicated by a neutrophil infiltration of the gastric tissue (Blaser *et al* 1992). Neutrophils are important histological markers for grading the activity of gastritis (Dixon *et al* 1996).

Although the exact mechanism how *H. pylori* infection contributes to the damage of the gastrointestinal mucosa is still unclear, the observation the extent of mucosal injury beeing related to the degree of *H. pylori* infection and neutrophil infiltration (Bayerdörffer *et al* 1992), supports the hypothesis that neutrophil products including neutrophil proteases, and/or the effect of *H. pylori* bacteria on the gastric epithelium may play an essential role in this process.

Neutrophils contain the serin proteases elastase, cathepsin G, and proteinase 3, as well as other enzymes. Under inflammatory conditions these enzymes are found extracellularly in the gastric mucosa and if not inactivated by endogenous protease-inhibitors they may cause proteolytic damage. Furthermore, these specific inhibitors themselves may become a substrate for excessive unbalanced proteolytic activity. Specific inhibitors for neutrophil proteases are plasma-derived $\alpha_1$-antitrypsin, $\alpha_2$-macroglobulin and epithelial secretory leukocyte protease inhibitor (SLPI).

The major aims of the present study were to assess the influence of *H. pylori* infection on the proteolytic action of neutrophil elastase and the antiproteolytic action of its specific inhibitors SLPI and $\alpha_1$-antitrypsin.

# 2. MATERIAL AND METHODS

## 2.1 Subjects

Biopsy specimens of the gastric antrum of *H. pylori* infected and non-infected subjects were taken during routine gastroscopy. *H. pylori* infection was diagnosed by Helicobacter urease test (HUT), histology and *H. pylori* culture. Biopsy specimens were processed for immunohistochemistry (paraffin- and cryosectioning), ELISA and Westernblot (homogenization) and RT-PCR (mRNA extraction).

## 2.2 Immunohistology

Location of SLPI and Elastase in paraffin sections was done by immunostaining with specific rabbit anti human SLPI antibodies (custom made by EUROGENTEC) and rabbit anti human Elastase antibodies (DAKO) as primary antibodies. Detection was performed with biotinylated secondary antibodies and an alkaline phosphatase detection system (ABC-kit, VECTOR).

## 2.3 *In situ* Hybridization

SLPI-mRNA was localized in cryostat sections using 30-mer double FITC-labeled oligonucleotides. Controls included β-actin, vimentin, poly dT and SLPI-sense oligonucleotides. Specific binding was detected using alkaline phosphatase labeled anti-FITC-antibodies (ROCHE-DIAGNOSTICS) and NBT/BCIP substrate.

## 2.4 RT-PCR

SLPI-mRNA first was described into cDNA (First-strand cDNA-kit, BOEHRINGER) and than amplified with 30 cycles using Taq polymerase (ROCHE-DIAGNOSTICS) and SLPI-specific sense-and antisense oligonucleotid primers.

## 2.5 ELISA

Elastase, SLPI and $\alpha_1$-antitrypsin levels in tissue homogenates were measured using commercial ELISA (SLPI-ELISA, R & D; $\alpha_1$-antitrypsin ELISA, IBL; PMN-Elastase-ELISA, MERCK) according to the manufacturers recommendations. Concentrations measured were related to the protein concentrations of the homogenates (PIERCE-Protein-Assay).

## 2.6 Westernblot

Tissue homogenates and marker proteins (rSLPI, R&D; $\alpha_1$-antitrypsin, SIGMA) were separated on 4 – 20 % gradient polyacrylamide gels (NOVEX) using the Tris-Glycine buffer system. Proteins were blotted onto nitrocellulose sheets by semi-dry blotting using a discontinuous buffer system. Blocking was done with PBS-BSA for SLPI-blots and with 5 % non fat milk-PBS for $\alpha_1$-antitrypsin blots. Protein bands were developed using a commercial polyclonal goat-anti human SLPI antibodies (R & D) and rabbit-anti human $\alpha_1$-antitrypsin antibodies (DAKO) as primary antibodies. The detection system comprised a rabbit anti goat (SLPI) or swine anti rabbit ($\alpha_1$-antitrypsin) bridging antibody combined with a PAP goat or rabbit antibody. Antigen bands were developed by incubation of the sheets in diaminobenzidine as a substrate.

## 2.7 Statistical analysis

Statistical differences were analysed by student's T-test and Mann-Whitney-Test. Differences were considered to be significant when $p < 0.05$.

# 3. RESULTS

## 3.1 Immunohistochemical localization of elastase, SLPI and $\alpha_1$-antitrypsin

In *H. pylori* negative tissue only few neutrophils positive for elastase could be detected, whereas their number is markedly increased in *H. pylori* positive patients. The neutrophils are located in the epithelial region and around the gastric glands. In contrast, SLPI, in non-infected subjects, is located only within the epithelial cells of the gastric mucosa whereas a significant decrease is observed in *H. pylori* positive patients. In these patients SLPI also could be demonstrated within inflammatory cells. The endogenuous elastase-inhibitor $\alpha_1$-antitrypsin is equally, distributed within the gastric tissue (Fig 1).

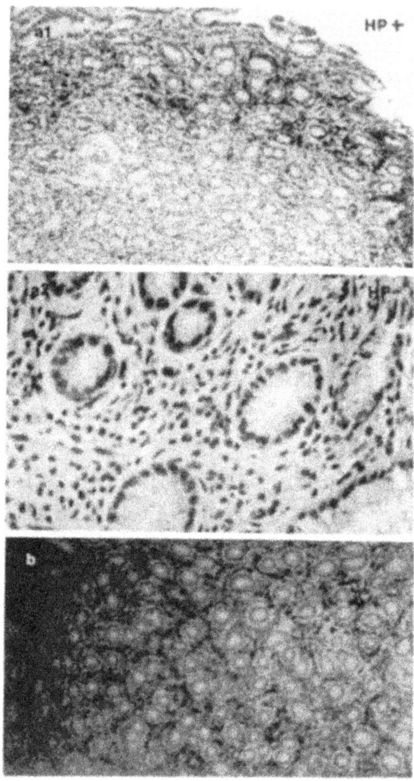


*Figure 1.* Immunohistological localization of proteases and antiproteases in gastric mucosa. a1) Elastase, in H.pylori positive patients, staining predominant in the epithelial region; a2) Elastase in H.pylori negative patients, only few cells are stained (arrows); . b) $\alpha_1$-antitrypsin, homogenous distribution in the tissue.

## 3.2 Effect of *H. pylori* infection on SLPI-transcription

SLPI-mRNA in detectable amounts was only seen in *H. pylori* negative subjects. *H. pylori* infected patients showed negligible amounts of mRNA. This could be demonstrated by *in situ* hybridization with specific oligonucleotides against SLPI-mRNA. Performing RT-PCR in tissue of infected and non-infected people, we were not able to demonstrate any amplification product for SLPI in *H. pylori* positives. These results indicate nearly total inhibition of SLPI-mRNA. (Fig 2).

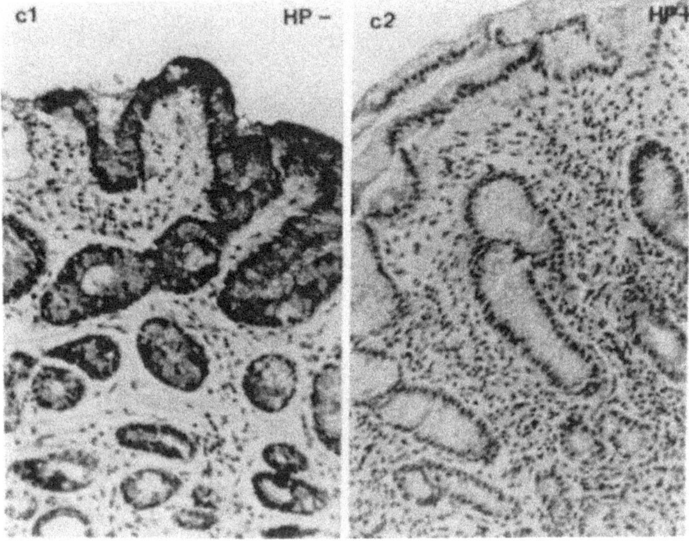


*Figure 1.* c1) SLPI in H.pylori negative patients, strong staining of the epithelium. .c2) SLPI in H.pylori positive patients, only few staining of the epithelial cells.

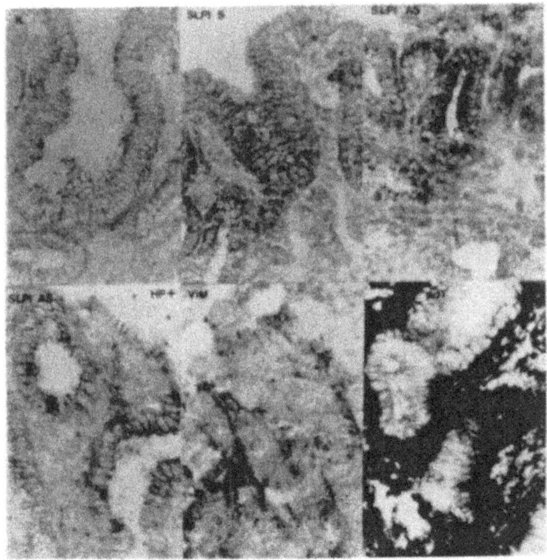

*Figure 2.* a) *In situ* Hybridisation of SLPI- mRNA C = Negative control, SLPI-S = Sense probe, SLPI-AS = Antisense probe, VIM = Vimentin control, pdT = poly dT control. No staining is seen in negative control and with SLPI-sense probe. Strong positive staining is seen in H.pylori negative tissue (HP-) , vimentin and pdT-controls. Only few staining is seen in H.pylori positive tissue (HP+).

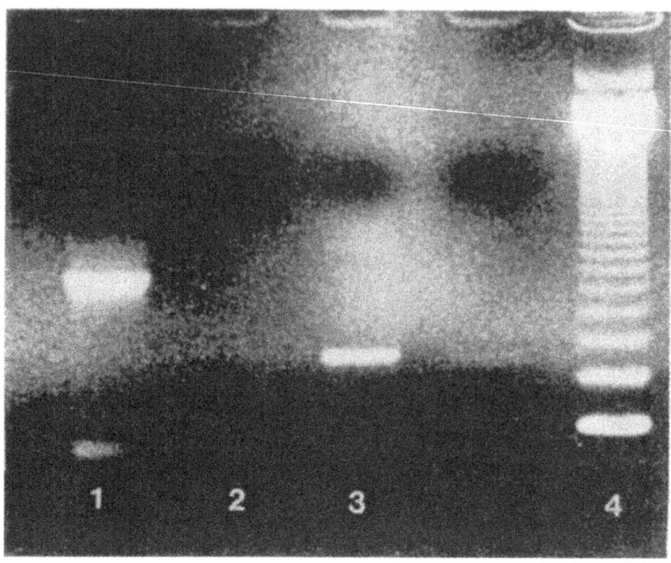


*Figure 2.* b) RT-PCR from patient tissue RNA. β-actin (1), HP-positive patient (2), HP-negative patient (3), 1kb-ladder (4).

## 3.3 Effect of *H. pylori* infection on protease and antiprotease production

Quantitative measurement of the protease and antiprotease concentrations by ELISA showed the following results: In the tissue homogenates of *H. pylori* infected subjects, elastase concentrations were significantly higher and SLPI concentrations were significantly lower than in non-infected subjects. The $\alpha_1$-antitrypsin concentrations were also higher in *H. pylori* infected patients without being statistically significant (Fig 3).

## 3.4 Degradation of SLPI and $\alpha_1$-antitrypsin

The SLPI protein, independent from infection or elastase concentration, remained unaffected when investigated for degradation of the molecule by Western immunoblot method. No complex formation indicating interaction with protease(s) could be demonstrated. In contrast, $\alpha_1$-antitrypsin showed both, degradation of the molecule and complex formation. In patients with high elastase levels, in majority being *H. pylori* positive, $\alpha_1$-antitrypsin degradation and complex formation is predominant. In patients with medium or low elastase tissue concentrations, native undegraded $\alpha_1$-antitrypsin is predominant. (Fig. 4).

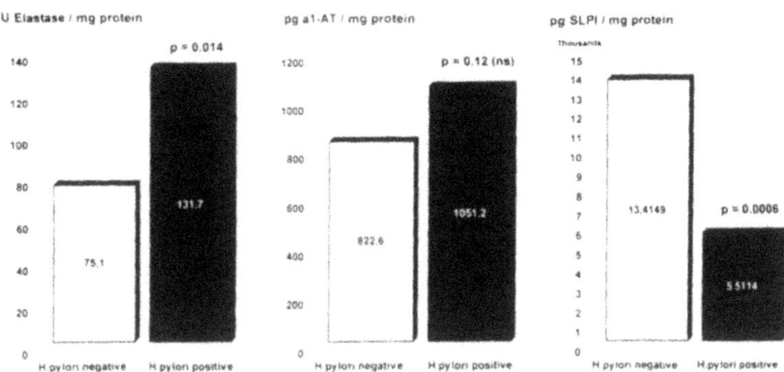


*Figure 3.* Concentrations of Protease and Protease inhibitor in tissue homogenate measured by ELISA. Elastase concentrations are significantly in H. pylori positives (1) $\alpha_1$-antitrypsin concentrations are also in H. pylori positives (2) SLPI concentrations are significantly decreased in H.pylori positives (3).

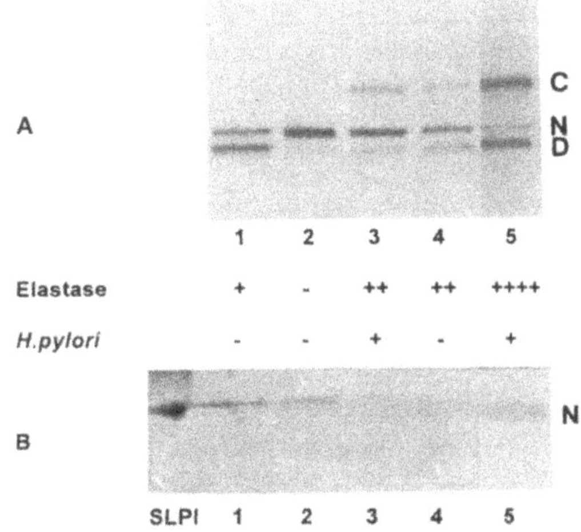


*Figure 4.* Integrity of protease inhibitor molecules tested by Westernblot A) native $\alpha_1$-antitrypsin is predominant in healthy controls (2) with higher elastase concentrations (cf. number of +signs) native band (N) decreases and Complex (C) and degradation products (D) are increased (3-5) B) Corresponding SLPI Westernblot of all patients only showed undegraded, uncomplexed native protein bands (N).

# 4. DISCUSSION

The correlation between neutrophil infiltration of the gastric mucosa following *H. pylori* infection is well known and is used also for grading the activity of inflammation. Therefore it is not surprising that the increasing amount of neutrophils may also be able to contribute to tissue damage by releasing proteolytic enzymes after activation. Passively released or actively secreted, these proteolytic enzymes, and in our case especially the predominant elastase, have been linked to the pathologic processes of a variety of inflammatory diseases, including rheumatoid arthritis, respiratory distress syndrom, and ulcerative colitis (Döring 1994, Weiss 1989). Uncontrolled proteolysis in tissue also occurs when the levels of regulating proteinase inhibitors are reduced by genetic disorders. A classic example is familial emphysema (Janoff 1985). Recently, it has also been shown that individuals presenting with genetic $\alpha_1$-antitrypsin deficiency more frequently are duodenal ulcer patients positive for *H. pylori* infection (Elzouki *et al* 1998).

This indicates that elastase if present in considerable concentrations outside of the neutrophils may contribute to mucosal damage during chronic inflammation, if not balanced by endogenous protease inhibitors such as $\alpha_1$-antitrypsin, the epithelial SLPI and the third important endogenous proteinase inhibitor, $\alpha_2$-macroglobulin, not investigated in this study. Furthermore it is imaginable that these inhibitors themselves may be degraded by uncontrolled proteolysis.

As our results show during *H. pylori* infection the protective shield of endogenous antiproteases is attacked in different ways. SLPI locally produced and secreted from the epithelial cells is significantly reduced. This loss in biologically active molecules is not due to a degradation of the molecule by proteolytic acitivity but by an inhibition of the mRNA. This is surprising, for other studies have shown that neutrophil elastase and pro-inflammatory cytokines (IL1,TNF$\alpha$) may induce SLPI-mRNA expression (Marchand *et al* 1997, Sallenave *et al* 1994). This indicates that SLPI mRNA production may be influenced directly by specific *H. pylori* factors or indirectly by *H. pylori* induced other factors such as immunoregulatory IFN-$\gamma$ or prostaglandin $E_2$. These factors are known suppressors of SLPI (Denison *et al* 1998, Jin *et al* 1998) and are also elevated in the inflammatory reaction following *H. pylori* infection (Ihan *et al* 1999, Lindholm *et al* 1998, Franco *et al* 1999, Fu *et al* 1999). Therefore it could be hypothesized that these factors during the inflammatory process may lead to a downregulation of SLPI-mRNA. It has also has been shown, that bacterial LPS and LPS-induced antiinflammatory cytokines such as IL-10 and IL-6 in a later stage after onset of bacterial infection induce SLPI expression in macrophages and neutrophils (Jin *et al* 1997). These findings are confirmed by our study showing SLPI positive inflammatory cells in *H. pylori* infected patients. The lack of degradation products and complex formation of SLPI indicates that SLPI may not be involved in the proteolytic process of the inflammatory reaction and also indicates that *H. pylori* itself may not possess any proteolytic activity as has been already demonstrated (Nilius *et al* 1996).

The other specific inhibitor of neutrophil elastase, $\alpha_1$-antitrypsin, is not reduced in *H. pylori* infection but shows a slight increase. This indicates that there is a mechanism which results in signalling for renewed and/or increased production of this inhibitor. Probably this is the same mechanism working by acute phase protein recruitment. It also has been shown that $\alpha_1$-antitrypsin transcription is regulated by neutrophil elastase increasing this molecule in peripheral blood monocytes and macrophages up to eightfold (Perlmutter *et al* 1988). The latter also may contribute to the higher levels measured in the gastric tissue of *H. pylori* infected patients. As our results show, the affection of this defensive mucosal factor by *H. pylori* infection is merely indirect. It seems to be dependent more directly on the level of proteolytic activity because *H. pylori* negative subjects also showed a distinct degradation rate of the native $\alpha_1$-antitrypsin molecule. At high elastase tissue levels a significant increase in complex formation and breakdown of the native protein is observed. But these high levels are only reached if *H. pylori* induced chemotactic factors, such as proinflammatory cytokines, provoke a massive neutrophil infiltration of the gastric tissue.

In conclusion, SLPI in partnership with $\alpha_1$-antitrypsin comprises an anti elastase shield in the gastric mucosa limiting the neutrophil induced inflammation. This shield in *H. pylori* infection is inactivated either by inhibition of SLPI-mRNA resulting in a decrease of mature biologically active SLPI protein or by saturation and degradation of $\alpha_1$-antitrypsin if elastase levels are high. The exact mechanism by which *H. pylori* infection causes this weakening of the antiproteolytic defense has to be elucidated in further studies.

## ACKNOWLEDGMENTS

The technical assistance of Ursula Stolz in immunohistology is kindly appreciated.

## REFERENCES

Bayerdörffer, E. *et al.*, 1992, Difference in expression of *Helicobacter pylori* gastritis in antrum and body. *Gastroenterology* **102:** 1575-1582.

Blaser, M.J., *et al.*, 1992, Hypothesis on the pathogenesis and natural history of *Helicobacter pylori*-induced inflammation. *Gastroenterology* **30:** 720-727.

Denison F.C., *et al.*, 1998, The action of prostaglandin $E_2$ on the human cervix: Stimulation of interleukin 8 and inhibition of secretory leukocyte protease inhibitor. *Am. J. Obstet. Gynecol.* **180:** 614-620.

Dixon, M.F. *et al.*, 1996, Classification and grading of gastritis. *Am. J. Surg. Pathol.* 20, 1161-1181.

Döring, G., 1994, The role of neutrophil elastase in chronic inflammation. *Am.. J. Respir. Crit. Care Med.* **150:** S114-S117.

Elzouki, A.N., *et al.*, 1998, *Helicobacter pylori* infection and $\alpha_1$-antitrypsin deficiency. *Gut* (Suppl. 2) **43:** A59.

Franco, L., *et al.*, 1999, Expression of COX-1, COX-2, and inducible nitric oxide synthase protein in human gastric antrum with *Helicobacter pylori* infection. *Prostaglandin Other Lipid Mediat.* **58:** 9-17.

Fu, S., *et al.*, 1999 Increased expression and cellular localization of inducible nitric oxide synthase and cyclooxygenase 2 in *Helicobacter pylori* gastritis. *Gastroenterology* **116:** 1319-1329.

Ihan, A., *et al.*, 1999, Diminished interferon-gamma production in gastric mucosa T lymphocytes after *H. pylori* eradication in duodenal ulcer patients. *Hepatogastroenterol.* **46:** 1740-1745.

Janoff, A., 1985, Elastases and emphysema: current assessment of the protease-antiprotease hypothesis. *Am. Rev. Respir. Dis.* **132:** 417-433.

Jin, F., *et al.*, 1997, Secretory leukocyte protease inhibitor: A macrophage product induced by and antagonistic to bacterial lipopolysaccharide. *Cell* **88:** 417-426.

Jin, F., *et al.*, 1998, Identification of genes involved in innate responsiveness to bacterial products by differential display. *Methods.* **16:** 396-406.

Lindholm, C., *et al.*, 1998, Local cytokine response in *Helicobacter pylori* infected subjects. *Infect. Immun.* **66:** 5964-5971.

Marchand, V., *et al.*, 1997, The elastase-induced expression of secretory leukocyte protease inhibitor is decreased in remodelled airway epithelium. *Eur. J. Pharmacol.* **336:** 187-196.

Nilius, M., *et al.*, 1996, *Helicobacter pylori* and proteolytic activity. *Eur. J. Clin. Invest.* **26:**1103-1106.

Perlmutter, D.H. ,*et al.*, 1988, Elastase regulates the synthesis of its inhibitor, $\alpha_1$-proteinase inhibitor, and exaggerates the defect in homozygous PiZZ $\alpha_1$ PI deficiency. *J. Clin. Invest.* **81:** 1774-1780.

Sallenave, J.M., *et al.*, 1994, Regulation of secretory leukocyte proteinase inhibitor (SLPI) and elastase-specific inhibitor (ESI/ELAFIN) in human airways epithelial cells by cytokines and neutrophilic enzymes. *Am. J. Respir. Cell Mol. Biol.* **11:** 733-741.

Weiss, S.J., 1989, Tissue destruction by neutrophils, *New Engl. J. Med.* **320:** 365-389.

# THE ROLE OF BACTERIAL AND HOST PROTEINASES IN PERIODONTAL DISEASE

James Travis, Agnieszka Banbula and Jan Potempa
*Dept. of Biochemistry & Molecular Biology*
*The University of Georgia*
*Athens, Georgia 30602*

## 1. INTRODUCTION

The development of periodontal disease is initiated by the growth of specific anaerobic bacteria, referred to as periodontopathogens, under the gum line. This occurs as a further consequence of initial plaque build-which results in the inflammatory condition referred to as gingivitis. While this latter disease can be reversed with proper dental treatment, periodontitis is essentially irreversible and results in massive tissue destruction and loss of tooth attachment. It is significant that the abnormal biochemical processes that are a hallmark of periodontitis are a result of both pathogen and host-derived proteinases (Potempa *et al* 1995a, 1995b). In this review we will first try to explain to the reader just how all of this can occur. This will then allow us to consider the possibility whether the abnormal biochemical reactions associated with the development of periodontitis might also be utilized by other infectious bacteria whose growth and proliferation is also associated with tissue destruction.

## 2. SOURCES AND GENERAL PROPERTIES OF BACTERIAL PROTEINASES

Examination of the various bacterial species residing within the plaque of individuals with adult chronic periodontitis has led to the suggestion that the opportunistic, anaerobic pathogen Porphyromonas gingivalis (P. gingivalis) may be primarily responsible for the progression of this disease (Socransky *et al* 1992). This organism produces enormous quantities of a variety of proteolytic activities, representing three of the four known classes of proteinases, many of which are described in subsequent sections. Various forms of many of these enzymes are either secreted, bound within the cell envelope, or shed in vesicles. A very important and significant property of all of these enzymes is that they are not inhibited by host proteinase inhibitors. Instead, virtually all of the members of the Serpin (SERine Proteinase INhibitor) class of inhibitors which are present in plasma are either inactivated and/or degraded by several of the P. gingivalis-derived proteinases (Meyer *et al* 1997, Lee *et al* 1997). This, of course, results in the dysregulation of several important cascade systems utilized by the human host to combat infection, including coagulation, fibrinolysis, complement activation, and kinin production. It is the lack of control of such systems that allows both P. gingivalis growth and proliferation, as well as host tissue destruction.

## 3. THE GINGIPAINS

Two of the major proteinase groups produced by P. gingivalis are the two classes of cysteine-proteinases referred to as the Arg-gingipains (Rgps) and Lys-gingipain (Kgp) (Potempa *et al* 1998). In a general sense, each group has an exquisite specificity for the cleavage of peptide bonds between either Arg-X (Rgps) or Lys-X (Kgp) peptide bonds, with X being any amino acid. Neither enzyme, however, degrades native proteins very well, instead acting primarily on those peptide bonds that can be easily accessed in apparently random coil structures but not in α-helices or β-sheets (Eichinger *et al* 1999).

The Rgps apparently exist in multiple forms, being the products of two different genes (Potempa *et al* 1998) (Fig 1). In one case the primary gene product (RgpB) contains only a single catalytic domain . In contrast, the major product (HRgpA) of the second gene is cleaved at multiple sites, thus yielding a heteromultimer of significantly higher Mr than RgpB (50 kdA vs 110 kDa). This enzyme also contains an additional domain which has the property of being able to bind a number of components, including fibrinogen, fibrinonectin, hemoglobin, and phospholipid (Rangarajan *et al* 1997). It is generally referred to as an adhesion domain. Both enzyme-types can be secreted; however, much of the HrgpA is found predominantly in the cell envelope fraction. Significantly, and despite the fact that both enzymes have virtually identical catalytic domains, HrgpA appears to be more efficient in peptide bond hydrolysis, and this is likely to be reflective of the importance of the binding function(s) of the adhesion domain.

Kgp, in contrast to the Rgps, is encoded by a single gene (Pavloff *et al* 1997). The mature form of the enzyme also exists as a heteromultimer

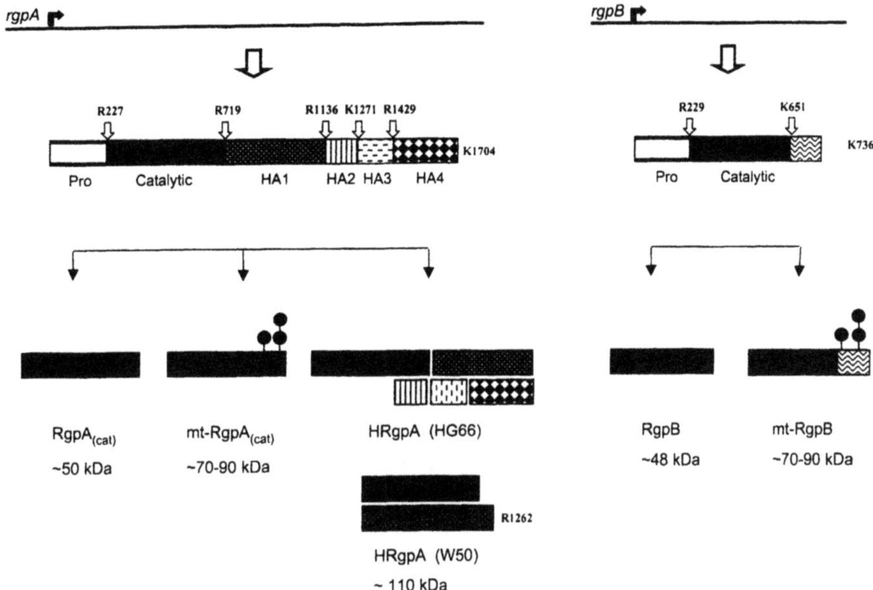


*Figure 1.* Schematic representation of the post-translational processing of the rgpA and rgpB gene products. The cleavages sites which generates mature gingipains are marked with arrows above progingipains. Full circles attached to protease catalytic domain indicate postranslational modifications.
Pro: propeptide; cat: catalytic domain; HA: haemagglutinating/adhesin domain; mt: membrane type

containing two domains that have both catalytic and adhesion properties. Significantly, there is a high degree of homology within the adhesion domain to that structure present in HrgpA; however, there is only limited identity to the Rgps, in general, within the catalytic domain, which is in agreement with the specificity of Kgp for cleavage between lys-X peptide bonds.

## 4. OTHER PROTEOLYTIC ACTIVITIES

P. gingivalis is a remarkable organism in that it is replete with a number of other proteolytic enzymes, each with an unusual specificity. One of the best characterized examples is a cysteine proteinase, referred to as periodontain (Nelson *et al* 1999), which occurs as a non-covalent heterodimer of Mr near 75 kDa with the heavy chain (55 kDa) containing the catalytic site. This enzyme has two significant properties. First, it can only degrade denatured proteins (e.g. gelatin) and polypeptides, being inactive on native proteins such as collagen and fibrinogen. Second, and perhaps most important peridontain rapidly inactivates plasma serpins, including α-1-proteinase inhibitor, apparently by attacking the random coil structure present within the reactive site loop of these proteinase inhibitors. This is perhaps its most important property since it renders these regulating proteins useless in controlling host-derived proteolytic events.

Another well characterized enzyme isolated from P. gingivalis is the exopeptidase referred to as prolyl tripeptidyl peptidase A (PtpA) (Banbula *et al* 1999). This enzyme, a member of the serine-class of peptidases, is cell-surface associated and liberates tripeptides from the amino terminus of polypeptide substrates in which the third residue from the amino-terminus is proline. It is important for two reasons. First, it can process polypeptides into smaller fragments on the cell surface of P. gingivalis , the products of which can then be taken up and utilized as both carbon and energy sources. Second, it is a logical enzyme to be utilized in the final processing of collagen/gelatin degradation products, these large connective tissue-derived substrates being present in the gingiva in exceedingly high concentrations.

In addition to these enzymes, a number of other proteolytic activities have also been reported to be either synthesized and/or secreted by P. gingivalis. In addition, genes encoding putative proteolytic activities have been found in analysis of the P. gingivalis genome. These include several di- and tri-peptidyl peptidases, a collagenase, a gelatinase, and a poorly characterized metalloproteinase active on multiple protein substrates. Thus, it is clear that this organism has more than enough proteolytic activity (Table 1) for both the random and specific cleavage of protein and peptide substrates with which it comes into contact within the human host.

## 5. P. GINGIVALIS PROTEINASES AND HOST DEFENSE

It is well known that any pathogen invading the human body will become a target for a response by the host immune system. This involves both a direct response by professional phagocytes which are mobilized by activation of the complement pathway, together with an inducible response due to antibody binding in concert with cytotoxic T cell and macrophage activation. The result is the development of an inflammatory reaction and may include tissue swelling due to edema formation, redness, and elevated temperature, all of which normally culminate in the elimination of the intruding organism.

However, pathogens have developed mechanisms for evading host defense mechanisms, and P. gingivalis is certainly one of these organisms, utilizing its array of proteolytic activities as a major part of its strategy for survival. In addition, this bacterial pathogen has gone one step farther so that it actually utilizes the attenuated host response to provide itself with the nutrients required for growth. This scenario is described in the following subset of steps, all of which involve synergistic roles played by P. gingivalis and host-derived proteinases.

*Table 1.* Porphyromonas gingivalis proteases

| Protease | Catalytic class | Accession No | References |
|---|---|---|---|
| RgpA | Cysteine | X82680 | Aduse-Opoku *et al.* 1995 |
| | | U15282 | Pavloff *et al.* 1997 |
| | | D26470 | Okamoto *et al.* 1995 |
| | | L27483 | Fletcher *et al.* 1997 |
| | | AF026946 | unpublished |
| RgpB | Cysteine | AF007124 | Slakeski *et al.* 1998 |
| | | U85038 | Nakayama *et al.* 1995 |
| | | D64081 | Nakayama *et al.* 1997 |
| Kgp | Cysteine | U75366 | Slakeski *et al.* 199 |
| | | U54691 | Pavloff *et al.* 1997 |
| | | U42210 | Barkocy-Gallagher *et al.* 1996 |
| | | AF017059 | Levis and Macrina 1998 |
| Periodontain | Cysteine | | Nelson *et al.* 1992 |
| PrtT | Cysteine | S75942 | Madden *et al.* 1995 |
| Tpr | Cysteine | M84471 | Bourgean *et al.* 1992 |
| | | AF020499 | Lu and McBride 1998 |
| Collagenase (PrtC) | Metallo and reducing agents dependent | M60404 | Kato *et al.* 1992 |
| | | AB006973 | Takahashi *et al.* 1999 |
| PtpA | Serine | | Banbula *et al.* 1999 |
| DPPIV | Serine | AF026511 | Kiyama *et al.* 1998 |
| PepO | Metallo | AB010400 | Unpublished |
| Collagenase | Metallo | | Birkedal-Hansen *et al.* 1998 |

## 5.1 Dysregulation of the Complement Pathway

Complement activation is one of the most important and effective defense strategies utilized by the human host to offset pathogen invasion. However, P. gingivalis has developed intriguing mechanisms for evading complement attack, primarily through the use of its vast array and high concentration of specific proteinases. In particular, the gingipains are very effective in utilizing the complement pathway for their own benefit. First, it has been found that the Rgps can readily attack and degrade the complement factor, C3 (Wingrove *et al* 1992). Normal activation of this protein to yield C3a and C3b has two important ramifications for pathogen elimination. First, C3b is required for opsonization of bacteria, so that its loss would effectively eliminate this process. Second, it is needed to form the C5 convertase complex, essential for the generation of the potent chemotactic factor, C5a. Elimination of both of these effects is obviously important for the survival of bacteria, in general. What is intriguing, however, is the fact that the Rgps ignore the second function of C3b and directly activate C5 to C5a and at a rate significantly faster than the host C5 convertase complex.

This process may, at first, seem lethal since it allows neutrophils to be attracted to P. gingivalis. However, the process is attenuated by the fact that vesicles shed by this bacterium have a massive arsenal of proteinases, most of which have been described previously, which are able to degrade the C5a receptor on neutrophil surfaces (Jagels *et al* 1996a, 1996b). While it is not clear which enzyme on vesicles is responsible for receptor degradation, it is obvious that the final result would be loss of a "homing" signal for the neutrophils and an inability to phagocytize the bacterium. Ultimately, these cells would die and/or degranulate near the site of infection, releasing high concentrations of hydrolytic enzymes. Since it is well known that the controlling inhibitors of neutrophil proteinases, α-1-proteinase inhibitor for elastase and proteinase III, and α-1-antichymotrypsin for cathepsin G, are inactivated by periodontain and other P. gingivalis-derived enzymes, the ultimate result is the complete loss of control of host neutrophil proteolytic activities. Since such enzymes are very effective in degrading collagen, elastin, and proteoglycan, it is almost certain that they play major roles in the destruction of periodontal connective tissue (Travis *et al* 1994).

In the context described above, it should also be noted that other chemotactic factors, such as IL-8, are also utilized to attract neutrophils towards the invading P. gingivalis bacterium. Thus, it has been found that truncation of IL-8 by the Rgps and Kgp provides a markedly more effective chemokine for neutrophil migration (Mikolajczyk-Pawlinska *et al* 1998). Significantly, shed vesicles rapidly degrade IL-8 indicating once again that there is a protective halo proximal to the bacterium which destroys not only chemotactic receptors (e.g., C5a receptor) but also IL-8 and possibly C5a. Clearly, this is a fascinating mechanism for inducing neutrophil chemotaxis and yet halting the process before phagocytosis can occur. In this way the pathogen can take advantage of the supply of general proteolytic activity now provided by the host. In essence, the bacterium lives in a sea of phagocytes and utilizes its enzymes for nutrient processing and, indirectly, for its own growth and proliferation.

## 5.2 Dysregulation of the Kallikrein/Kinin Pathway

Although it is clear that connective tissue degradation within the periodontal pocket is the most important factor in tooth loss, it is also apparent that the primary purpose of this effect from the point of view of P.

gingivalis is that it serves as a source of nutrients. Nevertheless, there is a second, striking characteristic of periodontitis which is likely to result in a far more important source of nutrients, and this is the development of edema together with increased crevicular flow. Again, it is apparent that proteolytic enzymes from P. gingivalis play an integral part in this process through the uncontrolled activation of the kallikrein/kinin system. Bradykinin (BK) has powerful biological activities, and at the site of infection is not only responsible for pain but also for local extravasation leading to edema. Normally, this important mediator is released from high molecular weight kininogen (HMWK) by the action of plasma kallikrein, which is, in turn, generated from prekallikrein by activated Hageman factor (XIIa). However, the Rgps have been found to be very effective in directly activating prekallikrein and thus, indirectly, causing the production of BK (Imamura *et al* 1995). Significantly, when Kgp is added to the Rgps there is a synergistic effect in that the pair of enzymes directly releases BK from HMWK. The ultimate result is an increased flow of gingival fluid (plasma exudate) at periodontitis sites infected with P. gingivalis. Clearly, this provides a continuous supply of nutrients necessary for bacterial growth and virulence. It should be pointed out, however, that this is not the only effect BK may be having during the progression of periodontitis. In fact, it is likely to be involved in alveolar bone loss through activation of prostaglandin synthesis in osteoblasts and periodontal-ligament cells.

## 5.3 Dysregulation of the Coagulation/Fibrinolysis Cascade Pathway

One of the hallmarks of chronic periodontal disease is the bleeding effect, which is found to occur on dental probing. Such an abnormal physiological response strongly suggests a local abnormality in either the coagulation and/or the fibrinolytic pathways within the human host Once again, all available information would suggest an abnormal role for P. gingivalis proteinases in interfering with both of these tightly controlled systems. Indeed, there is compelling data which indicates that the RGPs can rapidly activate both factor X and prothrombin (Imamura *et al* 1997) at rates which are comparable to those which occur physiologically. In addition, KGP and the RGPs, together, can also degrade fibrinogen. What is not clear, however, is how the interference of these two opposing systems (coagulation vs fibrinolysis) is manifested pathophysiologically. In the one case, where bleeding occurs, it would seem apparent that the degradation of fibrinogen is paramount and that clot formation is hindered. However, there is now ample evidence to indicate a strong relationship between severe periodontitis and the onset of myocardial infarctions. In this case it is likely that localized

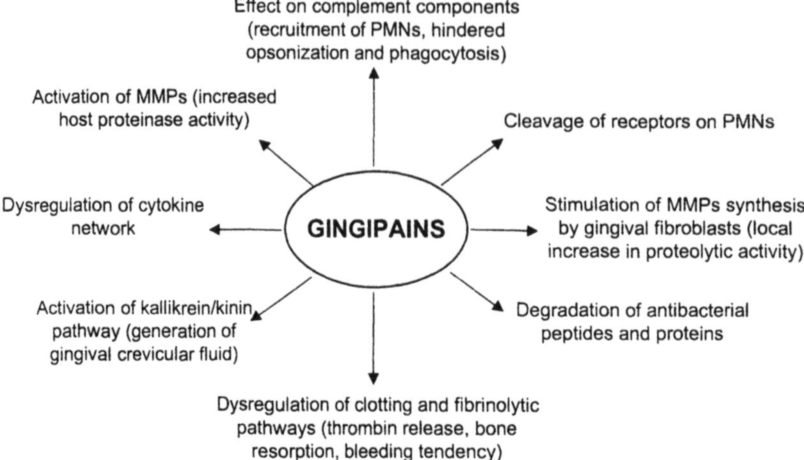


*Figure 2.* Roles of gingipains in the pathogenesis of periodontitis.

microemboli are forming and that the activation of coagulation factors may be important. When one considers the fact that there is a considerably higher concentration of fibrinogen in the blood in comparison to that present at periodontal sites, it seems likely that the effects of the RGPs to produce active thrombin and, therefore, fibrin clot formation far outweigh the ability of all of the P. gingivalis proteinases to degrade fibrinogen.

## 6. SUMMARY

It is abundantly obvious that the uncontrolled degradation and/or activation of host defense pathways is the major pathway by which the periodontal pathogen P. gingivalis promotes its growth and proliferation. By being able to shed host receptors, degrade cytokines, and activate coagulation, complement, and kallikrein/kinin pathways it is clear that this organism has found a mechanism(s) to evade host defense and at the same time develop a system for cannibalizing host proteins for its own nutritional usage (Fig 2). Thus, it seems only logical that the development of inhibitors against these bacterial proteinases would be a useful method for negating their activities and making such pathogens more susceptible to attack by host phagocyte cells. In this respect, the structure of the truncated form of RGP has just been elucidated. Thus, it should only be a question of time before inhibitors to this enzyme will be developed and, hopefully, be used to reduce the pathologies associated with the development of periodontitis and/or eliminate the disease altogether.

## REFERENCES


Aduse-Opoku, J., Muir, J., Slaney, J.M., Rangarajan, M., Curtis, M.A., 1995, Characterization, genetic-analysis, and expression of a protease antigen (PrpRI) of Porphyromonas gingivalis W50. *Infect. Immun.* **63:** 4744-4754.

Banbula, A., Mak, P., Bugno, M., Silberring, J., Dubin, A., Nelson, D., Travis, J., and Potempa, J., 1999, Prolyl tripeptidyl peptidase from Porphyromonas gingivalis. A novel enzyme with possible pathological implications for the development of periodontitis. *J. Biol. Chem.* **274:** 9246-9252.

Barkocy-Gallagher, G.A., Han, N., Patti, J.M., Whitlock, J., Progulske-Fox, A., and Lantz, M.S., 1996, Analysis of the prtP gene encoding porphypain, a cysteine proteinase of Porphyromonas gingivalis. *J. Bacteriol.* **178:** 2734-2741.

Birkedal-Hansen, H., Taylor, R.E., Zambon, J.J., Barwa, P.K., Neiders, M.E., 1988, Characterization of collagenolytic activity from strains of Bacteroides gingivalis. *J. Periodontal. Res.* **23:** 258-264.

Bourgeau, G., Lapointe, H., Peloquin, P., and Mayrand, D., 1992, Cloning, expression, and sequencing of a protease gene (tpr) from Porphyromonas gingivalis W83 in Escherichia coli. *Infect. Immun.* **60:** 3186-3192.

Eichinger, A., Beisel, H.G., Jacob, U., Huber, R., Medrano, F.-J., Banbula, A., Potempa, J., and Travis, J., 1999, Bode W Crystal Structure of Gingipain R: an Arg-specific bacterial cysteine proteinase with a caspase-like fold. *EMBO J.* **18:** 5453-5462.

Fletcher, H.M., Schenkein, H.A., and Macrina, F.L., 1994, Cloning and characterization of a new protease gene (prtH) from Porphyromonas gingivalis. *Infect. Immun.* **62:** 4279-4286.

Imamura, T., Potempa, J., Tanase, S., and Travis, J., 1997, Activation of blood coagulation factor X by arginine-specific cysteine proteinases (gingipain-Rs) from Porphyromonas gingivalis. *J. Biol. Chem.* **272:** 16062-16067.

Imamura, T., Potempa, J., Pike, R.N., and Travis, J., 1995, Dependence of vascular permeability enhancement on cysteine proteinases in vesicles of Porphyromonas gingivalis. *Infect. Immun.* **63:** 1999-2003.

Jagels, M.A., Travis, J., Potempa, J., Pike, R., and Hugli, T.E., 1996a, Proteolytic inactivation of the leukocyte C5a receptor by proteinases derived from Porphyromonas gingivalis. *Infect. Immun.* **64:** 1984-1991.

Jagels, M.A., Ember, J.A., Travis, J., Potempa, J., Pike, R., and Hugli, T.E., 1996b, Cleavage of the human C5A receptor by proteinases derived from Porphyromonas gingivalis: cleavage of leukocyte C5a receptor. *Adv. Exp. Med. Biol.* **389:** 155-164.

Kato, T., Takahashi, N., and Kuramitsu, H.K., 1992, Sequence analysis and characterization of the Porphyromonas gingivalis prtC gene, which expresses a novel collagenase activity. *J. Bacteriol.* **174:** 3889-3895.

Kiyama, M., Hayakawa, M., Shiroza, T., Nakamura, S., Takeuchi, A., Masamoto, Y., and Abiko, Y., 1998, Sequence analysis of the Porphyromonas gingivalis dipeptidyl peptidase IV gene. *Biochim. Biophys. Acta* **1396:** 39-46.

Lee, H.M., Golub, L.M., Chan, D., Leung, M., Schroeder, K., Wolff, M., Simon, S., and Crout R., 1997, Alpha-1-Proteinase inhibitor in gingival crevicular fluid of humans with adult periodontitis: serpinolytic inhibition by doxycycline. *J. Periodontal Res.* **32:** 9-19.

Levis, J.P., and Macrina, F.L., 1998, IS195, an insertion sequence-like element associated with protease genes in Porphyromonas gingivalis. *Infect. Immun.* **66:** 3035-3042.

Lu, B., and McBride, B.C., 1998, Expression of the tpr protease gene of Porphyromonas gingivalis is regulated by peptide nutrients.*Infect. Immunol.* **66:** 5147-5156.

Madden, T.E., Clark, V.L., and Kuramitsu, H.K., 1995, Revised sequence of the

Porphyromonas gingivalis prtT cysteine protease/hemagglutinin gene: homology with streptococcal pyrogenic exotoxin B/streptococcal proteinase. *Infect. Immun.* **63:** 238-247.

Meyer, J., Guessous, F., Huynh, C., Godeau, G., Hornebeck, W., Giroud, J.P., and Roch-Arveiller, M., 1997, Active and alpha-1 proteinase inhibitor complexed leukocyte elastase levels in crevicular fluid from patients with periodontal diseases. *J. Periodontol.* **68:** 256-261.

Mikolajczyk-Pawlinska, J., Travis, J., and Potempa, J., 1998, Modulation of interleukin-8 activity by gingipains from Porphyromonas gingivalis: implications for pathogenicity of periodontal disease. *FEBS Lett.* **4:** 282-286.

Nakayama, K., Kadowaki, T., Okamoto, K., and Yamamoto, Y., 1995, Construction and characterization of arginine-specific cysteine proteinase (Arg-gingipain)-deficient mutants of Porphyromonas gingivalis - Evidence for significant contribution of arg-gingipain to virulence. *J. Biol. Chem.* **270:** 23619-23626.

Nakayama, K., 1997, Domain-specific rearrangement between the two Arg-gingipain-encoding genes in Porphyromonas gingivalis: possible involvement on nonreciprocal recombination. *Microbiol. Immunol.* **41:** 185-196.

Nelson, D., Potempa, J., Kordula, T., and Travis, J., 1999, Purification and characterization of a novel cysteine proteinase (periodontain) from Porphyromonas gingivalis. Evidence for a role in the inactivation of human $\alpha_1$-proteinase inhibitor. *J. Biol. Chem.* **274:** 12245-12251.

Okamoto, K., Misumi, Y., Kadowaki, T., Yoneda, M., Yamamoto, K., and Ikehara, Y., 1995, Structural characterization of argingipain, a novel arginine-specific cysteine proteinase as a major periodontal pathogen from Porphyromonas gingivalis. *Arch. Biochem. Biophys.* **316:** 917-925.

Pavloff, N., Pemberton, P.A., Potempa, J., Chen, W.C.A., Pike, N.R., Prochazka, V., Kiefer, M.C., Travis, J., and Barr, P., 1997, Molecular cloning and characterization of Porphyromonas gingivalis Lys-gingipain. A new member of an emerging family of pathogenic bacterial cysteine proteinases. *J. Biol. Chem.* **272:** 1595-1600.

Potempa, J., Pike, R., and Travis, J., 1995a, Host and Porphyromonas gingivalis proteinases in periodontitis: a biochemical model of infection and tissue destruction. *Prospect Drug Discovery Design* **2:** 445-458.

Potempa, J., Pavloff, N., and Travis, J., 1995b, Porphyromonas gingivalis: a proteinase/gene accounting audit. *Trends. Microbiol.* **3:** 430-434.

Potempa, J., and Travis, J., 1998, Gingipain R and gingipain K. In: *Handbook on Proteinases* (A.J. Barrett, N.D. Rawlings, F.Woessner, eds.) Academic Press, Amsterdam, pp. 762-768.

Rangarajan, M., Aduse-Opoku, J., Slanet, J.M., Young, K.A., and Curtis, M.A.., 1997, The prpR1 and the prR2 arginine-specific protease genes of Porphyromonas gingivalis W50 produce five biochemically distinct enzymes. *Mol. Microbiol.* **23:** 855-965

Slakeski, N., Bhogal, P.S., O'Brien-Simpson, N.M., and Reynolds, E.C., 1998, Characterization of a second cell-associated Arg-specific cysteine proteinase of Porphyromonas gingivalis and identification of an adhesion-binding motif involved in association of the prtR and prtK proteinases and adhesins into large complexes. **144:** 1583-1592.

Slakeski, N., Cleal, S.M., Bhogal, P.S., and Reynolds, E.C., 1999, Characterization of a Porphyromonas gingivalis gene prtK that encodes a lysine-specific cysteine proteinase and three sequence-related adhesins. *Oral Microbiol. Immunol.* **14:** 92-97.

Socransky, S.S., and Haffajee, A.D., 1992, The bacterial etiology of destructive periodontal disease: current concepts. *J. Periodontol.* **63:** 322-331.

Takahashi, N., and Kato, T., 1991, Kuramitsu HK. Isolation and preliminary characterization of the Porphyromonas gingivalis prtC gene expressing collagenase activity. *FEMS Microbiol. Lett.* **68:** 135-138.

Travis, J., Pike, R., Imamura, T., Potempa, J., 1994, The role of proteolytic enzymes in the development of pulmonary emphysema and periodontal disease. *Am. J. Respir. Care Med.* **150:** 150143-150146.

Wingrove, J.A., DiScipio, R.G., Chen, Z., Potempa, J., Travis, J., and Hugli, T.E., 1992, Activation of complement components C3 and C5 by a cysteine proteinase (gingipain-1) from Porphyromonas (Bacteroides) gingivalis. *J. Biol. Chem.* **267:** 18902-18907.

# MULTIFUNCTIONAL ROLE OF PROTEASES IN RHEUMATIC DISEASES

Jörn Kekow, Thomas Pap* and Susanne Zielinski
*Clinic of Rheumatology, Otto-von-Guericke-University of Magdeburg, D-39245 Vogelsang, Germany, *Center for Experimental Rheumatology, Department of Rheumatology, University Hospital, CH-8091 Zürich, Switzerland*

## 1. INTRODUCTION

Proteases have long been recognized only as matrix degrading effector proteins. Their complex nature is exemplified by the increasing number of proteases found in synovial fluid (SF) and tissues. Besides high levels of proteases, a broad spectrum of immunomodulating cytokines is detectable in inflammation. Many of these cytokines are secreted in a latent, non-bioactive form. There is now strong evidence that these cytokines are activated by proteases, shedding new light on the mechanisms of cytokine regulation itself and the biological role of proteases. Moreover, it can be anticipated that proteases are able to abrogate cytokine effects by proteolysis of the bioactive cytokine. Recent data indicate that proteases are also involved in cytokine expression at the cellular level. Knowledge of these proteolytic processes is an important key to understanding the pathogenesis of rheumatic diseases and the development of new therapeutic strategies. In this article we review important findings on the multifunctional role of proteases in rheumatic diseases.

## 2. PROTEASES IN RHEUMATIC DISEASES

Progressive joint destruction distinguishes rheumatoid arthritis (RA) from other inflammatory joint diseases. It is mediated by various proteinases, the most prominent being matrix metalloproteinases (MMPs) and cathepsins

(Müller-Ladner et al 1996). The MMP family consists of at least 20 structurally related members (Nagase and Okada 1997). They include collagenase (MMP-1), gelatinases (MMP-2 and MMP-9), and stromelysin (MMP-3). Collagenase 3 (MMP-13) is a novel member of the MMP family that was cloned from mammary carcinoma tissue and, subsequently, from osteoarthritic and rheumatoid synovial tissue (Mitchell et al 1996). Recently discovered membrane-type MMPs (MT-MMP) also belong to the MMP family (d'Ortho et al 1997). They are characterized by a transmembrane domain and act on the surface of cells. MMPs are secreted as inactive proenzymes and are activated proteolytically by various enzymes such as trypsin, plasmin, and other proteases. The MMPs differ with respect to their substrate specificities. Whereas MMP-1 degrades collagen types I, II, III, VII, and X only when they are arranged in a triple helical structure, MMP-2 can also cleave denatured collagen. MMP-3 is able to activate MMP-1 as well as to degrade proteoglycans. Several reports have implicated MMPs in rheumatoid joint destruction. MMP-1 and MMP-3 have been found to be elevated in SF of patients with RA as compared to osteoarthritis (OA) and are released in large amounts by synovial fibroblast-like cells in culture (Firestein and Paine 1992). MMP overexpression, however, is not a specific feature of RA. As shown in Fig 1, high levels of MMP-3 are detectable also in other inflammatory joint diseases (e.g. psoriatic arthritis, reactive arthritis, gout).

Using *in situ* hybridization techniques, MMPs have been localized in the RA synovial membrane. Synovial-fibroblast-like cells within the lining layer or at the site of cartilage invasion have been identified as the major source of MMPs. MMP-13 is also found expressed at the mRNA and protein level (Lindy et al 1997), especially in the lining layer of rheumatoid synovium. Due to this localization, its substrate specificity for collagen type II, and its relative resistance to known MMP inhibitors, MMP-13 might play an important role in joint destruction. MT-MMPs are also abundantly expressed in cells aggressively destroying cartilage and bone in RA (Pap et al 1998). This is of particular importance because MT1-MMP degrades extracellular matrix components and can activate other disease relevant MMPs such as MMP-2 and MMP-13 (Kauper et al 1996).

Normally MMP activity is balanced by the naturally occurring tissue inhibitors of metalloproteinase (TIMP-1 and TIMP-2), which interact irreversibly with MMPs such as MMP-1 and MMP-3 and are synthesized and secreted by chondrocytes, synovial fibroblasts, and endothelial cells (Nagase and Okada 1997, Firestein and Paine 1992). *In situ* hybridization studies have demonstrated striking amounts of TIMP-1 mRNA in the synovial lining of patients with RA (Firestein and Paine 1992). However, the molar ratio of MMPs to TIMP rather than the absolute levels of TIMP are crucial for joint

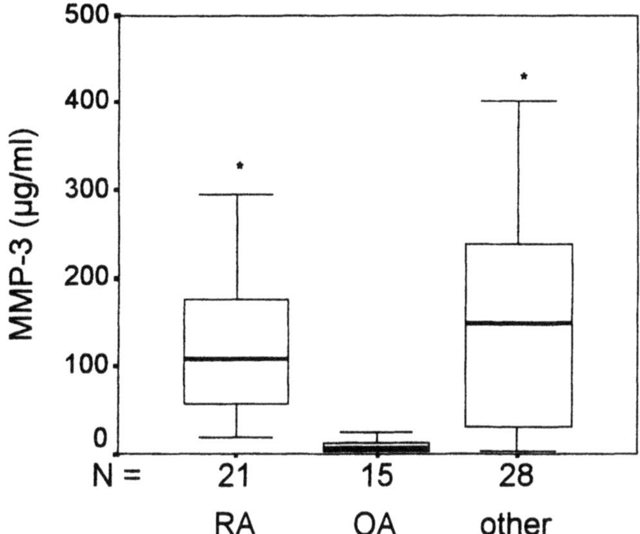


*Figure 1.* MMP-3 concentrations in SF (data are presented as median, quartiles and extreme values). The MMP-3 levels in the RA (and the non-RA or non-OA) patients group are significantly elevated as compared with those of the OA group ($p < 0.01$). Non-RA or non-OA patients (*other*) suffered from spondylarthropathy ($n = 11$), psoriatic arthritis ($n = 5$), reactive arthritis ($n = 5$), gout ($n = 4$), or juvenile rheumatoid arthritis ($n = 3$). ELISA for MMP-3 was performed using the MMP-3 kit obtained from Amersham Pharmacia Biotech.

destruction. In RA the amount of MMPs produced by far outweighs that of TIMPs, allowing destruction to take place (Firestein and Paine 1992).

Cathepsins (B, D, and L) are the other major group of proteases involved in joint destruction (Müller-Ladner et al 1996). They are classified by their catalytic mechanism and cleave cartilage types II, IX, and XI as well as proteoglycans (Müller-Ladner et al 1996). Cathepsin K, a novel cysteine proteinase, is thought to play an important role in osteoclast-mediated bone resorption. Recently RA synovial fibroblasts and macrophages were reported to express cathepsin K, especially at the site of synovial invasion into articular bone, suggesting that it participates in bone destruction in RA (Hummel et al 1999).

Besides their role in matrix degradation proteases serve as autoantigens. Systemic vasculitides share the phenomenon that proteases serve as autoantigens. The patient's autoantibodies recognize e. g. proteinase 3 (PR3), elastase or cathepsins. The mechanism for the induction of autoantibodies is unknown. Interestingly healthy individuals also carry proteases on cell surfaces but do not develop the corresponding autoantibodies. Nevertheless numerous data point to the pathogenic role of these autoantibodies. Cytokines may enhance disease activity by translocating proteases captured in granula to the cell surface (Csernok et al 1996).

## 3. REGULATION OF PROTEASE EXPRESSION

Both proteases and protease inhibitors are regulated in their expression by cytokines. For example, upregulation of tissue inhibitor of metalloproteinase 1 (TIMP-1) can be achieved by transforming growth factor ß (TGFß), interleukin (IL) 6 and IL 11 (Brennan et al 1997). New therapeutic strategies employing tumor necrosis factor α (TNFα)-antibodies can profoundly reduce MMP-1 and MMP-3 levels, as reported recently (Brennan et al 1997). Local reduction of MMP-3 levels may also contribute to the anti-inflammatory effect of intraarticular administered glucocorticoids (Taylor et al 1994). Quite recently TGFß1 and TGFß2 were identified as important triggers of MMP-1 and MMP-13 production by human osteoarthritic chondrocytes and fibroblasts (Tardif et al 1999).

Proto-oncogenes are deeply involved in the activation of joint destroying cathepsins and MMPs (Müller-Ladner et al 1996, Ishidoh et al 1997). The *c-fos* proto-oncogene, which is known to be co-expressed with egr-1, has also been found in RA synovium (Asahara et al 1997, Dooley et al 1996, Brahn et al 1998, Combe 1998). Interestingly, it encodes for a basic leucine zipper transcription factor and is part of the transcriptional activator AP-1 (jun/fos). The promoters of several MMPs (e. g. MMP-9) contain consensus binding sites for the transcription factor AP-1, and the AP-1 site has been shown to be involved in tissue-specific expression of MMPs (Benbow and Brinckerhoff 1997). The cysteine proteinase cathepsin L, which has been shown to be the major ras-induced protein in *ras*-transformed murine NIH 3T3 cells, was detected in 50 % of the RA cases, predominantly in synovial cells (Trabandt et al 1990). Interestingly, cathepsin L in these cases was co-localized with *ras* and *myc*. Most recent data indicate that some of these proto-oncogenes are directly involved in the up-regulation of different MMPs. Gelatinases (MMP-2 and MMP-9) together with MT1-MMP are probably regulated by growth factors that mediate their effects through the *ras* proto-oncogene, and c-Ras plays a critical role in the increased expression and proteolytic activation of MMPs in fibroblasts (Korzus et al 1997, Gum et al 1997).

The cysteine proteases cathepsin B and L are up-regulated in RA synovium, especially at sites of cartilage invasion (Nagase and Okada 1997, Lemaire et al 1997). Cathepsins are activated by proto-oncogenes in a fashion similar to MMPs. In addition, several studies have shown that proinflammatory cytokines such as IL 1 and TNFα can stimulate the production of cathepsins B and L by synovial fibroblast-like cells (Lemaire et al 1997, Huet et al 1993).

# 4. PROTEOLYSIS OF CYTOKINES

Many cytokines are present in a premature form and serve as protease substrates. TGFß can be used as a typical example of cytokine activation by proteolytic processes. There is strong evidence that a variety of proteases are capable of regulating TGFß-mediated biological effects by activating latent TGFß at sites of inflammation. Applying different detection assays, we found a high percentage of bioactive TGFß2 in SF of RA patients (Fig 2) (Lotz et al 1990, Szymkowiak et al 1995).

In vitro studies could identify plasmin, leukocyte elastase, cathepsin G and proteinase 3 as potential activators of latent TGFß (Csernok et al 1996, Kekow et al 1997). Polymorphonuclear cells (PMN) may be one important source of these proteases. It is likely that this predominant cell subset in SF of RA patients contributes to the high levels of TGFß2 in SF. PMN thus constitute a good example of the comprehensive network of cytokine production and activation (Csernok et al 1996). Our inhibition experiments

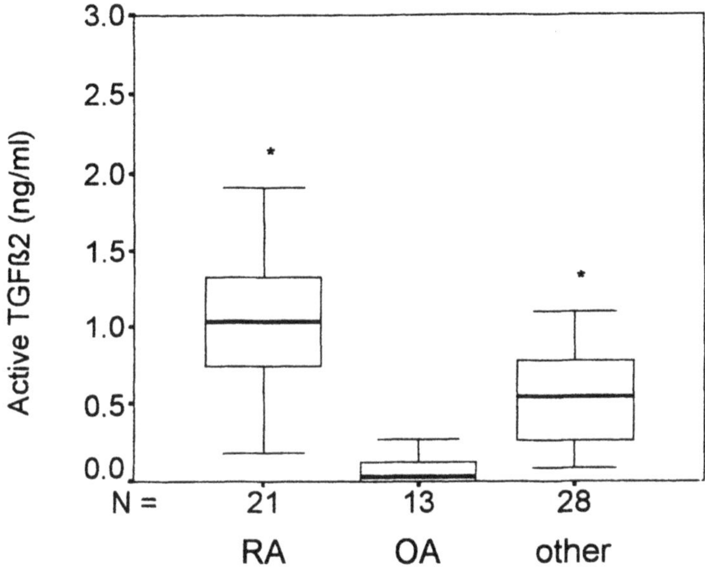


*Figure 2.* Active TGFß2 levels in SF (Data are presented as median, quartiles and extreme values). The TGFß2 levels in the RA (and the non-RA or non-OA) patients group are significantly elevated as compared with those of the OA group ($p < 0.01$). For patients refer to Figure 1.

with aprotinin and phenanthroline and the detection of high levels of MMP-3 in SF from arthritic joints indicate that serine proteases and metalloproteinases are important candidates for the activation of latent TGFβ in SF. TGFß activation due to low pH is unlikely because the pH remains rather constant at not less than 6.8 (Farr et al 1985). In addition matrix-metalloproteinases like MMP-3 can degrade TGFß-binding proteins and thereby increase the levels of active TGFß (Imai et al 1997). Similar proteases can change the TGFß-binding affinity of immunoglobulins (Kaminishi et al 1995, Kekow et al 1998, Stach and Rowley 1993). Non-bioactive precursor cytokine forms requiring proteolytic processing are also described for IL 1, IL 2, IL 8, IL 16, IL 18, and TNFα (Flad et al 1997, Thomson 1998). PR3 was recently described as an alternative mechanism for the cleavage and release of biologically active forms of TNFα and IL 1ß (Coeshott et al 1999). Regarding IL 8, aminopeptidase N was found to degrade this cytokine and to inactivate its chemotactic activity (Kanayama et al 1995). In turn this membrane peptidase is regulated by cytokines such as IL 4 and TGFß (Riemann et al 1998). The particular cytokine effect may also be modulated by the presence of corresponding signaling cellular receptors, which have also been described as a target for proteolytic processes (e. g. IL 2R and IL 6R) (Bank et al 1997).

## 5. CLINICAL APPROACHES

Proteases have become an important target of drug therapy. Established disease-modifying antirheumatic drugs (DMARDs, e. g methotrexate) ameliorate disease activity and joint destruction. They are also protease inhibitors. New therapeutic strategies in RA include neutralizing antibodies to TNFα or TNF-receptor fusion proteins. Their disease controlling effect may include the downregulation of protease expression in the synovium.

Many other drugs used in inflammatory joint diseases, such as gluco-corticoids, NSAIDS, and cyclosporin A, also affect MMP activity. Their effects are mediated via nitric oxide synthase or prostaglandins ($PGE_2$) (39). Tetracyclines have long been examined with regard to their antibiotic effectiveness in controlling arthritides secondary to bacterial infection. The observation of MMP-1 and gelatinase inhibition sheds new light on the use of tetracyclines in rheumatic diseases and extends their indication to include OA (Smith et al 1998).

Rheumatic diseases have also been targeted for new therapeutic strategies, e.g. gene therapy. Among the new approaches for controlling autoimmune processes, stimulation of systemic and even more importantly local overexpression of TGFß is important. Overexpression of TGFß in its latent form is sufficient to effectively treat collagen-induced arthritis in mice and may be superior to the application of bioactive TGFß, which can have several side effects. It is well known that transfected T cells migrate to the joints and that latent TGFß is activated at inflammation sites (Chernajovsky et al 1997).

More direct approaches for controlling matrix degradation are protease inhibiting drugs. Examples include MDL 101,146 and FK 706 as neutrophil elastase inhibitors (Jannsz and Durham 1997) and (Shinguh et al 1997) Ro 32-3555 as an orally active collagenase inhibitor (Lewis et al 1997). Inhibitors of the exoprotease dipeptidylpeptidase IV (DP IV) also can suppress arthritis, as shown in the model of collagen-induced arthritis (Tanaka et al 1997). The underlying mechanisms have not been fully elucidated. They do though include modulation of T- and B-cell activation. Moreover, the inhibition of DP IV suppresses IL 1 production in monocytes and upregulates TGFß1 production, which have been shown to be beneficial in animal models of RA (Reinhold et al 1998).

## 6. CONCLUSION

Further in vitro and in vivo experiments are necessary to fully elucidate the mechanisms of cartilage and bone destruction. The identification and regulation of particular proteases involved in matrix degradation and cytokine activation or even in deactivation will lead to new therapeutic approaches in rheumatology. Experiments carried out in knock out mice and SCID mice have demonstrated the complexity of cartilage and bone degrading enzymes in erosive arthritis (Mudgett et al 1998). It can be anticipated therefore that effective treatment will consist of more than one compound. This is already obvious from the high rate of non-responders to biologicals (e. g. TNFα blocker) as well as the observation that classical DMARDS have a number of effects on the immune system, including the modulation of protease and cytokine expression.

## ACKNOWLEDGMENTS

This work was supported by the Deutsche Forschungsgemeinschaft SFB 387.

## REFERENCES

Amin, A.R., Attur, M., and Abramson, S.B., 1999, Nitric oxide synthase and cyclooxygenases: distribution, regulation, and intervention in arthritis, *Curr.Opin.Rheumatol.* **11**: 202-209.

Asahara, H., Hasunuma, T., Kobata, T., Inoue, H., Muller-Ladner, U., Gay, S., Sumida, T., and Nishioka, K., 1997, In situ expression of protooncogenes and Fas/Fas ligand in rheumatoid arthritis synovium, *J.Rheumatol.* **24:** 430-435.

Bank, U., Reinhold, D., Kunz, D., and Ansorge, S., 1999, Selective proteolytical cleavage of the ligand-binding chains of the IL-2-receptor and Il-6-receptor by neutrophil-derived proteases, *Adv.Exp.Med.Biol.* **421:** 231-242.

Benbow, U., and Brinckerhoff, C.E., 1997, The AP-1 site and MMP gene regulation: what is all the fuss about?, *Matrix Biol.* **15:** 519-526.

Brahn, E., Banquerigo, M.L., Firestein, G.S., Boyle, D.L., Salzman, A.L., and Szabo, C., 1998, Collagen induced arthritis: reversal by mercaptoethylguanidine, a novel antiinflammatory agent with a combined mechanism of action, *J.Rheumatol.* **25:** 1785-1793.

Brennan, F.M., Browne, K.A., Green, P.A., Jaspar, J.M., Maini, R.N. and Feldmann, M. 1997, Reduction of serum matrix metalloproteinase 1 and matrix metalloproteinase 3 in rheumatoid arthritis patients following anti- tumour necrosis factor-alpha (cA2) therapy, *Br.J.Rheumatol.* **36:** 643-650.

Chernajovsky, Y., Adams, G., Triantaphyllopoulos, K., Ledda, M.F., and Podhajcer, O.L., 1997, Pathogenic lymphoid cells engineered to express TGF beta 1 ameliorate disease in a collagen-induced arthritis model, *Gene Ther.* **4:** 553-559.

Coeshott, C., Ohnemus, C., Pilyavskaya, A., Ross. S., Wieczorek, M., Kroona, H., Leimer, A.H., and Cheronis, J., 1999, Converting enzyme-independent release of tumor necrosis factor alpha and IL-1beta from a stimulated human monocytic cell line in the presence of activated neutrophils or purified proteinase 3, *Proc.Natl.Acad.Sci.U.S.A* **96:** 6261-6266.

Combe, B., 1998, Inflammation and joint destruction during rheumatoid polyarthritis: what relation?, *Presse Med.* **27:** 481-483.

Csernok, E., Szymkowiak, C.H., Mistry, N., Daha, M.R., Gross, W.L., and Kekow, J., 1996, Transforming growth factor-beta (TGF-beta) expression and interaction with proteinase 3 (PR3) in anti-neutrophil cytoplasmic antibody (ANCA)-associated vasculitis, *Clin.Exp.Immunol.* **105:** 104-111.

Dooley, S., Herlitzka, I., Hanselmann, R., Ermis, A., Henn, W., Remberger, K., Hopf, T., and Welter, C., 1996, Constitutive expression of c-fos and c-jun, overexpression of ets-2, and reduced expression of metastasis suppressor gene nm23-H1 in rheumatoid arthritis, *Ann.Rheum.Dis.* **55:** 298-304.

d'Ortho, M.P., Will, H., Atkinson, S., Butler, G., Messent, A., Gavrilovic, J., Smith, B., Timpl, R., Zardi, L., and Murphy, G., 1997, Membrane-type matrix metalloproteinases 1 and 2 exhibit broad-spectrum proteolytic capacities comparable to many matrix metalloproteinases. *Eur. J. Biochem.* **250:** 751-757.

Farr, M., Garvey, K., Bold, A.M., Kendall, M.J., and Bacon, P.A., 1985, Significance of the hydrogen ion concentration in synovial fluid in rheumatoid arthritis, *Clin.Exp.Rheumatol.* **3:** 99-104.

Firestein, G.S., and Paine, M. M., 1992, Stromelysin and tissue inhibitor of metalloproteinases gene expression in rheumatoid arthritis synovium, *Am.J.Pathol.* **140:** 1309-1314.

Flad, H.D., Härter, L., Petersen, F., Ehlert, J.E., Ludwig, A., Bock, L., and Brandt, E., 1999, Regulation of neutrophil activation by proteolytic processing of platelet-derived alpha-chemocines, *Adv.Exp.Med.Biol.* **421:** 223-230.

Gum, R., Wang, H., Lengyel, E., Juarez, J., Boyd, D., 1997, Regulation of 92 kDa type IV collagenase expression by the jun aminoterminal kinase- and the extracellular signal-regulated kinase-dependent signaling cascades, *Oncogene* **14:** 1481-1493.

Huet, G., Flipo, R.M., Colin, C., Janin, A., Hemon, B., Collyn-d'Hooghe, M., Lafyatis, R., Duquesnoy, B., and Degand, P., 1993, Stimulation of the secretion of latent cysteine proteinase activity by tumor necrosis factor alpha and interleukin-1, *Arthritis Rheum.* **36:** 772-780.

Hummel, K.M., Franz, J.K., Petrow, P.K., Müller-Ladner, U.,.Aicher, W.K., Gay, R.E., Brömme, D., and Gay, S., 1999, Cathepsin K mRNA expression in normal synovial cells and cells from patients with rheumatoid arthritis (RA) and osteoarthritis (OA)., *J.Rheumatol.*: In Press

Imai, K., Hiramatsu, A., Fukushima, D., Pierschbacher, M.D., and Okada, Y., 1997, Degradation of decorin by matrix metalloproteinases: identification of the cleavage sites, kinetic analyses and transforming growth factor-beta1 release, *Biochem.J.* **322:** 809-814.

Ishidoh, K., Taniguchi, S., and Kominami, E., 1997, Egr family member proteins are involved in the activation of the cathepsin L gene in v-src-transformed cells, *Biochem.Biophys.Res.Commun.* **238:** 665-669.

Janusz , M.J., and Durham, S.L., 1997, Inhibition of cartilage degradation in rat collagen-induced arthritis but not adjuvant arthritis by the neutrophil elastase inhibitor MDL 101,146, *Inflamm.Res.* **46:** 503-508.

Kaminishi, H., Miyaguchi, H., Tamaki, T., Suenaga, N., Hisamatsu, M., Mihashi, I., Matsumoto, H., Maeda, H., and Hagihara. Y., 1995, Degradation of humoral host defense by Candida albicans proteinase, *Infect.Immun.* **63:** 984-988.

Kanayama, N., Kajiwara, Y., Goto, J., el Maradny, E., Maehara K., Andou, K., and Terao, T., 1995, Inactivation of interleukin-8 by aminopeptidase N (CD13), *J.Leukoc.Biol.* **57:** 129-134.

Kekow, J., Csernok, E., Szymkowiak, C., and Gross, W.L., 1997, Interaction of transforming growth factor beta (TGF beta) with proteinase 3, *Adv.Exp.Med.Biol.* **421:** 307-313.

Kekow, J., Reinhold, D., Pap, T., and Ansorge, S., 1998, Intravenous immunoglobulins and transforming growth factor beta, *Lancet* **351**: 184-185.

Knauper, V., Will, H., Lopez-Otin, C., Smith, B., Atkinson, S.J., Stanton, H., Hembry, R.M., and Murphy, G., 1996, Cellular mechanisms for human procollagenase-3 (MMP-13) activation. Evidence that MT1-MMP (MMP-14) and gelatinase a (MMP-2) are able to generate active enzyme, *J.Biol.Chem.* **271:** 17124-17131.

Korzus, E., Nagase, H., Rydell, R.,and Travis, J.,1997, The mitogen-activated protein kinase and JAK-STAT signaling pathways are required for an oncostatin M-responsive element-mediated activation of matrix metalloproteinase 1 gene expression, *J.Biol.Chem.* **272:** 1188-1196.

Lemaire, R., Huet, G., Zerimech, F., Grard, G., Fontaine, C., Duquesnoy, B. and Flipo, R.M., 1997, Selective induction of the secretion of cathepsins B and L by cytokines in synovial fibroblast-like cells, *Br.J.Rheumatol.* **36:** 735-743.

Lewis, E.J., Bishop, J., Bottomley, K.M., Bradshaw, D., Brewster, M., Broadhurst, M.J., Brown, P.A., Budd, J.M., Elliott, L., Greenham, A.K., Johnson, W.H., Nixon, J.S., Rose, F., Sutton, B., and Wilson, K., 1997, Ro 32-3555, an orally active collagenase inhibitor, prevents cartilage breakdown in vitro and in vivo, *Br.J.Pharmacol.* **121:** 540-546.

Lindy, O., Konttinen, Y.T., Sorsa, T., Ding, Y., Santavirta, S., Ceponis, A., and Lopez-Otin, C., 1997, Matrix metalloproteinase 13 (collagenase 3) in human rheumatoid synovium, *Arthritis Rheum.* **40:** 1391-1399.

Lotz, M., Kekow, J., and Carson, D.A., 1990, Transforming growth factor-beta and cellular immune responses in synovial fluids, *J.Immunol.* **144:** 4189-4194.

Mitchell, P.G., Magna, H.A., Reeves, L.M., Lopresti-Morrow, L.L., Yocum, S.A., Rosner, P.J., Geoghegan, K.F., and Hambor, J.E., 1996, Cloning, expression, and type II collagenolytic activity of matrix metalloproteinase-13 from human osteoarthritic cartilage. *J. Clin. Invest.* **97:** 761-768.

Mudgett, J.S., Hutchinson, N.I., Chartrain, N.A., Forsyth, A.J., McDonnell, J., Singer, I.I., Bayne, E.K., Flanagan, J., Kawka, D., Shen, C.F., Stevens, K., Chen, H., Trumbauer, M., and Visco, D.M., 1998, Susceptibility of stromelysin 1-deficient mice to collagen-induced arthritis and cartilage destruction, *Arthritis Rheum.* **41:** 110-121.

Müller-Ladner, U., Gay, R.E., and Gay, S., 1996, Cysteine proteinases in arthritis and inflammation. *Perspectives in drug discovery and design* **6:** 87ff.

Nagase, H., and Okada, Y., 1997, Proteinases and matrix degradation. In *Textbook of rheumatology*, (W.N. Kelly, J. Harris, S. Ruddy and C.B. Sledge, eds.), Saunders, Philadelphia.

Pap, T., Kuchen, S., Hummel, K.M.,.Franz, J.K, Gay, R.E., and Gay, S., 1998, In-situ expression of membrane-type matrix metalloproteinase 1 (MT1-MMP) and MT3-MMP mRNA in rheumatoid arthritis (RA), *Arthritis Rheum.* **41:** 317ff.

Reinhold, D., Bank, U., Bühling, F., Täger, M., Born, I., Faust, J., Neubert, K., and Ansorge, S., 1997, Inhibitors of dipeptidyl peptidase IV (DP IV, CD26) induces secretion of transforming growth factor-beta 1 (TGF-beta 1) in stimulated mouse splenocytes and thymocytes, *Immunol.Lett.* **58:** 29-35.

Riemann, D., Kehlen, A., Kühne, C., Kekow, J., and Langner, J., 1998, IL4 and TGFß regulate the expression of membrane peptidase CD10 on human fibroblast like synoviocytes, *Arthritis Rheum.* **41:** 37ff.

Rowley, D.A., and, Stach, R.M., 1993, A first or dominant immunization. II. Induced immunoglobulin carries transforming growth factor beta and suppresses cytolytic T cell responses to unrelated alloantigens, *J.Exp.Med.* **178:** 835-840.

Shinguh, Y., Imai, K., Yamazaki, A., Inamura, N., Shima, I, Wakabayashi, A., Higashi, Y., and Ono, T., 1997, Biochemical and pharmacological characterization of FK706, a novel elastase inhibitor, *Eur.J.Pharmacol.* **337:** 63-71.

Smith, G.N., Jr., Yu, L.P., Jr., Brandt, K.D., and Capello, W.N., 1998, Oral administration of doxycycline reduces collagenase and gelatinase activities in extracts of human osteoarthritic cartilage, *J.Rheumatol.* **25:** 532-535.

Szymkowiak, C.H., Mons, I., Gross, W.L., and Kekow, J., 1995, Determination of transforming growth factor beta 2 in human blood samples by ELISA, *J.Immunol.Methods* **184:** 263-71.

Tanaka, S., Murakami, T., Horikawa, H., Sugiura, M., Kawashima, K., and Sugita, T., 1997, Suppression of arthritis by the inhibitors of dipeptidyl peptidase IV, *Int.J.Immunopharmacol.* **19:** 15-24.

Tardif, G., Pelletier, J.P., Dupuis, M., Geng, C., Cloutier, J.M., and Martel-Pelletier, J., 1999, Collagenase 3 production by human osteoarthritic chondrocytes in response to growth factors and cytokines is a function of the physiologic state of the cells, *Arthritis Rheum.* **42:** 1147-1158.

Taylor, D.J., Cheung, N.T., and Dawes, P.T., 1994, Increased serum proMMP-3 in inflammatory arthritides: a potential indicator of synovial inflammatory monokine activity, *Ann.Rheum.Dis.* **53:** 768-772.

Thomson, A.W., 1998, "The cytokine handbook", 3 Ed., Academic press, London.

Trabandt, A., Aicher, W.K., Gay, R.E., Sukhatme, V.P., Nilson, H.M., Hamilton, R.T., McGhee, J.R., Fassbender, H.G., and Gay, S., 1990, Expression of the collagenolytic and ras-induced cysteine proteinase cathepsin L and proliferation-associated oncogenes in synovial cells of MRL/I mice and patients with rheumatoid arthritis, *Matrix* **10:** 349-361.

# EVIDENCE OF PROTEOLYTIC ACTIVATION OF TRANSFORMING GROWTH FACTOR ß IN SYNOVIAL FLUID

Susanne Zielinski, Kathleen Bartels, Kristin Cebulski, Cornelia Kühne and Jörn Kekow
*Clinic of Rheumatology, Otto-von-Guericke-University of Magdeburg, D-39245 Vogelsang, Germany*

## 1. INTRODUCTION

Transforming growth factor β (TGF-β) is a pleiotrophic cytokine primarily found in a latent form due to noncovalent association with a latency-associated peptide that cannot interact with its high affinity receptors. This holds true for all three isoforms found in human. Activation of the latent complex is therefore a critical regulatory step for the biological effect of TGF-β. TGF-β can be activated by extreme pH, temperature, ionizing radiation, chaotropic agents, thrombospondin, and proteases (Miyazono *et al* 1991). Proteases are also thought to be responsible for the formation of active TGFβ under physiological conditions. Elastase, plasmin, proteinase 3 and cathepsin G have already been described as potential activators of TGF-β1 (Lyons *et al* 1988, Csernok *et al* 1996). In addition, matrix metalloproteinases have been shown to release active decorin-bound TGFβ by digestion of this proteoglycan (Imai *et al* 1997). In this report we describe the presence of proteolytic activity in synovial fluid (SF) and its association with active forms of TGF-β.

*Cellular Peptidases in Immune Functions and Diseases 2*
Edited by Langner and Ansorge, Kluwer Academic/Plenum Publishers, 2000 

# 2. MATERIALS AND METHODS

## 2.1 Patients

SF was obtained from 53 patients with rheumatoid arthritis (RA) (n = 21), spondyloarthropathy (n = 14), osteoarthritis (n = 6), psoriatic arthritis (n = 5), gout (n = 4), and juvenile rheumatoid arthritis (n = 3). SFs were centrifuged twice to remove cells and stored at –20 °C prior to analysis. Means of 32.1 mg/l (range 5 - 150 mg/l) C-reactive protein (CRP) and 39.6 mm/h (range 1 - 151 mm/h) erythrocyte sedimentation rate (ESR) indicated moderate to high disease activity in the patients investigated.

## 2.2 Cytokine Assays

TGF-β1 and 2 were determined using two isoform-specific ELISAs. Levels of total TGF-β were ascertained after transient acidification as described elsewhere (Szymkowiak *et al* 1995, Reinhold *et al* 1997).

## 2.3 Determination of Proteolytic Activity

Overall proteolytic activity in cell-free SF was measured with the Enzchek™ Protease assay kit (Molecular Probes). Casein was digested in a 10 mM Tris-HCl buffer at pH 7.8 for 24 hours at 37 °C. Inhibition experiments were performed with aprotinin at a final concentration of 1 μg/ml, E-64 at 10 μg/ml and 1,10-phenanthroline at 0.25 mM.

## 2.4 Matrix Metalloproteinase-3 (MMP-3)

ELISA for MMP-3 was performed using the MMP-3 kit obtained from amersham pharmacia biotech according to the manufacturers instructions.

# 3. RESULTS

The SFs were found to contain low concentrations of total TGF-β1 and only traces of spontaneously active TGF-β1, but high levels of total TGF-β2 and active TGF-β2 (table 1). Active TGF-β2 levels correlated with disease activity markers such as CRP and ESR ($r = 0.354/p < 0.05$ and $r = 0.336/p < 0.05$, respectively).

The proteolytic activity of diluted SF ranged from 45 to 220 arbitrary units (AU) and revealed a statistically significant association with ESR and

*Table 1.* TGF-β concentration in synovial fluids

| | total TGF-β1 (ng/ml) | active TGF-β1 (ng/ml) | total TGF-β2 (ng/ml) | active TGF-β2 (ng/ml) |
|---|---|---|---|---|
| Mean (n = 53) | 1.129 | 0.037 | 25.070 | 0.790 |
| SE | 0.130 | 0.010 | 0.700 | 0.070 |
| Min | 0.120 | 0.000 | 16.110 | 0.079 |
| Max | 4.462 | 0.360 | 41.260 | 2.424 |

CRP ($r = 0.336$ and $r = 0.404$, respectively, both at $p < 0.05$). A good correlation was also found between proteolytic activity and concentrations of spontaneously active TGF-β2 (Fig 1).

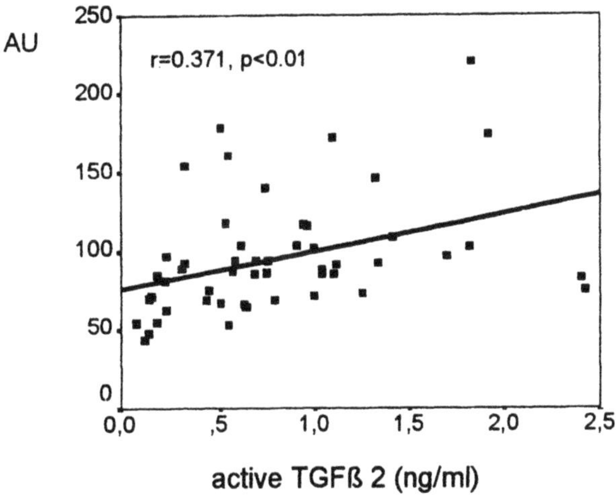


*Figure 1.* Correlation of spontaneously active TGF-β2 and proteolytic activity (AU) in synovial fluid

In 4 selected SF samples exhibiting high proteolytic activity, the addition of inhibitors for serine proteases (aprotinin) and metalloproteinases (1,10-phenanthroline) resulted in a loss of activity. Pretreatment with aprotinin or phenanthroline decreased the proteolytic activity by 20 % and 50 %, respectively. In contrast, the cystein protease inhibitor E64 did not cause a significant reduction in proteolytic activity.

The strong inhibitory effect of phenanthroline prompted us to closely study MMP-3 expression in SF. A mean concentration of 134.2 μg/ml (SE: 14.1 μg/ml) MMP-3 (range 2.3 to 400 μg/ml) was detected in SF. MMP-3 concentrations correlated well with the overall proteolytic activity (Fig 2).

MMP-3 levels also showed a strong association with CRP and ESR ($r = 0.360$, $p < 0.05$ and $r = 0.380$, $p < 0.01$, respectively), and a good correlation with levels of active TGF-β2 ($r = 0.405$, $p < 0.01$).

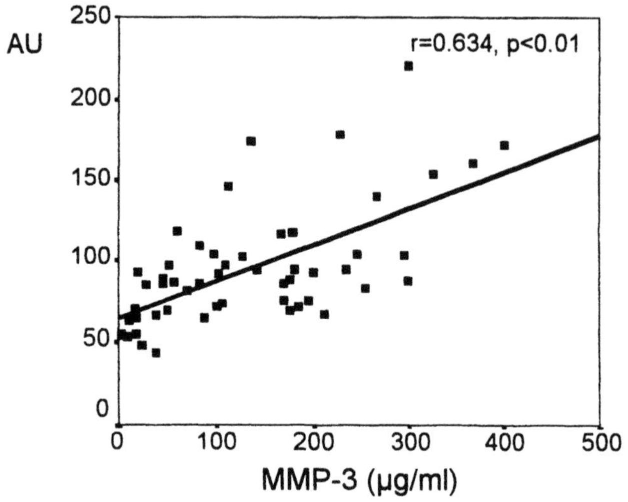


*Figure 2.* Correlation of MMP-3 levels and proteolytic activity (AU) in synovial fluid

# 4. CONCLUSION

TGF-β2 was the predominant isoform of TGF-β found in SF of the arthritic joint. This result agrees with earlier reports (Lotz *et al* 1990). Concentrations of spontaneously active TGF-β2 correlated well with CRP and ESR, suggesting an association between TGF-β activation and disease activity.

SF of RA patients is characterized by an overexpression of proteolytic enzymes produced by synoviocytes, mast cells and neutrophils (Moore *et al* 1993, Bresnihan 1999). Our data indicate that the overall proteolytic activity in SF depends on disease activity. The association of high levels of active TGF-β with increased proteolytic activity suggests that proteolysis could be an important mechanism for the activation of TGF-β in SF.

Earlier reports have shown that serine proteases, collagenases and other proteolytic enzymes are detectable in SF of RA patients (Larbre *et al* 1994, Nakano *et al* 1999). Our inhibition experiments with aprotinin and phenanthroline and the detection of high levels of MMP-3 in SF from arthritic joints indicate that serine proteases and metalloproteinases are potential candidates for the activation of latent TGF-β in SF.

Further experiments are in progress to substantiate the present findings as well as to identify the proteases involved in the activation of TGF-β within the inflamed joint and to characterize the proteolytic sites.

## ACKNOWLEDGMENTS

This work was supported by the Deutsche Forschungsgemeinschaft (SFB 387/B9).

## REFERENCES

Bresnihan, B., 1999, Pathogenesis of joint damage in rheumatoid arthritis. *J. Rheumatol.* **26:** 717-719.

Csernok, E., Szymkowiak, C. H., Mistry, N., Daha, M. R., Gross, W. L. and Kekow, J., 1996, Transforming growth factor-beta (TGF-β) expression and interaction with proteinase 3 (PR3) in anti-neutrophil cytoplasmic antibody (ANCA)-associated vasculitis. *Clin. Exp. Immunol.* **105:** 104-111.

Larbre, J.-P., Moore, A. R., Da Silva, J. A. P., Iwamura, H., Ioannou, Y. and Willoughby, D. A., 1994, Direct degradation of articular cartilage by rheumatoid synovial fluid: contribution of proteolytic enzymes. *J. Rheumatol.* **21:** 1796-1801.

Lotz, M., Kekow, J. and Carson, D. A., 1990, Transforming growth factor-β and cellular immune response in synovial fluid. *J. Immunol.* **144:** 4189-4194.

Lyons, R.M., Keski-Oja, J. and Moses, H.L., 1988, Proteolytic activation of latent transforming growth factor-β from fibroblast-conditioned medium. *J. Cell Biol.* **106:** 1659-1665.

Miyazono, K. and Heldin, C.-H., 1991, Latent forms of TGF-β: molecular structure and mechanisms of activation. *Ciba Found. Symp.* **157:** 81-89.

Moore, A. R., Iwamura, H., Larbre, J. P., Scott, D. L. and Willoughby, D. A., 1993, Cartilage degradation by polymorphnuclear leucocytes: in vitro assessment of the pathogenic mechanisms. *Ann. Rheum. Dis.* **52:** 27-31.

Nakano, S., Ikata, T., Kinoshita, I., Kanematsu, J. and Yasuoka, S., 1999, Characteristics of the protease activity in synovial fluid from patients with rheumatoid arthritis and osteoarthritis. *Clin. Exp. Rheumatol.* **17:** 161-170.

Imai, K., Hiramatsu, A., Fukushima, D., Pierschbacher, M. D. and Okada, Y., 1997, Degradation of decorin by matrix metalloproteinases: identification of the cleavage sites, kinetic analyses and transforming growth factor-β1 release. *Biochem. J.* **322:** 809-814.

Reinhold, D., Bank, U., Bühling, F., Junker, U., Kekow, J., Schleicher, E. and Ansorge, S., 1997, A detailed protocol for the measurement of TGF-β1 in human blood samples. *J. Immunol. Meth.* **209:** 203-206.

Szymkowiak, C., Mons, I., Gross, W.L. and Kekow, J., 1995, Determination of transforming growth factor β 2 in human blood samples by ELISA. *J. Immunol. Meth.* **184:** 263-271.

# MATRIX METALLOPROTEINASES AND TACE PLAY A ROLE IN THE PATHOGENESIS OF ENDOMETRIOSIS

Constanze Gottschalk[1], Kurt Malberg[1], Marco Arndt[1], Johannes Schmitt[2], Albert Roessner[2], Dagmar Schultze[3], Jürgen Kleinstein[3], Siegfried Ansorge[1]
*[1]Institute of Immunology, [2]Institute of Pathology, [3]Department of Reproductive Medicine and Gynecological Endocrinology, Otto-von-Guericke-University, Magdeburg, Germany*

Key words MMP, TACE, endometriosis

Abstract Endometriosis, a benign gynecologic disorder, occurs in about 10% of women in reproductive age and in up to 50 % of women with infertility. The basic etiologic factors causing this disease are unknown as yet. Matrix metalloproteinases (MMP) are involved in degradation of the extracellular matrix (ECM). Their proteolytic activity is regulated by tissue inhibitors of metalloproteinases (TIMPs). Tumor necrosis factor-α converting enzyme (TACE) is a membrane-bound disintegrin metalloproteinase that processes the membrane-associated cytokine proTNF-α to its mature soluble form. TNF-α induces the secretion of several MMPs. In order to study the expression of MMP-1, -2, -3 and -9, TIMP-1 and -2, TACE and TNF-α in endometrium and endometriotic tissue, we investigated formalin-fixed paraffin sections of endometriotic tissues and normal endometrium with immunohistochemical techniques and *in situ* hybridisation. Furthermore, quantitative PCR was used for quantification of TACE-mRNA in fresh tissue. We found in this study significant higher protein expression of MMP-1 and TACE and significant lower protein expression of TIMP-1 and -2 in endometriotic tissue compared to endometrium. This data may suggest that high TACE expression causes the increased conversion of membrane-bound proTNF-α into its soluble form, which stimulates the increased secretion of MMP-1. The simultaneous deficiency of TIMP-1 and -2 in endometriotic tissue suppose an additional proteinase inhibitor imbalance in endometriosis.

## 1. INTRODUCTION

Endometriosis is defined as the presence of endometrial glandular and stromal cells outside their normal location in the uterine cavity. This disorder is affecting many women during their reproductive life. Endometriosis is a cause of acquired dysmenorrhoea, dyspareunia, intermenstrual bleeding, menorrhagia and pelvic pain and occurs in up to 50 % of women with infertility.

The basic etiologic factors causing endometriosis are unknown as yet. While it is not a malignant disorder, endometriosis exhibits cellular proliferation, cellular invasion and neoangiogenesis. Endometriosis can invade tissues and surfaces with as much aggression as malignancy, and yet it is a benign disorder. Mechanisms underlying cellular invasion include an important role for matrix metalloproteinases (MMPs). MMPs are a family of enzymes involved in extracellular matrix (ECM) remodeling (Nagase 1994). These proteases have been implicated in endometrial remodeling, which occurs during the cellular proliferative phase and as participants in tissue desquamation at the time of menstrual bleeding. MMPs are a multigene family of enzymes that share sequence homology, require a heavy metal, zinc, in the active site and can each degrade at least one component of ECM (Bode *et al* 1999).

The proteolytic activity of MMPs can be inhibited by TIMPs (Gomez *et al* 1997). The synthesis of MMPs is regulated by cytokines, e. g. TNF-α, (Mauviel 1993), which is synthesized in endometrium and stimulates the secretion of MMP-1,-3 and -9 in endometrial stromal cells (Rawdanowicz *et al* 1994). It is processed from the membrane-associated proTNF-α to its soluble mature form by tumor necrosis factor-α converting enzyme (TACE) that is a membrane-bound disintegrin metalloproteinase (Black *et al* 1997).

An imbalance between inhibitors, cytokines and enzymes, resulting in excessive degradation of the ECM has been implicated in tumor invasion and metastasis. The aim of the present study was to investigate the expression of MMP-1, -2, -3 and -9, TIMP-1 and -2, TNF-α and TACE in glandular epithelial cells of endometriotic tissue and normal endometrium to clarify the role of MMPs and TACE in the development and progression of endometriosis.

## 2. MATERIALS AND METHODS

In order to study the protein expression of MMP-1, -2, -3 and -9, TIMP-1 and -2, TACE and TNF-α in endometrium and endometriotic tissue, we investigated sections of formalin-fixed and paraffin-embedded endometriotic tissues from 33 infertile women with endometriosis and normal endometrium from 52 women without endometriosis. For detection of the protein expression of these enzymes and TNF-α the immunoenzyme staining by Vector-AP-ABC-method (Vector Laboratories, Burlingame,

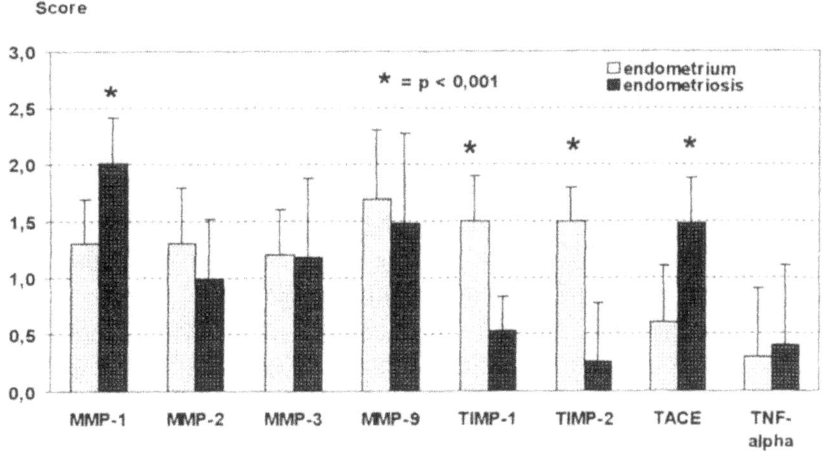


*Figure 1.* Immunohistochemical analysis of protein expression of MMP-1,-2,-3 and -9, TIMP-1 and -2, TACE and TNF-α in glandular epithelial cells of endometriotic tissue compared to eutope endometrium

USA) was used. The expression of the antigen was classified by intensity of immunoreactive staining into negative (0), low (0.5 and 1), intense (1.5 and 2) and high-intense (2.5 and 3). The evaluation of the sections was performed by two independend observers. For calculation of statistical significances the $\chi 2$- test was used, a p-value of 0.001 was set as significant difference. The detection of TACE-mRNA was performed by *in situ* hybridisation with digoxigenin labeled sense- and antisense-probes. For quantification of TACE-mRNA in fresh endometriotic tissue and normal endometrium the quantitative RT-PCR (Lightcycler) was used.

## 3. RESULTS

We found significant higher protein expression of MMP-1 and TACE and significant lower protein expression of TIMP-1 and -2 in glandular epithelial cells of endometriotic tissue compared to endometrium by immunohistochemistry. There were no significant different results in protein expression for MMP-2, -3 and -9 in glandular epithelial cells. Interestingly, there was no significant difference in protein expression of TNF- α in both tissues, too (Figure 1). Using *in situ* hybridisation, we could show for the first time that TACE is expressed in normal endometrium and in endometriotic tissue. Surprisingly, quantitative RT-PCR did not show different results of TACE-mRNA expression in fresh endometriotic tissue compared to normal endometrium (data not shown).

# 4. CONCLUSIONS

In conclusion, TACE and MMP-1 are involved in genesis and progression of endometriosis. We could show that there is a high protein expression of MMP-1 and TACE in glandular epithelial cells of endometriotic tissue in contrast to normal endometrium. Our results suggest that high TACE protein expression causes the increased conversion of membrane-bound proTNF-α in its soluble form. This cytokine stimulates the increased secretion of MMP-1 in endometriotic tissue.

Moreover, the simultaneous deficiency of TIMP-1 and -2 in endometriotic tissue suppose an additional proteinase inhibitor imbalance in this disease.

# REFERENCES

Black, R.A. *et. al.*, 1997, A metalloproteinase disintegrin that releases tumour-necrosis factor-alpha from cells. *Nature* **385**: 729-733.

Bode, W. *et. al.*, 1999, Structural properties of matrix metalloproteinases. *Cel. Mol. Life Sci.* **55**: 639-652.

Gomez, D.E. *et. al.*, 1997, Tissue inhibitors of metalloproteinases: structure, regulation and biological functions. *Eur. J. Cell Biol.* **74**: 111-122.

Mauviel, A., 1993, Cytokine regulation of metalloproteinase gene expression. *J. Cell. Biochem.* **53**: 288-295.

Nagase H., 1994, Matrix metalloproteinases. A mini-review. *Contrib. Nephrol.* **107**: 85-93.

Rawdanowicz, T.J., *et. al.*, 1994, Matrix metalloproteinase production by cultured human endometrial stromal cells: identification of interstitial collagenase, gelatinase-A, gelatinase-B, and stromelysin-1 and their differential regulation by interleukin-1 alpha and tumor necrosis factor-alpha. *J. Clin. Endocrinol. Metab.* **79**: 530-536.

# INFLUENCE OF PROLIFERATION, DIFFERENTIATION AND DEDIFFERENTIATION FACTORS ON THE EXPRESSION OF THE LYSOSOMAL CYSTEINE PROTEINASE CATHEPSIN L (CL) IN THYROID CANCER CELL LINES

Alexander Plehn[1], Dagmar Günther[1], Hendryk Aurich[1], Chuong Hoang-Vu[2], Ekkehard Weber[1]
[1]*Institute of Physiological Chemistry*
[2]*AG Exp. & Chirurgische Onkologie, Klinik für Allgemeinchirurgie, Martin Luther University Halle-Wittenberg*

## 1. INTRODUCTION

In order to develop into a metastasizing tumor transformed cells have to employ proteolytic enzymes. All stages of this development as local infiltration, penetration through neighboring structures such as basement membranes, intra- and extravasation, embedding into an unfamiliar environment and metastasis depend on the degradation of a variety of proteins. Tumor cells use proteinases of several classes and secrete most of them as inactive precursors. The immature proenzymes are processed to the active enzymes by autoactivation, other proteinases or by charged surface structures.

Lysosomal cysteine proteinases are participating in these processes (Chauhan *et al* 1991, Denhardt *et al* 1987, Gottesman 1993, Sivaparvathi *et al* 1996, Jean *et al* 1996). They are able to degrade proteins of the extracellular matrix, plasma membrane receptors and adhesion proteins. In addition, one of them, procathepsin L, the precursor of cathepsin L a potent endoproteinase was reported to have particular functions as growth factor (Kasai *et al* 1993) and in cell differentiation (Homma *et al* 1994).

Overexpression of procathepsin L was reported for many transformed cells and its enhanced secretion is supposed to be correlated to the malignancy of tumor cells and their ability to metastasize. Blocking the activity of procathepsin L by specific antibodies (Weber *et al* 1994, Jean 1996) or low molecular weight inhibitors (Yagel *et al* 1989) and the down regulation of its expression by antisense mRNA were shown to suppress tumor development. Nevertheless, only few details are known about a specific role of extracellular procathepsin L in these processes and about regulatory mechanisms of its expression. It is conceivable that its action may facilitate one or all of the steps necessary to metastasize. It could also help in supplying the tumor cell with amino acids or peptide substrates for survival or it may help in escaping defense mechanisms of the host (Jean, 1996).

We investigated the expression of cathepsin L on the mRNA and on the protein level in monolayer cultures of two thyroid cancer cell lines, the follicular carcinoma cell line FTC 133 and the anaplastic thyroid cancer line 8505C under the influence of five regulators of thyroid proliferation, differentiation and dedifferentiation. Both cell lines were developed from carcinomas of the follicular epithelium of the thyroid gland. Despite their identical origin those carcinomas differed in malignancy. FTC 133 was selected from a tumor with a relatively benign prognosis and the undifferentiated cell line 8505C came from an aggressive tumor with a pronounced tendency to metastasize.

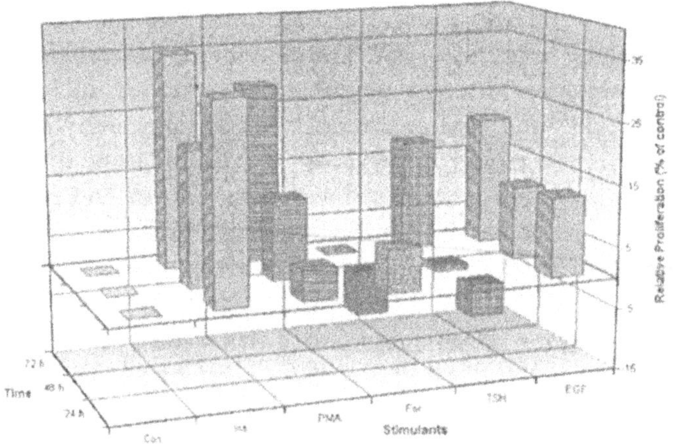


*Figure 1.* Cell line 8505C

Relative proliferation of cell lines 8505C and FTC 133 after 24, 48 and 72 hours of stimulation by insulin (ins), phorbol-12-myristate-13-acetate (PMA), forskolin (for), thyroid stimulating hormone (TSH) and epidermal growth factor (EGF). Cells were harvested using trypsin and counted in a cell counter. After averaging, proliferation is expressed as percentage of the proliferation of the control group (Con).

# 2. MATERIALS AND METHODS

## 2.1 Cell culture

Cells of the thyroid cancer cell lines FTC 133 and 8505C were cultured in DMEM-Ham's F-12 medium supplemented with 10 % FCS. After reaching confluency cells were kept in serum-free medium for another 72 hours. Cells were treated for 24, 48 or 72 hours with thyroid stimulating hormone (TSH), 100 μU/ml; forskolin (For), 10 - 5 M; epidermal growth factor (EGF), 5 ng/ml; insulin (Ins), 5 μg/ml; or phorbol-12-myristate-13-acetate (PMA), 10 ng/ml. Cells were harvested after 24, 48 or 72 hours; counted and homogenized. The cytoplasmic fraction of the homogenate was used for Western blot and ELISA analyses. The stimulating medium was replaced and collected for analysis every 24 hours.

## 2.2 Amplification of procathepsin L mRNA

The total RNA was isolated from stimulated cells transcribed into DNA and amplified by the Polymerase Chain Reaction (PCR) with specific primers and according to a standard protocol.

## 2.3 Enzyme Activity Determination

Procathepsin L activities were determined according to Heidtmann (1993). 40μl of cell culture supernatant were added to 300 μl citrate buffer (125 mM), pH 3.5, containing DTT (2 mM) and EDTA (1 mM) and 200 μl Z-F-R-MCA (25 μM) at 37 °C for 15 minutes. The increase in fluorescence due to the release of MCA was measured.

## 2.4 Western blotting

The monoclonal anti-human cathepsin L antibody 33/1 (Weber *et al* 1997), detecting amino acids 238 to 244 of the procathepsin L sequence was used for Western blot analyses of cell lysates and cell culture supernatants.

## 2.5 Enzyme linked immunosorbent assay (ELISA)

For the determination of cathepsin L and procathepsin L an ELISA was established using the epitope specific mouse anti-human cathepsin L/procathepsin L monoclonal antibody 33/1 as catcher antibody on the microtiter plate and a polyclonal rabbit anti-cathepsin L antibody for detecting bound cathepsin L/procathepsin L. The lowest detectable concentration of cathepsin L was 2 ng/ml.

# 3. RESULTS

## 3.1 Proliferation

Our proliferation assay revealed insulin to be the most potent of all factors stimulating proliferation in cell line 8505C, whereas it does not seem to have a mitogenic effect on the follicular thyroid cell line FTC133. PMA appears to have a constantly increasing stimulating impuls on proliferation in 8505C reaching that of insulin after 72 hours. In the same cell line TSH, initially inhibiting mitosis, is showing a late onset mitogenic effect after 72 hours, while EGF is increasing proliferation at constant levels.

Compared to the control group PMA, TSH and EGF resemble in their mitogenic effect on cell line FTC133. After an initial significant stimulation, proliferation decreased markedly and reached nearly control levels at the end of the stimulation period. Forskolin also seems to be a potent trigger of proliferation in FTC 133 but after 48 hours and particularly after 72 hours this effect inverted into a considerable inhibition. Forskolin, however, did not induce proliferation in cell line 8505C at all.

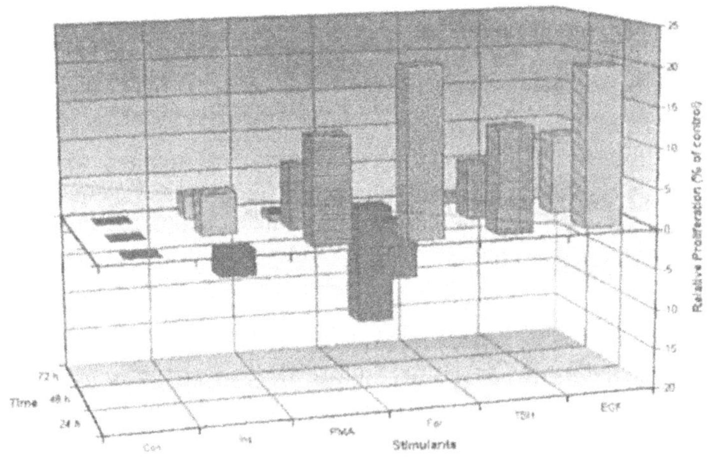


*Figure 2.* Cell line FTC 133
Relative proliferation of cell lines 8505C and FTC 133 after 24, 48 and 72 hours of stimulation by insulin (ins), phorbol-12-myristate-13-acetate (PMA), forskolin (for), thyroid stimulating hormone (TSH) and epidermal growth factor (EGF). Cells were harvested using trypsin and counted in a cell counter. After averaging, proliferation is expressed as percentage of the proliferation of the control group (Con).

## 3.2 RT-PCR analysis

RT-PCR analysis showed Cathepsin L to be constitutively expressed on m-RNA levels in both cell lines but could not reveal any factor-specific differences in its expression during stimulation.

## 3.3 Intracellular Cathepsin L

All factors (PMA, TSH, EGF) which promoted the proliferation of both cell lines caused an increase in the intracellular concentration of cathepsin L. Factors without influence on proliferation did not change cathepsin L expression. The pronounced proliferation effect shown by insulin in 8505C was accompanied by an increase in cytoplasmic cathepsin L. In FTC 133 on which insulin has no proliferation stimulating effect the cytoplasmic cathepsin L concentration remains unchanged as compared to untreated cells. Forskolin is initially enhancing cell proliferation in FTC 133 and an upregulation of the cathepsin L expression on the protein level is observed. Both effects were not seen in cells of cell line 8505C.

## 3.4. Procathepsin L secretion

The enzyme linked immunosorbent assay of the culture medium revealed a considerable difference between cell lines 8505C and FTC133. Basic procathepsin L secretion was much higher in the anaplastic carcinoma cell line 8505C compared to the follicular thyroid carcinoma cell line FTC133. Procathepsin L secretion in 8505C was increased only by PMA while FTC133 stayed completely unaffected by all agents (Fig 3). Activity determination of procathepsin L and Western Blot analysis of CL confirmed the results found by ELISA in the culture medium.

## 4. Discussion and Conclusions

Overexpression and secretion of proteolytic enzymes by cancer cells are presumably a basic mechanism in tumor progression and metastasis. We investigated the influence of five different factors of thyroid differentiation and dedifferentiation on the expression of the lysosomal cysteine proteinase cathepsin L in two thyroid cancer cell lines and on the proliferation of these cell lines. We found that unstimulated cells of the anaplastic thyroid carcinoma cell line 8505C are secreting more procathepsin L than those of the follicular thyroid cancer cell line FTC 133 what may correlate with the higher malignancy of anaplastic thyroid carcinomas the origin of cell line 8505C. Stimulation by TSH, EGF or PMA promotes cell proliferation and

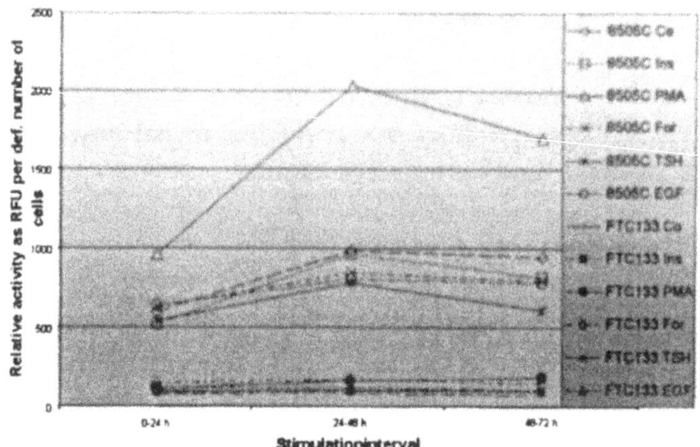


*Figure 3.* Procathepsin L (pro-CL) activity in cell culture supernatants
As described in "Material and Methods" procathepsin L activity determinations were performed in cell culture supernatants of cell lines FTC 133 and 8505C after the first, second and third 24 hour intervals with no stimulation (Con) or stimulation by insulin (ins), phorbol-12-myristate-13-acetate (PMA), forskolin (for), thyroid stimulating hormone (TSH) or epidermal growth factor (EGF). The relative pro-CL activity is given as RFU of pro-CL secreted by a defined number of cells.

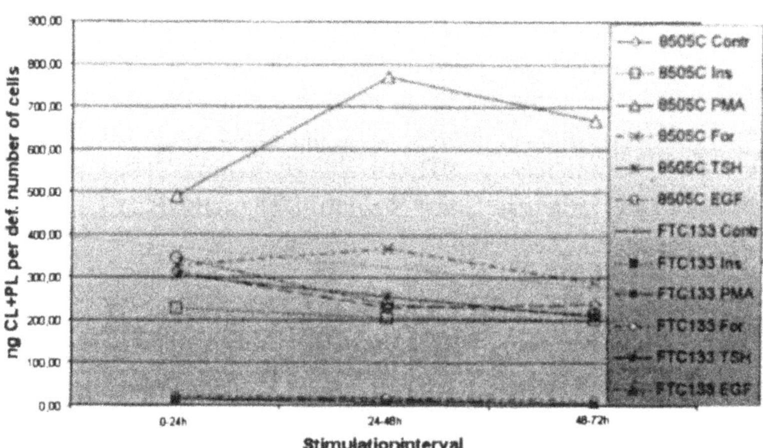


*Figure 4.* Secreted cathepsin L and procathepsin L (pro-CL) in cell culture supernatants determined by ELISA.
After the first, second and third 24 hours with no stimulation (Con) or stimulation by insulin (ins), phorbol-12-myristate-13-acetate (PMA), forskolin (for), thyroid stimulating hormone (TSH) or epidermal growth factor (EGF), 50 µl of cell culture supernatnat of cell lines FTC 133 and 8505C were analyzed in a sandwich ELISA using two cathepsin L specific antibodies. Amounts of determined CL and pro-CL are given as relative amounts of CL secreted by a defined number of cells.

increases the expression of cathepsin L in both cell lines but to different extents. Insulin is only enhancing CL expression and cell proliferation in 8505C, forskolin, however, only in FTC 133.

We were also able to show that PMA is a strong activator of procathepsin L secretion in cell line 8505C but not in FTC 133. Since PMA is also promoting the proliferation of 8505C we assume that the protein kinase C system may be important for the expression of malignant characteristics of at least some human thyroid tumors. Earlier evidence indicated that the TSH-triggered cAMP release might be the main stimulatory mechanism to induce growth in thyroid benign tumors while in thyroid cancers TSH seems to activate the phospholipase C pathway, too (Goretzki, 1989, Schatz *et al* 1989, Müller-Gärtner, 1989). Since the stimulants used here act via different transduction pathways (Clark 1989, Raspé 1989) it may be concluded that probably several mechanism can be similar pathway is only increasing proliferation and CL expression in 8505C. For the adenylate cyclase pathway tumor specific differences may exist since TSH and forskolin increase CL expression in FTC 133 cells but they have only a minor influence on its expression in 8505C.

Our results indicate that different transduction pathways may be open in the two investigated thyroid cancer cells lines and that a specific combination of their activities could be responsible for defining the characteristics of each cell line. Only TSH and EGF exert similar effects on both cell lines while the other factors showed differing effects depending on the cell line. In general the stimulation of proliferation resulted in an increase in the expression of cathepsin L on the protein level. Its amount relative to the overall cytoplasmic protein level is increasing (data not shown), but only PMA is stimulating the secretion of procathepsin L.

## ACKNOWLEDGMENTS

This work was supported by grants No. 2792A/0087H and 2795A/0087H of the Kultusministerium of LSA

## REFERENCES

Chauhan, S.S., Goldstein, L.J., and Gottesman, M.M., 1991, Expression of cathepsin L in human tumors. *Cancer Res.* **51:** 1478 – 1481.

Clark, O. H., Siperstein, A., Miller, R., and van Raavenswaay-Classen, H. ,1989, Receptors and receptor-transducing systems in normal and neoplastic human thyroid tissue. In *Growth regulation of thyroid gland and thyroid tumors.* (P.E. Goretzki, H.D. Röher, eds.) Karger, Basel, pp. 14 – 27.

Denhardt, D.T., Greenberg, A.H., Egan, S.E., Hamilton, R.T., and Wright, J.A., 1987, Cysteine proteinase cathepsin L expression correlates closely with the metastatic potential of H-ras-transformed murine fibroblasts. *Oncogene* **2:** 55 – 59.

Frade, R., Rodrigues-Lima, F., Huang, S., Xie, K, Guillaume, N. and Bar-Eli, M., 1998, Procathepsin L, a proteinase that cleaves human C3, the third component of complement, confers high tumorigenic and metastatic properties to human melanoma cells. *Cancer Research* **58:** 2733 – 2736

Goretzki, P. E., Frilling, A., Simon, D., Rastegar, M., Ohmann, C., 1989, Growth regulation of human thyrocytes by thyrotropin, cyclic adenosine monophosphate, epidermal growth factor and insulin-like growth factor. *In Growth regulation of thyroid gland and thyroid tumors.* (eds.) Karger, Basel, pp. 56 – 80.

Gottesman, M.M., 1993, Cathepsin L and cancer. In *Proteolysis and protein turnover.* Portland Press Proceedings, pp. 248 – 251.

Heidtmann, H.H., Salge, U., Havemann, K., Kirschke, H., and Wiederanders, B., 1993, Secretion of a latent, acid activatable cathepsin L precursor by human non-small lung cancer cell line. *Oncology Res.* **5:** 441 – 451.

Homma, K., Kurata, S., and Natori, S., 1994, Purification, characterization, and cDNA cloning of procathespin L from the culture medium of NIH-Sape-4, an embryonic cell line of Sarcophaga peregrina (flesh fly), and its involvement in the differentiation of imaginal discs. *J.Biol.Chem.* **269:** 15258 – 15264.

Jean, D., Bar-Eli, M., Huang, S., Xie, K., Rodrigues-Lima, F., Hermann, J., and Frade, R, 1996, A cysteine proteinase, which cleaves human C3, the third component of complement, is involved in tumorigenicity and metastasis of human melanoma. *Cancer Res.* **56:** 254 – 258.

Kasai, M., Shirakawa, T., Kitamura, M., Ishido, K., Kominami, E., and Hirokawa, K., 1993, Proenzyme form of cathepsin L produced by thymic epithelial cells promotes proliferation of immature thymocytes in the presence of IL-1, IL-7, and anti-CD3 antibody. *Cell. Immunol.* **150:** 124 – 136.

Müller-Gärtner, H.W., Baisch, H., Garn, M., de la Roche, J., 1989, Individually different proliferation responses of differentiated thyroid carcinomas to thyrotropin. In *Growth regulation of thyroid gland and thyroid tumors.* (P.E. Goretzki, H.D. Röher,eds.) Karger, Basel, pp. 137 – 151.

Raspé, E., Reuse, S., Maenhaut, C., Roger, P., Corvilain, B., Laurent, E., Mockel, J., Lamy, F., van Sande, J., and Dumont, J. E., 1989, Importance and variability of transducing systems in the control of thyroid cell function, proliferation and differentiation. *In Growth regulation of thyroid gland and thyroid tumors.* (P.E. Goretzki, H.D. Röher,eds.) Karger, Basel, pp. 11-13.

Schatz, H., Freiberger, R., Richter, C., Wiss, F., & Weber, K., 1989, Influence of thyroid-stimulating hormone, epidermal growth factor, and insulin-like growth factor I on growth of thyroid cells in vitro. In *Growth regulation of thyroid gland and thyroid tumors.* (P.E. Goretzki, H.D. Röher,eds.) Karger, Basel, pp. 88-97.

Sivaparvathi, M., Sawaya, R., Wang, S.W., Rayfor, A., Yamamoto, M., Liotta, L.A., Nicolson, G.L., and Rao, J.S., 1996, Expression and immunohistochemical localization of cathepsin L during the progression of human gliomas. *Clin. Experim. Metastasis* **14:** 27-34.

Weber, E., Bahn, H. and Günther, D., 1997, Monoclonal antibodies against cathepsin L and procathepsin L of different species. *Hybridoma* **16:** 159 – 166.

Weber, E., Günther, D., Laube, F., Wiederanders, B., and Kirschke, H., 1994, Hybridoma cells producing antibodies to cathepsin L have greatly reduced potential für tumor growth. *J. Cancer Res. Clin. Oncol.* **120:** 564 – 567.

Yagel, S., Warner, A.H., Nellans, H.N., Lala, P.K., Waghorne, C., and Denhardt, D. T., 1989, Suppression by cathepsin L inhibitors of the invasion of amnion membranes by murine cancer cells. *Cancer Res.* **49:** 3553 – 3557.

# CONTRIBUTORS

**Catherine A. Abbott**
A.W. Morrow Gastroenterology and Liver Centre, Centenary Institute of Cell Biology and Cancer Medicine, Royal Prince Alfred Hospital and The University of Sydney, NSW, Australia

**Reinhard Andreesen**
Department of Hematology and Oncology, University of Regensburg, 93042 Regensburg, Germany

**Siegfried Ansorge**
Institute of Experimental Internal Medicine, Center of Internal Medicine, Otto von Guericke University Magdeburg, D-39120 Magdeburg, Germany

**Petra Arck**
Department of Internal Medicine, Charité Campus Virchow-Klinikum, Humboldt-Universität zu Berlin, Germany

**Marco Arndt**
Institute of Immunology, Center of Internal Medicine, Otto von Guericke University Magdeburg, D-39120 Magdeburg, Germany

**Koen Augustyns**
Laboratory of Pharmaceutical Chemistry, University of Antwerp, Wilrijk, Belgium

**Hendryk Aurich**
Institute of Physiological Chemistry, Martin Luther University, D-06097 Halle, Germany

**Gunther Bal**
Laboratory of Pharmaceutical Chemistry, University of Antwerp, Wilrijk, Belgium

**Agnieszka Banbula**
Dept. of Biochemistry & Molecular Biology, The University of Georgia, Athens, Georgia 30602

**Ute Bank**
Institute of Immunology, Otto-von-Guericke-Universität, Magdeburg, D-39120 Magdeburg

**Kay Barnes**
School of Biochemistry and Molecular Biology, University of Leeds, Leeds LS2 9JT, U.K.

**Kathleen Bartels**
Clinic of Rheumatology, Otto-von-Guericke-University of Magdeburg, D-39245 Vogelsang, Germany

**Eckart Bartnik**
Biochemical Research, Hoechst Marion Roussel Deutschland GmbH, Frankfurt/M., Germany

**Ilona Born**
Martin-Luther-University Halle-Wittenberg, Department of Biochemistry and Biotechnology, Institute of Biochemistry, 06120 Halle, Germany.

**Alan Boyde**
Zentrum Biochemie und Molekulare Zellbiologie, Abteilung Biochemie II, Universität Göttingen, 37073 Göttingen, Germany

**Wolfgang Brandt**
Department of Biochemistry and Biotechnology, Institute of Biochemistry, Martin-Luther-University Halle-Wittenberg, 06120 Halle, Germany

**Stefan Brocke**
Neuroimmunology Branch, NINDS, National Institutes of Health, Bethesda, USA

**Carolyn D. Brown**
School of Biochemistry and Molecular Biology, University of Leeds, Leeds LS2 9JT, U.K.

**Dieter Brömme**
Department of Human Genetics, Mount Sinai School of Medicine, New York, NY 10029, USA

**Josef Brunner**
Department of Biochemistry, Swiss Federal Institute of Technology, Zürich, Swizerland

**Frank Bühling**
Institute of Experimental Internal Medicine, Otto-von-Guericke Universität Magdeburg, 39120 Magdeburg

**Frank H. Büttner**
Biochemical Research, Hoechst Marion Roussel Deutschland GmbH, Frankfurt/M., Germany

**Julie A. Carson**
School of Biochemistry and Molecular Biology, University of Leeds, Leeds LS2 9JT, U.K.

**Kristin Cebulski**
Clinic of Rheumatology, Otto-von-Guericke-University of Magdeburg, D-39245 Vogelsang, Germany

**W. Bret Church**
Garvan Institute of Medical Research, Sydney

**Lisa Demchik**
Barbara Ann Karmanos Cancer Institute, Detroit, Michigan 48201, USA

**Ingrid De Meester**
Laboratory of Clinical Biochemistry, University of Antwerp, Wilrijk, Belgium

**Hans-Ullrich Demuth**
probiodrug GmbH, Biocenter, 06120 Halle, Germany

**Jan Duessing**
First Department of Internal Medicine, Albert Ludwigs University, Freiburg, Germany

**Jonathan S. Duke-Cohan**
Tumor Immunology, Department of Cancer Immunology and AIDS, Dana-Farber Cancer Institute and Department of Medicine, Harvard Medical School, Boston, MA, USA

**Christine Durinx**
Laboratory of Clinical Biochemistry, University of Antwerp, Wilrijk, Belgium

**Vincent Everts**
Dept Periodontology and Dept. Cell Biology and Histology, Academic Centre for Dentistry, Amsterdam; University of Amsterdam, Amsterdam, The Netherlands

**Jürgen Faust**
Martin-Luther-University Halle-Wittenberg, Department of Biochemistry and Biotechnology, Institute of Biochemistry, 06120 Halle, Germany

**Annett Fengler**
Department of Biochemistry and Biotechnology, Institute of Biochemistry, Martin Luther University Halle-Wittenberg, 06120 Halle, Germany

**Karin Frank**
Institute of Experimental Internal Medicine, D-39120 Magdeburg, Germany

**Ivar Friedrich**
Klinik für Herz- und Thoraxchirurgie, Martin Luther University, D-06097 Halle

**Ameer Gado**
National Institute of Neurological Disorders and Stroke, National Institutes of Health, Bethesda,MD, USA

**R. Gelling**
Department of Physiology, University of British Columbia, Vancouver, BC., Canada V6T 1Z3

**Annegret Gerber**
Inst. of Immunology, Otto-von-Guericke University Magdeburg, Magdeburg, Germany

**Harald Gollnick**
Dept. of Dermatology and Venereology, Otto-von-Guericke University, D-39120 Germany

**Filip Goossens**
Laboratory of Clinical Biochemistry, University of Antwerp, Wilrijk, Belgium

**Mark D. Gorrell**
A.W. Morrow Gastroenterology and Liver Centre, Centenary Institute of Cell Biology and Cancer Medicine, Royal Prince Alfred Hospital and The University of Sydney, NSW, Australia

**Constanze Gottschalk**
Institute of Immunology, Center of Internal Medicine, Otto von Guericke University Magdeburg, D-39120 Magdeburg, Germany

**Henner Graeff**
Klinische Forschergruppe der Frauenklinik der Technischen Universität München, Klinikum rechts der Isar, D-81675 München, Germany

**Bruno Gran**
Neuroimmunology Branch, NINDS, National Institutes of Health, Bethesda, USA

**Dagmar Günther**
Institute of Physiological Chemistry, Martin Luther University, D-06097 Halle, Germany

**Carsten Häckel**
Institute of Pathology, Otto-von-Guericke Universität Magdeburg, 39120 Magdeburg

**Walter Halangk**
Division of Experimental Surgery, Otto-von-Guericke-Universität Magdeburg

**Bettina Hause**
Institute of Plant Biochemistry, Martin Luther University Halle, D-06120 Halle, Germany

**Michael Heinrich**
Institute of Medical Microbiology and Hygiene, University of Regensburg, Germany

**Bernhard Hemmer**
Clinical Neuroimmunology group, University of Marburg, Germany

**Martin Hildebrandt**
Department of Internal Medicine, Charité Campus Virchow-Klinikum, Humboldt-Universität zu Berlin, Germany

**Oliver Hiller**
University Bielefeld, Faculty of Chemistry, Department of Biochemistry, Bielefeld, Germany

**Simon A. Hinke**
Department of Physiology, University of British Columbia, Vancouver, BC., Canada V6T 1Z3

**Chuong Hoang-Vu**
Klinik für Allgemeinchirurgie, AG Exp. & Chirurgische Onkologie, Martin Luther University Halle, D-06097 Halle

**Torsten Hoffmann**
probiodrug GmbH, Biocenter, 06120 Halle, Germany,

**Henk C. Hoogsteden**
Departments of Pulmonary Medicine, Erasmus University and University Hospital Rotterdam-Dijkzigt, Rotterdam, The Netherlands

**Nigel M. Hooper**
School of Biochemistry and Molecular Biology, University of Leeds, Leeds LS2 9JT, UK

**Ernst Hunziker**
Universität Bern, M.E. Müller-Institut für Biomechanik, Murtenstrasse 35, CH-3010 Bern, Switzerland

**Leonid Izikson**
National Institute of Neurological Disorders and Stroke, National Institutes of Health, Bethesda,MD, USA

**Paul Janowitz**
Dept. Internal Medicine, Teaching Hospital Burg Otto-von-Guericke-University Magdeburg, D-39120 Magdeburg, Germany

**Sheila Jones**
Department of Anatomy and Developmental Biology, University College London, London WC1E 6BT, UK.

**Thilo Kähne**
Institute of Experimental Internal Medicine, Department of Internal Medicine, Otto-von-Guericke University Magdeburg, D-39120 Magdeburg, Germany

**Astrid Kehlen**
Institute of Medical Immunology, Martin Luther University Halle, D-06097 Halle, Germany,

**Jörn Kekow**
Clinic of Rheumatology, Otto-von-Guericke-University of Magdeburg, D-39245 Vogelsang, Germany

**Burghard F. Klapp**
Department of Internal Medicine, Charité Campus Virchow-Klinikum, Humboldt-Universität zu Berlin, Germany

**Jürgen Kleinstein**
Department of Reproductive Medicine and Gynecological Endocrinology, Otto-von-Guericke-University, Magdeburg, Germany

**Stefan W. Krause**
Department of Hematology and Oncology, University of Regensburg, 93042 Regensburg, Germany

**Martin Krönke**
Institute of Medical Microbiology and Hygiene, University of Cologne, Germany

**Burkhard Krüger**
Division of Medical Biology, Department of Pathology, Rostock University, Germany

**Sabine Krüger**
Institute of Pathology, Otto-von-Guericke Universität Magdeburg, 39120 Magdeburg

**Kerstin Kühn-Wache**
probiodrug GmbH, Biocenter, 06120 Halle, Germany,

**Cornelia Kühne**
Clinic of Rheumatology, Otto-von-Guericke-University of Magdeburg, D-39245 Vogelsang, Germany

**Bernadette Küpper**
Institute of Immunology, Otto-von-Guericke-Universität, Magdeburg, D-39120 Magdeburg

**Jürgen Langner**
Martin Luther University, Institute of Medical Immunology, D-06097 Halle, Germany

**Uwe Lendeckel**
Institute of Experimental Internal Medicine, Center of Internal Medicine, Otto von Guericke University Magdeburg, D-39120 Magdeburg, Germany

**Markus M. Lerch**
Department of Medicine B, Westfälische Wilhelms-Universität Münster

**Brynn Levy**
Department of Human Genetics, Mount Sinai School of Medicine, New York, NY 10029, USA

**Miriam T. Levy**
A.W. Morrow Gastroenterology and Liver Centre, Centenary Institute of Cell Biology and Cancer Medicine, Royal Prince Alfred Hospital and The University of Sydney, NSW, Australia

**Andrea Lichte**
University Bielefeld, Faculty of Chemistry, Department of Biochemistry, Bielefeld, Germany

**Harald Loppnow**
Martin-Luther-Universität Halle-Wittenberg, Klinik und Poliklinik für Innere Medizin III, Forschungslabor, D-06097 Halle

**Susan Lorey**
Martin-Luther-University Halle-Wittenberg, Institute of Biochemistry and Biotechnology, D-06120, Halle, Germany

**Thomas Luther**
Institut für Pathologie der Technischen Universität Dresden, D-01307 Dresden, Germany

**Ulla Magdolen**
Klinische Forschergruppe der Frauenklinik der Technischen Universität München, Klinikum rechts der Isar, D-81675 München, Germany

**Viktor Magdolen**
Klinische Forschergruppe der Frauenklinik der Technischen Universität München, Klinikum rechts der Isar, D-81675 München, Germany

**Jianxin Mai**
Department of Pharmacology, Wayne State University School of Medicine

**Kurt Malberg**
Institute of Immunology, Otto-von-Guericke-University, Magdeburg, Germany

**Peter Malfertheiner**
Center of Internal Medicine, Dept. Gastroenterology, Hepatology and Infectious Diseases Otto-von-Guericke-University Magdeburg, D-39120 Magdeburg, Germany

**Susanne Manhart**
probiodrug GmbH, Biocenter, 06120 Halle, Germany,

**Roland Martin**
Neuroimmunology Branch, NINDS, National Institutes of Health, Bethesda, USA

**Christine Mayr**
Department of Internal Medicine, Charité Campus Virchow-Klinikum, Humboldt-Universität zu Berlin, Germany

**Geoffrey W. McCaughan**
A.W. Morrow Gastroenterology and Liver Centre, Centenary Institute of Cell Biology and Cancer Medicine, Royal Prince Alfred Hospital and The University of Sydney, NSW, Australia

**Christopher H.S. McIntosh**
Department of Physiology, University of British Columbia, Vancouver, BC., Canada V6T 1Z3

**Robert Ménard**
Biotechnology Research Institute, National Research Council of Canada, Montréal, Québec, Canada H4P 2R2,

**Werner Mögelin**
Martin-Luther-University Halle-Wittenberg, Department of Biochemistry and Biotechnology, Institute of Biochemistry, 06120 Halle, Germany.

**Kamiar Moin**
Department of Pharmacolog, Wayne State University School of Medicine, Detroit, Michigan 48201, USA

**Carmen Mrestani-Klaus**
Department of Biochemistry and Biotechnology, Institute of Biochemistry, Martin-Luther-University Halle-Wittenberg, 06120 Halle

**Bernd Muehlenweg**
Klinische Forschergruppe der Frauenklinik der Technischen Universität München, Klinikum rechts der Isar, D-81675 München, Germany

**Brigitta A.E. Naber**
Departments of Immunology Erasmus University and University Hospital Rotterdam-Dijkzigt, Rotterdam, The Netherlands

**Dorit K. Nägler**
Institute of Pathology, Otto-von-Guericke University Magdeburg, 39120 Magdeburg, Germany.

**Alexander Navarrete Santos**
Institute of Medical Immunology, Martin Luther University, D-06097 Halle

**Klaus Neubert**
Martin-Luther-University Halle-Wittenberg, Department of Biochemistry and Biotechnology, Institute of Biochemistry, 06120 Halle, Germany.

**Manfred Nilius**
Center of Internal Medicine, Dept. Gastroenterology, Hepatology and Infectious Diseases Otto-von-Guericke-University Magdeburg, D-39120 Magdeburg, Germany

**Ove Norén**
Department of Medical Biochemistry and Genetics, Biochemistry Laboratory C, The Panum Institute, University of Copenhagen, Copenhagen, Denmark

**André Oberpichler**
University Bielefeld, Faculty of Chemistry, Department of Biochemistry, Bielefeld, Germany

**Jørgen Olsen**
Department of Medical Biochemistry and Genetics, Biochemistry Laboratory C, The Panum Institute, University of Copenhagen, Copenhagen, Denmark

**Shelley E. Overbeek**
Departments of Pulmonary Medicine, Erasmus University and University Hospital Rotterdam-Dijkzigt, Rotterdam, The Netherlands

**Thomas Pap**
Center for Experimental Rheumatology, Department of Rheumatology, University Hospital, CH-8091 Zürich, Switzerland

**Edward T. Parkin**
School of Biochemistry and Molecular Biology, University of Leeds, Leeds LS2 9JT, UK

**Raymond A. Pederson**
Department of Physiology, University of British Columbia, Vancouver, BC., Canada V6T 1Z3

**Christoph Peters**
First Department of Internal Medicine, Albert Ludwigs University of Freiburg, Freiburg, Germany

**Alexander Plehn**
Institute of Physiological Chemistry, Martin Luther University, D-06097 Halle, Germany

**Jan Potempa**
Dept. of Biochemistry & Molecular Biology, The University of Georgia, Athens, Georgia 30602

**Nuria Arroyo de Prada**
Klinische Forschergruppe der Frauenklinik der Technischen Universität München, Klinikum rechts der Isar, D-81675 München, Germany

**Paul Proost**
Laboratory of Molecular Immunology, Rega Institute for Medical Research, University of Leuven, Leuven, Belgium

**Enrico O. Purisima**
Biotechnology Research Institute, National Research Council of Canada, Montréal, Québec, Canada H4P 2R2

**Laura Quigley**
National Institute of Neurological Disorders and Stroke, National Institutes of Health, Bethesda, MD, USA

**Michael Rehli**
Department of Hematology and Oncology, University of Regensburg, 93042 Regensburg, Germany

**Dirk Reinhold**
Institute of Experimental Internal Medicine, Center of Internal Medicine, Otto von Guericke University Magdeburg, D-39120 Magdeburg, Germany

**Georg Reiser**
Otto-von-Guericke Universität Magdeburg, Medizinische Fakultät, Institut für Neurobiochemie, D-39120 Magdeburg, Germany

**Ina Repper**
Center of Internal Medicine, Dept. Gastroenterology, Hepatology and Infectious Diseases Otto-von-Guericke-University, D-39120 Magdeburg, Germany

**Ute Reuning**
Klinische Forschergruppe der Frauenklinik der Technischen Universität München, Klinikum rechts der Isar, D-81675 München, Germany

**Werner Reutter**
Institut für Molekularbiologie und Biochemie, Freie Universität Berlin, Germany

**Dagmar Riemann**
Institute of Medical Immunology, Martin Luther University Halle, D-06097 Halle, Germany,

**Jana Röntsch**
Institute of Medical Immunology, Martin Luther University Halle D-06097 Halle;

**Albert Roessner**
Institute of Pathology, Otto-von-Guericke Universität Magdeburg, 39120 Magdeburg

**Matthias Rose**
Department of Internal Medicine, Charité Campus Virchow-Klinikum, Humboldt-Universität zu Berlin, Germany

**Paul Saftig**
Zentrum Biochemie und Molekulare Zellbiologie, Abteilung Biochemie II, Universität Göttingen, 37073 Göttingen, Germany

**Abdulgabar Salama**
Department of Internal Medicine, Charité Campus Virchow-Klinikum, Humboldt-Universität zu Berlin, Germany

**Simon Scharpé**
Laboratory of Clinical Biochemistry, University of Antwerp, Wilrijk, Belgium

**Stuart F. Schlossman**
Tumor Immunology, Department of Cancer Immunology and AIDS, Dana-Farber Cancer Institute and Department of Medicine, Harvard Medical School, Boston, MA, U.S.A

**Johannes Schmitt**
Institute of Pathology, Otto-von-Guericke-University, Magdeburg, Germany

**Manfred Schmitt**
Klinische Forschergruppe der Frauenklinik der Technischen Universität München, Klinikum rechts der Isar, D-81675 München, Germany

**Wulf Schneider-Brachert**
Institute of Medical Microbiology and Hygiene, University of Regensburg, Germany

**Cora Schüler**
Institut für Anatomie, Med. Akademie Carl Gustav Carus der Technischen Universität Dresden

**Stefan Schütze**
Institute of Immunology, University of Kiel, Germany

**Dagmar Schultze**
Department of Reproductive Medicine and Gynecological Endocrinology, Otto-von-Guericke-University, Magdeburg, Germany

**Rolf-Edgar Silber**
Klinik für Herz- und Thoraxchirurgie, D-06097 Halle

**Hans Sjöström**
Department of Medical Biochemistry and Genetics, Biochemistry Laboratory C, The Panum Institute, University of Copenhagen, Copenhagen, Denmark

**Bonnie F. Sloane**
Department of Pharmacolog, Wayne State University School of Medicine, Detroit, Michigan 48201, USA

**Armin Sokolowski**
Dept. Clinical Chemistry Otto-von-Guericke-University Magdeburg, D-39120 Magdeburg, Germany

**Stefan Sperl**
Klinische Forschergruppe der Frauenklinik der Technischen Universität München, Klinikum rechts der Isar, D-81675 München, Germany

**Antje Spiess**
Institute of Immunology, Institute of Experimental Internal Medicine, Center of Internal Medicine, Otto von Guericke University Magdeburg, D-39120 Magdeburg, Germany

**Andreas Steinbrecher**
National Institute of Neurological Disorders and Stroke, National Institutes of Health, Bethesda, MD, USA

**Beate Stiebitz**
Martin-Luther-University Halle-Wittenberg, Department of Biochemistry and Biotechnology, Institute of Biochemistry, 06120 Halle, Germany.

**Angela Stöckel-Maschek**
Martin-Luther-University Halle-Wittenberg, Department of Biochemistry and Biotechnology, Institute of Biochemistry, 06120 Halle, Germany.

**Sofie Struyf**
Laboratory of Molecular Immunology, Rega Institute for Medical Research, University of Leuven, Leuven, Belgium

**Traian Sulea**
Biotechnology Research Institute, National Research Council of Canada, Montréal, Québec, Canada H4P 2R2,

**Anke Suter**
Zentrum Biochemie und Molekulare Zellbiologie, Abteilung Biochemie II, Universität Göttingen, 37073 Göttingen, Germany

**Wendy Tam**
Biotechnology Research Institute, National Research Council of Canada, Montréal, Québec, Canada H4P 2R2,

**Wen Tang**
Tumor Immunology, Department of Cancer Immunology and AIDS, Dana-Farber Cancer Institute and Department of Medicine, Harvard Medical School, Boston, MA, USA

**James Travis**
Dept. of Biochemistry & Molecular Biology, The University of Georgia, Athens, Georgia 30602

**Nancy Tresser**
National Institute of Neurological Disorders and Stroke, National Institutes of Health, Bethesda, MD, USA

**Alison J. Trew**
School of Biochemistry and Molecular Biology, University of Leeds, Leeds LS2 9JT, UK

**Harald Tschesche**
University Bielefeld, Faculty of Chemistry, Department of Biochemistry, Bielefeld, Germany

**Anthony J. Turner**
School of Biochemistry and Molecular Biology, University of Leeds, Leeds LS2 9JT, UK

**Joachim J. Ubl**
Otto-von-Guericke Universität Magdeburg, Medizinische Fakultät, Institut für Neurobiochemie, D-39120 Magdeburg, Germany

**Vincent H.J. van der Velden**
Departments of Pulmonary Medicine, Erasmus University and University Hospital Rotterdam-Dijkzigt, Rotterdam, The Netherlands

**Peter Th. W. van Hal**
Departments of Pulmonary Medicine, Erasmus University and University Hospital Rotterdam-Dijkzigt, Rotterdam, The Netherlands

**Thomas Vahldieck**
Dept. Internal Medicine, Teaching Hospital Burg Otto-von-Guericke-University Magdeburg, D-39120 Magdeburg, Germany

**Marjan A. Versnel**
Departments of Immunology Erasmus University and University Hospital Rotterdam-Dijkzigt, Rotterdam, The Netherlands

**Robert Vetter**
Dept. of Dermatology and Venereology, Otto-von-Guericke University, D-39120 Germany

**Kurt von Figura**
Zentrum Biochemie und Molekulare Zellbiologie, Abteilung Biochemie II, Universität Göttingen, 37073 Göttingen, Germany

**Nadine Waldburg**
Inst. of Immunology, Otto-von-Guericke University Magdeburg, Magdeburg, Germany

**Ekkehard Weber**
Institute of Physiological Chemistry, Martin Luther University, D-06097 Halle, Germany

**Thomas Weber**
Memorial Sloan-Kettering Cancer Center, New York, NY-10021, USA
Department of Biochemistry, Swiss Federal Institute of Technology, Zürich, Swizerland

**Olaf Wehmeyer**
Zentrum Biochemie und Molekulare Zellbiologie, Abteilung Biochemie II, Universität Göttingen, 37073 Göttingen, Germany

**Tobias Welte**
Department of Pneumology and Critical Care, Otto von Guericke University Magdeburg, Magdeburg, Germany

**Karl Werdan**
Martin-Luther-Universität Halle-Wittenberg, Klinik und Poliklinik für Innere Medizin III, Forschungslabor, D-06097 Halle

**Elena Westphal**
Martin-Luther-Universität Halle-Wittenberg, Klinik und Poliklinik für Innere Medizin III, Forschungslabor, D-06097 Halle

**Heike Wex**
Department of Human Genetics, Mount Sinai School of Medicine, New York, NY 10029, USA

**Thomas Wex**
Department of Human Genetics, Mount Sinai School of Medicine, New York, NY 10029, USA

**Marc Wickel**
Institute of Medical Microbiology and Hygiene, University of Regensburg, Germany

**Bernd Wiederanders**
Friedrich-Schiller-Universität Jena, Klinikum, Institut für Biochemie, D-07743 Jena

**Olaf G. Wilhelm**
Klinische Forschergruppe der Frauenklinik der Technischen Universität München, Klinikum rechts der Isar, D-81675 München, Germany

**Supandi Winoto-Morbach**
Institute of Medical Microbiology and Hygiene, University of Regensburg, Germany

**Sabine Wrenger**
Department of Internal Medicine, Institute of Experimental Internal Medicine, Otto-von-Guericke-University Magdeburg, 39120 Magdeburg, Germany

**Denise Yu**
A.W. Morrow Gastroenterology and Liver Centre, Centenary Institute of Cell Biology and Cancer Medicine, Royal Prince Alfred Hospital and The University of Sydney, NSW, Australia

**Rulin Zhang**
Biotechnology Research Institute, National Research Council of Canada, Montréal, Québec, Canada H4P 2R2

**Susanne Zielinski**
Clinic of Rheumatology, Otto-von-Guericke-University of Magdeburg, D-39245 Vogelsang, Germany

# INDEX


ACE: *see* Angiotensin converting enzyme
Acute pancreatitis, 403
Acute phase proteins, 358
Acylaminoacyl peptidases (ACPH), 103
Adamalysin, 382
ADAMs: *see* A disintegrin and metalloproteinase
Adenylyl cyclase, 73
A disintegrin and metalloproteinase
  family members, 19, 382
  function, 15
  role in Alzheimers disease, 383
  role in amyloid processing, 382
  structure, 15
Airways, *see also* Asthma
  occurrence of peptidases, 413
  peptidergic innervation, 413
  soluble peptidases in, 419
  tachykinin containing nerves, 413
Alzheimer's disease, 355, 379–390
Aminopeptidase B, substrates
  β-casomorphin-5, 69
  endomorphin-2, 69
  substance P, 68
Aminopeptidase N
  in asthmatic airways, 415
  calcium signals, 43
    dependence on enzyme activity, 43–47
  circulating isoform, 420
  classification, 25
  dendritic cells, 416
  distribution in bronchial tissues, 416
  domain structure, 26
  eosinophils, 416
  expression on synoviocytes, 57–66
  post-translational modifications, 31
  regulation by cell-cell contact, 57–66
  signaling via APN, 36–41, 43–47
  soluble form in airways, 419
  vascular endothelial cells, 416
  Wnt-5a protooncogene activation, 36
Aminopeptidase N/review, 3, 25
Aminopeptidase P/APP, 9
  distribution, 9
  substrates, 10
Amyloid
  deposits, 379
  islet amyloid polypeptide (IAPP), 78
β-amyloid peptide, 379
Amyloid precursor protein, proteolytic processing, 379–390
Amyloidogenesis, 379–385
Amyotrophic lateralsclerosis, 355
Anaerobic bacteria
  peridontopathogenesis, 455
  Porphyromonas gingivalis (P. gingivalis), 455
Angelman syndrome, 353
Angiotensin converting enzyme, release by secretase, 381, 383
Anorexia nervosa, 197
  DPIV activity in patients, 200
α1-antichymotrypsin, 358
APn: *see* Aminopeptidase n
Apoptotic signalling, via ceramide, 306
APP: *see* Aminopeptidase N
APP: *see* Amyloid precursor protein
Ascites, Il-6 determination, 432
Aspartatic proteases, 241
Asthma
  origin via neurogenic inflammation, 413
  peptidases in bronchial lavage fluid, 419
  role of peptidases, 413
    modulation by glucocorticoids, 423
Ataxin-3, 355
Attractin, 104, 173–185
  CUB domain, 177
  gene product of mahogany gene, 180
  relation to DPIV, 175
  role in energy metabolism, 182
  role in melanin metabolism, 181
  structural characterization, 176
  T cell activation antigen, 175
Autoantigens in rheumatic diseases, 469

BAL: *see* Bronchial lavage fluid
Batimastat, 381
Bendless, 353
Blood glucose homeostasis, 73

Bone resorption, osteoclasts, 293
Bronchial lavage fluid, peptidase content, 419
  influence of smoking, 420
BT 20 human breast carcinoma cells, 393
Bulimia, 197
  DPIV activity in patients, 200

$Ca^{2+}$-signals
  following APN ligation, 43–48
  PAR-1 evoked, 324
  in rat astrocytes, 324
Caerulein, 405
Calcitonin gene family as DPIV substrates, 77
  CALC-1 gene, 77
  calcitonin, 77
  preprocalcitonin, 77
  procalcitonin gene related peptide mRNA, 77
CALLA: *see* Neprilysin; Neutral endopeptidase 24.11
Carboxypeptidase M, 205–216
  monoclonals MAX, 208
  natural substrates, 210–212
  regulation of expression, 210
  sites of expression, 207
Carboxypeptidases
  Carboxypeptidase D/review, 14
  Carboxypeptidase M/review, 14
β-casomorphin substrates, 72
Caspase-1
  in PMA leukocytes, 346
  in vascular smooth muscle cells, 343
Cathepsins, 287
  papain family, 271
  thiol-dependent, 271
  tumor progression, 366
Cathepsin A, 241
Cathepsin B, 241, 271, 261, 282, 367
  antisense technology, 439
  colocalization with trypsinogen in pancreas, 407
  different isoforms, 392
  Elisa, 440
  enzyme expression, 440
  expression in lung epithelial cells, 287–292
  expression in lung/lung tumor, 284
  inhibition by cystatin C, 321
  inhibitors, CA074, 443
  localization in tumors, 391–401
  mRNA expression, 288
  northern blot, 439
  in osteoclasts, 299
  pro-uPA conversion, 332
  quantification, 440
  regulation by IL-6, 287
  regulation by TGF-β, 287
    A549 cells, 287
    BEAS-2B cells, 287
Cathepsin B (*cont.*)
  tumor cell trafficking, 392
  tumor invasion, 442
Cathepsin B transfected cell lines
  MNNG/HOS, 439
  osteosarcoma, 439
Cathepsin C, 241, 271
  dipeptidylpeptidase I, 271
Cathepsin D, 241
  ceramide-binding protein, 305–315
  isoforms, 309
  subcellular compartmentation, 309
Cathepsin E, 241
Cathepsin F, 271–280, 261
  chromosomal localisation, 272
  cloning, 272
  homology based models, 255
Cathepsin F/review, 244
  cystatin-like domain, 244, 279
  expression, 244
  signal sequence, 244
Cathepsin F-like proteases, 276
Cathepsin G, 241, 357
  role in IL-6 proteolysis, 431
Cathepsin H, 241, 261, 271, 282
  homology based models, 255
Cathepsin K, 271, 261
  cathepsin K deficient mice, 293–303
    osteopetrotic changes, 293
  collagen cleavage, 282
  elastinolytic activity, 282
  expression, 294
    in lung tissue, 281–286
    in lung tumors, 284
  gene mutations, 295
  homology based models, 255
  immunohistochemistry, 283
  matrix degradation, 282
  osteoporosis, 295, 301
  quantitative RT-PCR, 282
  role in bone resorption, 293
  role in other tissues, 295
Cathepsin K/review, 245
  enzymatic activity, 245
  expression, 245
  gene, 245
  substrates, 245
  Xray structure, 246
Cathepsin-K deficient osteoclasts, 296
Cathepsin L, 241, 261, 271, 282, 287, 367
  ELISA detection
    in normal lung parenchyma, 289
    in tumor cells, 289
  expression in lung/lung tumor, 284
  gene regulating factors, 287
  inhibition by cystatin C, 321

Cathepsin L (*cont.*)
osteoclasts, 299
protein level, 289
pro-uPA conversion, 332
in thyroid cancer cell lines, 488
Cathepsin O, 271
Cathepsin O/review, 247
enzymatic activity, 247
expression, 247
gene, 247
Cathepsin S, 241, 261, 271, 282
homology based models, 255
inhibition by propeptide domain, 266
osteoclasts, 299
propeptide as chaperone, 266, 268
Cathepsin V (L2), 271
Cathepsin V/review, 247
activation, 248
cloning, 247
expression, 248
structure, 248
substrate specificity, 248
Cathepsin W, 271–280
chromosomal localisation, 272
cloning, 272
Cathepsin W/review, 248
expression, 248
lymphopain, 271
Cathepsin X
characterization as new cysteine protease, 317–322
inhibition by cystatin C, 321
kinetic characterization, 319
molecular modelling, 319
primary structure, 319
Cathepsin X (2), 271
Cathepsin X/review, 249, 317–322
activity, 250
cloning, 249
expression, 249
inhibition, 250
molecular model, 250
CD10: *see* Neutral endopeptidase 24.11
CD13: *see* Aminopeptidase N
CD26: *see* Dipeptidyl peptidase IV
CD87, uPAR, 332
cDNA array technique, 36
Cell-cell-contact, 57–66
Ceramide
apoptosis signal, 306
binding to cathepsin D, 305–315
second messenger, 305
Cerebrospinal fluid, Il-6 determination, 432
CFTR: *see* Cystic fibrosis transmembrane conductance regulator
Chemokines as DPIV substrates, 80
chemokine receptors, 80
Chemokines as DPIV substrates (*cont.*)
eotaxin, 80
family C, 80
family CC, 80
family CX3C, 80
family CXC, 80
heptahelical G-protein-coupled receptors, 80
RANTES, 80, 181
SDF-1, 80
Cholecystokinin, 405
Cigarette smoke, influence on activity of neutral endopeptidase 24.11, 418, 422
Coagulation/Fibrinolysis Cascade Pathway, 461
bleeding, 461
fibrinogen, 461
myocardial infarction, 461
prothrombin, 461
Coculture of lymphocytes with synoviocytes, 57–66
Collagenases, 218, 246
Complement pathway
complement factor C3; C5, 459
Controlling inhibitors, 460
α1-antichymotrypsin, 460
α1-proteinase inhibitor, 460
Creutzfeldt–Jacob disease, 355
CSF: *see* Cerebrospinal fluid
Cystatins, 244
Cystatin M, 244
Cysteine peptidases
catalytic domain, 262
lysosomal, 261
propeptide domain, 261–270
Cystein proteinases
papain familiy, 255
papain family of/review, 241–254, 261–270
phylogenetic tree, 242, 277
proenzymes, 243
tissue degeneration, 282
tumour progression, 282, 366
Cystic fibrosis transmembrane conductance regulator, degradation via UMPS, 354
Cytokines
DPIV inhibition and cytokin synthesis, 139–143
regulation of cathepsin gene expression, 287–291
by IL-6, 290
by TGF-β1, 290

Dipeptidyl aminopeptidase long forms (DPPX-L), 104
α/β hydrolase fold, 91, 104
β propeller fold, 91, 104

Dipeptidyl aminopeptidase short forms (DPPX-S), 104
Dipeptidyl peptidase II (DPP II), 106
Dipeptidyl peptidase IV/review, 10
active site, 97
activity in patients with anorexia nervosa, 197
activity in patients with bulimia, 197
activity on MBP specific T cells, 150
association with CD45 and ADA, 11
in asthmatic airways, 417
autoimmune diseases, 155–160
binding of inhibitors, 97–101
cell cycle arrest, 131
costimulatory signal in T cells, 12
distribution in bronchial tissue, 417
other organs, 10
effects of inhibition by actinonin, 12
effects of inhibitors, 12
effects of TGF-β, 13
fluorogenic substrates, 111–115
cell associated fluorescence, 112
flow cytometry, 115
fluorescence microscopy, 115
kinetic constants
rhodamine 110 (R110), 111
Xaa-Pro-R110-Y, 112
function, 11
HIV-1 Tat binding, 100
hydrolase domain, 91
adenosine deaminase binding protein (ADAbp), 91
inflammatory CNS diseases, 145–153, 155–160
inhibitor actions, 131
inhibitors, 117–123
amino acid thiazolidides, 119
HIV-1 Tat protein peptides, 125–129
Ile-morpholide, 120
Ile-piperidide, 120
influence on cytokine synthesis, 139
influence on DPIV mRNA, 139
influence on EAE-T cells, 145–153
inhibition of DNA synthesis in keratinocytes, 167–171
inhibition of DNA synthesis in T cells, 152
irreversible, 117
Lys[Z($NO_2$)]-piperidide, 118
monoclonal antibody binding, 90
point mutations, 91
pyrrolidine derivative inhibitors, 118
(S)-(+)-1-(2-pyrrolidinylmethyl)-pyrrolidine, 118
reversible, 117
role in immunosuppression, 161
thioxo amino acids, 121
THO and TH1 cells, 418
MAP kinase activation, 131
Dipeptidyl peptidase IV (*cont.*)
molecular model, 89–95, 97–101
natural substrates, 67–87
β-casomorphin-5, 69
endomorphin-2, 69
neuropeptide Y (NPY) substrates, 71
μ-opioid receptor
pancreatic polypeptide family substrates, 71
pancreatic polypeptide-fold (PP-fold) substrates, 71
procolipase substrates, 69
prolyl oligopeptidase (PO), 98
substance P, 68
β-propeller domain, 89–95
regulation by cell-cell-contact
regulation, 12
role in T cell activation, 155–160
S1-binding site, 99
signal transduction by, 12, 13, 131–137
soluble form in airways, 419
structurally conserved regions (SCRs), 98
structurally variable regions (SVRs), 98
substrates, 11, 67–87, 111–115, 187–195
tertiary structure, 89, 97
TGF-β1 mediated effects on cells, 131
vascular endothelial cells, 418
Dipeptidylpeptidase IV-β, 104
α/β hydrolase fold, 91, 104
β-propeller fold, 91, 104
Disease-modifying anti-rheumatic drugs (DMARDs), 472
DPIV like proteins, 103–109
DPIV-β: *see* Dipeptidylpeptidase IV-β
DPIV: *see* Dipeptidyl peptidase IV
DPIV-related gene family, 104

E-64d, 409
E6-associated protein, 353
Elafin, 358
Elastase, 246, 357
PMN-derived, role in IL-6 proteolysis, 431
Enamelysin, 218
Endometriosis, 483–486
Endometrium, 484
Endothelin converting enzymes, 230
isoforms, 230
localization, 230
Endothelins, 229
Enterokinase, 407
Enterostatin, 69
ERFNAQ motif, 276
ERFNIN motif, 263, 271
Eumelanin, 181
Experimental autoimmune encephalomyelitis, 145

Extracellular matrix, 332
Extracellular matrix degradation, 357, 361

FAP-α: *see* Fibroblast-activation protein-α
Fibrinogen, 217–228
  degradation by MMPs, 218
Fibroblast-activation protein-α (FAP), 104
  relation to attractin, 174
  relation to DPIV, 174
Fluticasone (glucocorticoid), 425

G protein-coupled receptors (GPCR), 324
Gastric mucosa, 445–454
Gelatinases, 218
GFP: *see* Green Fluorescent protein
Gingipains, 456
  cysteine-proteinases, 456
    arg-gingipains (Rgps), 457
    lys-gingipains (Kgp), 457
Gingivitis, 455
Glucagon superfamily as DPIV substrates, 73
  gastric inhibitory peptide (GIP), 73, 74, 75
  gastric-releasing peptide, 76
  glicentin, 74
  glucagon-like peptide 1 (GLP-1), 74
  glucagon-like peptide 2 (GLP-2), 74
  glucose-dependent insulinotropic peptide, 74, 75
  growth hormone releasing factor (GRF), 74, 76
  oxyntomodulin, 74
  proglucagon, 74
  secretin, 73
  vasoactive intestinal peptide (VIP), 73
  vasoactive peptide, 76
Glucagon-like peptide-1, 187–195
  degradation by DPIV, 187
Glucocorticoids, modulation of peptidases in asthmatics by, 423
Glucose-dependent insulinotropic polypeptide, 187–195
  biological half-life, 188
  DPIV resistant analogs, 188
    synthesis of, 188
  DPIV substrate, 190
Glutamate carboxypeptidases, 106
  NAALADase I, 106
  NAALADase II, 106
  NAALADase L, 106
Green Fluorescent Protein, Cathepsin B fusion proteins, 393
GSK-3β, glycogen synthase kinase 3β, 35

Hagemann factor
  degradation by kallikrein, 226
  degradation by MMPs, 217–228
    scheme of molecular destruction, 227
Hect domain, 354
Helicobacter pylori infection, 359, 445
  neutrophil infiltration, 445
HIV-1 Tat protein, 125, 161
Hormones
  chromogranin A, 76
  growth hormone releasing hormone (GRF), 76
  insulin-releasing hormone, 75
  parathyroid hormone secretion, 77
  preprocalcitonin, 77
  prohormone convertase, 76
  procalcitonin (PCT), 77
  vasostatin I, II, 76
Huntingtin, 355

IL-6, regulation of cathepsin synthesis by, 287–291
Inflammation, local proteolytic potential, 431
Inflammatory diseases
  antiprotease balance, 357, 363
  role of proteases, 357
Inhibitors, *see also* Protease inhibitors
  cystein proteases, E64, 479
  metalloproteinases, phenanthroline, 479
  serine proteases, aprotinin, 479
Insulin, 76
Interferon-γ-inducing factor: *see* Interleukin-18
Interleukin-1, vascular cells, 343
Interleukin 6
  analysis of proteolysis products, 432
  determination in body fluids, 432
  inactivation by neutrophil proteases, 431
  soluble IL-6 receptor, influence on IL-6 proteolysis, 431
Interleukin-18
  precursor, 343
  processing in PMN/monocytes, 345
  processing in vascular smooth muscle cells, 343–348
  vascular cells, 343
Intracellular stores of $Ca^{++}$, 73
$^{125}$I-TID ceramid, 307

Kallikrein/Kinin pathway, 460
  bradykinin (BU), 461
  Hageman factor (XIIa), 461
  kallikrein, 460
  kininogen (HMWK), 461
  prekallikrein, 460
Katacalcin, 77
Kell, member of neprilysin family, 231
Keratinocytes, 167–171
  DNA synthesis in, 167
    influenced by DPIV inhibition, 167
  DPIV activity in, 167
  HACAT cell line, 167

Knock-out cells, Cathepsin B, MEF-I-mouse fibroblast, 393
Knock-out mice
 notch, 385
 presenilin-1, 385
Knockout mouse models
 cathepsin K knockout mice, 299
  model for human pycnodysostosis, 299
 PAI-1 and -2-knockout mice, 334

Liddle syndrome, 354
Lipid rafts, 385
L-valinyl-L-boroproline (Vbp), 105
Lysosomal enzyme trafficking, in living tumor cells, 391–401

α2-macroglobulin, 358
Malignant transformation, 392
Mannose-6-phosphate receptor, 392
Mannose-6-phosphate receptor pathway, 242
Matrix metalloproteinases, 382
 activation cascades, 363
 inhibitors, 357
 interaction with clotting system, 217
  degradation of factor XII, 223
  degradation of fibrinogen, 218
matrilysin, 218
 MMP-13, 217
 MMP-14, 217
 MMP-7, 218
 MMP-8, 217
 review, 217–228
 role in endometriosis, 483–486
 TIMPs, 357, 363
 inactivation of TIMPs, 364
Matrixins: *see* Matrix metalloproteinases
Membrane microdomains: *see* Lipid rafts
Mental eating disorders, 197–204
 DPIV activity in patients, 197–204
Metastasis, 332
MMPs: *see* matrix metalloproteinases
Mobilization of $Ca^{++}$, 73
Mononuclear phagocyte system, 206
 macrophage differentiation, 206
 role of carboxypeptidase M, 205–216
Multiple sclerosis, 145, 155
Myelin basic protein, 150, 155

Natriuretic peptides, 229
Nedd 4, 354
NEP: *see* Neprilysin; Neutral endopeptidase 24.11
Neprilysin family/review, 229–240
 natural substrates, 229
 processing of regulatory peptides, 229
Neurogenic inflammation, 413
Neurokinin A, 413
Neuropeptides, role in asthma, 413
Neutral endopeptidase 24.11
 aerosol administration as therapy, 414
 bronchial tissue, 419
 glucocorticoid influence, 423
 increased serum activity
  in miners, 419
  in sarcoidosis, 419
 role in asthma, 414, 418
 shedding, 422, 424
 soluble form in airways, 419
 vascular endothelial cells, 418
Neutrophils
 C5 a receptor, 460
 proteinases
  cathepsin G, 460
  elastase, 460
  proteinase 3, 460
 serin proteases
  elastase, 445
  cathepsin G, 445
  proteinase 3, 445

Opioide, 229
μ-Opioid receptor, 69
Osteoclasts, 294
 TRAP-stained, 295

p53, 353
Pancreas
 serine protease activation, 403–411
 zymogen granules, 404
  colocalization trypsinogen/cathepsin B, 407
Pancreatic polypeptide family substrates, 71
 neuropeptide Y (NPY) substrates, 71
Pancreatic polypeptide-fold (PP-fold) substrates, 71
Papain-like cysteine proteinases, 261
Parkinson's disease, 355
Peptidases
 CD antigens, 2
 role in immune function, 1
 substrates, 2, 67, 111, 431
Peptidases in serum
 of allergic asthmatics, 420
 release from leukocytes, 422
 after shedding from tissues, 421
Periodontain, 459
Periodontitis, 455
Phaeomelanin, 181
Pick's disease, 355
Plasma kallikrein, degradation of Hagemann factor, 226
Plasmin, 332
Plasminogen, 332
Pleural effusion, Il-6 determination, 432

PMN degranulation, 431
PMN leukocyte derived proteases, 431
  membrane rebinding, 357, 359
  elastase, inactivation of IL-6, 431
  cathepsin G, inactivation of IL-6, 431
  proteinase 3, inactivation of IL-6, 431
  substrates
    Cytokines, 431
    non matrix proteins, 431
  subcellular fractionation, 431
Post-proline peptidases, 105
Polyubiquitination, 351, 353
Presenilins, 383
  N-terminal fragment, 384
  C-terminal fragment, 384
  mutations, 384
  presenilinase, 384
Procathepsin B, 392
Procathepsin K, 282
Procathepsin L, 265
  3-D-structure, 265
  crystal structure, 265
Procathepsin S, 262
  homology modelling, 265
  processing, 263
  structure, 263
  site directed mutagenesis, 262
Procolipase substrates, 69
Progelatinases, 218
Progressive supranuclear palsy, 355
Proliferation of thyroid cancer cell lines, 488
  under insulin stimulation, 488
  under stimulation
    by PMA, 488
    by TSH, 488
    by EGF, 488
    by forskolin, 488
Prolyl endopeptidases (PEP), 103
Prolyl tripeptidyl peptidase A (PtpA), 458
Prolyloligopeptidase, 103
Protease-antiprotease imbalance, 357, 363
Protease inhibiting drugs, anti-rheumatics, 472
α1-protease inhibitor, 358
Protease inhibitors
  $\alpha_1$-antitrypsin, 445
  batimastat, 381
  BB2116, 381
  complex formation, 450
  compound 1, 381
  compound 4, 381
  degradation, 450
  E64d, 381
  genetic variants, 360
  hydroxamate derivatives, 381
  immunohistology, in situ hybridization, 446
  inactivation, 364
Protease inhibitors (*cont.*)
  $\alpha_2$-macroglobulin, 445
  marimatat, 381
  1,10 phenanthroline, 381
  SLPI: *see* Secretory leukocyte proteinase inhibitor
    suppressors, 452
    antiproteolytic action, 452
  therapeutics, 364
Protease expression, regulation
  by cytokines, 470
  by protooncogenes, 470
Proteasome, 356
Proteinase 3, 357
  role in IL-6 proteolysis, 431
Proteinase receptor activation, 323
Proteinase-activated receptors
  PAR-1, 323–329
    rat astrocytes, 232–329
    signal termination and desensitisation, 324
  PAR-2, -3, -4, 323
Proteolysis of cytokines
  IL-6, 431
  TGF-β, 471
Pycnodysostosis, 293–303
  characteristic feature, 299
    osteopetrosis, 295
  clinical picture, 299

Quiescent peptidyl peptidase (QPP), 106

Receptor phosphorylation and internalisation, 327
Receptors
  antagonists, 73
  NPY, 73
  vasoconstrictive Y1, 73
  Y2-, Y5-receptors, 73
Restrained molecular dynamics calculations (MD calculations), 126
Rheumatoid arthritis, 477
  synovial fluid, 477
  synovial fluid proteinase content, 469
  synovial membrane, 469
Rheumatic diseases, 467
  proteinases
    autoantigens, 469
    cathepsin B, 469
    cathepsin K, 469
    cathepsin L, 469
    matrix-metalloproteinases (MMPs), 467
    membrane-type MMPs (MT-MMPs), 467
    stromelysin (HHP-S), 467

Secretase α, β, γ, 380
Seprase, relation to DPIV, 174

Serine protease inhibitor superfamily: *see* Serpins
Serine protease uPA, 331
  inhibitors PAI-1, PAI-2, 333
  receptor uPAR
Serine proteases, 241
  thrombin as ligand for PARs, 323–329
  trypsin as ligand for PARs, 323–329
Serpins, 358, 456
  PAI-1 and PAI-2 as, 335
    reaction
    with activated protein C, 333
    with plasmin, 333
    with protein C inhibitor, 333
    with proteinase nexin-1, 333
    with thrombin, 333
    with trypsin, 333
Secretory leukocyte proteinase inhibitor (SLPI), 358
Smads, 50
Smad 7, quantification of mRNA expression, 50
Soluble secreted endopeptidase, 230
Soybean trypsin inhibitor (SBTI), 327
Sphingomyelinase, 306
  plasma membrane bound, 306
  endo lysosoma, 3061
Stromelysins, 218
Substance P, 413
Synovial fluid, Il-6 determination, 432
Synoviocytes, 57
Synthetic thrombin receptor agonist peptide (TRag), 324
  rat astrocytes, 325

TACE: *see* TNF-α converting enzyme
Tachykinin, 415, 229
  containing nerves, 413
Tat nonapeptides, 125–129, 161–165
  conformational analysis, 125
  influence on T cell DNA synthesis, 161
Tau protein, 355
Tetracyclines, 472
TGF-β: *see* Transforming growth factor-β
Thyroid cancer cell lines
  anaplastic thyroid cancer line 8505C, 488
  follicular carcinoma cell line FTC133, 488
  thyroid cancer cell lines, 488
TIMPs: *see* Tissue inhibitors of metalloproteinases
Tissue inhibitors of metalloproteinases (TIMPs), 357, 363, 468
  inactivation of TIMPs, 364
  members of the family, 218
  role in endometriosis, 483–486
Tissue type plasminogen activator, 332
TNF-α convertase (TACE) , 362, 382, 483
  matrix metalloproteinases, 382
  presenilinase, 384
  reprolysin, 382
tPA, 332; *see also* Tissue type plasminogen activator

Transduction pathways for cathepsin L stimulation in thyroid cells
  adenylate cyclase pathway, 491
  cAMP release, 491
  Phospholipase C pathway, 491
  Protein kinase C system, 491
  tyrosin kinase pathway, 491
Transforming growth factor-β
  activation, 479
  activators, 479
  isoforms, 478
  regulation of APN expression, 49–56
  regulation of cathepsin synthesis, 287–291
  synovial fluid, 477–481
Transmembrane signalling
  via aminopeptidase N, 36–41, 43–47
  via DPIV, 131–137
  via proteinase activated receptors, 323–329
Trypsinogen activation, 404
Tumor cell invasion, 332
Type 2 diabetes, 75

Ubiquitin-mediated proteolytic system
  and cell cycle control, 352
  defects in UMPS, 353
    Angelman syndrome, 353
    cystic fibrosis, 354
    Liddle syndrome, 354
    muscle diseases, 355
  enzyme cascade, 350
  and mitosis, 352
  substrates, 351
    antigens, 352
    cyclin-dependent kinase, 352
    cyclins, 351
    membrane channels, 352, 353
    NFκB, 352
    oncoproteins, 352
    signalling receptors, 352
    T cell receptor ζchain, 352
    transcription factors, 352
  ubiquitin conjugating enzymes, 350
  viral tumorigenesis, 356
UMPS: *see* Ubiquitin mediated proteolytic system
Urokinase-type plasminogen activator (uPA), 331–341, 439
  synthetic inhibitors, 336, 338
  tumor antigen, 334
  tumor cell surface-associated, 331

Wagner's disease
Wnt-5a proto-oncogene, 35
WW domains, 354

X-converting enzyme, 230

www.ingramcontent.com/pod-product-compliance
Ingram Content Group UK Ltd.
Pitfield, Milton Keynes, MK11 3LW, UK
UKHW051325070726
13610UKWH00014B/88